Congratulations!

You're using
ChemCom: Chemistry in the Community
A Project of the American Chemical Society (ACS)

Now join the ChemCom family!

By completing and returning this reply card, you'll be added to the list of enthusiastic Kendall/Hunt educators who are interested in receiving information about workshops, teaching methods, new editions, and other helpful resources. We'll also ask for information about how you use *ChemCom* in your classroom, send you supplementary materials, and connect you with user support groups in your part of the country.

NAME_____ SCHOOL/
ORGANIZATION_____

ADDRESS_____ CITY_____

STATE_____ ZIP_____ PHONE (   )_____ EXT._____

SUMMER ADDRESS_____ CITY_____

STATE_____ ZIP_____ PHONE (   )_____ EXT._____

Please mark all boxes that are appropriate.

1. I currently teach the following level(s):

   ☐ middle/junior high   ☐ high school   ☐ college/university

2. If high school, what level(s)?: ☐ 9   ☐ 10   ☐ 11   ☐ 12 Number of students in class:____

3. I have responsibilities as a:

   ☐ science supervisor (K-12)   ☐ administrator   ☐ department chair

   ☐ science consultant   ☐ other_____

4. How did you learn about *ChemCom?*

   ☐ convention   ☐ colleague   ☐ advertisement   ☐ ACS   ☐ other_____

5. What text did *ChemCom* replace?_____

6. Which of the following influenced your decision to teach this program?

   ☐ colleague   ☐ administrator   ☐ Kendall/Hunt   ☐ ACS   ☐ other_____

7. Please share with us your first impressions of *ChemCom.* _____
_____

8. Would you be willing at a later date to complete an evaluation questionnaire based on your experiences using *ChemCom?*   ☐ Yes   ☐ No

9. Would you be interested in more information on Kendall/Hunt programs in:

   ☐ math   ☐ language arts   ☐ physical education and health

KENDALL/HUNT PUBLISHING COMPANY
4050 Westmark Drive P.O. Box 1840 Dubuque, Iowa 52004-1840

Attn.: Teacher Services Department
Kendall/Hunt Publishing Company
4050 Westmark Drive, P.O. Box 1840
Dubuque, IA 52004-1840

- -
(fold)

(tape here)   (fold)   (tape here)
- -

KENDALL/HUNT PUBLISHING COMPANY
4050 Westmark Drive P.O. Box 1840 Dubuque, Iowa 52004-1840

At Kendall/Hunt Publishing Company, we develop innovative materials in Science, Language Arts, Physical Education, Health, and Social Studies that offer teachers a real alternative to traditional material. We work in partnership with universities, school districts, and curriculum development groups such as the American Chemical Society to develop *and* implement curricula rather than just sell textbooks.

Because our programs are on the leading edge of educational reform, we are committed to extensive teacher training and inservice. At Kendall/Hunt you don't become just a sales figure after you make your selection. Instead you become part of the growing community of dedicated educators using Kendall/Hunt programs. And the ultimate benefactors are the students. Thank you for selecting a Kendall/Hunt program.

ChemCom
Chemistry in the Community

A Project of the American Chemical Society

Teacher's Guide
Second Edition

KENDALL/HUNT PUBLISHING COMPANY
4050 Westmark Drive   Dubuque, Iowa 52002

ChemCom
Second-Edition Revision Team

Project Manager:
Keith Michael Shea

Chief Editor:
Henry Heikkinen

Assistant to Chief Editor:
Wilbur Bergquist

Editor of Teacher's Guide:
Jon Malmin

Second Edition Editorial Advisory Board:
Diane Bunce, Henry Heikkinen (*ex officio*), S. Allen Heininger, Donald Jones (chair), Jon Malmin, Paul Mazzocchi, Bradley Moore, Carolyn Morse, Keith Michael Shea (*ex officio*), Sylvia Ware (*ex officio*)

Teacher Reviewers of First Edition:
Vincent Bono, New Dorp High School, New York; Charles Butterfield, Brattle Union High School, Vermont; Regis Goode, Spring Valley High School, South Carolina; George Gross, Union High School, New Jersey; C. Leonard Himes, Edgewater High School, Florida; Gary Hurst, Standley Lake High School, Colorado; Jon Malmin, Peninsula High School, Washington; Maureen Murphy, Essex Junction Educational Center, Vermont; Keith Michael Shea, Hinsdale Central High School, Illinois; Betsy Ross Uhing, Grand Island Senior High School, Nebraska; Jane Voth-Palisi, Concord High School, New Hampshire; Terri Wahlberg, Golden High School, Colorado.

Safety Consultant:
Stanley Pine

Editorial:
The Stone Cottage

Design:
Bonnie Baumann

Design of Teacher's Guide:
James Rhoades
Robert Burns

Art:
Additional art for this edition by Seda Sookias Maurer

This material is based upon work supported by the National Science Foundation under Grant No. SED-88115424 and Grant No. MDR-8470104. Any opinions, findings, and conclusions or recommendations expressed in this publication are those of the authors and do not necessarily reflect the views of the National Science Foundation. Any mention of trade names does not imply endorsement by the National Science Foundation.

Thanks to the faculty, administration, and students of Hinsdale Central High School (Illinois) for use of their chemistry laboratory facilities in photographs depicting laboratory set-ups and procedures in this textbook.

Copyright © 1988, 1993 by American Chemical Society

ISBN 0-8403-5506-8

Printed in the United States of America

10 9 8 7 6 5 4

CONTENTS

PETROLEUM: TO BUILD? TO BURN?   72

Introduction   74

IMPORTANT NOTICE

ChemCom is intended for use by high school students in the classroom laboratory under the the direct supervision of a qualified chemistry teacher. The experiments described in this book involve substances that may be harmful if they are misused or if the procedures described are not followed. Read cautions carefully and follow all directions. Do not use or combine any substances or materials not specifically called for in carrying out experiments. Other substances are mentioned for educational purposes only and should not be used by students unless the instructions specifically so indicate.

The materials, safety information, and procedures contained in this book are believed to be reliable. This information and these procedures should serve only as a starting point for laboratory practices, and they do not purport to specify minimal legal standards or to represent the policy of the American Chemical Society. No warranty, guarantee, or representation is made by the American Chemical Society as to the accuracy or specificity of the information contained herein, and the American Chemical Society assumes no responsibility in connection therewith. The added safety information is intended to provide basic guidelines for safe practices. It cannot be assumed that all necessary warnings and precautionary measures are contained in the document and that other additional information and measures may not be required.

PREFACE

The United States is a world leader in science, technology, and the education of scientists and engineers. Yet, overall, U.S. citizens are barely literate in science. In responding to this situation, our government and many professional groups have assigned high priority to improving the nation's science literacy.

Chemistry in the Community (ChemCom) represents a major effort to enhance science literacy through a high school chemistry course that emphasizes chemistry's impact on society. Preliminary work on *ChemCom* began in 1980 with the formation of a steering committee, staff, and groups of writing teams consisting of high school, college, and university teachers, assisted by chemists from industry and government. Developed by the American Chemical Society (ACS) with financial support from the National Science Foundation and several ACS funding sources, the themes of *ChemCom* were developed, emphasizing the application of chemistry to societal issues utilizing student-centered activities. During the summers of 1982 and 1983 the first drafts of the eight *ChemCom* units were written and carefully evaluated by content specialists, social science consultants, and field-tested in local communities. In 1984 the overall syllabus was developed, and the national field test version was written by a team of writers under the direction of Henry Heikkinen.

In 1985 the field test began in 13 centers around the country, in which 61 teachers used the materials with 2900 students. In several locations the teachers met monthly to share ideas and iron out difficulties, proposing alterations in the text and laboratory material. *ChemCom*'s showing was very favorable, as indicated by the increase in requests for textbooks by field test teachers, and after final revision the book was published by the Kendall/Hunt Publishing Company in 1988. Since then, *ChemCom* has been successfully implemented by chemistry teachers in thousands of classrooms. Many teachers report that the program offers a motivational, engaging approach to the study of chemistry for a remarkably wide range of students. The second edition was released in 1992.

Briefly, *ChemCom* is designed to help students:
- realize the important roles that chemistry will play in their personal and professional lives.
- use chemistry knowledge to think through and make informed decisions about issues involving science and technology.
- develop a lifelong awareness of the potential and limitations of science and technology.

Each of *ChemCom's* eight units focuses on a chemistry-related technological issue currently confronting our society and the world. The issue serves as a basis for introducing the chemistry needed to understand and analyze it. The setting for each unit is a community. This may be the school community, the town or region in which the students live, or the world community–Spaceship Earth.

The major *ChemCom* topics are: *Supplying Our Water Needs*; *Conserving Chemical Resources*; *Petroleum: To Build? To Burn?*; *Understanding Food*; *Nuclear Chemistry in Our World*; *Chemistry, Air, and Climate*; *Health: Your Risks and Choices*; and *The Chemical Industry: Promise and Challenge*.

The eight units include the major concepts, vocabulary, thinking skills, and laboratory techniques expected in any introductory chemistry course. However, the program contains a greater number and variety of student-oriented activities than is customary. In addition to numerous laboratory exercises, including many developed especially for *ChemCom*, each unit contains three levels of decision-making activities and several types of problem-solving exercises.

The first four *ChemCom* units are designed to be studied in sequence. Each new unit builds on the previous ones. For example, considering water as a resource leads to consideration of resources in general, which leads to information about another very special resource, petroleum. The two uses of petroleum, for energy production and building petrochemicals, mimic the two main uses of food, for energy production and building components. The first four units also involve systematic development of basic chemical concepts, problem-solving skills, and decision-making abilities. By contrast, the last four units have been designed independently, so that they can be studied in any sequence, based on the interests and needs of your students.

This new edition of *ChemCom*, while maintaining the overall structure and approach of the first edition, provides up-dated information on many *ChemCom* topics as well as detailed improvements based on suggestions from classroom experience. A new feature, appropriately named *Chemistry in the Community*, describes how individuals in widely diverse careers find uses for chemistry in their daily work. All second-edition changes are intended to make *ChemCom* even more "user-friendly" for both teachers and their students.

Dozens of professionals from all segments of the chemistry community contributed their talents and energies to create *ChemCom*. Their hope is that its impact will be substantial and lasting, and that those who study *ChemCom*, will find chemistry interesting, captivating, and useful.

CREDITS

ChemCom is the product of teamwork involving individuals from all over the United States over the past decade. The American Chemical Society is pleased to recognize all who contributed to *ChemCom*.

The team responsible for this new edition of *ChemCom* is listed on the copyright page. Those individuals who contributed to the initial development of *ChemCom*, and to the release of the program's first edition in 1988, are listed below.

Principal Investigator:
W. T. Lippincott

Project Manager:
Sylvia Ware

Chief Editor:
Henry Heikkinen

Contributing Editor:
Mary Castellion

Assistant to Contributing Editor:
Arnold Diamond

Editor of Teacher's Guide:
Thomas O'Brien

Revision Team:
Diane Bunce, Gregory Crosby, David Holzman, Thomas O'Brien, Joan Senyk, Thomas Wysocki
Editorial Advisory Board: Glenn Crosby, James DeRose, Dwaine Eubanks, W. T. Lippincott (ex officio), Lucy McCorkle, Jeanne Vaughn, Sylvia Ware (ex officio)

Writing Team:
Rosa Balaco, James Banks, Joan Beardsley, William Bleam, Kenneth Brody, Ronald Brown, Diane Bunce, Becky Chambers, Alan DeGennaro, Patricia Eckfeldt, Dwaine Eubanks (dir.), Henry Heikkinen (dir.), Bruce Jarvis (dir.), Dan Kallus, Jerry Kent, Grace McGuffie, David Newton (dir.), Thomas O'Brien, Andrew Pogan, David Robson, Amado Sandoval, Joseph Schmuckler (dir.), Richard Shelly, Patricia Smith, Tamar Susskind, Joseph Travello, Thomas Warren, Robert Wistort, Thomas Wysocki

Steering Committee:
Alan Cairncross, William Cook, Derek Davenport, James DeRose, Anna Harrison (ch.), W. T. Lippincott (ex officio), Lucy McCorkle, Donald McCurdy, William Mooney, Moses Passer, Martha Sager, Glenn Seaborg, John Truxall, Jeanne Vaughn

Consultants:
Alan Cairncross, Michael Doyle, Donald Fenton, Conard Fernelius, Victor Fratalli, Peter Girardot, Glen Gordon, Dudley Herron, John Hill, Chester Holmlund, John Holman, Kenneth Kolb, E. N. Kresge, David Lavallee, Charles Lewis, Wayne Marchant, Joseph Moore, Richard Millis, Kenneth Mossman, Herschel Porter, Glenn Seaborg, Victor Viola, William West, John Whitaker

Synthesis Committee:
Diane Bunce, Dwaine Eubanks, Anna Harrison, Henry Heikkinen, John Hill, Stanley Kirschner, W. T. Lippincott (ex officio), Lucy McCorkle, Thomas O'Brien, Ronald Perkins, Sylvia Ware (ex officio), Thomas Wysocki

Evaluation Team:
Ronald Anderson, Matthew Bruce, Frank Sutman (dir.)

Field Test Coordinator:
Sylvia Ware

Field Test Workshops:
Dwaine Eubanks

Field Test Directors:
Keith Berry, Fitzgerald Bramwell, Mamie Moy, William Nevill, Michael Pavelich, Lucy Pryde, Conrad Stanitski

Pilot Test Teachers:
Howard Baldwin, Donald Belanger, Navarro Bharat, Ellen Byrne, Eugene Cashour, Karen Cotter, Joseph Deangelis, Virginia Denney, Diane Doepken, Donald Fritz, Andrew Gettes, Mary Gromko, Robert Haigler, Anna Helms, Allen Hummel, Charlotte Hutton, Elaine Kilbourne, Joseph Linker, Larry Lipton, Grace McGuffie, Nancy Miller, Gloria Mumford, Beverly Nelson, Kathy Nirei, Elliott Nires, Polly Parent, Mary Parker, Dicie Petree, Ellen Pitts, Ruth Rand, Kathy Ravano, Steven Rischling, Charles Ross, Jr., David Roudebush, Joseph Rozaik, Susan Rutherland, George Smeller, Cheryl Snyder, Jade Snyder, Samuel Taylor, Ronald Tempest, Thomas Van Egeren, Gabrielle Vereecke, Howard White, Thomas Wysocki, Joseph Zisk

Field Test Teachers:
Vincent Bono, Allison Booth, Naomi Brodsky, Mary D. Brower, Lydia Brown, George Bulovsky, Kay Burrough, Gene Cashour, Frank Cox, Bobbie Craven, Pat Criswell, Jim Davis, Nancy Dickman, Dave W. Gammon, Frank Gibson, Grace Giglio, Theodis Gorre, Margaret Guess, Yvette Hayes, Lu Hensen, Kenn Heydrick, Gary Hurst, Don Holderread, Michael

Ironsmith, Lucy Jache, Larry Jerdal, Ed Johnson, Grant Johnson, Robert Kennedy, Anne Kenney, Joyce Knox, Leanne Kogler, Dave Kolquist, Sherman Kopelson, Jon Malmin, Douglas Mandt, Jay Maness, Patricia Martin, Mary Monroe, Mike Morris, Phyllis Murray, Silas Nelsen, Larry Nelson, Bill Rademaker, Willie Reed, Jay Rubin, Bill Rudd, David Ruscus, Richard Scheele, Paul Shank, Dawn Smith, John Southworth, Mitzi Swift, Steve Ufer, Bob Van Zant, Daniel Vandercar, Bob Volzer, Terri Wahlberg, Tammy Weatherly, Lee Weaver, Joyce Willis, Belinda Wolfe

Field Test Schools:
California: Chula Vista High, Chula Vista; Gompers Secondary School, San Diego; Montgomery High, San Diego; Point Loma High, San Diego; Serra Junior-Senior High, San Diego; Southwest High, San Diego; Colorado: Bear Creek Senior High, Lakewood; Evergreen Senior High, Evergreen; Green Mountain Senior High, Lakewood; Golden Senior High, Golden; Lakewood Senior High, Lakewood; Wheat Ridge Senior High, Wheat Ridge; Hawaii: University of Hawaii Laboratory School, Honolulu; Illinois: Project Individual Education High, Oak Lawn; Iowa: Linn-Mar High, Marion; Louisiana: Booker T. Washington High, Shreveport; Byrd High, Shreveport; Caddo Magnet High, Shreveport; Captain Shreve High, Shreveport; Fair Park High, Shreveport; Green Oaks High, Shreveport; Huntington High, Shreveport; North Caddo High, Vivian; Northwood High, Shreveport; Maryland: Charles Smith Jewish Day School, Rockville; Owings Mills Junior-Senior High, Owings Mills; Parkville High, Baltimore; Sparrows Point Middle-Senior High, Baltimore; Woodlawn High, Baltimore; New Jersey: School No. 10, Patterson; New York: New Dorp High, Staten Island; Texas: Clements High, Sugar Land; Cy-Fair High, Houston; Virginia: Armstrong High, Richmond; Freeman High, Richmond; Henrico High, Richmond; Highland Springs High, Highland Springs; Marymount School, Richmond; Midlothian High, Midlothian; St. Gertrude's High, Richmond; Thomas Dale High, Chester; Thomas Jefferson High, Richmond; Tucker High, Richmond; Varina High, Richmond; Wisconsin: James Madison High, Madison; Thomas More High, Milwaukee; Washington: Bethel High, Spanaway; Chief Sealth High, Seattle; Clover Park High, Tacoma; Foss Senior High, Tacoma; Hazen High, Renton; Lakes High, Tacoma; Peninsula High, Gig Harbor; Rogers High, Puyallup; Sumner Senior High, Sumner; Washington High, Tacoma; Wilson High, Tacoma

Safety Consultant:
Stanley Pine

Social Science Consultants:
Ross Eshelman, Judith Gillespie

Art:
Rabina Fisher, Pat Hoetmer, Alan Kahan (dir.), Kelly Richard, Sharon Wolfgang

Copy Editor:
Martha Polkey

Administrative Assistant:
Carolyn Avery

Student Aides:
Paul Drago, Stephanie French, Patricia Teleska

ACS also offers thanks to the National Science Foundation for its support of the initial development of ChemCom, and to NSF project officers Mary Ann Ryan and John Thorpe for their comments, suggestions, and unfailing support

SAFETY IN THE LABORATORY

In *ChemCom* you will frequently perform laboratory activities. While no human activity is completely risk free, if you use common sense and a bit of chemical sense, you will encounter no problems. Chemical sense is an extension of common sense. Sensible laboratory conduct won't happen by memorizing a list of rules, any more than a perfect score on a written driver's test ensures an excellent driving record. The true "driver's test" of chemical sense is your actual conduct in the laboratory.

The following safety pointers apply to all laboratory activities. For your personal safety and that of your classmates, make following these guidelines second nature in the laboratory. Your teacher will point out any special safety guidelines that apply to each activity.

If you understand the reasons behind them, these safety rules will be easy to remember and to follow. So, for each listed safety guideline:
- Identify a similar rule or precaution that applies in everyday life—for example in cooking, repairing or driving a car, or playing a sport.
- Briefly describe possible harmful results if the rule is not followed.

Rules of Laboratory Conduct

1. Perform laboratory work only when your teacher is present. Unauthorized or unsupervised laboratory experimenting is not allowed.
2. Your concern for safety should begin even before the first activity. Always read and think about each laboratory assignment before starting.
3. Know the location and use of all safety equipment in your laboratory. These should include the safety shower, eye wash, first-aid kit, fire extinguisher, and blanket.
4. Wear a laboratory coat or apron and protective glasses or goggles for all laboratory work. Wear shoes (rather than sandals) and tie back loose hair.
5. Clear your benchtop of all unnecessary material such as books and clothing before starting your work.
6. Check chemical labels twice to make sure you have the correct substance. Some chemical formulas and names may differ by only a letter or a number.
7. You may be asked to transfer some laboratory chemicals from a common bottle or jar to your own test tube or beaker. Do not return any excess material to its original container unless authorized by your teacher.
8. Avoid unnecessary movement and talk in the laboratory.
9. Never taste laboratory materials. Gum, food, or drinks should not be brought into the laboratory.
10. If you are instructed to smell something, do so by fanning some of the vapor toward your nose. Do not place your nose near the opening of the container. Your teacher will show you the correct technique.
11. Never look directly down into a test tube; view the contents from the side. Never point the open end of a test tube toward yourself or your neighbor.
12. Any laboratory accident, however small, should be reported immediately to your teacher.
13. In case of a chemical spill on your skin or clothing rinse the affected area with plenty of water. If the eyes are affected water-washing must begin immediately and continue for 10 to 15 minutes or until professional assistance is obtained.
14. Minor skin burns should be held under cold, running water.

15. When discarding used chemicals, carefully follow the instructions provided.

16. Return equipment, chemicals, aprons, and protective glasses to their designated locations.

17. Before leaving the laboratory, ensure that gas lines and water faucets are shut off.

18. If in doubt, ask!

COMBINED EXPENDABLE ITEMS SUPPLY LIST

Quantities listed are for a class of 24 students working in pairs and completing all recommended *ChemCom Laboratory Activities*. For information concerning the actual quantities needed for specific activities, see the individual supply lists found at the end of each unit of the Teacher's Guide.

CODE:
- W → Supplying Our Water Needs
- R → Conserving Chemical Resources
- P → Petroleum: To Build? To Burn?
- F → Understanding Food
- N → Nuclear Chemistry in Our World
- A → Chemistry, Air, and Climate
- H → Health: Your Risks and Choices
- I → The Chemical Industry: Promise and Challenge

Item	Unit(s)	Quantity
Acetic acid, concentrated	W/P/F	61 mL
Acrylic sheets	H	12
Albumin powder or 200 mL of egg whites	H	20 g
Alcohol, 2-propanol (isopropyl)	N/H	300 mL
Algae, filamentous *Spirogyra* (demo)	F	200 strands
Algae water (demo)	F	200 mL
Alka Seltzer	A	40 tablets
Ammonia, concentrated (demo)	A	100 mL
Ammonium chloride, crystals	W	25 g
Ammonium hydroxide, concentrated (demo)	A	15 mL
Ammonium molybdate	I	10 g
Ammonium nitrate	I	5 g
Amyl alcohol, pentanol (optional)	P	10 mL
Amylase enzyme powder	H	2 g
Apples	A	2
Ascorbic acid, vitamin C, crystals	F/I	21 g
Asphalt, pieces	P	15
Balloons, small (demo)	F	10
Balloons, large (demo)	A	10
Barium chloride dihydrate	W/F/I	85 g
Benedict's qualitative solution	H	300 mL
Benzoic acid (optional)	P	10 g
Beverage samples	F	400 mL each
Biuret reagent	F	50 mL
Boiling chips	F	100 g
Boric acid (demo)	H	15 g
Bottle, empty soft drink (demo)	A	1
Bottle, plastic, with lid (demo)	A	1
Boxes, cigar or shoe	N	8
Boxes, cardboard with lids	N	8
Bromthymol blue indicator	A	100 mL
Calcium chloride	W	3 g
Calgon (sodium hexametaphosphate)	W	1 box
Can, aluminum, soft drink	P/A	13
Candles, birthday (demo)	A	24
Candles, paraffin	P/A	35
Capillary melting point tubes	A	24
Carbon, graphite sticks or charcoal briquets	R	70 g
Carbon dioxide gas, cylinder (optional)	A	1
Cardboard or waxpaper pieces (5 cm × 5 cm × 0.1 cm)	N/A	17

Item	Unit(s)	Quantity
Charcoal, decolorizing	W	200 g
Charcoal, wood, granular (optional)	R	90 g
Cigarettes, unfiltered and filtered	H	2 packs
Citric acid (demo)	H	6 g
Clay or cornstarch	W	200 g
Cloth, pieces	N	12
Coffee grounds, used	W	1/2 cup
Copper(II) chloride dihydrate	R	14 g
Copper filings	A	70 g
Copper foil approximately 0.1 mm thick	R/I	150 cm^2
Copper(II) nitrate	R/I	370 g
Copper(II) nitrate trihydrate	I	75 g
Copper (II) oxide (optional)	R	50 g
Copper(II) sulfate pentahydrate	W/I	450 g
Copper wire, 24 gauge (0.5 mm diameter)	A/H	200 cm
Cotton balls	F/A	100
Cups, paper	W	20
Detergent, liquid dish (demo)	A	30 mL
Dry Ice (demos)	N/A	5.5 lbs.
Eggs	H	2
Elodea plant (demo)	A	4
Ethanol, denatured	W/P/F	230 mL
Euglena culture	H	230 mL
Filter paper, Whatman #1, 11 cm diameter	W/F	230
Food samples	F	5 kinds
Garlic powder	W	1 tsp.
Gas lantern mantles	N	12
Glass tubing, 6 mm diameter (optional)	R	5 m
Gloves, disposable	N/A	36 pairs
Glycerin	A	30 mL
Graph paper	N	50 sheets
Gravel, fine grind	W	2 kg
Hair strands	H	500
Hard water, Ca^{2+}(*aq*)	W	1 L
Helium gas, cylinder (demo)	A	1
Hexane	W/P	900 mL
Hydrion paper AB, full scale	A/H	3 packages
Hydrochloric acid, concentrated	R/F/A/H/I	500 mL
Hydrogen gas, cylinder (demo)	A	1
Hydrogen peroxide, 3% solution	F/A/H	2 L
Iodine crystals	A	0.5 g
Ion exchange resin, sodium, Dowex 50W-8X, 50-100 mesh	W	100 g
Iron, filings or thin wire	R	50 g
Iron strips	I	150 cm^2
Iron(III) chloride hexahydrate	W/F	60 g
Iron(III) nitrate nonahydrate	W/I	130 g
Iron(II) sulfate heptahydrate	W/I	20 g
Kerosene	P	750 mL
Lighter fluid (optional)	W	100 mL
Limewater	A	60 mL
Litmus paper, red and blue	A/I	2 packages each
Liver (raw)	H	2 pieces

Item	Unit(s)	Quantity
Magnesium, ribbon	R/F/A/I	600 cm
Magnesium nitrate hexahydrate	R/I	125 g
Manganese dioxide	A/H	10.5 g
Marble chips, calcium carbonate	A	100 g
Marshmallows, large (demo)	A	2
Methanol	P	40 mL
Milk, nonfat, (dry milk powder)	F	300 mL
Milk, whole	W	1 L
Naphthalene flakes	W	25 g
1-Naphthol	F	5 g
Nichrome or platinum wire	I	300 cm
Nitric acid, concentrated	R/A	140 mL
Nitrogen gas, cylinder (demo)	A	1
Octanol, octyl alcohol (optional)	P	10 mL
Oil, automotive	P	750 mL
Oil, household lubricating	P	750 mL
Oil, mineral (light)	P	750 mL
Oil, vegetable	W/A	400 mL
Paper bags, small brown	N	24
Paraffin wax	P	50 g
Paramecium (or mixed) culture	A	1
Pepsin enzyme powder	H	2 g
Perfume (demo)	A	1 bottle
Permanent wave solution	H	180 mL
Permanent wave neutralizer	H	90 mL
Phenolphthalein (demo)	A	5 mL
Potassium dihydrogen phosphate (demo)	F	1 g
Potassium iodate	F	1 g
Potassium iodide	F	10 g
Potassium monohydrogen phosphate	I	2 g
Potassium nitrate	I	5 g
Potassium thiocyanate	W/F	5 g
Propanoic acid (optional)	P	10 g
Rubber bands, orthodontist's type	H	225
Rubber bands, 10-cm diameter	P	20
Rubber bands, small	A	324
Salicylic acid	P	20 g
Salt (table), sodium chloride	W	50 g
Salt, low-sodium (Lite™)	N	1 box
Sand	W	1 kg
Silicon, lump	R	50 g
Silver nitrate	W/R	4 g
Soap, Ivory (liquid)	W	1 bottle
Sodium carbonate monohydrate	W	15 g
Sodium hydrogen carbonate (baking soda)	H	15 g
Sodium hydroxide	R/H/I	90 g
Sodium nitrate	I	5 g
Sodium oxalate monohydrate	W	3 g
Sodium phosphate dodecahydrate	I	10 g
Sodium sulfate decahydrate	I	10 g
Sodium sulfite (demo)	F	2 g
Splints, wooden	A/H	60
Starch, water soluble	F/H	3 g
Steel wool pads	R/A/I	35 pads

Item	Unit(s)	Quantity
Straws, drinking	H	24
Sulfur hexafluoride gas, cylinder (demo)	A	1
Sulfur, lump roll	R	50 g
Sulfur powder	A	40 g
Sulfuric acid, concentrated	P/F/I	600 mL
Sun-sensitive paper	N/H	14 pieces
Suntan oil (SPF #2, #6, #10 and #15)	H	1 bottle of each
Tin, mossy or foil	R	50 g
Tygon tubing (optional)	R	1 m
Trisodium phosphate	H	50 g
Universal indicator solution	A/H	95 mL
Urea Crystals	W	25 g
Zinc, finely granulated	R	10 g
Zinc, mossy (demo)	R	300 g
Zinc, strips (0.25 mm thick)	I	150 cm^2
Zinc nitrate hexahydrate	R/I	115 g

COMBINED NONEXPENDABLE ITEMS SUPPLY LIST

Quantities listed are for a class of 24 students working in pairs. Common nonexpendable items such as glassware are not listed.

CODE:
- W → Supplying Our Water Needs
- R → Conserving Chemical Resources
- P → Petroleum: To Build? To Burn?
- F → Understanding Food
- N → Nuclear Chemistry in Our World
- A → Chemistry, Air, and Climate
- H → Health: Your Risks and Choices
- I → The Chemical Industry: Promise and Challenge

Item	Unit(s)	Quantity
Alligator clips with leads	I	12
Assorted small objects for blackbox	N	12 of each
Batteries, 6-V lantern or DC power sources	I	12
Batteries, 9-V, and connectors	R	12
Beads, 7-mm metal	P	72
Calipers, vernier (optional)	N	1
Cloud chamber kits	N	12
Coins, assorted	N	12
Conductivity apparatus (optional)	R	12
Cups, porous, 75-mL	I	12
Culture tubes with caps (20 × 150 mm)	P	72
Dominoes	N	12 sets
Forceps	N	12
Glass plate (5 cm × 5 cm × 0.1 cm)	N	144
Graphite rods	R/I	24
Hammer	R	1
Lead plates (5 cm × 5 cm × 0.1 cm)	N	48
Light source, intense	N	12
Motors, low current (optional)	I	12
Pennies, (pre-1982, post-1982)	N	1400
Radiation sources (alpha, beta, and gamma)	N	12 sets
Ratemeter	N	12
Sand paper	R	12 pieces
Spot plates	I	12
Squeeze bottles	N	12
Syringe, plastic 50-mL (demo)	A	1
Timers, digital, or stop watches	P	12
Tongs, crucible	N	12
Watch faces, luminescent (optional)	N	12

CHEMICALS AND SCIENTIFIC EQUIPMENT SUPPLIERS

The following representative firms provide general supplies and equipment for all areas of science teaching as well as specific items for chemistry teaching.

Aldrich Chemical Company, P.O. Box 355, Milwaukee, WI 53201. 414-273-3850, 800-558-9160.

Carolina Biological Supply Company, 2700 York Rd., Burlington, NC 27215. 919-584-0381, 800-334-5551. Request Physical Sciences and Biological Sciences Catalogs.

Central Scientific Company, 11222 Melrose Ave., Franklin Park, IL 60131-1364. 708-451-0150, 800-262-3626.

Connecticut Valley Biological Supply Company, Inc., 82 Valley Rd., Southampton, MA 01073. 413-527-4030, 800-628-7748.

Edmund Scientific Company, 101 East Gloucester Pike, Barrington, NJ 08007. 609-573-6250.

Fisher Scientific Company, Educational Materials Division, 4901 W. LeMoyne St., Chicago, IL 60651. 312-378-7770, 800-955-1177.

Flinn Scientific Inc., P.O. Box 219, 131 Flinn St., Batavia, IL 60510. 708-879-6900, 800-452-1261. The *Flinn Chemical Catalog* also provides information on chemical safety, storage, and disposal.

Frey Scientific Company, 905 Hickory Lane, Mansfield, OH 44905. 419-589-9000, 800-225-FREY.

Hach Company, P.O. Box 389, Loveland, CO 80539. 800-227-4224. Test kits for environmental studies.

Lab Safety Supply, P.O. Box 1368, Janesville, WI 53547-1368. 608-754-2345. Specializes in safety equipment and supplies.

LaMotte Chemical Products, Box 329, Chestertown, MD 21620. 410-778-3100, 800-344-3100. Test kits for environmental studies.

Micro Mole Scientific, 1312 North 15th Street, Pasco, WA 99301.

Nalgene Labware Division, P.O. Box 20365, Rochester, NY 14602-0365. 716-586-8800. Specializes in transparent and translucent plastic laboratory equipment.

NASCO, 901 Janesville Ave., Ft. Atkinson, WI 53538. 414-563-2446, 800-558-9595.

Ohaus Scale Corp., 29 Hanover Rd., Florham Park, NJ 07932. 201-377-9000, 800-672-7722.

Sargent-Welch Scientific Company, 7400 North Linder, Skokie, IL 60076. 708-677-0600, 800-727-4368.

Science Kit & Boreal Laboratories, Inc., 777 E. Park Dr., Tonawanda, NY 14150-6782. 716-874-6020, 800-828-7777.

Wards Natural Science Establishment, Inc., 5100 West Henrietta Rd., P.O. Box 92912, Rochester, NY 14692. 716-359-2502, 800-962-2660.

CHEMICAL SAFETY, STORAGE, AND DIS-POSAL REFERENCES

While no laboratory exercise is without some potential risk, most accidents can be prevented by basing laboratory practices and procedures on established chemical safety guidelines. The following references offer valuable background information. Since risk assessments are subject to change with new studies, it is important to keep up to date by reading professional journals (the *Journal of Chemical Education, The Science Teacher*, etc.) and learning about any guidelines and regulations issued in your local school district or state.

Anon., *Aldrich Catalog Handbook of Fine Chemicals*, Aldrich Chemical Co., P.O. Box 355, Milwaukee, WI 53201. 414-273-3850, 800-558-9160.

Anon., *Chemical Catalog/Reference Manual*, Flinn Scientific, Inc., P.O. Box 219, 131 Flinn St., Batavia, IL 60510. 708-879-6900, 800-452-1261.

Anon., *Fire Protection Guide for Laboratories Using Chemicals*, National Fire Protection Association, Battery March Park, Quincy, MA 00269. Current edition. National safety code as applied to laboratory fire prevention and protection.

Anon., *The Merck Index*, 11th Edition (1989); Merck & Co., Inc., Professional Handbooks Dept., P.O. Box 2000, Rahway, N.J. 07065-0901. Authoritative, brief descriptions of properties (including health hazards) of many substances.

Anon., *Pocket Guide to Chemical Hazards*, U.S. Department of Health and Human Services, Public Health Service, Center for Disease Control, National Institute for Occupational Safety and Health Administration, Washington, D.C., 1980. A pocket summary of hazardous properties of substances regulated by OSHA.

Anon., *School Science Laboratories: A Guide to Some Hazardous Substances*, Council of State Science Supervisors and U.S. Consumer Products Safety Commission, Washington, D.C. 20207.

Council Committee on Chemical Safety (ACS), *Safety in Academic Chemistry Laboratories*, 5th edition. American Chemical Society, Washington, D.C. 20036. 1990.

Gerlovich, J. A. *et al. School Science Safety*, Flinn Scientific, Inc., Batavia, IL 60510. 708-879-6900, 800-452-1261. 1988. Two volumes; one for elementary science and one for secondary science.

Steere, N. V., ed., *CRC Handbook of Lab Safety*, Flinn Scientific, Inc., P.O. Box 219, 131 Flinn St., Batavia, IL 60510. 708-879-6900, 800-452-1261.

Wood, Claire. *Safety in School Science Labs*, Flinn Scientific, Inc., P.O. Box 219, 131 Flinn St., Batavia, IL 60510. 708-879-6900, 800-452-1261.

COMPOSITIONS OF CONCENTRATED ACIDS AND BASES

	Molar mass of reagent, g/mol	Approximate strength of concentrated reagent[a]	Molarity of concentrated reagent (M)	Milliliters of concentrated reagent needed to prepare one liter of 1 M solution[b]
Acetic acid, $HC_2H_3O_2$	60.05	99.8	17.4	57.5
Hydrochloric acid, HCl	36.46	37.2	12.1	82.5
Nitric acid, HNO_3	63.01	70.4	15.9	63.0
Sulfuric acid, H_2SO_4	98.07	96.0	18.0	55.5
Ammonium hydroxide, $NH_3(aq)$	17.03	28.0	14.5	69.0

[a] Representative value, mass percent
[b] Rounded to nearest 0.5 mL

INTRODUCTION

The goal of the Teacher's Guide is to provide you with (1) the history, philosophy and assumptions upon which this *ChemCom* course is based; (2) day-to-day teaching support; and (3) a variety of optional activities, supplemental references, and resource materials for you and your students to explore.

ChemCom is a student-centered, activity based, issues-oriented chemistry curriculum that encourages small group learning. Developed by the American Chemical Society, it includes many traditional chemical concepts and laboratory skills, as well as more bio-, industrial, organic, and nuclear chemistry than is normally encountered in a standard high school chemistry text. Thus, any prior teaching experience with chemistry will prove invaluable.

Perhaps the most unique feature of *ChemCom* is that chemistry is taught on a "need-to-know" basis, with societal and technological issues/problems determining the depth and breadth of chemical concepts taught. If your previous background in "real world" chemistry and chemical technology is limited, we invite you to explore the Optional Background Information and Supplemental Readings provided in this guide. Additionally, you'll find that articles about *ChemCom* topics in local newspapers, magazines, and TV news will help you to assist students in seeing chemistry in the context of their own communities.

Most of the complex and perplexing issues and problems facing our nation involve more than scientific concepts. They also involve individual and social values and group decision-making processes. Accordingly, *ChemCom* aims to prepare students for informed, effective citizenship through stimulating and engaging their hands, hearts (values and attitudes), and heads (higher cognitive abilities). As such, *ChemCom* places much of the responsibility for learning *on students themselves* as they explore how chemical concepts apply to their everyday lives. In addition to hands-on laboratory activities (on average, every four or five days), *ChemCom* includes a variety of student "decision-making" activities.

The decision-making focus, central to the *ChemCom* approach, requires you to engage in some teaching approaches not typical in traditional chemistry courses. The following material outlines the premises underlying this focus and highlights similarities and differences among various types of decision-making activities, especially the varied roles you will be asked to play. It is our hope that this guide will help make teaching *ChemCom* as enjoyable for you as it will be for your students.

A: Premises Underlying *ChemCom*'s Decision-Making Focus

1. Problem-solving and decision-making challenges are common in a variety of personal, social, and work-related settings.
2. Problem-solving and decision-making require higher level cognitive skills which can be improved with appropriate experience and practice.
3. Academic problem-solving as presented in traditional school settings has not the kinds of real-life decision-making students will encounter later as voting citizens and professionals.
4. Science curricula should be designed to include opportunities for students to practice real-life decision making strategies.
5. A science curriculum featuring an issues focus (such as *ChemCom*) provides a productive setting to develop students' decision-making skills.

6. Student application and use of chemical knowledge are at least as important educationally as their initial acquisition of that knowledge.
7. Modern society depends on group cooperation as much as on individual competition. Schools should provide opportunities for students to work together on cooperative ventures.
8. Academic preparation for life is at least as important as preparation for a subsequent course of study.

B: Characteristics of Problem-solving

Academic	Real-Life
One result expected	Multiple alternatives
Fully defined	Imperfectly defined
Discipline focus	Multidisciplinary
Right/wrong	Burdens/benefits
Judged immediately	Judged later
Algorithmic	Heuristic
Driven by knowledge	Constrained by missing knowledge
Objectively oriented	Value-laden
No solution is not a solution	No decision is itself a decision (!)
Dull (at worst)	Agonizing (at worst)

Source: Heikkinen, H. "Decision Making and *ChemCom*." *Chemunity*, Vol. 8, No. 2 (Summer 1987), p. 7.

C: Overview of Types of *ChemCom* Decision-making Activities

YOUR TURN

ChemQuandary

You Decide

Putting It All Together

Instructional time

10 minutes — 2 days

Open endedness

A "right" answer — Alternatives

Small group learning

Low — High

Student autonomy

Teacher-directed — Student-run

Problem-solving/decision-making

Academic — Real-life

YOUR TURN

Goal: to give students practice and reinforcement on basic chemical concepts, skills, and calculations in the context of applied, "real world" chemistry problems.

Time: typically assigned as homework, followed by brief (5-15 minutes) class discussion.

Instructional group: work performed by individual students—"checking" may be done in a variety of formats, but most commonly through whole class, teacher-led discussions.

Teacher's role: to insure students have acquired prerequisite skills before assignment and to provide subsequent in-class feedback and reinforcement.

Sample activities: paper and pencil activities on the metric system, nomenclature, computations, balancing equations, graphing and graphical analysis, etc., typically leading to single "right" answers.

Typical number per unit: 9.

ChemQuandary

Goal: to motivate and challenge students to think about chemical applications and societal issues, which are often open-ended and may generate additional questions beyond a specific "right" answer.

Time: may be assigned as either homework or in-class class start-up activities with brief (10-15 minutes) follow-up discussion.

Instructional group: individual or small groups of students (3-5) followed by class discussion.

Teacher's role: to encourage students to use scientific concepts to help frame and answer relevant questions; at times it may be important to avoid giving students "the answer", allowing them to wrestle with and discover answers on their own.

Sample activities: discrepant-event type information; indirect water use benefit and burden analyses; chlorination of water; developing a scenario of life without abundant fuel; using the concept of limiting reactant as a metaphor to account for world hunger; comparing and contrasting industrial and automotive pollution control.

Typical number per unit: 3-5 (plus other options in the Teacher's Guide).

You Decide

Goal: an inquiry activity where students are presented aspects of a societal/technological problem, asked to collect and/or analyze data for underlying patterns, and challenged to develop and support or refute hypotheses/solutions based on scientific evidence and clearly-stated opinions.

Time: one-half to one class period (some involve at-home work).

Instructional group: individual and/or small group (3-5 students) assignments leading to class discussions.

Teacher's role:
1. to teach students prerequisite communication and research skills (library, mathematical, etc.) and conceptual knowledge.
2. to hold a short pre-activity briefing (equivalent to a pre-lab presentation), forming student groups and clarifying objectives and general procedures of the activity (in some cases this may be a week or more in advance of the actual class activity to allow time for student data gathering).

3.   to arrange the physical environment to support small group and/or whole-class communication, as appropriate.

4.   to monitor and encourage students to:

   a. use their individual talents and strengths to contribute to group learning

   b. separate misconception/myth from fact, fact from opinion, and to use scientific concepts to clarify the issue and their own opinions

   c. work cooperatively, receiving and respecting the opinions of others

   d. deal productively with conflict and controversy by evaluating ideas and issues, not people

   e. stay on task to give students the autonomy they need to discover their own solutions, not dominating or obtrusively managing the discussion by trying to "solve" all group-dynamic problems (you should not lead the activity any more or less than you would lead a conventional laboratory activity).

6.   to evaluate individual and/or group work based on written reports, resource notes, and actual interactions/discussions.

7.   to provide any needed review and closure.

Sample activities: collect and analyze data on personal use of resources; interpret implications of environmental data; survey public understanding of nuclear issues; prepare personal diet and health analyses; interview the elderly on the "good old days."

Typical number per unit: 5.

Putting It All Together

Goal: to give students the opportunity to sum up, review and apply what they have learned in a unit in the context of a hypothetical or real societal/ technological, chemically-related community issue. Students are expected to develop and defend scientifically sound positions that acknowledge the roles of economics, politics, and personal/social values (this typically involves weighing benefits and burdens and developing workable solutions).

Time: two days of class time preceded by individual or group research.

Instructional group: individual assignment within small groups (3-5 students) which then interact in a whole-class format (simulation, debate, etc.).

Teacher's role: similar to those listed for a *You Decide*, but also:

1. to provide timely and clear pre-assignments with periodic progress checks as needed.

2. to conduct a more extensive debriefing of the activity, insuring that any scientific misconceptions are corrected and providing feedback to students that encourages improvements in their small group learning skills.

Sample activities: simulation of a town meeting on implications of a fish kill; evaluating the success of air pollution controls; forecasting alternative depletion curves for metal reserves in light of various assumptions; evaluating statements on nuclear issues in terms of validity and underlying values.

Typical number per unit: 1 (end of unit activity).

D: *ChemCom* Problem-solving Activities

Laboratory Activities

Goal: to motivate student interest in chemistry and its applications; to introduce and develop key chemical concepts; and to provide practice with laboratory skills.

Time: 50-minute class period (a few require two days).

Instructional group: student pairs

Teacher's role: similar to that for conventional chemistry laboratory activities:

1. to conduct a pre-lab discussion to clarify objectives, demonstrate new techniques, and emphasize safety procedures.
2. to monitor student laboratory work at all times (never leave the students unattended) to:
   a. insure safety regulations are followed;
   b. insure that students are actively engaged in the laboratory procedures;
   c. provide needed assistance.
3. to direct a post-lab discussion, pulling together and analyzing class data, clearing up any misconceptions, and connecting previous and subsequent learning activities.
4. to evaluate student work.
5. to encourage students to identify additional everyday applications of the concepts they discover.

Note: if you elect to substitute or add some of your own laboratory activities, try to maintain the "real-life" focus by strategies such as using household and grocery store chemicals and supplies.

Sample activities: purifying "foul" water; making a petrochemical (ester); simulating half-life; comparing sunscreens; qualitative and quantitative analysis of fertilizers, foods, etc.; electroplating.

Typical number per unit: 5.

Summary Questions

Goal: to help students review important concepts and skills developed in each major part of a unit and to introduce additional practical applications of chemistry.

Time: typically assigned as homework with an in-class review of usually no more than one-half a class period.

Instructional group: individual and/or group.

Teacher's role: to provide appropriate class settings (teacher-led recitations, review games, computer data bases, posted answer keys, etc.) that enable students to check their responses and receive needed remediation (it may be desirable to use some items for grading purposes).

Sample activities: specific items may involve naming compounds, listing key points, performing calculations, balancing equations, answering essay questions, etc. In many cases students are asked to apply their knowledge in new (but still familiar) contexts.

Typical number per unit: a typical *ChemCom* unit contains five sections; each section is supported by 3-12 summary questions.

Extending Your Knowledge

Goal: to stimulate and challenge able students and/or arouse interest among all students; to apply chemistry to new situations; to encourage students to seek additional information; may be open ended or not and may involve review, application, and extension of concepts previously learned. These are optional activities to be used at teacher discretion.

Time: highly variable (out-of-class time).

Instructional group: individual or group.

Teacher's role: to encourage students to pursue selected issues in greater depth; to direct students to appropriate resources, and (if desired) to allow class time for student reports.

Sample activities: optional library research on oil spills, origins of chemical symbols, DDT, history of the metric system, alternative automotive fuels and engines, cryogenics, Boyle's Law and scuba diving, etc., optional at-home "laboratory" activities on chemical indicators, acid rain, etc.

Typical number per unit: most units contain 1-3 (the Teacher's Guide contains additional project ideas).

E: Chemical Concepts Grid

Concept	*ChemCom* Units							
	Water	Resources	Petroleum	Food	Nuclear	Air	Risk	Industry
Metric (SI) measurement	I	U	E	U	U	U	U	U
Scale and order of magnitude	I	U	U	U	U	U	U	U
Physical and chemical properties	I	E	E	U	E	E	E	E
Solids, liquids, and gases	I	U	E		U	E	U	U
Solutions and solubility	I	E	U	U	U	U	E	U
Elements and compounds	I	E	E	E	E	U	U	U
Nomenclature	I	E	E	E	E	U	U	U
Formula and equation writing	I	E	E	E	U	U	U	U
Atomic structure	I	E	E		E			
Chemical bonding	I	U	E	E		U	E	U
Shape of molecules	I		E	U			E	U
Ionization	I	U	E		E	E	E	E
Periodicity		I/E/U						
Mole concept		I	E	U	E	E	U	U
Stoichiometry		I	E	E		U	U	U
Energy relationships		I	E	E	E	E	E	E
Acids, bases, and pH	I			E	E		E	U
Oxidation-reduction		I		U		U	U	E
Reaction rate/kinetics				I	E	U	E	U
Gas laws						I/E/U		
Equilibrium								I/U
Chemical analysis	I	E	E	E		U	U	U
Chemical synthesis			I			U		E
Biochemistry				I	U		E	
Industrial chemistry	I	E	E	E	E	E	E	E
Organic chemistry			I	E			E	
Nuclear chemistry					I/E/U			

CODE:	I	→	Introduced
	E	→	Elaborated
	U	→	Used

F. CONCEPTS AND ASSUMPTIONS RELATED TO *CHEMCOM* ISSUES

The following concepts and assumptions underlie and pervade the entire *ChemCom* program. They are listed separate from individual unit objectives to highlight their importance and to distinguish them from more straightforward, objective chemical concepts and skills. Though they may not be as readily testable (or in some cases even appropriate for testing purposes), students should be given many opportunities to examine and work with them.

Between the activities listed in the student text, options suggested in the Teacher's Guide, and your own resourcefulness in connecting local community events to chemistry, your students will undoubtedly examine and wrestle with many of these ideas. This process is a least as important as coverage of more specific science content objectives.

1. Citizen participation in public policy decision-making can often be enhanced by understanding the underlying scientific and technological concepts.
2. Students may become better citizens if they are a part of a cooperative learning environment where they are given opportunities to (a) learn and apply fundamental scientific concepts and methods and (b) practice group decision-making skills.
3. Many complex problems can be made manageable by breaking them into simpler subproblems or components.
4. Most everyday, "taken-for-granted" conveniences of our technological society did not exist a century ago and even today are considered luxuries (often unattainable) in many other societies.
5. Our technological society/economy is heavily dependent upon basic chemical resources found in the lithosphere, hydrosphere, and atmosphere. The indirect, "hidden" use of such resources far exceeds more direct, visible use.
6. Increasing population and consumerism place increasing demands on our chemical and societal resources, and increasing stress on the environment.
7. The ability to detect various pollutants often exceeds our capabilities to assess long term health effect and/or to control emissions.
8. Expert agreement upon underlying scientific and technological facts related to a given societal or technological issue does not necessarily imply that the experts will agree on a particular "fix." Social, political, economic, and ethical values influence the opinions and advice of experts.
9. Experts sometimes do not even agree on the underlying scientific and technological facts.
10. Public decision-making in a democratic government involves a balancing of differing viewpoints and interests with the intent of producing the greatest good for the greatest number (without denying the rights of minorities).
11. Citizen involvement in public policy is both a right/privilege and responsibility. Not to decide (or to decide not to decide or act) is also a decision.
12. Wise resource management on "Spaceship Earth" aims at maximizing both short and long term benefits while minimizing environmental and human health burdens/risks/costs.

13. Societal laws/regulations can either be written in congruence with natural laws or in opposition to them. In the end, nature's limitations and boundaries prevail.

14. Individual actions that may seem insignificant considered alone, can have major societal and ecological impact when multiplied by similar actions of many individuals.

15. The nonuniform distribution of chemical resources on our planet and the demand for a great diversity of goods necessitate a world import /export market economy.

16. All sources of information are not equally reliable. The best, most up-to-date scientific information is typically found in professional journals or obtained from scientific associations.

17. Cost-benefit analysis may be imprecise because: (a) the costs/risks/ burdens may be distributed over time, geography, and socioeconomic levels; (b) it is difficult to place a specific weight on many aesthetic values or environmental resources; (c) we often lack conclusive data on long term impacts; and (d) calculations may lead to different conclusions for different societal groups.

18. Pollution can be produced by natural events as well as by human activity. Nature has mechanisms to cleanse itself of a wide variety of natural and manmade pollutants.

19. "Zero" pollution is an economic and scientific impossibility. A reasonable goal is to act in ways that do not overload nature's mechanisms and capabilities.

20. Human technological systems (pollution control) can be designed to work with and/or supplement natural systems to maintain a high quality environment.

21. Pollution control approaches include (in decreasing order of desirability): reducing emissions, capturing/converting emissions at the source, or cleaning up after dispersal.

22. Pollutants and drugs/poisons, and even foods, can act in combination (synergistically) on biological systems to produce effects greater that the sum of their individual contributions.

23. All technological benefits are associated with some level of risks/costs/ burdens.

24. Short term technological "fixes" may produce unanticipated and unwanted side effects. Resolution of one problem may produce another.

25. Individual and societal health and well-being are largely dependent upon human decisions and actions.

26. Individuals are often willing to accept higher risks in activities of their own choosing than they are to accept smaller, involuntary risks imposed by pollutants or technologies they cannot avoid.

27. Government regulations are often designed to lower unavoidable and/or involuntary risks associated with various technologies.

28. A social partnership exists between scientific/technological industries and society as a whole. Ideally, such industries deliver useful products and services at a reasonable profit with minimal negative human health and environmental impacts.

29. Thoughtful and informed group debate will not necessarily produce "the answer" or unanimity of opinion about a particular issue/problem. Compromise is sometimes essential.

30. Our present state of knowledge about any given societal/technological issue is likely to contain some imprecision, inaccuracy, and uncertainty. Society must act upon the best available information with the understanding that additional information may call for subsequent reevaluation of a given issue and/or previous solution.

31. Scientific knowledge is an essential though not sufficient ingredient for successful resolution of societal/technological issues. Science alone will not provide "the answer" but can help clarify the underlying issues, pose relevant questions, and raise the level of public concern and debate.

32. Scientific and technological knowledge are capable of being used for both constructive or destructive purposes. Technologies must be developed and applied with compassion and respect for both human and environmental life and dignity if they are to serve humanity well.

AUDIOVISUAL/COMPUTER SUPPLEMENTS

The following audiovisuals and computer software are possible supplements for *ChemCom* units. Citation of a specific product does not imply endorsement. Instead, the listings are meant only to suggest types of materials that may prove useful in the classroom. In using such materials, it is important to remind students that learning is an active process. To guard against passivity and to reinforce active learning, utilize such techniques as overviews that tie an audiovisual to previous learning or potential discussion/debate questions that may arise from the subject matter.

GENERAL SOURCES

Bureau of Mines, Cochrans Mill Road, P.O. Box 18070, Pittsburgh, PA 15236.

Cambridge Development Laboratory, Inc., 1696 Massachusetts Ave., Cambridge, MA 02138. 800-637-0047. Offers a variety of software and interfacing devices.

Coronet Films and Video, 108 Wilmot Road, Deerfield, IL 60015. 800-621-2131. Offers a wide selection science-related programs on video. Discussion guides are available for most titles.

Hawkhill Associates, Inc., P.O. Box 1029, Madison, WI 53701. 800-422-4295. Offers a variety of science/technology/society programs on video. With every order you get a free year-long subscription to the *Hawkhill Science Newsletter for Scientific Literacy*.

International Film Bureau (IFB) Inc., 332 South Michigan Avenue, Chicago, IL 60604. 312-427-4545.

Modern Talking Picture Service, 5000 Park Street North, St. Petersburg, FL 33709. 813-541-7571. Videos produced by corporations, civic organizations, and government agencies are available. Videos will be shipped from the nearest library to minimize shipping time and return mail cost.

National Audiovisual Center, National Archives and Records Administration, Customer Services Section PY, 8700 Edgewood Drive, Capitol Heights, MD 20743-3701. 301-763-1896.

NOVA, WGBH Educational Foundation, Box 2222-TG, South Easton, MA 02375. Weekly STS-oriented 60-minute shows broadcast on public television. Semi-annual teacher's guides (containing a program synopsis, viewing goals, discussion questions, suggested activities, vocabulary, and recommended readings) and program transcripts are available. Include NOVA-program title desired, Box 322, 125 Western Avenue, Boston, MA 01234. Inquire with your school audiovisual coordinator concerning fair use, legal videotaping, or rent/purchase the programs through Coronet Films and Video.

PBS Video, 1320 Braddock Place, Alexandria, VA 22314-1698. 800-344-3337.

Project SERAPHIM, John W. Moore, Director, Department of Chemistry, University of Wisconsin, Madison, WI 53706. 608-263-2837. A National Science Foundation-funded clearinghouse for instructional computer information and software in chemistry. Request a current catalog of software titles, the *Teacher's Guide to SERAPHIM Software* (keys software to six popular chemistry books), and *Teaching Tips*.

CAREERS IN SCIENCE/TECHNOLOGY

Hawkhill Associates, Inc., P.O. Box 1029, Madison, WI 53701. 800-422-4295.

"People in Science Today." Focuses on individual scientists from the "Scientists at Work" series. Relevant titles include: "A Nuclear Physicist," "An Astrophysicist," "A Toxicologist," "An Industrial Chemist," "A Microbiologist," "A Biotech Entrepreneur," "A Molecular Biologist," "A Biochemist," and "A Geologist."

"Scientists at Work." Titles include: "Chemists at Work": Part 1 presents a short history of chemistry, from the alchemists to the present; Part 2 profiles the work of four living chemists; and Part 3 challenges students to consider careers in chemistry and their impact on economic, political, and social issues. Also available are the titles "Physicists at Work," "Biologists at Work," and "Earth Scientists at Work."

"Women in Science." "Part 1. Women in Science Yesterday" and "Part 2. Women in Science Today" highlight five outstanding young women scientists in biology, geology, physics, and chemistry.

SUPPLYING OUR WATER NEEDS

"Lake Study." Project SERAPHIM. Intended to involve students in scientific exploration by allowing them to collect data and identify the pollutant that is killing fish in a lake. "Pond Study" is a less rigorous version.

"Waqual." Project SERAPHIM. Computer simulation of a wastewater treatment plant. Students choose the fraction of waste that will receive primary, secondary, and tertiary treatment and note the effect on the pollution level in a river.

"Pollute." Cambridge Development Laboratory, Inc. Interactive simulation which examines the impact of various pollutants on typical bodies of water. Students consider the impact on oxygen content and fish survival of water temperature, waste type, waste treatment type, rate of waste dumping, and type of receiving body of water. Generates both tabular and graphical data, and enables students to test various hypotheses.

"Toxic Turmoil: The Silicon Valley Story." Modern Talking Picture Service. Describes how leaks from underground storage facilities contaminated groundwater, tarnished the clean image of the electronics industry, and caused widespread public concern. This well-balanced documentary examines the sources of the contamination, industry's response to the problem, and the potential health risks involved. Producer: Clean Water Task Force. Available only in California.

"The New Williamette." Modern Talking Picture Service. Documents the joint efforts of industry, citizens, and government to clean up the once polluted Williamette River in Oregon. It serves as a model for the cleanup task ahead for other rivers of the world. Producer: U.S. Army Corp of Engineers.

"Water—It's What We Make It." Modern Talking Picture Service. Focuses on the need for pure water and filtering technologies that can be used to obtain it. Producer: Water Quality Association.

"Water." Hawkhill Associates, Inc. Introduction to the physics, chemistry, and biology of water. Concepts include the shape of the water molecule, heat of fusion and vaporization, function in erosion, role as the Earth's thermostat and circulatory system, as well as the mystery and beauty of flowing water.

"Water: More Precious than Oil." PBS. Examines the issues of water management and its conservation on a global basis. Producer: KTCA, Minneapolis-St. Paul, MN.

CONSERVING CHEMICAL RESOURCES

"Mineral Resources." Project SERAPHIM. A simulation emphasizing energy and entropy. Acquaints students with factors related to the maintenance of mineral resources.

"Copper." Bureau of Mines/U.S. Dept. of Interior. Depicts both the past and present uses of copper, the methods devised for winning it from the Earth, and the importance of recycling.

"The Minerals Challenge." Bureau of Mines/U.S. Dept. of Interior. Shows the key role of minerals in today's society and how the ever increasing need for fuels, metals, and other mineral materials is being met by advanced technologies to find and mine thinner deposits, lower grade ores, and deeper and less accessible veins. Specific sequences include surface and underground mining of metals, nonmetals, and coal; ore preparation and transportation; petroleum and natural gas operations; and research to extract oil from oil shale.

"Wealth Out of Waste." Bureau of Mines/U.S. Dept. of Interior. Identifies the variety of wastes generated in the United States, and tells the story of recycling and waste utilization research aimed at turning the rubbish heaps into valuable resources.

"Mining and the Environment." Modern Talking Picture Service. Describes copper and silver mining in Troy, Montana, and how the ecology of the land and its abundant wildlife are being preserved while providing a livelihood for 350 families. An example of enlightened management working to provide the minerals needed by modern society while preserving the environment. Producer: Asarco Inc.

"Natural Resources: Doomsday or Utopia." Hawkhill Associates, Inc. Program on natural resources and their role in past, present, and future societies. Concepts include: resource "value," exploitation, and limitations.

"Population and Pollution." International Film Bureau Inc. Demonstrates how air (industry, planes, and automobiles), water (sewage, detergents, and poisonous chemicals), and land (littering) pollution results from the misuse of the environment and the demands of a constantly growing population. Emphasizes the need for a change in attitudes, and offers specific suggestions for individuals.

"Waste: The Penalty of Affluence." International Film Bureau Inc. Presents the environmental hazards of improper disposal of household, commercial, and industrial waste. Demonstrates sanitary landfills, composting, and incineration as well as techniques for converting waste to raw materials. Producer: Dr. Heinrich Feuter for the World Wildlife Fund.

PETROLEUM: TO BUILD? TO BURN?

"Octane." Project SERAPHIM. Tutorial and game on gasoline, octane rating, engine knock and compression ratio, and gasoline additives. Game involves a simulated cross-country drive for which gasoline must be blended to match characteristics of the car.

"Refiner." Project SERAPHIM. Tutorial: explains the various operations of a modern refinery. Game: the player (an operations manager of a refinery) makes decisions about operating a refinery in order to maximize profits.

"After Oil What?" International Film Bureau Inc. Discusses the limited lifespan of fossil fuels and considers the advantages and disadvantages of the alternatives of solar, wind, wave, tidal, and geothermal energies. Considers hydroelectric power, nuclear fission, fusion, and breeder reactors.

"Fields of Fuel: The Ethanol Debate." International Film Bureau Inc. Presents evidence from World War II and from present day U.S. and Brazilian farms and research laboratories that demonstrates that alcohol fuels are a practical and economic alternative for vehicles. Producer: Iowa State University.

"Energy—A Family Album." National Audiovisual Center—Energy Research and Development Center. Presents a history of energy use in America and highlights future options.

"Rethinking Tomorrow." National Audiovisual Center—Department of Energy. Depicts the growing national concern and mobilization for energy conservation. Flashbacks show how cheap oil fueled the turn of the century industrial revolution. Highlights energy programs of three cities.

"Plastics: The World of Imagination." Modern Talking Picture Service. Emphasizes the role of plastics in aeronautics, architecture, textiles, packaging, transportation, energy conservation, and medicine.

"Refiner." Modern Talking Picture Service. Highlights both the scientific and economic principles behind refining. Extensive animated sequences show how molecules are changed and reformed. Producer: Exxon Corp.

"The Refinery Film." Modern Talking Picture Service. How science works to produce hundreds of useful products from the hydrocarbon soup that is crude oil. Students learn a little about chemistry, a bit of physics, and a lot about what goes on inside the tanks and pipes of a large, modern oil refinery. Producer: Chevron USA.

"What Price Coal?" PBS. Focuses on environmental health and safety concerns associated with increased reliance on coal as a source of energy. Producer: NOVA.

UNDERSTANDING FOOD

"The Food Processor." Cambridge Development Laboratory, Inc. A dietary analysis program built around a database of over 1500 brand name and generic foods that is capable of analyzing individual foods and whole menus for nutritional content, and comparing these values to RDA recommendations.

"Biotechnology: Unlocking Nature's Secrets." Modern Talking Picture Service. Shows how scientists at Pioneer Hi-Bred International are combining traditional plant breeding procedures with new genetic engineering techniques to create improved crop seeds.

"Food for the Future." Modern Talking Picture Service. Visits the Land Pavilion at EPCOT Center to see Kraft scientists working to discover new ways to grow food.

"Food: Its Science and Your Future." Modern Talking Picture Service. Discusses the various career options (such as product development, food processing, quality control, regulatory compliance, and package design) within the field of food science and technology. Producer: Institute of Food Technologists.

"The Insect Challenge." Modern Talking Picture Service. Uses macrophotography to portray the complexity and sophistication of insects and man's efforts to control them. Producer: Chevron Chemical Co.

"Insects and Agriculture." Modern Talking Picture Service. Explores the scientific research conducted in the struggle humans wage with insects for production of food and fiber. Producer: Chevron Chemical.

"Soybeans: Food, Feed and Future." Modern Talking Picture Service. Describes how Pioneer Hi-Bred International breeders are trying to improve the soybean varieties. This title is not available in the states of Arkansas or Hawaii.

NUCLEAR CHEMISTRY IN OUR WORLD

"Nuclear Chemistry." Project SERAPHIM. Contains two programs: "Nuke 4" is a simulation of first-order radioactive decay involving a dice-throwing experiment, and "Decay" is a drill program on radioactive decay.

"Rutherford." Project SERAPHIM. Part I provides an illustrated overview of the classic alpha particle scattering experiment. Part II is an experiment simulation allowing students to experiment creatively with scattering phenomena.

"The Atom." Hawkhill Associates, Inc. "Part 1: How We Found Out About Atoms." traces the history of the atomic theory from Democritus to Bohr. "Part 2: What Is an Atom?" presents the modern view of the atom, including the work being done with the supercooled accelerator at Fermilab on quarks and an understanding of its relevance to the ordinary citizen.

"Nuclear Power." Hawkhill Associates, Inc. "Part 1: The History of Nuclear Power" and "Part 2: Nuclear Power—Today and Tomorrow" explain how chain reactions work, the difference between fission and fusion, and the difference between controlled nuclear reactions in a power plant, and explosive ones in a nuclear bomb. Concludes with a summary of the pros and cons of nuclear power and challenges students to debate the issue.

"Radiation." Hawkhill Associates, Inc. "Part 1: How Radiation Was Discovered" traces discoveries from Isaac Newton to Marie Curie. "Part 2: What Radiation Is, and What Radiation Can Do" distinguishes ionizing and nonionizing radiation, as well as electromagnetic radiation and nuclear radiation.

"Conquest of the Atom." International Film Bureau Inc. Recreates the experiments of J. J. Thomson, Ernest Rutherford, Sir John Cockcroft, Sir James Chadwick, and Hahn and Strassman. Points out that, in an atomic pile, the chain reaction is controlled so that a continuous supply of energy can be produced.

"The Discovery of Radioactivity." International Film Bureau Inc. Surveys the work of Roentgen, Becquerel, Curie, Elster, Geitel, and Rutherford.

"The Day Tomorrow Began." National Audiovisual Center—Atomic Energy Commission. Traces the history of the building and testing of the first atomic pile and the team which achieved the first sustained chain reaction on Dec. 2, 1942.

CHEMISTRY, AIR, AND CLIMATE

"Solids/Liquids/Gases." Project SERAPHIM. Simulations of Charles' Law, Boyle's Law, a ride in a hot air balloon, and a tutorial on the gas laws.

"The Earth's Atmosphere." Coronet Films and Video. Discusses the layered structure of the Earth's atmosphere and its interaction with solar radiation.

"The Energy Balance." Coronet Films and Video. Examines Earth's delicate energy balance and how humans consciously or unconsciously affect it.

"The Hole in the Sky." Coronet Films and Video. Discusses the role of chlorofluorocarbons in creating a "hole" in the Earth's ozone layer and the possible effects. Producer: NOVA.

"Air." Hawkhill Associates, Inc. Explores the history of how the Earth came to have an atmosphere, its present composition, and its life-sustaining roles.

"Air is Life." International Film Bureau Inc. Using visuals from around the world, stresses a global perspective and a need for cooperative international efforts as well as prudent individual fuel consumption. Producer: Condor Film.

"The CFC Story." Modern Talking Picture Service. Discusses the uses of chlorofluorocarbons and their possible effects on the stratospheric ozone layer. Producer: DuPont.

"The Climate Factor." National Audiovisual Center—National Oceanic and Atmospheric Administration. Discusses how climate has affected both animals and humans over the last 100 million years.

"National Center for Atmospheric Research." PBS. Focuses on the research of scientists and technicians at the NCAR, especially as connected to the effects of increased carbon dioxide levels. Producer: Creativity with Bill Moyers.

"To Catch a Cloud: A Thoughtful Look at Acid Rain." Modern Talking Picture Service. Seeks to dispel the propaganda and rhetoric surrounding this controversial issue with factual information about the atmosphere and research results. Producer: Alliance for Balanced Environmental Solutions.

HEALTH: YOUR RISKS AND CHOICES

"Health Awareness Games." Cambridge Development Laboratory, Inc. Five interactive programs: "Coronary Risks," "Why Do You Smoke?" "Exercise Weight," "Life Expectancy," and "Lifestyle" that relate student personal health and lifestyles to how well and how long they'll live.

"Health Risk Appraisal." Cambridge Development Laboratory, Inc. Using data from the Centers for Disease Control in Atlanta, this program helps students assess their health risks. By answering 40 health-related questions, students can determine mortality risks and means of lowering them.

"Disease and Health." Hawkhill Associates, Inc. "Part 1: The Agony and the Pride—a History of Disease and Health" highlights the work of Jenner, Pasteur, Koch, and Nightingale. "Part 2: Avoiding Disease, Achieving Health" emphasizes personal choice issues like smoking, diet, and drugs and presents an overview of the consensus and the disputes within the medical and scientific communities.

"Cancer and the Environment." National Audiovisual Center—National Institute of Health. Discusses pollution, food additives, smoking, and diet as possible causes of cancer, problems of the industrial environment, drug and chemical testing, prevention, and the mathematics of risk assessment.

"Cancer—What Is It?" National Audiovisual Center—National Institute of Health. Overview of the nature of cancer, including definitions of biomedical terms and illustrated examples of cancer growth.

"Enzymes: Nature's Magic." Modern Talking Picture Service. Examines the functions and roles of enzymes in nature and their potential use in combatting pollution and world hunger. Producer: Novo Industries.

THE CHEMICAL INDUSTRY: PROMISE AND CHALLENGE

"BCTC." Project SERAPHIM. Allows students to gather data on a fictitious, potential industrial pollutant. It illustrates the contribution science can make to a social problem for which the solution is not obvious.

"Toxic Wastes." Hawkhill Associates, Inc. "Part 1: A History of Toxic Wastes in the Biosphere" points out many striking examples of toxic wastes in the "good old days." "Part 2. Toxic Wastes Today and Tomorrow" discusses the nature of chemicals and their cycling in the biosphere, food chains, tolerance levels, radiation, and environmental controls.

"The Development of Electrochemistry." International Film Bureau Inc. Surveys work on the production of an electric current by chemical action as conducted by Volta, Davy, Oersted, Faraday, Van't Hoff, and contributions of Ostwald, Arrehenius, Heroult, and Hall. Illustrates some modern technological applications of electrochemistry.

"Doing Something." Modern Talking Picture Service. Shows what the chemical industry is doing to insure safe production, transportation, use, and disposal of chemicals. Producer: Chemical Manufacturers' Association.

"The Electrolysis of Brine." Modern Talking Picture Service. Demonstrates the electrolysis of brine into chlorine and sodium hydroxide and the importance of these chemicals in daily life. Producer: PPG Industries, Inc.

"How Much is Too Much?" Modern Talking Picture Service. Explains how chemical companies (such as Dow) test products to insure that they can be made, transported, and used safely. Concepts such as risk analysis, risk vs. hazard, and toxicity are discussed using everyday examples.

"The Need to Know." Modern Talking Picture Service. Shows what the chemical industry is doing about the issue of hazardous waste disposal. Producer: Chemical Manufacturers' Association.

"The Right Chemistry." Modern Talking Picture Service Explores the three major divisions of Morton Thiokel, Inc. (salt) and their functions.

ChemCom NEWSLETTER

An important part of the *ChemCom* package is *Chemunity News*, the free newsletter that highlights the educational programs and resources of the American Chemical Society as well as movements and trends in science education including *ChemCom*. *Chemunity News* is published five times a year: January, March, May, September, and November. If you wish to be added to the mailing list, send your name and address to: American Chemical Society, *Chemunity News*, Room 806, 1155 Sixteenth Street, NW, Washington, DC 20036.

SUPPLYING OUR WATER NEEDS
PLANNING GUIDE

Section	Laboratory Activity	You Decide
A. The Quality of Our Water		
Introduction A.1 Measurement and the Metric System A.4 Water and Health A.5 Water Uses A.6 Back through the Water Pipes A.7 Where Is the Earth's Water?	A.2 Foul Water	A.3 Information Gathering A.6 Water Use Analysis A.9 Riverwood Water Use
B. A Look at Water and Its Contaminants		
Introduction B.1 Physical Properties of Water B.2 Mixtures and Solutions B.4 Molecular View of Water B.5 Symbols, Formulas, and Equations B.6 The Electrical Nature of Matter B.7 Pure and Impure Water B.10 What Are the Possibilities?	B.3 Mixtures B.8 Water Testing	B.9 The Riverwood Mystery
C. Investigating the Cause of the Fish Kill		
C.1 Solubility C.2 Solution Concentration C.3 Oxygen Supply and Demand C.6 Acid Contamination C.7 Ions and Ionic Compounds C.8 Dissolving Ionic Compounds C.9 Heavy Metal Ion Contamination C.11 Molecular Substances in the River	C.12 Solvents	C.5 Too Much, Too Little? C.10 Heavy Metal Ions
D. Water Purification and Treatment		
Introduction D.1 Natural Water Purification D.3 Hard Water and Water Softening D.4 Municipal Water Purification D.5 Chlorine in Our Water	D.2 Water Softening	D.6 Chlorination and THMs
E. Putting It All Together: Fish Kill-Who Pays?		
Introduction E.1 Directions for Town Council Meeting E.2 Looking Back and Looking Ahead		

TEACHING SCHEDULE

	DAY 1	DAY 2	DAY 3	DAY 4	DAY 5
Class Work	Intro to *ChemCom* Read pp. 2-10	Discuss pp. 5-7		CQ p. 7	YT pp. 12-13
Laboratory		Pre-lab LA pp. 8-10	LA pp. 8-10	Post-lab LA pp. 8-10	
Homework	YT pp. 6-7	Read pp. 10-11	YD pp. 10-11	Read pp. 11-14	Read pp. 15-19

	DAY 6	DAY 7	DAY 8	DAY 9	DAY 10
Class Work	Discuss pp. 15-17	YDs pp. 17-19	Review Part A Quiz Part A		YT p.29
Laboratory			Pre-lab LA pp. 25-26	LA pp. 25-26	Pre-lab LA pp. 32-35
Homework	YD pp.10-11	SQ p.19 Read pp. 20-26	Read pp. 26-35	YT pp. 27-28	Read pp. 36-38

	DAY 11	DAY 12	DAY 13	DAY 14	DAY 15
Class Work		YD pp. 36-37	Review Part B Quiz Part B	Discuss pp. 42-44 YT pp. 43-44	YD pp. 47-51
Laboratory	LA pp. 32-35				
Homework		SQ pp. 37-38	Read pp. 39-44 YT 41-42	Read pp. 44-51	Read pp. 51-54

	DAY 16	DAY 17	DAY 18	DAY 19	DAY 20
Class Work	Discuss pp. 51-54 YT p. 54	YD pp. 58-59 Discuss pp. 59-60		Review Part C Quiz Part C	
Laboratory		Pre-lab LA pp. 60-61	LA pp. 60-61	Pre-lab pp. 65-67	LA pp. 65-67
Homework	Read pp. 55-62	SQ pp. 61-62	Read pp. 63-67		Read pp. 67-75

	DAY 21	DAY 22	DAY 23	DAY 24	DAY 25
Class Work	YD pp. 74-75	Review Part D Quiz Part D PIAT pp. 78-82	PIAT pp. 78-82	Review Unit	Exam
Laboratory					
Homework	SQ p. 75 Read pp. 76-83		PIAT Letter		

LA = Laboratory Activity; **CQ** = ChemQuandary; **YT** = Your Turn; **YD** = You Decide; **PIAT** = Putting It All Together.

This first *ChemCom* unit focuses on the fictional town of Riverwood and problems with its water supply. From that begining, it goes on to explore properties of water and aqueous mixtures, and to examine procedures for obtaining adequate supplies of water pure enough for our needs. Fundamental concepts related to formulas, nomenclature, and solubility are introduced and developed along with the Riverwood story.

The beginning of the unit exemplifies the *ChemCom* approach: first create a need-to-know in the minds of the students via a science-technology-society (STS) problem or issue, and then develop the chemical concepts that shed light on the issue. This is a fairly radical departure from traditional chemistry, where students often merely memorize answers to questions that are remote from their lives and concerns.

OBJECTIVES

Upon completion of this unit the student will be able to:

1. List and use the units of the modernized metric system (SI) in measurements of length, volume, mass and density. [A.1]
2. Discuss direct and indirect water use and their importance for water conservation. [A.5, A.8, A.9]
3. Describe the function and operation of the hydrologic cycle and indicate the primary storage reservoirs of the Earth's water supply. [A.6, A.7]
4. Discuss the impact of water's unusual physical properties on Earth's life forms. [B.1]
5. Define the terms solution, solvent, and solute and apply them in an example. [B.2]
6. Classify matter in terms of elements, compounds, and mixtures; and distinguish among different types of mixtures (solutions, colloids, and suspensions) in a laboratory setting. [B.2-B.4]
7. Interpret the symbols and formulas in a balanced chemical equation in terms of atoms and molecules. [B.5]
8. Describe the three basic subatomic particles (proton, neutron, and electron) and their connection to the polarity and solubility of a compound. [B.6, C.1]
9. Define the terms insoluble, unsaturated, saturated, and supersaturated, and calculate solution concentration as a percentage. [C.1, C.2]
10. Use solubility curves to describe the effect of temperature on solubility, and calculate percent saturation. [C.1, C.2, C.4]
11. Demonstrate the ability to organize and interpret environmental and other data in graphs or tables. [C.1, C.5, C.10]
12. Given the pH of a solution, classify it as acidic, basic, or neutral. [C.6]
13. Determine the formula and name of simple ionic compounds when provided with the anion's and cation's names and charges. [C.7]
14. Evaluate the risks of contaminants in our water supply, with particular attention to heavy metal ions of lead, mercury, and cadmium. [C.10]
15. Compare and contrast natural and artificial water purification systems, and assess the risks and benefits of water softening and chlorination. [D.1, D.3–D.5]

A: THE QUALITY OF OUR WATER

The newspaper articles on pages 2-4 ("Fish Kill..." and "Townspeople React...") introduce students to a troubling water-quality problem in Riverwood, setting the stage for the major "story line" of the unit. Students will follow the progress of Riverwood citizens in dealing with this issue. Encourage your students to note the difference between reported fact and opinion, and to identify the kind of scientific information useful in solving such issues. It's also worth noting that a large problem can often be solved more easily if it is first reduced to a series of smaller sub-problems. Invite your students to identify some of the sub-problems involved in the Snake River water-quality crisis.

ChemCom units provide the framework for student involvement in various STS issues. You may want to have students collect relevant newspaper articles and carry out individual projects. (The *Extending Your Knowledge* sections provide suggestions.)

A.1 MEASUREMENT AND THE METRIC SYSTEM (pages 5-7)

Emphasize that the modernized metric system (SI) is used worldwide, and serves as an important international "language." Encourage students to consider items that are already measured exclusively in metric (photographic equipment, skis, ammunition, 2-L plastic soft drink containers, etc.) versus items that do not allow for easy conversion (football fields, automotive tools, etc.). Try to avoid referring to our customary U.S. (or English) units as a "system"—the term grants these units more coherence and order than they actually merit. The only real *system* of units is the metric system. Both the text and the student activity focus on "thinking metric" rather than emphasizing English/metric or metric/English conversions.

YOUR TURN

Meters and Liters (pages 6-7)

This activity provides drill and practice in metric manipulation and conversions, and encourages students to use metric measurements in their everyday lives. It also illustrates why scientists prefer metric over U.S. customary units.

Answers to Questions:

1. a. 18 mm (penny); 21 mm (nickel); 17 mm (dime); 23 mm (quarter).
   b. 1.8 cm (penny); 2.1 cm (nickel); 1.7 cm (dime); 2.3 cm (quarter).
2. a. 1000 cm^3;
   b. 1000 mL;
   c. 1 L;
   d. 1000 sugar cubes.
3. a,b. Individual student answers here. Most common beverage containers now report volumes in both U.S. customary units and metric units. One liter is slightly larger than a quart (1 qt = 0.947 L or 0.947 dm^3). One fluid ounce is approximately 30 times larger than one milliliter. To be more precise, 1 mL = 0.034 oz, or 1 oz = 29.6 mL.
4. a. The conversion of the magnitude of metric units, based on the number 10, is accomplished by decimal point changes. The units of volume are related to the length by 1 L = 1 dm^3, 1 mL = 1 cm^3, etc. For conversion in the U. S. customary units one must memorize factors such as 5,280 feet = 1 mile, 16 fluid ounces = 1 quart, etc.
   b. At the present time in the United States we sometimes need to translate metric units into U. S. customary units, contributing to one apparent disadvantage. Such conversions often have the effect of making metric–based units seem unnecessarily awkward. SI units are also less familiar to many U. S. citizens and require retooling for manufacturers (practical disadvantages of SI) and are probably the source of some resistance to going metric in this country.

Water, Water Everywhere (page 7)

To challenge students, consider closing a class session by stating the puzzle described in the unit, but do not discuss the explanation. Underscore the problem by pointing out that the water-use figures shown for one egg and a can of juice mean that eating two eggs and drinking 1.3 L of juice would involve enough water for taking 10 showers, washing 10 loads of clothes, or 25 loads of dishes, or flushing a toilet 100 times. Why?

In discussing this *ChemQuandary*, you might begin by asking students to list processes involved in obtaining eggs or cans of juice: raising feed for chickens, providing drinking water for the animals, washing eggs, watering fruit trees, washing fruit, etc.. Then trace the history of the water used in the juice can as follows. Washing the can after production probably required a small amount of water, an example of direct water use. However, digging the iron ore from the ground and moving it to the steel mill required an additional 4 L of water per can. The steel mill that produced the original metal in the can used about 76 L of water. The rolling mill that fabricated the metal for the can consumed about 46 L of additional water. Thus, the total indirect volume of water used to make the juice can is about 126 L. Thus, drinking canned juice may help solve the short-term problem of water in Riverwood, but does not contribute to long-term water conservation.

OPTIONAL BACKGROUND INFORMATION

Indirect Water Use

Like energy, water is an essential "behind-the-scenes" ingredient in nearly every human endeavor. In the United States, where overall water use is approximately 6230 L/person/day, industrial processes, such as electrical energy production, primary metals production, chemical products, petroleum refining, pulp and paper manufacturing, and food processing, account for about 51% of the use, agriculture for about 40%, and the municipal sector for about 9%. The point to make with students is that not only do they use large quantities of water directly (325 L/person/day), but even larger quantities are used for them behind the scenes.

Water Used on a Daily Per-Person Basis in the U.S. in 1980

Use or product	Average volume in L (gal)	% of total
Direct personal use	325 L (86 gal)	5.1%
Indirect use	6000 L (1584 gal)	94.9%
TOTAL	**6325 L (1670 gal)**	**100.0%**
Subcategories of indirect use		
Irrigating land and growing food	2600 L (693 gal)	41.7%
Producing energy (especially electricity)	2400 L (632 gal)	38.5%
Mining and manufacturing	700 L (183 gal)	11.2%
Commercial (jobs and services)	340 L (89 gal)	5.4%

Note: Estimates of water use vary with the source and with the year. It is not important that students remember exact figures, but only the order of magnitude and a sense that: (a) We all use directly a very large quantity of water, with little awareness of the fact, and (b) by far, the largest portion of our water use is hidden or indirect. The *ChemQuandary* in the text provides just two examples of indirect water use. Other examples include: automobile (379,000 L); Sunday newspaper (1,050 L), loaf of bread (570 L); glass of milk (300 L), refining one gallon of gasoline from crude oil (42 L).

You may wish to assign students an additional activity, depending on where in the United States you reside.

Some suggestions:

1. Since 90% of the water use in the West is for irrigation, investigate the source of the irrigation water, examining the watershed, the Continental Divide, and concerns of farmers using irrigation downriver. Students may be interested in discussing problems encountered at different sites along the same river.

2. Investigate the uses of cooling water in electrical plants, and the relative numbers and types in the West and East. Also explore the difference in water use between a coal–fired or oil–fired plant and a nuclear power plant. Students may think that the presence of a cooling tower indicates a nuclear power plant, but this is not necessarily true.

A.2 LABORATORY ACTIVITY: FOUL WATER (pages 8-10)

This laboratory activity is designed to stimulate student interest in some of the practical problems associated with water quality. The goal of the activity is for students to purify a sample of highly impure "foul" water to the point that it could be used to wash their hands. **Do not extend the challenge to making the water suitable for drinking,** since no laboratory-based materials should be swallowed. The activity also introduces students to the techniques of filtration, absorption, and separation.

Since this is the first laboratory activity in the course, emphasize safety procedures, identification and use of safety equipment supplied in your classroom (such as the eye wash, shower, or fire blanket), as well as the general rules of safe laboratory conduct. Be sure to discuss the *Safety in the Laboratory* section in the preliminary pages of the book, pages *xiv-xv*. You should also discuss the potential hazards of wearing contact lenses in the laboratory—*we strongly recommend that students remove contacts prior to any laboratory activity*. Any reagent splashed into a contact-wearer's eye is held directly onto the eye's surface by the contact lens; even an eye-wash fountain cannot rinse this trapped reagent away.

Explain the meaning of the protective goggle symbol found at the beginning of each laboratory activity. The goggle sign indicates that students should wear goggles for the activity, be alert, and that the activity may require particular attention to minimize danger. Highlight all of your own school's laboratory safety policies.

Time

One class period for student work; one additional day for post-lab distillation demonstration.

Materials (for a class of 24 working in pairs)

    2-3 L of foul water (100 mL/pair) (see recipe below)
    12 100-mL graduated cylinders
    24 150-mL beakers
    12 funnels (73 mm inside diameter)
    12 ring stands
    12 ring clamps
    12 clay triangles
    12 cups (paper preferred)
    12 paper clips (for punching holes in cups)
    12 filter papers (Whatman #1, 11 cm diameter)
    12 125-mL Erlenmeyer flasks (for mixing charcoal and water)
    150 g decolorizing charcoal
    2 kg sand, cleaned
    2 kg gravel, cleaned (small grind; roughly 5 mm diameter)
    1 distillation apparatus
    1 conductivity meter
    rubber tubing

Foul Water Recipe

Add the following ingredients to an empty two-liter soft-drink bottle (All specified quantities can be estimated.)

    1/2 tablespoon table salt (NaCl)
    1/2 teaspoon garlic powder
    1/2 cup used coffee grounds
    1/3 cup vegetable oil

Fill the bottle with tap water, cap tightly, and shake well. Save the excess for Section B.2 and the discussion of mixtures.

Advance Preparation

If you wish, encourage students to compete with other lab teams in obtaining the largest volume of "pure" water. Set up a water-distillation apparatus, which you will demonstrate at the close of the activity. (See Post-Lab Activities below.) You should also assemble a conductivity apparatus to be used to test the student purified water samples.

Pre-Lab Discussion

Provide general safety orientation. Make an initial display of the foul-water mixture, perhaps by pouring into a large beaker. Possibly discuss pure and potable water, and introduce the problem of knowing how pure a water sample actually is. Review the methods of purification used in the activity.

Point out that the data table should be larger than the sample versions shown on page 8 in the text—the tables are purposely reduced in size in the unit to discourage students from writing in their books.

Lab Tips

Dispense about 100 mL of the foul water to each pair, shaking the mixture before dispensing to ensure homogeneity between samples.

Some forms of decolorizing charcoal are quite messy and can be irritating if inhaled; it is helpful to prepare a thick slurry of the charcoal with water beforehand. Have your students use a spoonful of the slurry for this step. Granular charcoal, used in aquaria, is a cheaper alternative to the powdered charcoal.

Do not reveal the composition of the original foul-water sample until your post-lab discussion.

Provide containers for students to place used sand, gravel, and charcoal.

Filtration will not remove the dissolved table salt; the final sample will conduct electricity. Compare the conductivity of distilled water with the student purified water.

During the post-lab distillation demonstration, the liquid should not be boiled away to dryness, since the residue will be etched into the flask, preventing cleaning, and an overheated flask may break. About 1/4 of the original liquid should be left in the flask at the end of the distillation. *Do not permit students to drink any of the water samples used in this activity.*

Calculations

A range of 40-60% recovery is to be expected. Students should see that water purification has a cost, and there is also a loss of usable water in the process.

Expected Results

	Color	Clarity	Odor	Presence of oil	Presence of solids	Volume
Before treatment	dark	cloudy	strong	yes	yes	80-100 mL
Oil-water separation	dark	cloudy	strong	less	yes	75-95 mL
Sand fitration	dark	clearer	strong	none	none	70-90 mL
Charcoal filtration	lighter	clearer	less	none	none	65-85 mL

Post-Lab Activities

Suggested sequence: (1) Check the electrical conductivity of student purified-water samples and compare to that of tap water or distilled water. (2) Distill a sample of student purified water. Discard the initial 20% of the distillate since it may contain dissolved gases, and discard the final 20% because it may contain dissolved salts; only the middle 60% is considered "pure." (3) Compare the electrical conductivity of the distillate with your previous tests. Discuss the implications of these observations.

Many students will think their samples are pure because they look clear and have no odor. The conductivity test should lead students to think about the presence of contaminants that cannot be detected with the unaided senses. Point out the use of distilled water as a control in the conductivity test. Students may wish to ponder the relatively high cost involved in distillation (in terms of equipment, energy, and time) in order to obtain "pure" water. In our society, this is an impractical way to purify municipal water supplies.

Answers to Questions:

1. Color, clarity, odor, presence of immiscible liquids, presence of ions as shown by the conductivity tests.
2. Distillation is too expensive for large volumes of water due to the time, energy, and equipment required.

A.3 YOU DECIDE: INFORMATION GATHERING (pages 10-11)

Throughout *ChemCom*, students will have many opportunities to explore the role of chemistry in their own lives. This activity, although requiring participation of family members, generally receives a positive response and can be beneficial. The activity explores water use and promotes awareness of the large quantities of water used routinely in a household. The activity also emphasizes information gathering as an important first step in making decisions. The data collected will be analyzed later in *You Decide* A.8.

The time needed for data collection is three days. If possible, include one weekend day, in order to obtain a more typical sample period.

A.4 WATER AND HEALTH (page 11)

This brief section provides a rationale for why we should be concerned with the quality and quantity of our water. We tend to take our continuing supply of clean water for granted—something that is only a "faucet twist" away. However, this is only a recent luxury, as the text points out. In addition, the lack of adequate water in drought-stricken areas of the world presents us with a continuing succession of tragic headlines. The water-related issues for all nations are associated both with adequate quantity and quality. In addition, the effect of too much water and flooding can also be explored.

OPTIONAL ACTIVITY

You may wish to initiate an ongoing student project of maintaining a "Chemistry-in-the-News" bulletin board where students will assemble local, regional, and national newspaper or magazine articles, cartoons, etc., that relate to the main issue of the unit being covered. While many variations are possible, you may wish students to use four different color highlighters to mark statements that refer to (A) the issue; (B) relevant chemical facts; (C) proposed or hypothesized solution(s); and (D) political implications.

Activities like this help insure that students see that Riverwood-type scenarios are not just fictions used for teaching. Chemistry-related issues consistently appear in both national and local news, and *ChemCom* topics produce a wide variety of information far beyond the usual collection of articles about chemical spills and explosions.

OPTIONAL BACKGROUND INFORMATION/DISCUSSION TOPIC

On a worldwide basis enough rain and snow fall each year to cover Earth's total land area to a depth of 83 centimeters. And yet, uneven distribution and inadequate technology make quality fresh water a major constraint on economic growth, food production, and general health in many nations. This is especially true of Asia and Africa, where periodic droughts lead to the deaths of millions through famines. Even when water is available in adequate quantities, it may be of insufficient quality. It has been estimated that 80% of all the world's sickness is caused by contaminated water, and that persons with water-borne diseases occupy half of the world's hospital beds and die at a rate of 25,000 per day.

It is hard for students from industrialized nations, who directly use between 100-325 L of water/day and have never seen the results of cholera and other water-borne diseases, to appreciate conditions in developing nations, where individuals may have access to only a public faucet or, even worse, must constantly search to obtain the biological minimum of 2-5 L of water/day. It has been estimated that less than 10% of the world's population has access to sufficient quantities of clean water. Because of time constraints, this global, humanitarian issue is not addressed in the unit. However, it certainly deserves mentioning and would also make an excellent extra-credit report option.

A.5 WATER USES (pages 11-13)

We consume much more water than many persons imagine in growing, preparing, and manufacturing foods and consumer goods. Students may have the impression that direct home water use represents the major water demand of our society. The fact is that the manufacturing of goods far exceeds the more visible (direct) use of water.

YOUR TURN

U. S. Water Use (page 12)

Answers to Questions:

1. a. manufacturing
   b. irrigation
   c. irrigation
   d. irrigation
   e. mining
   f. irrigation.

2. Possible reasons would include land forms (mountains, etc.), available precipitation, population, climate for grow—ing, location of factories, etc.
3. Since the East is more densely populated, 32% of the total water used in the East is for homes and offices com—pared to 5% in the West. Since the East contains more manufacturing plants, 40% of the total water used in the East is used for manufacturing as compared to 2% in the west. 13% of the water use in the East is for steam-electrical plant generating, while the West, with extensive hydroelectric resources, uses only 1% for generating.

A.6 BACK THROUGH THE WATER PIPES (pages 15-16)

This section traces tap water "back through the pipes" to its source in the hydrologic cycle, supporting the fact that the Earth has a finite (though recyclable) supply of water. Students can investigate the source of water in their own commu—nity: reservoirs, water towers, pumping stations, wells, rivers, or other site-specific locations. A field trip can be highly motivational and informative.

A.7 WHERE IS THE EARTH'S WATER? (page 17)

Students will not be surprised to learn that most of the Earth's water supply (97%) is in the oceans. However, they may be puzzled to find that the next largest store of the Earth's water is locked in glaciers and ice caps. Neither of these sources of water is readily available—it would be very costly to change either to a useful form.

We have been fortunate to have a generally abundant water supply in the United States, but how long will this situation continue? Remind students that even though this country's total water-supply profile is favorable, local and regional problems still are quite visible. You can use the fish-kill problem in Riverwood as one typical example. What water-use priorities are needed when water is in short supply? *You Decides* A.8 and A.9 will challenge students to make thoughtful decisions about how water should be used during such a shortage.

A.8 YOU DECIDE: WATER USE ANALYSIS (pages 17-18)

This activity is intended to inform students about the large quantities of water their families use daily. Students will gain useful experience transforming raw data into more interpretable forms. The data processing includes finding the total volume of water used by the family during the survey, and then calculating the total water use per person as well as the daily water use. Students will compare their results with individual water-use information provided in the unit.

The activity takes about 25-35 minutes. A data sheet for reporting the results may be helpful, and group work on calculations may help some students over the rough parts. These calculations provide a good opportunity to stress the importance of keeping track of units associated with any quantities being computed.

OPTIONAL BACKGROUND INFORMATION FOR YOU DECIDES A.8 AND A.9

Although household and other municipal water use accounts for about only 9% of the total water use in the United States, delivering adequate quantities of water of sufficient quality for this purpose is becoming increasingly expensive for individuals and communities. Investments in household water-saving devices tend to pay off through lower water, sewer, and energy costs within a few months to a year, as shown in the following chart:

Effect of Water Saving Devices on Water Use

Activity	% of total indoor water use	% W/O conservation (10^3 L/person/year)	% With conservation	% Savings with conservation
Toilet flush	38	34.5	16.4	52
Bathing	31	27.6	21.8	21
Laundry and dishes	20	18.0	13.1	27
Drinking and cooking	6	5.5	5.5	0
Miscellaneous	5	4.1	3.7	10
TOTAL	100	89.7	60.5	33

Source: Lester R. Brown (ed.), *State of the World—1985: A Worldwatch Institute Report on Progress toward a Sustainable Society*, Chapter 3 (W.W. Norton & Co.).

Using fairly simply conservation devices that families can find in hardware stores and install themselves, water consumption can be reduced by as much as a third, saving money without real hardship. You may wish to encourage students to visit hardware or plumbing stores to examine such devices.

A.9 YOU DECIDE: RIVERWOOD WATER USE (pages 18-19)

This activity simulates typical water-use decisions individuals might make during a water shortage. Students are challenged to plan how to use a limited supply (375 L) of water for maximum benefit. More specifically, the objectives are to determine what general actions could be taken to reduce water use in a family household; to decide which uses could be eliminated; to decide which uses have priority; and, to determine what repeated savings might help conserve this resource.

This activity lends itself to group work and will take 15-20 minutes to complete. Students may have experienced some actual short-term community emergency, such as a waterline break, that would increase the realism of this exercise.

Answers to Questions:

1. a. Eliminate washing cars, windows, pets, clothes, watering indoor and outdoor plants, watering lawn. Some may answer bathing, showers, washing hair!
   b. Dirty car, dirty pet; plants may die, lawn turn brown; odor level would increase!
2. a. Reduce water for bathing, showering, washing clothes and dishes, flushing toilet.
   b. Shorter showers, lower bath water level; wash dishes by hand; don't flush toilet every time; low-flow toilets and shower heads.
3. a. Use impure water for washing car, flushing toilet, possibly watering plants and lawn.
   b. From washing hands, preparing food.

It has been estimated that 30-50% of the water used in the United States is wasted. In the discussion, lead students to see how individual acts of conservation can lead to significant savings at the household level, and even more at the community or national level, where the impact is multiplied by 240 million.

OPTIONAL CHEMQUANDARY

Explain the relevance of the Benjamin Franklin quotation (1746) "When the well is dry, we learn the worth of water."

Answer: (a) It is in the nature of economic systems to place a higher value on scarce commodities. (b) Due to uneven distribution, water has differing values in different parts of the world— in oil-rich, water-poor Kuwait a barrel of water is more precious than a barrel of oil. (c) Even though water is a renewable resource, it is possible to deplete and pollute water at a rate that exceeds natural chemical recycling processes.

PART A: SUMMARY QUESTIONS (page 19)

1. a. dime = 1 mm; b. glass of water = 250 cm³; c. pencil diameter = 7mm; d. gallon = 3.8 L.
2. a. 735 cm³ = 735 mL = 0.735 L; b. 10.7 mm = 1.07 cm.
3. In general the implications of lack of water can be explored in both short-term and long term situations. Students will probably have several suggestions for family and town in the short-term, but encourage them to consider the long-term situation, which is easier to apply to the region or the United States.
4. a. Indirect water use refers to water used in the production or processing of foods and goods.
   b. Water for grain, water for mixing dough, growing yeast, washing pans, packaging materials, grinding and trans– porting grain and flour.
5. Water is very stable, and the Earth's gravity prevents losses to outer space. Our planet is a closed system, keeping the amount of water on the Earth constant.
6. Cholera, typhoid fever, dysentery (possibly others).
7. a. There is concern over the availability of pure water, since more than 97% of the total water is in the oceans and is unfit for human use. Due to increasing population we are using more and polluting more.
   b. We could use more sea water by distillation (using solar power) or purification (by using reverse osmosis). Plans for desalinating ocean or brackish water may become economically competitive with the cost of transporting fresh water over long distances.
   c. Ocean water could be used in coastal cities for less critical tasks, such as toilet flushing.

EXTENDING YOUR KNOWLEDGE (page 19)

- The hydrologic cycle works because liquid water evaporates to form a lower density gas which rises into the atmosphere to cool and form clouds. Evaporation also leaves behind the dissolved particles present in impure water. Water in clouds may freeze, producing higher density liquid and solid states that fall back to the Earth as precipitation.

Planet	Approximate temperature (°C)	Atmospheric Pressure (atm)	Composition
Earth	−40 to +45	1	N_2, O_2
Moon	No atmosphere	none	none
Venus	480	90	CO_2

The large amount of water on Earth's surface helps to stabilize the surface temperature during the Earth's rotation and revolution around the sun. Compared to other common substances, large quantities of heat energy are involved when the temperature of water rises or goes down. Also, a large quantity of energy is used or released when water changes from one state to another.

Heat of fusion is the thermal energy needed to melt (convert to liquid) a specified quantity of a solid at its melting point. The reverse process can be demonstrated by melting about 100 g of sodium acetate, allowing to cool slowly, then dropping a seed crystal into the supersaturated mixture: the temperature rises notably as solid crystals form. (Note that the word denoting a fuse—an inexpensive device that prevents overloaded electrical circuits—is based on the fact that a solid metal melts, or fuses, when the electric current exceeds safe levels.)

Heat of vaporization is the thermal energy needed to vaporize (convert to gas) a specified quantity of liquid at its boiling point.

Heat capacity refers to the quantity of thermal energy needed to change the temperature of a given sample of matter by one degree Celsius.

- Compared with other common substances, water has an unusually high heat of fusion, heat of vaporization, and heat capacity. For instance, the next best candidate for a life fluid is ammonia.

Formula	Melting point °C	Boiling point °C	Specific heat capacity joules/g	Heat of fusion joules/g	Heat of vaporization joules/g
H_2O	0	100	4.18	333	2.26
NH_3	−78	−33	4.48	341	1.37

Ammonia has a much lower freezing level point and boiling point than water has. Thus, even if ammonia were as good a solvent, an ammonia-based organism would have to function within a much narrower temperature range (-78°C to -33°C).

B: A LOOK AT WATER AND ITS CONTAMINANTS

Part B returns to Riverwood's problem with another news report about the fish kill. To understand the reported issues, students need to know more about the chemistry of water and how substances interact with water. Some properties of water are given, and students learn to classify mixtures as suspensions, colloids, or solutions in a laboratory activity. To explain these observations, a particulate model of matter is invoked. The universal symbolic language of chemistry is also introduced. Students then learn the chemical meaning of the terms "contamination" and "pure water", which leads them to laboratory testing for dissolved ions in water samples.

With this new information students go back to the newspaper articles with a more scientific viewpoint. Part B stresses that observations are explained by models, which permit reliable predictions regarding the behavior of matter. Chemical theory provides a framework for asking questions about societal problems that may lead to solutions.

MEETING RAISES FISH KILL CONCERNS (page 20)

Chemists and other authorities report their progress in the investigation of the Snake River fish kill, using such terms as "dissolved oxygen," "suspended," and "solubility." In order to understand the fish-kill issue, students need to understand basic chemical concepts regarding the behavior of water. The activity in *You Decide* B.9 (pages 36-37) will return to this article for additional analysis.

B.1 PHYSICAL PROPERTIES OF WATER (pages 21-23)

This section introduces students to directly observable properties of all matter, as well as focusing on water specifically. Concepts such as density, states of matter, evaporation/boiling, temperature, surface tension, boiling and freezing points, and the solvent characteristics of water are covered here. Some practical consequences of some of water's physical properties are discussed.

Water is probably the most familiar substance on Earth, but it is also the most unusual in terms of its properties and behavior. Water is a scarce commodity in our solar system, and perhaps in the universe. You may wish to display and contrast pictures of Earth and various other planets/moons—the presence of large amounts of water on our planet makes it unique in the solar system in its capability to support life as we know it.

DEMONSTRATION IDEAS

1. **Surface Tension:** Consider using the floating pin effect illustrated on page 23 in the student text—or show that you can add more water to a full glass of water by allowing the water to bulge over the top of the glass rim.
2. **Relative Densities of Liquid and Solid States:** Contrast the behavior of ice in liquid water by freezing a sample of ethylene glycol (melting point of –13°C) with dry ice or liquid nitrogen. Place the solid samples in a cooled (but not frozen) container of their own liquid. 2-propanol, M.P. – 89.5°C, could be used with liquid nitrogen.
3. **A Density Column:** Place the following materials in a large graduated cylinder (density in terms of g/ mL) in this order: lead sinker (d = 11.3), Karo syrup (d = 1.36), glycerin (d = 1.26), rubber stopper (d = 1.2), ethylene glycol (d = 1.11), water (d = 1.0), plastic (d = 0.9), oil (cooking or mineral d = 0.9), oak (d = 0.9), alcohol (either ethanol or 2-propanol, d = 0.79), and cork (d = 0.2). To obtain discrete layers, pour the materials slowly down the side of the tipped graduated cylinder, or add each layer with a long-stemmed funnel, if available.
4. **Discrepant event:** Place an ice cube in a beaker of alcohol.
5. **Relative Densities of Various Metals:** Place equal sized blocks of various metals on a double pan balance, and/or pass samples around: Al (2.7 g/mL), Pb (11.3 g/mL), Cu (8.94 g/mL), Fe (7.86 g/mL), and Zn (7.14 g/mL).

B.2 MIXTURES AND SOLUTIONS (pages 23-25)

This section introduces water mixtures and their classification. A sample of the original foul-water sample, plus the purified product, could illustrate the three types of mixtures presented in the text. The dissolving of atmospheric gases in water may be included as an example of a solution.

B.3 LABORATORY ACTIVITY: MIXTURES (pages 25-26)

This laboratory activity will provide students with the experience of applying chemical knowledge to the classification of mixtures. The objective is to classify each of four mixtures as a suspension, colloid, or solution.

The underlying concept is that the three kinds of mixtures are differentiated based on particle size, with solute size decreasing from suspensions, to colloids, to solutions. Suspensions contain particles which settle out or can be filtered; neither colloids nor solutions settle or can be filtered. Colloids, in turn can be differentiated from solutions by using the Tyndall effect.

Time

One class period

Materials (for a class of 24 working in pairs)

24 beakers (100 or 150 mL)
48 filter papers
12 funnels
12 ring stands
12 ring clamps
12 clay triangles
12 glass rods
Several light beam sources (flashlight, slide projector, overhead projector, laser)

"Unknown" solutions:

1.0 L milk-water colloidal mixture— #1
1.0 L 0.1 M $CuSO_4$ solution (25 g $CuSO_4 \cdot 5H_2O$/1.0 L)— #2
1.0 L clay-water suspension mixture— #3
1.0 L 0.1 M $FeCl_3$ solution (27 g $FeCl_3 \cdot 6H_2O$/1.0 L)— #4

Advance Preparation

1. The water-milk colloidal mixture should be quite dilute— add milk dropwise to the water (with stirring) until a very faint cloudiness is evident. Test the colloid for an observable Tyndall effect before dispensing the mixture to students.

2. The actual amount of clay used to prepare the clay suspension is not important. Adding a small amount of powdered charcoal may increase the visibility of the clay in the filtering experiment.

3. The two solutions should be freshly prepared, as they may form colloidal dispersions (due to the formation of hydroxides under alkaline conditions) upon standing. If the solutions appear cloudy, add a drop of hydrochloric acid or sulfuric acid to clarify them. The amount of salt used is not crucial— the minimum needed to produce distinct colors is sufficient if chemicals are scarce. It is acceptable to substitute blue food coloring for copper(II) sulfate, and brown or yellow food coloring for the iron(III) chloride. This would lower the cost for the experiment, and reduce problems of disposal.

Lab Tips

It is often convenient to borrow a laser from the physics department and demonstrate the Tyndall effect. A piece of cardboard with a pencil-size hole in it may be placed over a filmstrip or slide projector for students to use. You may also try using a piece of cardboard with a small hole in it on the overhead projector, but do not leave the projector on too long, or overheating problems may occur.

The clay mixture may be classified as both a suspension and a colloid, if the filtrate shows the Tyndall effect.

Pre-Lab Discussion

The pre-lab discussion should review the text material on solutions, suspensions, and colloids. Note that the color of a solution does not necessarily indicate the presence of a colloid.

1. If time is short, consider doing the Tyndall effect purely as a demonstration.

2. Another way to demonstrate the difference between a colloid and a solution is to use dialysis. Fill one piece of dialysis tubing, usually available from the biology department, with copper(II) sulfate solution and another with regular milk. Put each in a separate beaker of water and allow to stand: then compare the colors of the water in the beakers. The ions of the copper(II) sulfate solution are small enough to pass readily through the semiporous wall of the tubing, turning the beaker water the characteristic blue color of copper(II) ions. By contrast, the colloidal particles in milk are larger and cannot pass through the wall. The liquid in the beaker will remain clear.

Post-Lab Discussion

After the students have classified the mixtures, tell them the contents of each beaker. Milk was discussed in the previous section, and copper(II) sulfate and iron(III) chloride will be discussed as ionic compounds in Section B.6.

The post-lab discussion can also tie in the foul-water separations and mixtures involved in that lab as well.

Expected Results

Original Sample

Beaker	Color	Clarity	Settle Out?	Tyndall effect	Classification
1	white	very slight	no	yes	colloid
2	blue	clear	no	no	solution
3	light tan	cloudy	yes	yes	suspension
4	yellow	clear	no	no	solution

Filtrate

Beaker	Color	Clarity	Settle Out?	Tyndall effect	Filter paper	Classification
1	white	slightly cloudy	no	yes	nothing	colloid
2	blue	clear	no	no	nothing	solution
3	light tan	cloudy	no	yes	residue	colloid
4	yellow	clear	no	no	nothing	solution

Materials needed per student

4 Beral pipets, unlabeled
4 Beral pipets, labeled 1-4
1 24-well microplate, to hold pipets
2 12-well microplates
4 Pasteur pipets
cotton ball
wooden stick to fit into Pasteur pipet (for packing cotton)

Procedure

1. Prepare data tables as in textbook (pp. 25-26)
2. Fill four labeled Beral pipets with the following mixtures supplied by your teacher:
   pipet 1: milk and water
   pipet 2: copper(II) sulfate and water
   pipet 3: clay and water
   pipet 4: iron(III) chloride and water
3. Place 5-10 drops of each sample into separate wells of the 12-well microplate, and complete the following observations:
   a. Is the sample colored?
   b. Is the sample clear?
   c. If the sample is cloudy, are any particles settling out?
   d. Perform a light-scattering test, called the Tyndall effect, on each mixture in the 12-well microplate. Shine the light source through the side of the well, and observe the liquid from the top. If the beam is visible as it passes through the liquid, the Tyndall test is positive.
4. Filter each sample using a Pasteur pipet that has been packed with cotton. Use the wood sticks to pack a small amount of cotton in the wide end of the pipet; *do not pack the cotton too tightly!* Collect enough of each filtrate in a separate well to repeat the observations in step 3.
5. Observe the cotton used for filtration; are any particles visible? Do not confuse a discoloration of the cotton for a residue. A residue is a solid.
6. Compare the fitrate with the original liquid samples. Record any differences or similarities in their appearances.
7. Decide whether the original sample was a solution, colloid, or suspension. Record your decisions in the table.
8. Compare your observations and conclusions with those of other students in the laboratory.

Lab Tips

Caution students to shake each mixture before filling their Beral pipets and before dispensing liquid either into the microplates or when filtering.

B.4 MOLECULAR VIEW OF WATER (pages 26-28)

Moving beyond directly observable properties of water, this section builds a model based on the idea of atoms and molecules. The analogy between the alphabet, with 26 letters used to make words, and chemical elements, with 90 some elements used to make compounds, may be helpful here. Additional analogies focusing on the extremely small size of atoms and molecules may be appropriate. For instance: cesium, the largest (volume, not mass) atom known, has a diameter of 0.524 nanometers (0.524×10^{-9} meters)—it would take a million of these atoms lined up edge to edge to equal one half the thickness of a dime (1 mm thick). Hydrogen and oxygen atoms are much smaller still (0.074 nm and 0.132 nm diameters, respectively).

DEMONSTRATION IDEAS ON ELEMENTS, MIXTURES, AND COMPOUNDS

1. The classic (though smelly) demonstration to illustrate the difference between elements, compounds and mixtures is to examine the properties of iron and sulfur, noting the color, magnetic character, etc., of the two elements. Physically mix samples of the two and note that they still retain their individual characteristic properties and can be separated quite easily by physical means (magnet, density, etc.). Place a mixture of 4 g of sulfur and 7 g of iron (steel wool works better than iron filings in the reaction) in a large test tube and heat vigorously *under a fume hood*. Crack the test tube (plunging the hot tube in a beaker of cool water makes a striking change!) and note the properties of the product iron(II) sulfide (FeS): non-magnetic, different color, crumbles instead of being malleable, and cannot be separated back to its component elements by physical means.

YOUR TURN

Matter at the Microlevel (pages 27-28)

In this activity, students practice matching a theory with observations that provide support for it. Note that additional detail on aspects of the atomic theory will be forthcoming in subsequent units. At this point it is enough for students to account for certain aspects of the macroscopic world in terms of submicroscopic particles.

This activity may also be completed as a classroom inquiry exercise. The three observations presented in this activity can be introduced and supported as classroom demonstrations, then solved within each group. Regardless of how you organize this activity, it is needed to acquaint students with the use of scientific theory to explain observations.

Answers to Questions:

1. a. Molecules of gas in a balloon can escape slowly through small holes in the balloon wall. The helium-filled balloon shrinks more rapidly than other gas-filled balloons because the helium gas particles are smaller and faster-moving than those of most other gases.
    b. Metal-foil (Mylar) balloons do not shrink as quickly as plastic balloons because of smaller-size holes in the metal foil wall. (This can be set up as a two-day demonstration: Use two identical balloons filled equally, one with helium and the other with human breath; then use two same-size balloons (one rubber, one Mylar) filled with the same gas.) The point is that all matter is porous.
2. The decreasing density of a substance as a solid, liquid and gas may be explained by the close packing of particles at the atomic/molecular level.
3. When the perfume is opened, molecules evaporate from the liquid and join the gas phase, where they are free to move by convection throughout the room.

B.5 SYMBOLS, FORMULAS, AND EQUATIONS (pages 28-29)

Students are introduced to the notion of chemical equations and asked to identify the number of atoms of each element in a formula. Point out the importance of subscripts. For example, the different properties of H_2O (water) and H_2O_2 (hydrogen peroxide) may be explained, plus the fact that HO_2 does not exist. Also, emphasize the importance of a common, worldwide chemical language.

This section is an introduction; the topic will be further developed in the Resources unit. It is **NOT** necessary for students to memorize symbols or balance equations at this point. Instead, show students where to find a listing of symbols and elements and alert them to the idea of a balanced chemical equation by using the analogy of a recipe: it takes exact proportions of various ingredients to bake a cake.

Symbols, Formulas, and Equations (page 29)

This activity is designed to introduce students to chemical symbols and their use in representing chemical compounds. The objective is to interpret the given formulas by naming the elements in each compound and counting the number of atoms of each element shown in the formula.

Answers to Questions:

1. a. phosphorus; b. nickel; c. nitrogen; d. cobalt; e. bromine; f. potassium; g. sodium; h. iron (note the Latin-name symbols for the last three elements).

2. a. 2 hydrogens, 2 oxygens b. 1 calcium, 2 chlorines (c) 1 sodium, 1 hydrogen, 1 carbon, 3 oxygens;
    (d) 2 hydrogens, 1 sulfur, 4 oxygens.

    Extra items for optional practice:
    ethylene glycol (car antifreeze): $C_2H_6O_2$ — 2 carbons, 6 hydrogens, 2 oxygens;
    ammonium phosphate (a fertilizer): $(NH_4)_3PO_4$ — 3 nitrogens, 12 hydrogens, 1 phosphorus, 4 oxygens;
    iron(III) oxide (common rust): Fe_2O_3 — 2 irons, 3 oxygens

3. a. One molecule of nitrogen reacts with one molecule of oxygen to form two molecules of nitrogen oxide.
    b. Two atoms of nitrogen react with two atoms of oxygen to form two molecules of nitrogen oxide, each containing one atom of nitrogen and one of oxygen. Note that there are two atoms of nitrogen and two atoms of oxygen on each side of the equation.

B.6 THE ELECTRICAL NATURE OF MATTER (pages 29-31)

The electrical nature of matter is discussed in terms of the attraction between charges, and in terms of neutrons, protons, and electrons. Water is described as a polar molecule whose structure helps explain certain observations, such as water's relatively high boiling point. It is appropriate now to discuss the ionic compounds encountered earlier and the conductivity results obtained in the Foul Water laboratory activity.

B.7 PURE AND IMPURE WATER (pages 31-32)

The properties of water that make it uniquely suited for supporting life (such as its ability to dissolve a variety of materials) also make it easy to contaminate. In common usage the word "contaminant" means a poison or something harmful. To a chemist, the term means that one substance is mixed with another so that neither is pure. Certain contaminants, such as dissolved minerals and atmospheric gases, give water a more pleasing taste. Dissolved oxygen is essential for the survival of aquatic life. By contrast, contaminants such as ions of lead, cadmium, mercury, and arsenic, as well as pesticides and commercial and industrial wastes, are dangerous to human and other forms of life. We must prevent unwanted, potentially harmful substances (pollutants) from entering our water supply. A second challenge is to remove toxic substances already found in water, a costly proposition. These concerns should lead students to realize the importance of methods for detecting the presence of water contaminants—the subject of the next laboratory activity.

B.8 LABORATORY ACTIVITY: WATER TESTING (pages 32-35)

Students test different water samples for the presence of dissolved ions: iron(III), calcium, chloride, and sulfate. The qualitative analysis procedures in this activity are based on a "confirming test" approach. Students should realize that a positive result indicates that a particular ion is present, but a negative result does not prove the **total** absence of the ion. The actual concentration of the ion may just be below the sensitivity level of the confirming test.

Time

One class period

Materials (for a class of 24 working in pairs)

12 wash bottles
36 test tubes
12 10-mL graduated cylinders
12 test tube racks
12 marking pencils
48 dropping bottles

4-5 burets for dispensing reference solutions
4 L distilled water

Test reagent solutions in dropper bottles: [important ion]

50 mL 0.5 M KSCN (potassium thiocyanate, 2.4 g/50 mL) [SCN⁻ ions]
50 mL 0.1 M $Na_2C_2O_4$ (sodium oxalate, 0.67 g /50 mL) [$C_2O_4^{2-}$ ions]
50 mL 0.1 M $HC_2H_3O_2$ (acetic acid, 0.3 mL conc. acetic acid/50 mL)
50 mL 0.1 M $AgNO_3$ (silver nitrate, 0.85 g/50 mL) [Ag^+ ions]
50 mL 0.1 M $BaCl_2$ (barium chloride, 1.2 g $BaCl_2$· $2H_2O$/50 mL) [Ba^{2+} ions]

If dropper bottles are in short supply, several students may share these solutions from a central location.

Reference ion solutions:

200 mL 0.1 M $Fe(NO_3)_3$ (iron(III) nitrate, 8.0 g $Fe(NO_3)_3$· $9H_2O$/200 mL) [Fe^{3+} ions]
400 mL 0.1 M $CaCl_2$ (calcium chloride, 4.4 g/400 mL) [Ca^{2+} ions and Cl⁻ ions]
200 mL 0.1 M $FeSO_4$ (iron(II) sulfate, 5.6 g $FeSO_4$· $7H_2O$/200 mL) [SO_4^{2-}]

Note: The key ions have been noted for each solution to enable you to make substitutions in accord with your stockroom supplies of reagents.

Advance Preparation

You may want to dispense the reference-ion solutions from burets located centrally in the laboratory. The test-reagent dropper bottles may also be located in a central location, or a set of the seven dropper bottles can be placed on each lab bench. Assuming that four students work at each table, you will need to prepare seven or eight sets of reagents.

All the solutions should be prepared at least one day in advance. When making solutions remember to dissolve the solid or liquid solute in a water volume somewhat less than that of the final solution. Then add water to the desired total volume for the solution.

Note that the $CaCl_2$ solution is used as the source of both Ca^{2+} and Cl⁻ reference ions.

Pre-Lab Discussion

Repeat the philosophy behind this confirming test approach. Also emphasize the need for a control in tests of this type. Demonstrate how students should obtain their samples, whether from burets or dropper bottles. You may also decide to demonstrate how very faint colors can be judged by using the reference solution and by comparing the control samples. Remind students to exercise care in handling all chemicals in the laboratory. (Note: oxalates are toxic.)

If your tap water is particularly clean, you may wish to "manufacture" your own tap water sample, which will contain at least one of the ions being tested.

Lab Tips

Remind students that rinsing used test tubes with distilled water is important, so that stray ions are not left to contaminate any solutions added to the tube.

Tests on a reference solution may produce a dark color or a heavy precipitate. The iron thiocyanate complex may appear black if too concentrated. Suggest that students add a little water to see the actual color of the complex. If the same ion is present in tap water, the results may be fainter due to the lower concentration of the ion. Even a faint red or pink color indicates the presence of iron(III) ions in the Fe^{3+} test.

The activity can be done more economically by using spot plates instead of test tubes. Plates with both white and dark backgrounds are needed, and students must be extra observant when using smaller quantities.

Expected Results

Reference solutions	Color	Precipitate or product
Fe^{3+}	Red	$Fe(SCN)^{2+}$, iron thiocyanate complex
Ca^{2+}	White	CaC_2O_4, calcium oxalate precipitate
Cl⁻	White	AgCl, silver chloride precipitate
SO_4^{2-}	White	$BaSO_4$, barium sulfate precipitate

Post-Lab Discussion

Point out that sensitive instruments are routinely used in chemistry laboratories to detect and analyze the intensity of color in chemical samples. You may wish to demonstrate the use of a spectrophotometer or colorimeter, if one is available, to detect and measure the intensity of colored solutions. It may also be helpful to demonstrate how intensity of color is proportional to concentration with a series of solutions from dilute to concentrated, with molarity clearly labelled.

Answers to Questions:

1. a. The control represents the absence of the ion the chemist is attempting to detect and therefore is useful in interpreting the experimental results.
   b. Distilled water is presumed to be essentially free of contaminating ions and other dissolved materials except small amounts of dissolved atmospheric gases—carbon dioxide, nitrogen, and oxygen.
2. The concentrations of some ions may be too low to detect. Or, if more than one ion is present, the test for one ion may interfere with or mask the behavior of another ion.
3. These tests can confirm the presence of an ion, but cannot prove the absence of ions due to the reasons listed in question 2.
4. Improper cleaning may leave stray ions behind that can contaminate and interfere with tests of solutions in the test tube.

MICROSCALE PROCEDURE

Materials needed per student

12-well plate
Beral pipets filled with $CaCl_2$, $Fe(NO_3)_3$, KSCN, $Na_2C_2O_4$, $AgNO_3$, $FeSO_4$, $HC_2H_3O_2$, $BaCl_2$. Use the solutions desribed in the materials section. These should be labelled Ca^{2+}/Cl^-, Fe^{3+}, SCN^-, $C_2O_4^{2-}$, Ag^+, SO_4^{2-}, $HC_2H_3O_2$, and Ba^{2+}, respectively. In addition, you will need Beral pipets for the distilled water and the tap water sample.

Procedure

Iron(III) Test (Fe^{3+})
1. Add three drops of Fe^{3+} solution to one well. Add one drop of SCN^- solution, record your observation.
2. Repeat step 1, using three drops of tap water in place of the Fe^{3+} solution.
3. Repeat step 1, using three drops of distilled water in place of the iron(III) solution.

Calcium ion test (Ca^{2+})
1. Add three drops of Ca^{2+} ion to one well. Add one drop of acetic acid $(HC_2H_3O_2)$ and one drop of sodium oxalate, $Na_2C_2O_4$ to the same well. Record your observation.
2. Repeat step 1, using three drops of tap water in place of Ca^{2+} ion.
3. Repeat step 1, using three drops of distilled water in place of the calcium ion.

Chloride Ion Test (Cl^-)
1. Add three drops of Cl^- ion to one well. Add one drop of Ag^+ test solution; record your observation.
2. Repeat step 1, using three drops of tap water in place of the Cl^- ion.
3. Repeat step 1, using three drops of distilled water in place of the chloride ion.

Sulfate ion test (SO_4^{2-})
1. Add three drops of SO_4^{2-} ion to one well. Add one drop of Ba^{2+} test solution; record your observations.
2. Repeat step 1, using three drops of tap water in place of SO_4^{2-} ion.
3. Repeat step 1, using three drops of distilled water in place of the sulfate ion.

Clean–up: empty well plate and rinse several times with tap water, then once with distilled water.

Reminder: microscale labs take *much* less time, due to ease of dispensing reagents, and using smaller amounts.

B.9 YOU DECIDE: THE RIVERWOOD MYSTERY (pages 36-37)

The purpose of this activity is to acquaint students with a scientific approach to solving problems. After distinguishing facts from opinion in the original newspaper articles, the students seek regularities or patterns among data, and suggest reasons that might account for these patterns.

As preparation, discuss the behavior of different Riverwood citizens and the comments of the experts. It should be clear that jumping to immediate conclusions is not regarded as a wise or a productive pathway in science.

Students should work in groups of four to complete this activity. When finished, the entire class will compare and discuss the groups' answers. It may help to assign the reading as homework.

From the newspaper articles, students should recognize that if the cause of the fish kill were water related, it would be due to something suspended or dissolved in the water. This at least narrows the range of possible fish-kill causes. Students should be encouraged to begin wondering what types of substances can be dissolved or suspended in water, and which ones are potentially harmful to aquatic life. These questions will be covered in Part C of the unit.

The post-activity discussion should focus on questions 1-3 since they represent a sequence of data gathering, finding patterns in the data, and hypothesizing reasons for the patterns found. Students should be encouraged to distinguish facts from opinions.

1. Potentially relevant facts:
   - Dead trout and other fish were found floating at the base of the Snake River Dam, five miles upstream from the Riverwood water intake.
   - Preliminary tests revealed no evidence of toxic substances.
   - The town council voted to shut down the water plant three days.
   - Further analysis of the fish revealed signs of biological trauma, including hemorrhaging and small bubbles under the skin along the lateral line.
   - Suspended particles were ruled out as a cause of the kill.

2. Help students pose questions that must be answered in finding a cause of the Riverwood emergency. Review the chemical knowledge that students already have in hand and suggest how this knowledge can be used to form questions regarding the fish kill. Examples:
   - What are possible emission sources upstream from the dam?
   - What happens to water as it goes over the dam? Do any additional substances become dissolved in it?
   - What is the nature of the bubbles found in the fish?
   - What substances might harm fish but not humans?
   - Could the fish have died from a deficiency or overdose of a naturally occurring and necessary substance?

3. a,b. There are many possible answers, which depend on the student questions.

At this point closure or a correct answer is not necessary— there is more information coming to help clarify the situation. Some students may read ahead to find the answer! The next section will focus on testing for dissolved substances in the water. One characteristic of any problem is the formation of hypotheses regarding observations, followed by systematic checking, and elimination of unproductive explanations based on further experiments. This is the pathway in the next section.

B.10 WHAT ARE THE POSSIBILITIES? (page 37)

The major idea to emphasize is that scientific approaches are based on a combination of systematic procedures and careful logic, as well as occasional hunches, guesses, and even dumb luck. A fundamental and difficult part of scientists' work is deciding what questions are worth asking in the first place. ("It's no trick to get the right answer when you have all the data. The real creative trick is to get the right answer when you have only half of the data in hand and half of it is wrong. And you don't know which half...." Melvin Calvin, 1961 Nobel laureate as quoted in *Chemistry in Britain*, December 1973.) Knowledge of scientific facts, theories, laws, and principles helps the working scientist to form questions or hypotheses that have a reasonable chance of yielding useful solutions.

PART B: SUMMARY QUESTIONS (pages 37-38)

1 a. Unusual properties of water include its lower density as ice [solid]; unusually high boiling point [liquid]; high surface tension [liquid]; excellent solvent [liquid].

  b. Importance to life: ice floats on a lake, insulating life below the surface; water remains a liquid at higher temperatures than do other substances of comparable size and mass; surface tension allows water to bead and form raindrops, and also be carried up into plants from the roots; being a good solvent enables water to transport materials essential for human life to all cells in an organism.

2. Gasoline.

3. a. liquid water; b. lead; c. AgCl precipitate; d. cold air.
4. a. suspension; b. solution; c. suspension; d. colloid; e. solution; f. suspension; g. colloid.
5. a. compound; b. element.
6. a. "100% chemical-free water" is not possible, because water itself is a chemical.
   b. "Pure" means water without dissolved ions, such as distilled water. Clean and highly desirable water samples may still contain dissolved gases and minerals that give water a pleasing taste.
7. a. phosphoric acid: 3 hydrogens, 1 phosphorus, 4 oxygens;
   b. sodium hydroxide: 1 sodium, 1 oxygen, 1 hydrogen;
   c. sulfur dioxide: 1 sulfur, 2 oxygens.
8. a. One molecule of methane reacts with two molecules of oxygen to form one molecule of carbon dioxide and two molecules of water.
   b. methane – one carbon atom, four hydrogen atoms; oxygen – four oxygen atoms; carbon dioxide – one carbon and two oxygen atoms; water – four hydrogen atoms, two oxygen atoms. Note that each side contains the same number of each type of atom — the equation is balanced.

EXTENDING YOUR KNOWLEDGE (page 38)

• On the Celsius scale there are 100 degree units between the freezing point of water (0°C) and its boiling point (100°C). On the Fahrenheit scale there are 180 degree units between the freezing point of water (32°F) and its boiling point (212°F). Since 100 degrees Celsius = 180 degrees Fahrenheit, then 1.8° F = 1°C or 1° F = 0.56°C. Thus the following expression can be written: ° F = (1.8 × °C) + 32; if this expression is solved for °C, we obtain: °C = (°F – 32)/1.8.

• Since the density of oil is less than the density of liquid water, the oil will float on the water. An oil spill floating on water can ignite and burn, and also block the sunlight from reaching the aquatic plants. It can slow the exchange of gases between the water and the atmosphere and can be blown by the winds and carried by the currents and tides to beaches. On the other hand, if oil were more dense than water, cleanup efforts would probably be more difficult.

• Element	Symbol	Historical origin
Gold	Au	Aurum; Latin for "shining dawn"
Iron	Fe	Ferrum; Latin
Lead	Pb	Plumbum; Latin
Mercury	Hg	Hydrargentum; Latin for "liquid silver"
Potassium	K	Kalium; Latin, element first isolated by electroysis in 1807
Silver	Ag	Argentum; Latin
Sodium	Na	Natrium; Latin, also first isolated by electrolysis in 1807
Tin	Sn	Stannum; Latin

C: INVESTIGATING THE CAUSE OF THE FISH KILL

Part B established that whatever caused the Snake River fish kill must have been dissolved in the water. Part C provides students with deeper insights into the nomenclature, structure, and chemistry of ionic and molecular substances, and with the concepts of solubility and solution concentration. All of this is introduced within the context of exploring possible reasons for the Riverwood fish kill.

Part C begins by introducing basic terms and concepts related to solubility. After gaining a working knowledge of these ideas, students systematically study several kinds of substances that can dissolve in water. They first consider the importance of dissolved oxygen (DO) for the survival of aquatic life. To examine the possibility that a shortage of dissolved oxygen caused the fish kill, students find out how much oxygen is available to aquatic life and how much dissolved oxygen fish need to survive. This leads to questions about the influence of temperature on solubility. After examining and graphing the DO data from the Snake River, students rule out the possibility of DO shortages induced by thermal pollution and move on to consider other possibilities.

Acid contamination is then considered as another possible cause of the fish kill. Students learn to recognize common acids and bases from their formulas. Acid contamination is ruled out in the text, and students next consider heavy metal ions. A decision-making activity leads them to rule out this as well, and Part C ends by examining molecular substances that dissolve in water.

C.1 SOLUBILITY (pages 39-42)

Section C.1 presents vocabulary and concepts related to solubility: solute, solvent, solution, saturation, and solubility curves. The temperature dependence of gas and solid solubility is explained, as well as the effect of pressure on gaseous solubility.

It may be useful to help students practice reading graphs, and to reemphasize that points on a solubility curve indicate saturated solutions.

DEMONSTRATION IDEAS

1. Holding a sponge under slowly running water can illustrate "unsaturated" and "saturated". Initially all of the water is absorbed by the sponge, but eventually water drips from the sponge when saturation is reached.
2. Supersaturation can be demonstrated with very concentrated sodium acetate solutions, or even better, the heat pouches sold in sporting good stores. Students never seem to lose their fascination with these!
3. The effect of temperature on gas solubility can be demonstrated as follows: chill a bottle of carbonated drink (colored beverages show the result better) to near 0°C. Prepare three beakers of water. Have one at 0°C (ice-water bath), one at room temperature, and the third at about 40°C. Open the bottle and pour soda gently into three test tubes until full. Close each tube with a one-hole stopper, and with your finger over the stopper hole invert the tube and immerse it in one of the water baths. Do the same with the other two. Having two students assist you so that all three tubes are inverted and immersed simultaneously is more effective. In the room temperature and warmer water carbon dioxide is released from the solution and is visible as bubbles in the test tube; the relative amounts indicate that the solubility of CO_2 is less at higher temperatures.

YOUR TURN

Solubility and Solubility Curves (pages 41-42)

The purpose of this activity is to have students work with data presented in Figures 21 and 22 regarding the effect of temperature on solubility. Make sure that students understand that they need to know the quantities of solute and solvent, and the temperature of the solvent, in order to answer questions about solubility. This activity could be completed in small groups and used to supplement a class discussion on these principles.

Answers to Questions:

1. a. 80 g
   b. 42 g
2. a. 17 g
   b. 42 g $KNO_3/100$ g H_2O = 25 g KNO_3/x g H_2O $x = 60$ g H_2O
3. a. 7.7 mg at 30°C
   b. 9.2 mg at 20°C
4. 9.2 mg $O_2/1000$ g $H_2O = x$ mg/100 g H_2O $x = 0.92$ mg

OPTIONAL BACKGROUND INFORMATION

The solubility of a solute in a solvent (water in the present discussion) depends on the relative strengths of solute-solute attractions versus solute-solvent attractions. In terms of energy, we can represent the dissolving process as an equation:

$$solid \ + \ solvent \ \rightarrow \ solution$$

For many solids, the forward reaction is endothermic because energy is needed to break down the crystal lattice. Therefore, adding heat by raising the temperature will aid in dissolving solids. This is reflected in the solubility curves in Figure 21. If the heat of solution is small, then temperature has less of an effect—for example, the $\Delta H_{solution}$ value for NaCl is only 3.9 kJ/mol. Pressure has no significant effect on the solubility of solids, and grinding and stirring merely change the rate of dissolving by increasing surface area, but do not affect how much can dissolve.

$$gas \ + \ solvent \ \rightarrow \ solution$$

For gases there is no crystal lattice to be broken down, so the dissolving reaction always produces energy; a higher temperature impedes the dissolving process. Also, in contrast to solids, the solubility of gases is directly proportional to the pressure: this is Henry's law. This is particularly evident in soft drinks, where carbon dioxide is dissolved under as much as 10 atmospheres of pressure.

C.2 SOLUTION CONCENTRATION (pages 42-44)

This section elaborates on the ideas inherent in the solubility curves of Section C.1. Point out to students that there are various ways of expressing solution concentration: the solubility curves express concentration in terms of g solute/100 g solvent. One of the more common ways, and the one that will be used for the remainder of this unit is: grams solute/ grams solution. Typically, one considers 100 g solution (as in pph), 1,000 g solution (ppt), or ppm or ppb for very dilute solutions. The more familiar units of molarity and molality are not covered here.

An interesting application of the concentration idea is in considering doses in medicines. Children typically require smaller amounts of medicines, such as cough syrups or aspirin, than do adults. However, when one considers relative body masses of children and adults, one can calculate that the medicine or drug concentration in the body is roughly the same. This also explains why illegal drugs or nicotine used by a pregnant woman are so harmful to a developing fetus: a tolerable amount for an adult often produces extremely high concentrations of the drug in the fetus, with tragic results.

YOUR TURN

Describing Solution Concentrations (pages 43-44)

This activity introduces the idea of concentration and makes use of simple calculations based on proportionality. Concentration is expressed as a percentage, which can also be expressed as parts per hundred (pph). This important concept is expanded to include parts per million (ppm).

Answers to Questions:

1. a. Sugar
   b. Water
2. a. (17 g sucrose/200 g solution) $\times$ (100%) = 8.5% or 8.5 pph
   b. (34 g sucrose/400 g solution) $\times$ (100%) = 8.5% or 8.5 pph
3. "Saturated solution" means that the solvent has dissolved as much solute as is possible at that temperature.
4. a. Less oxygen in the warmer water.
   b. The solubility of gases in liquids decreases with increasing temperature, according to Figure 22.

C.3 OXYGEN SUPPLY AND DEMAND (pages 44-46)

OPTIONAL CHEMQUANDARY

Introduction to Section C.3

One 40-L aquarium is kept at 25°C, while a second identical aquarium is kept at 20°C. Which aquarium can support a larger population of fish? Why?

Answer: Fish require dissolved oxygen to survive. The fish tank that is kept at the lower temperature (20°C) will support a larger fish population. This can be verified by calculation as well (refer to Figure 22). Tank # 1 contains 40 L or 40,000 mL of water. The density of water is 1 g/mL, therefore the mass of the water in the tank is 40,000 g or 40 kg. At 25°C, the solubility of O_2 is 8.4 mg/100 g water. So, in 40,000 g H_2O, 336 mg O_2 will dissolve. Tank # 2 also holds 40,000 g H_2O. The temperature is 20°C, the solubility of O_2 is 9.2 mg/100 g of water. So, in 40,000 g H_2O, 376 mg O_2 will dissolve.

Conclusion: The tank kept at a lower temperature will support a larger fish population, assuming the temperature is high enough for survival.

This section points out that the oxygen atoms contained in water molecules are not usable by aquatic life. Emphasize to students that dissolved molecular oxygen gas (O_2) is needed. Oxygen gas enters water by dissolving directly from the air (aeration), and as a product of photosynthesis by aquatic plants.

Aerobic bacteria compete with other aquatic life for dissolved oxygen. A minimum concentration of dissolved oxygen is needed for a given species of fish to survive, as schematically shown in Figure 23, page 46.

C.4 TEMPERATURE AND GAS SOLUBILITY (pages 46-47)

Water temperature affects both the amount of dissolved oxygen available and the amount of oxygen used by fish. Since fish are "cold-blooded" animals, their metabolism and use of oxygen are dependent upon the temperature of the surrounding water. The metabolism of aerobic bacteria being dependent on temperature as well, bacteria and fish compete for the available dissolved oxygen. This explains why fish kills may be found after long periods of hot weather, as the text points out. (Note that increased water temperature increases the fishes' need for dissolved oxygen at the same time it decreases the available DO.)

Analogies are important aids to student understanding when discussing concentrations as low as 4 ppm (minimum concentration of DO that will support any fish life). The text provides one such analogy. Another useful one: consider a baby goldfish swimming inside a blue whale.

C.5 YOU DECIDE: TOO MUCH, TOO LITTLE? (pages 47-51)

This activity illustrates the use of graphed data to discover regularities and patterns. Students evaluate the likelihood that a shortage of dissolved oxygen killed the fish in Riverwood.

This activity will take one and one-half to two class periods. The work can be done in groups of students. Each group will complete the questions and report their conclusions to the class. If students have been taught how to graph data, you may wish to assign the graph of DO vs. temperature as homework, allowing more class time for discussion of the implications of the data and for questions.

You may wish to point out that temperature is plotted on the x–axis, because temperature causes the change in the dissolved oxygen level. By convention, the so-called independent variable (temperature, in this case) is plotted on the x–axis and the dependent variable (dissolved oxygen level here) is plotted on the y–axis. Also, if necessary, review the meaning of parts per million (ppm) and the concept of saturation.

An interesting variation on the graphing procedure is to plot both temperature and disssolved oxygen versus time on the same grid, but to use different symbols for the data points. The inverse relationship between temperature and dissolved oxygen is easy to see, and the graph is more consistent than DO versus temperature.

Answers to Questions: (pages 48-49)

1. As temperature increases, dissolved oxygen decreases.
2. Student prepares the graph.
3. a. The average water temperature in June is 11°C and in December is 7°C. One would thus expect a lower oxygen ppm in June, which is verified by the data.
   b. Since March and November exhibit the same water temperature (7°C), one would expect the oxygen concentrations to be similar. In fact both months show 11.0 ppm of dissolved oxygen.
4. August of this year shows a higher concentration of dissolved oxygen, with a lower average temperature, than last year's. The difference in the concentrations of oxygen in water is probably explained by the 2°C difference in the average temperatures.

Answers to Questions: (pages 49-51)

1. a. On both graphs the levels rise during the day and fall towards evening.
   b. The temperature would be expected to rise due to the energy from the sun. However, the interesting observation is that the dissolved oxygen level in early afternoon is **higher**: at the higher temperature we would expect the dissolved oxygen to be lower. Encourage your students to develop explanations for this unexpected result. They may remember that photosynthesis is a major source of oxygen for the water, which would account for the increase during the day.
   c. Oxygen levels are higher during the day, probably because of photosynthesis.
2. Average water temperature = 21.2°C
    Average oxygen concentration = 9.13 ppm
3. It is unlikely that the one-day average chosen for this year could be representative of the average weather for last year—there may have been drastic changes in weather, cloud cover, temperature, etc.
4. For a given day, maximum and minimum values provide more useful information than does an average value. Events occurring during the course of the day, such as increased photosynthesis at midday, might be overlooked if only averages were examined.

5. Data for One-Day, Hourly Samples (Table 4)

Time	Temp.(°C)	O$_2$, ppm	sat. O$_2$, ppm	% of sat.
8 A.M.	21	9.1	9.0	101
9	21	9.1	9.0	101
10	21	9.1	9.0	101
11	21	9.1	9.0	101
12 noon	22	9.2	8.8	105
1 P.M.	23	9.3	8.7	107
2	23	9.3	8.7	107
3	23	9.2	8.7	106
4	23	9.2	8.7	106
5	23	9.2	8.7	106
6	23	9.2	8.7	106
7	23	9.2	8.7	106
8	22	9.2	8.8	105
9	22	9.2	8.8	105
10	22	9.2	8.8	105
11	21	9.1	9.0	101
12 midnight	21	9.1	9.0	101
1 A.M.	21	9.1	9.0	101
2	19	9.0	9.3	97
3	19	9.0	9.3	97
4	19	9.0	9.3	97
5	19	9.0	9.3	97
6	19	9.0	9.3	97
7	19	9.0	9.3	97

6. Based on the above data, and the information in the text about dissolved oxygen levels needed to support river-water fish, the oxygen concentration in the Snake River appears adequate for most fish species to survive.

7. The amount of dissolved oxygen in the Snake River is not a likely cause of the Riverwood fish kill. The levels of dissolved oxygen, as calculated from the collected data, are well within the acceptable range for most fish species to survive.

C.6 ACID CONTAMINATION (pages 51-52)

Acidity and basicity play a major role in water quality and the survival of aquatic life. Students are introduced to general properties of acids and bases, and given a list of some of the more common compounds. The concepts of neutralization and the pH scale are also discussed briefly. The textbook's treatment of pH as a scale for distinguishing acids and bases is generally easily understood by students, who have encountered the term in advertisements ("pH balanced shampoo or deodorant") or in hot-tub or swimming pool maintenance. The mathematical definition is covered later in the Air unit.

Students learn here that pH measurements taken in the Snake River near the the site of the fish kill were within the normal range at the time of the fish kill. These data eliminate another possible cause of the fish kill.

Acid molecules have one or more hydrogen atoms that can be detached rather easily due to their bonding and structure. Of the acids listed in the textbook (carbonic, sulfuric, nitric, hydrochloric, acetic, and phosphoric), the acidic hydrogen atom(s) are written first. Thus, although an acetic acid molecule contains four hydrogen atoms, only one hydrogen atom contributes to its acidity – $HC_2H_3O_2$. (If its formula is written CH_3COOH the acidic hydrogen is still distinguished from the remaining three hydrogen atoms. This also represents a case where what appears to be an –OH group contributes to *acidic* rather than basic character.)

Bases include the hydroxide ion (OH⁻) in their structures; examples in the text are KOH, Mg(OH)$_2$, NaOH, and Ca(OH)$_2$. The identified bases all contain metals as cations, while the acids include anions composed of nonmetals. Although the textbook has not yet formally distinguished metals from nonmetals, students may at least be able to distinguish the CH_3– and CH_2– structures from formulas for metals such as magnesium, sodium, and calcium. The –OH group in alcohols (such as methanol/CH_3OH and ethanol/CH_3CH_2OH) does not make them basic.

1. You may wish to distribute litmus paper to your students and ask them to test various household liquids (acidic toilet bowl cleaners, citric juices, carbonated soft drinks, foods, basic antacid medications, ammonia cleaners, and drain cleaners) in order to get a first-hand experience of acids and bases.

2. Test various concentrations of hydrochloric acid and sodium hydroxide solutions (you will need to express the concentrations using percentages, or pph) with Universal Indicator. The pHs and corresponding colors are as follows: 4.0/red; 5.5/orange; 6.5/yellow; 7.5/green; 8.5-9/blue; 9.5/violet; 10.0/red-violet.

3. The following can be done as either an in-class demonstration or student home assignment. Place small pieces of red cabbage in boiling water or use a blender and warm water to obtain the purple solution. Like many natural pigments, red cabbage juice can serve as an acid-base indicator (approximate pH/color: 1-3/red; 4-5/rose; 6-8/purple; 8-11/blue; 11-13/green; 14/yellow). Since the solution thus obtained is not as concentrated as commercial indicators, much more is needed. Also, unless refrigerated, the indicator will not last as long.

C.7 IONS AND IONIC COMPOUNDS (pages 52–54)

The naming of ionic compounds and some rules for writing formulas for ionic compounds are introduced and discussed briefly. An important student activity involving writing ionic compounds follows the discussion.

Be sure your students understand that an ion is an atom or group of atoms possessing an electrical charge. The relative numbers of electrons and protons determine the overall charge. The bookkeeping involved may need some further attention in class. A mnemonic to help remember the new terms: *cat*ion is *paws*-ative, or "cation" has its sign in the letter "t" (+).

Figure 28 (page 53) provides an opportunity to point out that ionic crystals such as sodium chloride are not composed of molecules, but rather regularly packed arrangements of positive and negative ions—an ionic lattice. The formula NaCl thus represents the relative number of cations and anions in the structure — its simplest (empirical) formula.

YOUR TURN

Ionic Compounds (page 54)

This activity provides practice in recognizing, naming, and writing the formulas of ionic compounds. You may decide to supplement this activity with additional items, if your students would benefit from more practice. (Note: the idea behind listing the uses of the chemicals is for students to get a sense that chemicals, both natural and synthetic, are all around us, not just in laboratories. Students should not be required to memorize the commercial uses. If you would like to give additional items, the *Merck Index*, the annual listing of the top 50 chemicals in a June issue of *Chemical and Engineering News* (see Industry unit, page 479), and industrial chemistry texts are useful sources of information.)

Answers to Questions:

Cation	Anion	Formula	Name
1. K^+	Cl^-	KCl	potassium chloride
2. Ca^{2+}	SO_4^{2-}	$CaSO_4$	calcium sulfate
3. Ca^{2+}	PO_4^{3-}	$Ca_3(PO_4)_2$	calcium phosphate
4. NH_4^+	NO_3^-	NH_4NO_3	ammonium nitrate
5. Fe^{3+}	Cl^-	$FeCl_3$	iron(III) chloride
6. Al^{3+}	SO_4^{2-}	$Al_2(SO_4)_3$	aluminum sulfate
7. Na^+	HCO_3^-	$NaHCO_3$	sodium hydrogen carbonate
8. Mg^{2+}	OH^-	$Mg(OH)_2$	magnesium hydroxide
9. Fe^{3+}	O^{2-}	Fe_2O_3	iron(III) oxide
10. Ca^{2+}	CO_3^{2-}	$CaCO_3$	calcium carbonate

Note: Some students may use the common names for some of the compounds (baking soda or rust). Point out that these names give less information than the more systematic system of naming both cation and anion.

C.8 DISSOLVING IONIC COMPOUNDS (pages 55-56)

Knowing what ions are and how they are formed is only part of the chemical story. The dissolving of solid substances in water is now given a qualitative molecular-level interpretation. Plan to spend some time discussing the features of Figure 29 with your students, particularly the ion-water interactions. Note that the positive ends (the hydrogen ends) of polar water molecules are attracted to negative Cl^- ions. Conversely, the negative (oxygen) end of water molecules is attracted to positive Na^+ ions. Note that each dissolved ion is surrounded by an envelope of water molecules. Emphasize the three-dimensional nature of the process and if large molecular models are available use them.

Do not leave students with the impression that all ionic substances are highly soluble in water, however. As the text notes, if the ion-ion attractions within the solid ionic crystal are sufficiently strong (compared to ion-water attraction in solution), the substance may be only slightly soluble in water. Common examples of water–insoluble ionic compounds include silver chloride, barium sulfate, and calcium carbonate. Students will encounter such substances (in the form of ionic precipitates) in later laboratory activities. You may want to demonstrate some of these from the previous water-testing laboratory activity.

C.9 HEAVY METAL ION CONTAMINATION (pages 56-58)

Heavy metal ion contamination is a possible candidate for killing the fish. Heavy metals such as lead (Pb), mercury (Hg), and cadmium (Cd) can appear in water samples as dissolved ions. These three metal ions are discussed in relation to their chemistry, commercial uses, and toxic effects on living organisms.

Stress the point that metal ions, rather than metal atoms, may be found dissolved in water. Also point out that while toxic effects of the heavy metals are fairly well documented, we are more able to detect trace contaminants (at the parts per billion and even parts per trillion range) than we are to document possible negative health effects. Also, detection does not simplify the removal of such trace amounts from our water supply, if that seems necessary. Yet, despite such uncertainties and technical limits, it makes sense to avoid or limit exposure to known, avoidable hazards (such as cadmium in cigarette smoke). Also, in the case where substitutes are not available for an essential use, controlling the extent of exposure can minimize potentially harmful effects.

C.10 YOU DECIDE: HEAVY METAL IONS (pages 58-59)

This decision-making activity provides students scientific facts from which they evaluate hypothesized explanations for an observed event. The objective is to determine whether excessive ion concentrations killed the fish in the Snake River. The estimated time for this activity is one class period. You may elect to have students work within assigned groups of four to five.

The pre-activity discussion should include a review of calculations involving parts per million (ppm) as a concentration unit. Also, consider having students practice the comparison of numbers such as 0.0004 ppm and 0.001 ppm— here, for example, 0.001 ppm represents a concentration 25 times higher than 0.0004 ppm.

Answers to Questions:

1. Hg^{2+}, Cl^-, NO_3^-
2. Cd^{2+}, Pb^{2+}, Se^{2-}, SO_4^{2-}
3. Cd^{2+} = 0.67, Pb^{2+} = 0.27, Se^{2-} = 0.031, SO_4^{2-} = no risk limit
4. Cd^{2+} has the highest risk factor for aquatic life.
5. Cd^{2+} = 0.1, Pb^{2+} = 0.4, Se^{2-} = 0.8, SO_4^{2-} = 0.14
6. Lead ion and selenium ion have the greatest risk factor for humans, but neither is yet over EPA limits. The suggestion may be made that the community continue to monitor these two ions, even though there is no immediate danger.

The post-activity discussion should focus on the questions provided. Emphasize how knowledge of ions in solution, concentration (ppm), and the concept of a risk factor enable citizens to seek appropriate information from experts. Students should conclude that dissolved ionic substances in the river water did not kill the fish. Point out that other kinds of non-ionic substances can also dissolve in water. This leads to the consideration of molecular substances as a suspect in the fish kill—the topic for the next section and laboratory activity.

C.11 MOLECULAR SUBSTANCES IN THE RIVER (pages 59-60)

Although several possible explanations for the fish kill have been eliminated, the role of molecular substances remains to be explored. The notion of polarity, the distribution of electrical charge within a molecule, is introduced. Polarity of water is reviewed. The solubility of a substance in a solvent such as water depends on the electrical properties of both substances. The general rule "like dissolves like" is introduced. Note that the rule can be extended beyond polar-nonpolar interactions. In the case of ionic solubility, the "like" property involves the presence of electrical charge on both solute and solvent—full ionic charges on the solute (ionic compound), and partial charges (due to molecular polarity) on the solvent (water). Oxygen is much less soluble in water because it is a nonpolar molecule, and is not attracted strongly to the polar water.

The question now arises: Did dissolved molecular contaminants cause the fish kill? Challenge students to identify the types of factual information needed to answer this question.

C.12 LABORATORY ACTIVITY: SOLVENTS (pages 60-61)

In this activity students will explore the concept of "like dissolves like" by testing the solubility of several substances in water and hexane, examples of polar and nonpolar solvents.

Time

This activity will take approximately 30 minutes to complete.

Materials (for class of 24 working in pairs)

24 test tubes (assuming tubes are reused), 13 × 100 mm (or smaller: 10 × 50 mm will use less material)
12 test tube racks
12 10-mL graduated cylinders
12 marking pencils
12 spatulas
12 medicine droppers

Solutes

25 g urea, $CO(NH_2)_2$
25 g iodine, I_2
25 g ammonium chloride, NH_4Cl
25 g sodium chloride (table salt, NaCl)
25 g naphthalene flakes, $C_{10}H_8$, or lighter fluid (25 mL)
25 g copper(II) sulfate, $CuSO_4 \cdot 5H_2O$
100 mL ethanol (denatured alcohol), C_2H_5OH

Solvents

1 L distilled water, H_2O
600 mL hexane, C_6H_{14}

Pre-Lab Discussion

Clarify the logic and general objective of the activity. The basic purpose is to allow students to associate different solvent properties with polar and nonpolar materials. Note that student interpretations are based on bench observations rather than on any analysis of the structures of substances involved. A discussion of "like dissolves like" will help students analyze their observations.

Lab Tips

The major problem to avoid in this activity is using too much solute. Using wood splints as dispensers of crystals might encourage students to "think small." To minimize student exposure to iodine, consider demonstrating the procedure. In this fashion you can show the appropriate amounts of solid solute to use at the same time.

Remind students to use the test-tube mixing procedure illustrated in Figure 18, p. 33, which avoids direct skin contact with the tube contents.

If naphthalene is not permitted in your laboratory (it has appeared on a few "don't use" lists as a suspected carcinogen), lighter fluid may serve as an alternate solute in this activity.

Post-Lab Discussion

The post-lab discussion should center on the questions posed at the end of the activity. Note that the newspaper clipping "Fish Kill Remains a Mystery" (page 63) eliminates organic compounds such as pesticides, fertilizers, and industrial waste as potential causes of the fish kill.

Expected Results

Solute	Solubility in:	
	Water	Hexane
Urea, $CO(NH_2)_2$	S	I
Ammonium chloride, NH_4Cl	S	I
Naphthalene, $C_{10}H_8$	I	S
Iodine, I_2	I	S
Copper(II) sulfate, $CuSO_4$	S	I
Ethanol, C_2H_5OH	S	S
Sodium chloride, NaCl	S	I

S = soluble, I = insoluble

Answers to Questions:

1. NH_4Cl, $CuSO_4$, NaCl, and $CO(NH_2)_2$, are all more soluble in water than in hexane. Note: Urea is often used as a fertilizer due to its high nitrogen content and solubility in water.

2. Naphthalene and iodine are more soluble in hexane than in water.

3. Ionic compounds tend to dissolve in polar water, and molecular compounds tend to dissolve in hexane, which is non-polar. This reinforces the concept of polar with polar, and non-polar with non-polar, the "like dissolves like" concept.

4. Ethanol dissolves both in water and hexane, suggesting its structure has both polar and nonpolar features. (Its structure—and structures of other common organic substances including hexane—will be considered further in a later unit.)

PART C: SUMMARY QUESTIONS (pages 61-62)

1. Carbon dioxide gas is dissolved in the liquid of a bottle of capped soda. Since the solubility of a gas in a liquid decreases with increasing temperature, one would expect more gas to come out of solution at warmer temperatures, producing more "fizz" than at colder temperature.

2. The phrase "like dissolves like" means that polar solvents will tend to dissolve polar solutes and many ionic substances, while nonpolar solvents will tend to dissolve nonpolar solutes.

3. Sodium chloride is an ionic substance. Its ions will be dissolved by polar solvents such as water. Table salt will not dissolve in nonpolar cooking oil, since no electrical interactions of ions and polar solvent molecules are possible here.

4. a. Rivers with rapids have a greater concentration of dissolved oxygen due to increased aeration.
   b. A lake in springtime will have a greater concentration of dissolved oxygen than in the summer due to lower spring temperatures. However, the argument of more photosynthesis occurring in summer may also be suggested by your students.
   c. A lake containing trout will have a higher concentration of dissolved oxygen. Trout need a higher dissolved oxygen level to survive than do catfish (see Figure 23).

5. a. Since the density of liquid water is about 1 g/mL, 100 mL of water has a mass of 100 g. This will dissolve about 80 g KNO_3, according to Figure 21 (page 40). Note that the graph line cannot be read with high accuracy.
   b. Slightly supersaturated. At 60°C the same solution will be unsaturated (capable of dissolving more potassium nitrate).

6. Parts per million.

7. Seawater is basic since its pH is greater than 7.

8. a. sodium nitrate; b. magnesium sulfate; c. aluminum oxide; d. barium sulfate.

9. Mass of solution is 35 g + 115 g = 150 g.
   $(35\ g/150\ g) = (x/100\ g)$ $x = 23\ g$

10. Heavy metals are metallic elements, such as lead, mercury, and cadmium, composed of atoms that are more massive than those of the essential human elements such as iron, calcium, and magnesium. Heavy metals can appear in water as dissolved ions. Small amounts of heavy metals are poisonous to humans and can accumulate in the body. Heavy metals can also be concentrated in the food chain to dangerous levels. The general effects of heavy metal poisoning are damage to the nervous system, kidneys and liver, causing mental retardation and even death.

EXTENDING YOUR KNOWLEDGE (page 62)

- If DDT dissolves in the fatty tissue of animals (composed of nonpolar substances) it is also probably nonpolar, since the tendency is for "like" to dissolve "like." Water-soluble molecules will generally possess polar character; these will have low solubility in nonpolar solutes.
- If a student report is prepared on the topic of biological magnification, have it shared with the full class—this subject involves a direct application of the solubility principles developed in this unit, and (in the case of DDT, particularly) provides a useful case study of how scientific and technological changes can have wide-ranging and sometimes unexpected consequences.

D: WATER PURIFICATION AND TREATMENT

In Part D the case for concern about water quality is brought back to a more personal level. This part features information needed to make intelligent decisions about water quantity and quality, even without the extra motivation of a water emergency. It addresses questions such as: How pure is our water? How can we gain the water quality required? How can waterborne wastes be treated?

The natural water purification system, the hydrologic cycle, is elaborated here. A key point is that nature needs considerable time to purify water via this natural cycle. Therefore, society's use of water must not overload the natural system.

The problem of hard water, caused by high concentrations of certain dissolved minerals, is explored. This leads to techniques of water softening, featured in a laboratory activity.

Two questions are then addressed: How does a community provide drinkable water? How is wastewater treated before it is returned to the environment? Artificial water purification is both a pre-cleaning and post-cleaning process. Chlorination associated with these processes help control harmful bacteria, but it can also create problems. As usual, wise decisions must be made on the basis of balancing benefits and burdens.

We return once again to Riverwood. The most common causes of river water contamination have, by now, been eliminated. Tests show that the water is safe for human use. The fish kill remains unexplained at the close of Part D. (An answer will appear—finally—in Part E!)

OPTIONAL ACTIVITY

Individual Research Project or Class Field Trip on Water Use in the Students' Community

1. What is (are) the major source(s) of the water supply?
2. How is water use divided among agricultural, industrial, and commercial and domestic uses?
3. How much does water cost (your home, school, etc.)? Has this price been relatively stable or has it been steadily increasing in recent years?
4. What types of water and wastewater treatment plants are used? What (and how much) chemicals are added during the treatment process?
5. Has the supply of water always been adequate? Have there been any shortages in recent years? Has it always been safe?
6. Is the water provided particularly hard or disagreeable in other ways? If so, what types of home water purification products are available and to what extent are they used?
7. Is water being wasted? If so, what conservation methods might be used?
8. What future water problems (in terms of quantity or quality) are projected?

D.1 NATURAL WATER PURIFICATION (page 64)

The hydrologic cycle is explored here. Three basic steps of the hydrologic cycle are defined so students can recognize later that human water purification and treatment methods resemble aspects of the hydrologic cycle. This section points out that the hydrologic cycle works better when not overloaded. Nature does not always deliver water pure enough for home use. The problem of high concentration of dissolved minerals in the water (hard water) is briefly described. This leads to the goal of removing these dissolved materials from water chemically, with water softening procedures. Students experience these techniques in the following laboratory activity, before details are provided in Section D.3.

D.2 LABORATORY ACTIVITY: WATER SOFTENING (pages 65-67)

In this laboratory activity students become aware that water softening represents applied chemistry. The objective is to compare the water-softening effectiveness of an ion-exchange resin, Calgon, and sand in the removal of a representative hard water ion (Ca^{2+}) from tap water. Students will also note that differences in the solubility of various substances make some water-softening techniques possible.

Students learn that hard water contains an excess concentration of calcium (Ca^{2+}), magnesium (Mg^{2+}), and/or iron (Fe^{3+} ions). Water softening thus involves the full or partial removal of such ions from water. The actual chemistry of these processes will be considered in greater detail in Section D.3; delay your full discussion of water softening until then.

Time

One class period.

Materials (for 24 students working in pairs)

12 10-cm diameter funnels
12 ring stands
12 ring clamps
12 clay triangles
72 pieces of 11-cm filter paper (Whatman No. 1)
48 test tubes (18 × 150 mm)
24 test tube racks
12 20-cm glass stirring rods
12 10-mL graduated cylinders
12 washing bottles
6 dropper bottles for sodium carbonate solution several burets for dispensing hard water
1.5 L hard water, containing 0.01 M Ca^{2+} (2.2 g $CaCl_2 \cdot 2H_2O$/1.5 L)
100 mL 0.1 M sodium carbonate in dropper bottles (1.24 g $Na_2CO_3 \cdot H_2O$ / 100 mL)
Ivory liquid soap, or some equivalent "standard" soap solution
12 wash bottles
1 L distilled water in wash bottles
1 lb sand
1 box Calgon (sodium hexametaphosphate)
100 g cation exchange resin, ionic form Na^+, Dowex 50W-X8 or equivalent, 50-100 mesh

Advance Preparation

The ion-exchange resin should be prepared for the activity as a water slurry. Add water to the resin in a beaker (2:1 volume ratio of water to resin). Decant excess water. The resin should be soaked overnight before using.

When purchasing Calgon or equivalent water softeners, make sure to obtain sodium hexametaphosphate for bath use, not sodium sesquicarbonate for dishes.

Pre-Lab Discussion

The pre-lab discussion should emphasize that hard water is caused by excess concentrations of calcium, magnesium, and/ or iron(III) ions in water. Point out that water softening is done in many homes, and will be discussed later.

Lab Tips

The procedure may be modified to use only one funnel and test tube rack per group to minimize the glassware needed. Using short stem funnels and filtering directly into the test tubes will eliminate the need for ring stands, etc. Be sure that students clearly label their test tubes if this minimal glassware approach is used. Alternately, funnels may be obtained by cutting the top off 1 L soda bottles (note that the other end makes a useful beaker).

The exchange resin contains 40-50% moisture and thus must be stored in sealed condition to avoid drying. The ion-exchange resin and sand can be reused from period to period.

The resin can be regenerated for future use by washing it three times in a beaker with 1 M NaCl, and then three times with distilled water.

Post-Lab Discussion

Organize your post-lab discussion around the questions provided. Students may be asked to interpret their observations. Consider mentioning that the ion-exchange process is used in most home water-softener systems. Such information provides a good opener for Section D.3.

Expected Results

	Filter paper	Filter paper and sand	Filter paper and Calgon	Filter paper and Ion- Exchange Resin
Reaction with Sodium carbonate (Na_2CO_3)	cloudy	cloudy	no reaction	no reaction
Degree of cloudiness (turbidity) with Ivory soap	very clear	very clear	blue and turbid	very turbid
Height of suds	none	none	1-2 mm	1-2 mm

Answers to Questions:

1. The ion–exchange resin should remove the hard-water ions best, but Calgon results may turn out better.
2. The higher the concentration of Ca^{2+} remaining in the filtrate (that is, the harder the water), the less soluble the soap.
3. The cleaning action of a soap depends on how dispersed or dissolved it is in the water. The more soap molecules that are bound to hard-water ions such as Ca^{2+}, the less cleaning action can occur.
4. Soaps such as Ivory will react with the hard-water ions to form a precipitate in the water. When the water drains, this precipitate sticks to the sides of the tub leaving a characteristic "ring". Softeners such as Calgon contain compounds that effectively remove some of the hard-water ions. In the activity this was evidenced by the appearance of soap suds. The box suggests Calgon be used in laundry, for bathing, handwashing, and general cleaning. Calgon (sodium hexametaphosphate) is a cation binder. Calgon will bind the troublesome positive ions in the hard water, but is still soluble and does not form a precipitate. These bound ions are unable to react with soap to form scum; thus the cleaning action of soap is increased.

D.3 HARD WATER AND WATER SOFTENING (pages 67-69)

This section identifies sources of hard-water ions. Chemical explanations are given for common problems that hard water causes in home water uses. Section D.3 ends with a discussion of ion-exchange resins currently used in some homes to soften water.

The following equation will help supplement and summarize the ion-exchange information given in the text. R stands for an ion-binding site in the resin:

$$2NaR + Ca^{2+}(aq) \rightarrow CaR_2 + 2\,Na^+(aq).$$

The resin is a polymeric material containing anion binding sites; the resin is prepared by rinsing thoroughly with solutions of sodium ions, which bind to the anion sites. When hard water containing calcium or magnesium ions is filtered through the resin, the higher-charged ions are attracted to the anion sites more strongly, replacing the sodium ions, which continue with the flow of the liquid through the resin. The resin can be regenerated by washing with a concentrated sodium ion solution. Some ion-exchange resins contain both cation and anion sites, and can be charged with hydrogen and hydroxide ions, which are replaced by *any* cation or anion in the water. Water of excellent purity, as far as dissolved ions go, can be prepared with these types of resins.

Students will need to understand the concept of charge balance illustrated in this equation if they are asked to answer the *Extending Your Knowledge* questions. The notion is simply that the total positive charge must remain constant on the resin and in the water— here we see that two Na^+ ions are freed for each Ca^{2+} bound to the resin.

OPTIONAL CHEMQUANDARY

Ion–exchange water-softening units provide an economical solution to a natural chemical problem. Does the use of such units create any new problems?

Answer: Yes, since the water now contains higher concentrations of sodium ions. People with high blood pressure needing low-sodium intake must find alternate sources of drinking water.

D.4 MUNICIPAL WATER PURIFICATION (pages 69-72)

Many waterways are both the source of municipal water and a place to dump wastewater. As a consequence, water from these sources must be both pre-cleaned and post-cleaned. Both processes model the steps of the natural hydrologic cycle, as this section suggests. You'll find that Figures 34 (page 70) and 35 (page 72) will help students see the 'big picture' as the details are developed.

D.5 CHLORINE IN OUR WATER (page 74)

Chlorination of water offers both benefits and possible problems. The reference to chlorinating agents in swimming pools may encourage some students to volunteer personal experiences with highly chlorinated pool water. Students that have tropical fish tanks may comment that chlorinated tap water must be boiled or pre-treated before fish are introduced to the tank. Encourage such contributions from students, since they help tie the chemistry to everyday experiences.

D.6 YOU DECIDE: CHLORINATION AND THMs (pages 74-75)

This *You Decide* points out that technological solutions to many problems create new problems. The formation of trihalomethanes (THMs), such as chloroform ($CHCl_3$) from the chlorination of water offers one example.

The modern system for nomenclature names compounds containing chlorine and carbon as derivatives of methane. A one-carbon compound with three chlorines on it would be called trichloromethane. Since other halogens behave almost identically to chlorine, the general name for this group of compounds is trihalomethanes.

Answers to Questions:

1. Activated charcoal filters would be too expensive to install and operate on a large scale. Ozone or ultraviolet radiation disinfects water, but does not provide added protection once the water leaves the plant. Elimination of the pre-chlorination step may be a preferable compromise. In this case, chlorine would be added only once, after the water had been filtered of unwanted organic material. The post-addition of chlorine to water would promote formation of some THMs, but to a much lesser extent. One disadvantage of this idea is that the decreased chlorine concentration might make the water more susceptible to bacterial growth.
2. Other alternatives to chlorination include the use of chlorine dioxide, ozone, chloramines, UV irradiation, iodination, or some combination of these. See *Envir. Sci. Tech.* **16**, (10) 1982.

PART D: SUMMARY QUESTIONS (page 75)

1. Thermal energy from the sun causes surface water from the ocean and other sources to evaporate, leaving behind dissolved minerals. The water vapor rises and condenses into tiny droplets (clouds). The water eventually falls as precipitation, snow, or rain. It then joins the surface water or seeps into the ground. The groundwater will eventually become surface water and the cycle begins again.
2. a. The screening of large objects, settling, and sand filtration of water purification and sewage treatment is accomplished in nature through evaporation and the seeping of surface into underground aquifers.
   b. The chlorination and flocculation steps in artificial water purification and treatment procedures do not have direct counterparts in the hydrologic cycle.
3. a. The chlorination of water provides some assurance that the water is bacteria-free.
   b. Conversely, chlorine in water presents a health risk. Chlorine may react with organic compounds to produce trihalomethanes (THMs). Chloroform ($CHCl_3$) is a representative trihalomethane, a suspected carcinogen (cancer-causing agent).

EXTENDING YOUR KNOWLEDGE (page 75)

* Increased population growth has raised the demand for goods and services. This ultimately increases the direct and indirect demand for so much "clean" water that it can no longer be provided in total by nature's recycling system. In addition, new knowledge of the adverse health effects of poor-quality water has increased the trend toward regulation of water quality by government agencies.
* Tertiary water treatment refers to a series of specialized chemical and physical processes to reduce the quantity of pollutants after primary and secondary treatment. The removal of organic wastes is typically the goal of tertiary water treatment. Sample procedures include activated carbon beds, chemical precipitation, ion exchange, electrodialysis, reverse osmosis, and distillation. The cost of tertiary water treatment is very high due to the large quantity of energy required. See *Living in the Environment* by G. Tyler Miller, Jr. (Wadsworth Publishing Co., 1990).

- $Fe^{3+}(aq) + 3\,NaR \rightarrow FeR_3 + 3\,Na^+(aq)$. In this equation, R represents a cation-exchange resin site containing fixed negative charges which interacts with mobile cations. Ion-exchange resins are macromolecules or insoluble polyelectrolytes having fixed charges distributed uniformly throughout the structure. During the ion-exchange reaction, the resins release sodium ions and bind the undersirable ion. Note that the charges are balanced as well as the numbers of atoms.

- About 50% of all Americans (and more than 95% of all rural Americans) rely on groundwater. Twenty large U.S. cities rely exclusively on groundwater and 12 states use it for more than half of their public supplies. The topic was not addressed in the student text due to lack of space— but it makes an excellent topic for optional student reports. Refer students to the supplemental references.

E: PUTTING IT ALL TOGETHER: FISH KILL—WHO PAYS?

FISH KILL CAUSE FOUND (page 76); EDITORIAL (page 77)

These articles contain the resolution of the Riverwood fish-kill mystery, as well as Riverwood reactions to the final answer. Gas bubble disease, caused by an excess of air dissolved in the river water, has been identified as the cause of the fish kill. The most likely source of this super-aerated water is the power company's release of water from a dam upstream from Riverwood. In reality, gas bubble disease is well-documented in the technical literature. Note that additional background regarding this effect is presented in the "Scientists" and "Engineers" briefing sections (pages 79-81) in the unit.

E.1 DIRECTIONS FOR THE TOWN COUNCIL MEETING (pages 78-82)

A town council meeting (to be simulated in your classroom) will address two questions: Who is responsible for the fish kill? Who should pay for the water trucked into Riverwood during the three-day water shutoff? Your students will participate in the final resolution of these questions.

The purpose of this culminating activity is to encourage students to use the chemistry they have learned to think through real-world problems similar to those they will be called upon to consider as voting adults. This activity will help students to integrate and apply the chemistry they have learned so far in this course.

Students will prepare briefs for presentation, develop a presentation that expresses the views of their group, and write an editorial and/or develop an interview about the Riverwood problem. Ultimately, each student must form a reasoned opinion regarding who is responsible for the fish kill and who should pay for water trucked to Riverwood during the three-day water shutoff.

Science does not provide answers to social problems, but can be used to help formulate possible solutions. Different people can reach different conclusions from the same facts, based upon their own perspectives or values. Science enables us to pose the right kinds of questions, thus setting useful boundaries on the problem. Scientific knowledge helps illuminate many of these problems, even though the larger decisions to take action must always involve additional considerations from social, economic, political, moral, and related perspectives.

Pre-Activity Preparation and Discussion

Read the unit's presentation of this culminating activity carefully, since many of the detailed procedures are directly summarized for students there. (This should save you some organizational time in class.)

Students should be assigned their roles in the town council meeting at least one day before this activity begins. The two town council members are central to organizing and administrating the debate. Consider the personalities of the students you choose for this role in the debate.

Briefly describe the various occupations that might actually be represented by members of each group. Specific occupational descriptions and qualifications may be found in the *Dictionary of Occupational Titles (DOT)* and other sources used by school guidance counselors.

It is possible to add other "special interest" groups to your meeting, if you wish to have smaller groups and more speakers. For example, there could be two groups of engineers: one as a source of information, and the other as original designers of the dam. The Riverwood Corporation, with some fictitious product, could be represented. The only limit to this is time and your imagination!

Encourage students to bring appropriate props such as hard hats or lab coats to increase the authenticity of group identities. Each interest group could design a logo to be displayed on a badge or name tag symbolizing their identity. Student talents in other areas such as art, graphics, computer science, business, drama, journalism, and vocational education can be harnessed to create the town meeting atmosphere. Tables and chairs can be rearranged by groups with a podium at the front of the room. A microphone (even if not connected to an amplifier) can add realism to the setting. Videotaping the meeting always lends an element of the theater to the meeting, and students thoroughly enjoy seeing themselves, as well as the other classes. This does increase the time for the activity, however!

Note that the activity is intentionally open-ended. There is no single right answer that the class is to obtain. The intent is that the students come away with a sense that: (1) chemistry is directly involved in many social issues, (2) in a democratic government, citizen involvement is both a privilege and a responsibility, and (3) decision making in such a government involves a balancing of differing viewpoints and interests with the intent of the greatest good for the greatest number (without denying the rights of minorities).

Time/Grouping

This activity will take at least two full class periods to complete. On the first day, the groups are chosen and the presentations organized. On the second day the town meeting is essentially run by the students according to this schedule:

Introduction of the rules	2 min
Group preparation of positions	20 min
Group presentations (2 min each)	12 min
Group rebuttals (1 min each)	6 min
Class discussion or preparation of letter to the editor	10 min

On the following day you need to decide whether students will conduct a simulated television interview in place of, or in addition to, the preparation of letters to the editor of the Riverwood newspaper. Alternatively, you could show the videotape of the meeting for at least part of the period.

If you elect to conduct the simulated television interview, each group should appoint a member to represent the views of that group. To maximize participation, this group member should probably be a different student than the one who represented the group during the first-day debate. The interview should solicit the response of each group to the stated positions of the other interest groups, as well as to the issues raised in the newspaper editorial.

Post-Activity Discussion

Following the completion of this activity, conduct a class discussion regarding students' experiences in this Riverwood simulation. At this point students should abandon their assigned roles and critically analyze the entire decision-making process. This discussion is central to the success of the activity. In this discussion consider focusing on points such as these:

1. Informed decisions concerning any problem should take into account opposing views before taking action. Contrast this view with the tactic of jumping to conclusions. You may want to ask students to identify social and chemical consequences of prematurely jumping to conclusions, particularly in the context of the Riverwood fish-kill story.
2. Chemical knowledge was involved in many aspects of the Riverwood fish-kill crisis and in its resolution. Use some of the discussion time to clarify any incorrect use of science concepts expressed by students during the class debate.
3. Point out the benefits and consequences of seeking consensus solutions to complex issues. Discourage students from viewing the debate in terms of winning or losing.
4. Highlight, with students' help, examples of factual, opinion-based, emotional, and logical arguments presented during the debate. What role(s) do these approaches play in group debates or discussions? What role(s) should they play?
5. Point out that each group had worthwhile viewpoints to contribute to the debate. The final resolution to the problem needs to be a compromise among the different group positions.

The evaluation of student performance in this activity may be done individually or by group grades. Group grades can be based upon the oral presentation or the simulated interview. The written letters to the editor permit you to evaluate individual performance.

E.2 LOOKING BACK AND LOOKING AHEAD (pages 82-83)

This section provides a transition from this initial unit to "Conserving Chemical Resources," which starts on page 84. You'll note that several important points are made in this section regarding the story of the Riverwood water crisis. You may elect to discuss these ideas briefly or to assign this as take-home reading.

SUPPLY LIST
Expendable Items

	Section	Quantity (per class)
Acetic acid, conc.	B.8	1 mL
Ammonium chloride	C.12	25 g
Barium chloride dihydrate	B.8	2 g
Calcium chloride	B.8	3 g
Calgon (sodium hexametaphosphate)	D.2	1 box
Charcoal, decolorizing	A.2	200 g
Clay or cornstarch	B.3	200 g
Coffee grounds, used	A.2	1/2 cup
Copper(II) sulfate pentahydrate	B.3, C.12	75 g
Cups, paper	A.2	20
Ethanol, denatured alcohol	C.12	100 mL
Filter paper, Whatman #1, 11 cm diameter	A.2, D.2	150
Garlic powder	A.2	1 tsp.
Gravel, fine grind	A.2	2 kg
Hard water, Ca^{2+}(aq)	D.2	1 L
Hexane	C.12	800 mL
Ion exchange resin, sodium, Dowex 50W-8X, 50-100 mesh	D.2	100 g
Iron(III) chloride hexahydrate	B.3	55 g
Iron(III) nitrate nonahydrate	B.8	10 g
Iron(II) sulfate heptahydrate	B.8	10 g
Milk	B.3	1 L
Naphthalene flakes	C.12	25 g
Vegetable oil	A.2	100 mL
Lighter fluid (optional)	C.12	100 mL
Potassium thiocyanate	B.8	3 g
Salt (table), sodium chloride	A.2, C.12	50 g
Sand	A.2, D.2	1 kg
Silver nitrate	B.8	2 g
Soap, Ivory (or equivalent)	D.2	1 bottle
Sodium carbonate monohydrate	D.2	15 g
Sodium oxalate monohydrate	B.8	3 g
Urea crystals	C.12	25 g

Nonexpendable Items

	Section	Quantity
Dropping bottles	B.8	48
Marking pencils	B.8	12
Paper cups	A.2	12
Paper clips	A.2	12
Distillation apparatus	A.2	1
Conductivity apparatus	A.2	1
Light beam sources	B.3	12

SUPPLEMENTAL READINGS

The articles and publications in this unit and in the following units may be used for several purposes: (1) background support material for the teacher—given the strong applications focus of *ChemCom* and the more theoretical nature of many college-level chemistry courses, you may wish to undertake a selected updating of your knowledge base with respect to science-technology-society (STS) issues; (2) background reading material for students involved in simulations and role playing activities, and (3) optional reading/reporting assignments. Of course, any such listing can become dated with respect to specific topics and should be updated and revised with newer materials periodically.

It is especially important that students realize that (1) no textbook contains all the "right" answers (or even all the right questions), (2) as our knowledge base expands, textbooks are constantly becoming somewhat out-of-date, and (3) some journals/magazines can be generally relied upon for valid scientific information (such as those published by scientific associations), while others (such as many of the popular newspapers and magazines found in grocery check-out lanes) cannot. Encourage your students to follow national and local news on STS issues and to bring relevant items to the attention of the class. "Chemistry in the News" bulletin boards, cartoon contests, etc., can be organized by students to help make the *Chemistry in the Community* text a relevant, living, ever-growing course of study.

"How Soon Will We Measure in Metric?" *National Geographic*, August 1977, 287-294. "Our Most Precious Resource." *National Geographic*, August 1980, 144-179.

"Megameanings." *CHEM MATTERS*, Dec., 1984.

"The Groundwater Terrarium: Geology on a Small Scale." *The Science Teacher*, May 1982, v. 49 #5, 44-46. Instructions for constructing a model and demonstrations to examine the water cycle.

"An Atomic Tour." *CHEM MATTERS*, Oct., 1983.

"Water in the Geosystem: Phase Relationships." *The Science Teacher*, May 1974, v. 45 #5, 39-43. Discusses the role of phase changes in the hydrologic cycle.

"Ape Antibiotic." *CHEM MATTERS*, Feb., 1987.

"What to Do about Water." *The Science Teacher*, January 1985, v. 52 #1, 34-37. Lists the causes of freshwater supply depletion and highlights several successful reclamation projects. Advocates a national water policy.

"Swimming Pool Chemistry." *CHEM MATTERS*, April 1983, 4-5. Issue focuses on acids and bases in everyday life.

"Acids and Bases: Chemical Principles Revisited." *Journal of Chemical Education*, July 1978, 459-464. "The Use of Oxygen in the Treatment of Sewage (The Real World of Industrial Chemistry)." *Journal of Chemical Education*, Feb 1980, 137-8. Discusses the advantages, problems, and costs of using oxygen rather than air in the second stage treatment of municipal wastewater.

"Crystal Growing" and "Breakfast of Crystals.", *CHEM MATTERS*, Oct., 1983.

Groundwater. Chemical Manufacturers Association, Washington, D.C. Free booklet explains in laymen's terms where groundwater comes from, its uses, and some principles of proper management and national policy.

"The Creeping Poison Underground." *Discover*, February 1985, 74-78. Hazardous chemical and metals affect on groundwater.

Cleaning Our Environment: A Chemical Perspective. American Chemical Society, Washington, D.C., 1978, 457 pp. Includes chapters on chemical analysis and monitoring, toxicology, air, water, solid wastes, pesticides, and radiation.

"Fox River Rish Kill.", *CHEM MATTERS*, Oct., 1990.

CONSERVING CHEMICAL RESOURCES
PLANNING GUIDE

Section	Laboratory Activity	You Decide
A. Use of Resources		
Introduction A.2 Using Things Up A.3 Tracking Atoms A.5 Resources and Waste A.6 Disposing of Things	A.1 Striking It Rich A.4 Using Up a Metal	A.7 Consuming Resources
B. Why We Use What We Do		
B.1 Properties Make the Difference B.2 The Chemical Elements B.4 The Periodic Table B.6 The Pattern of Atomic Numbers B.7 Chemical Reactivity B.9 What Determines Properties? B.10 Modifying Properties	B.3 Metal, Nonmetal? B.8 Metal Reactivities	B.5 Grouping the Elements B.11 Restoring Ms. Liberty
C. Conservation in Nature and the Community		
C.1 Sources of Resources C.2 Conservation Is Nature's Way C.3 Atom, Molecule, and Ion Inventory C.4 Conservation Must Be Our Way		C.5 Recycling Drive
D. Metals: Sources and Replacements		
D.1 Copper: Sources and Uses D.2 Evaluating an Ore D.3 Metal Reactivity Revisited D.4 Metals from Ores D.6 Future Materials	D.5 Producing Copper	
E. Putting It All Together: How Long Will the Supply Last?		
E.1 Metal Reserves: Three Projections E.2 Options and Opportunities E.3 Looking Back and Looking Ahead		

TEACHING SCHEDULE

	DAY 1	DAY 2	DAY 3	DAY 4	DAY 5
Class Work	Discuss pp. 86-87 CQ p. 88		Discuss pp. 91-93 YT pp. 93-94		Discuss pp. 97-99 YD pp. 99-100
Laboratory	Pre-lab pp. 89-91	LA pp. 89-91	Pre-lab LA pp. 94-95	LA pp. 94-95	
Homework		Read pp. 91-95		Read pp. 95-101	SQ p. 101 Read pp.102-105

	DAY 6	DAY 7	DAY 8	DAY 9	DAY 10
Class Work	Review Part A Discuss pp. 102-103 YT p.103	Quiz Part A Discuss pp.104-105 YT p.105		YD p.108	Discuss pp.109-112 YTs pp.109-112
Laboratory		Pre-lab LA pp. 105-107	LA pp. 105-107		
Homework			Read pp. 107-108	Read pp. 108-115	

	DAY 11	DAY 12	DAY 13	DAY 14	DAY 15
Class Work	Discuss pp.112-113	YD pp.116-117 CQ p.116	Review Part B Discuss pp.120-121	Quiz Part B Discuss pp.122-123 YT pp.123-124	Discuss pp.124-126 YT p.126
Laboratory	LA pp.113-115				
Homework	Read pp.115-119	SQ p. 119 Read pp.120-121	Read pp. 122-127		Read pp. 127-130

	DAY 16	DAY 17	DAY 18	DAY 19	DAY 20
Class Work	YD pp.128-130 CQ p.129	Review Part C Quiz Part C YT pp.132-133	Discuss pp.133-134 YT p.134	Discuss pp.135-138 CQ p.135 YT p.136	Discuss pp.139-143
Laboratory				Pre-lab LA pp. 138-139	LA pp. 138-139
Homework	SQ p.130 Read pp.131-133		Read pp. 135-143		YT p.143 SQ p.143 Read pp. 144-146

	DAY 21	DAY 22	DAY 23		
Class Work	PIAT pp.144-146	Review Unit	Exam		
Laboratory					
Homework					

LA = Laboratory Activity; **CQ** = ChemQuandary; **YT** = Your Turn; **YD** = You Decide; **PIAT** = Putting It All Together.

This second unit is intended to give students some insight into the worldwide problem of limited chemical resources. As the Earth's human population has grown, demand for chemical resources such as metal ores and fossil fuels has increased. Our developing technology has provided us a more comfortable way of life, but we face decreasing reserves of resources and increasing disposal problems associated with large quantities of "waste" materials. These related trends generate a host of economic, environmental, and political issues.

Chemical concepts covered in this unit include equation balancing, properties and reactivities of metals, and the periodic table; there is particular emphasis on copper as a resource, along with recycling and reusing resources. We must consider resource conservation strategies if we are to maintain a high standard of living. For some resources this means maximizing use/reuse (multiple use of a manufactured object for the same or similar purpose). For others, recycling (collecting and reprocessing a manufactured item to obtain its basic raw material) or substitution (filling a given need with an alternative resource) is more feasible. For many resources a combination of these conservation strategies is desirable. In any case, all three options require knowledge of the properties and behavior of matter—that is, familiarity with chemistry!

If your students are not already familiar with chemical reference books, while using this unit they should be exposed to one or more of the following: the *CRC Handbook of Chemistry and Physics*, Lange's *Handbook of Chemistry*, the *Merck Index*, the McGraw-Hill *Encyclopedia of Science and Technology*, and similar sources.

OBJECTIVES

Upon completion of this unit the student will be able to:
1. Compare and contrast science and technology. [Introduction]
2. State the law of conservation of matter, and apply the law by determining whether a given chemical equation is balanced. [A. 2, A. 3]
3. Describe the Spaceship Earth analogy, and apply it to the terms "throw away" and "using up." [A. 4, A. 5, A. 7]
4. List common types and sources of municipal waste, and describe attempts to reuse and recycle waste. [A. 5, A.6]
5. Define and give examples of renewable and nonrenewable resources. [A.5, A. 7]
6. Distinguish between chemical and physical changes and/or properties when given specific examples of each. [B. 1]
7. Classify selected elements as metals, nonmetals, or metalloids based on observations of their chemical and physical properties. [B.2, B. 3]
8. Use the periodic table to (a) predict physical and chemical properties of an element; (b) write formulas for various compounds; (c) identify elements by their atomic masses and atomic numbers; and (d) locate periods and groups (families) of elements. [B. 4, B. 6]
9. Construct a workable periodic table and explain its organization, given chemical and physical properties of a set of elements. [B. 5]
10. Compare the reactivities of selected elements, and explain the results in terms of the structure of their atoms. [B. 7–B. 9]
11. Discuss the development of new materials as substitutes for dwindling resources. [B. 10]
12. Explain from a chemical viewpoint the problems and solutions involved in restoring the Statue of Liberty. [B. 11]
13. List the three primary layers of our planet and some resources that are "mined" from each region. [C. 1]
14. Write balanced chemical equations and relate them to the law of conservation of matter. [C. 2]
15. Define the term mole, and calculate the molar mass of a compound when provided its formula and the atomic masses of its elements. [C. 3]
16. Outline the production of a metal from its ore (using copper as an example) and list four factors which determine the profitability of mining. [D. 1]
17. Calculate the percent by mass of a specified element in a given compound. [D. 2]
18. Define oxidation and reduction, and compare the three most common redox–reaction methods for separating metals from their ores. [D. 4]
19. Use supply and demand data to estimate the lifetime of a given resource, and discuss options such as reusing, recycling and substitution. [E. 1– E. 3]

A: USE OF RESOURCES

Part A examines (from a chemical perspective) what it means to "use" resources and how we dispose of "waste" products. Introducing the chemical concepts of the law of conservation of matter and of balanced equations (the chemist's accounting system) leads students to discover that "using things up" actually means atoms and molecules of resources become too widely dispersed to be efficiently collected again. Seen from this perspective, pollution can be regarded as resources "out of place." This in turn leads to the more personal question of what individuals do with the wastes they generate.

   To help make the concept of resource limits (and the importance of wise use of resources) more concrete, we have first directed student attention to recent changes in the composition of pennies. The U.S. Mint is now producing pennies composed of copper-plated zinc instead of essentially pure copper, because of the increased scarcity and cost of copper metal. In the first laboratory exercise, students gain a sense of how chemistry can be used to alter the properties of materials for meeting specific practical and economic needs. Part A ends with a student activity focusing on what resources individuals use up, and what happens to the discarded resources.

CHEMQUANDARY

A Penny for Your Thoughts (page 88)

How much is a penny really worth?   The mathematics is a bit lengthy, but interesting.

   Old pennies had a mass of 3.10 g and were 95% copper. This gave a mass of 2.95 g of copper, which at $1.14/pound is worth 0.74 cents. The 5% zinc had a mass of 0.15 g, and at a price of $0.55/pound, is worth 0.02 cents, making pre-1982 pennies worth 0.76 cents.

   New pennies have a mass of 2.52 g and are 97.6% zinc. This gives a zinc mass of 2.46 g, and at the price of $0.55/pound, is worth 0.30 cents. The mass of copper is 0.06 g, and its price of $1.14/pound makes this worth 0.02 cents. The newer pennies are thus worth 0.32 cents, less than half the value of the old pennies.

Answer to margin note (page 89): Hobbyists and coin collectors see pennies as having special worth depending on their condition and the number of pennies in circulation.   For example, a 1909 uncirculated penny is worth $3200 today. Even as recent a coin as an uncirculated 1950 penny is worth $80.

A.1 LABORATORY ACTIVITY: STRIKING IT RICH (pages 89-91)

This laboratory activity, in combination with the previous information about pennies, is intended to generate student interest in the chemistry of metallic resources—both how chemistry is involved in their transformation, and how substitutions may be made when supplies are limited.

   In this activity, students will produce brass, an alloy of zinc and copper. The shiny brass color may remind students of gold; an intermediate step may suggest the appearance of silver on the coin as well. Do not give away the anticipated results—let students discover the changing appearance of the penny for themselves.

Time

Less than a full class period. Try to schedule this activity during the first day of this unit to heighten interest.

Materials (for a class of 24 working in pairs)

   24 protective goggles
   12 150-mL beakers
   24 250-mL beakers
   12 25-mL graduated cylinders
   12 crucible tongs
   12 ring stands
   24 ring clamps
   12 watch glasses
   12 wire gauze
   12 Bunsen burners
   12 matches or strikers
   36 copper pennies

10 g mossy zinc (Zn) or zinc foil
300 mL 3 M sodium hydroxide (NaOH) solution
2 L distilled water

Advance Preparation

Prepare the 3 M sodium hydroxide solution (120 g NaOH/L solution). Be sure to have a supply of pennies, or ask students to bring their own. (Note: If they do this, they can preclean them by soaking overnight in vinegar with a little table salt to generate a weak solution of hydrochloric acid. As an alternative, the pennies can be cleaned in the laboratory in a 6M solution of HCl (concentrated HCl diluted 50/50 with water).

Pre-Lab Discussion

During a short pre-lab discussion, demonstrate how to operate a Bunsen burner, as presented on page 89 of the student text.

Lab Tips

Students must wear protective goggles (as they should for all laboratory activities). No contact lenses should be worn (either by teacher or students) during this activity, due to the caustic fumes released during the heating of the sodium hydroxide solution and the risk of splattering. The room **must** be well ventilated. Especially, draw student attention to step 5: remind them to heat the sodium hydroxide solution gently until the solution just starts to bubble—*do not allow them to boil the sodium hydroxide solution vigorously*. Hot sodium hydroxide is very caustic and students should be advised that if sodium hydroxide is spilled or splattered on any part of their skin, they should immediately wash with cold water and inform you. The vapors are also quite choking and should be avoided.

Some teachers find it useful to clean the pennies by rubbing with steel wool or Brillo pads just before starting the reaction sequence.

Either pre- or post-1982 pennies can be used. However, since post-1982 pennies contain a much higher percent of zinc than pre-1982 pennies (97.6% vs. 5%), they are more likely to warp if overheated (melting point of copper: 1083°C vs. zinc: 420°C). In either case, tell students to heat the penny gently by passing the penny back and forth over the outer cone of the flame. Holding the coin vertically over the flame serves to equalize heating and produces the most aesthetically pleasing coins.

You might be somewhat surprised to see that, in the first reaction, zinc plates onto the surface of the copper penny—this is contrary to some observations, where copper plates out on zinc. In this system, zinc reacts with aqueous sodium hydroxide to form a zincate ion, with the formula $Zn(OH)_4{}^{2-}$ or $Zn(OH)_3{}^- \cdot H_2O$. A zincate can react to plate zinc metal on the copper surface.

Post-Lab Discussion

The post-lab questions and discussion can guide students from their initial excitement or surprise to consider the importance of characteristic physical and chemical properties in identifying a substance. Each element has a unique combination of properties that is its own characteristic, just as humans have unique fingerprints for identification. Identification of substances is a large part of the work of chemists, whether they are analyzing water, minerals, oil, food, air, biochemical samples, industrial products on Earth, or even samples from other planets. This idea of chemists as "matter detectives" occurs repeatedly throughout the *ChemCom* course.

The concept of physical and chemical properties will be further developed in Part B, where students will consider the practical significance of physical and chemical properties. At this time, you may wish to simply point out that alloys such as brass (60-90% Cu/40-10% Zn) often allow for a blending of desirable properties. Since this activity is a highly motivational opening activity, do not kill enthusiasm by concept overloading. As in other units, the goal of the opening is to motivate the students to want to learn more (and generate their own questions), not to provide ready-made answers to questions they don't yet care about.

Answers to Questions:

1. a. Untreated — copper color; heated in Zn and NaOH solution — shiny "silver" color; heated in Zn and NaOH solution and by flame — "gold" color.
   b. Treated pennies resemble silver and gold.
2. Physical properties like density (Zn: 7.1, Cu: 9.0, Ag: 10.5, Au: 19.3 g/mL), melting point (pure Zn: 420, Cu: 1083, Ag: 962, Au: 1064°C) and chemical reactivity (for example: zinc, copper, and silver all react with nitric acid while gold does not). Note: In the pre-lab discussion you may wish to introduce your students to one or more chemical reference books where such data can be found.

3.   Inexpensive substitute for silver or gold appearance.
4. a. A solid solution of brass was produced: the copper atoms were merely mixed with zinc.
   b. Yes, allowing the zinc to react with acid would remove the zinc and leave the copper layer.

A.2 USING THINGS UP (page 91)

Students encounter the idea that although things sometimes seem to disappear when used, matter is neither created nor destroyed at the atomic level. A better definition of "use" is the rearrangement of matter at the atomic level. You may wish to contrast a natural law like the law of conservation of matter to a societal law, such as speed limits or criminal law. The natural laws have been discovered by careful scientific inquiry. Society is free to make, and change, laws governing human behavior.

A.3 TRACKING ATOMS (pages 91-94)

Chemists acknowledge the law of conservation of matter each time they balance chemical equations. Counting atoms is the first step in balancing such equations— later, coefficients will be used to balance the number of atoms on each side of the reaction symbol. Two examples are worked out. The concept of balancing equations will be covered in a subsequent section, an excellent example of spiral learning.

YOUR TURN

Balanced Equations (pages 93-94)

This activity establishes a connection between properly written chemical equations and the law of conservation of matter. Balancing is not necessary here, but it is wise to provide further support for how atoms are counted in formulas involving parentheses, such as in $Al(OH)_3$.

Answers to Questions:

1. Answered in the text.

2. One formula unit of cellulose reacts with six molecules of oxygen gas, to form six molecules of carbon dioxide gas, five molecules of water, and energy.

Reactants:	Products:
6 C atoms	6 C atoms
10 H atoms	10 H atoms
17 O atoms	17 O atoms

This equation is balanced.

3. Four molecules of liquid nitroglycerin decompose to form six molecules of nitrogen gas, one molecule of oxygen gas, twelve molecules of carbon dioxide gas and ten molecules of gaseous water (water vapor).

Reactants:	Products:
12 C atoms	12 C atoms
20 H atoms	20 H atoms
12 N atoms	12 N atoms
36 O atoms	36 O atoms

This equation is balanced.

4. Four atoms of solid silver react with four molecules of hydrogen sulfide gas and one molecule of oxygen gas to form two molecules of solid silver sulfide and four molecules of liquid water. This is the reaction behind the tarnishing of silver and explains why silver vessels should not be used to heat sulfur-containing food such as proteins (eggs, etc.).

Reactants:	Products:
4 Ag atoms	4 Ag atoms
8 H atoms	8 H atoms
4 S atoms	2 S atoms
2 O atoms	4 O atoms

This equation is not balanced as written.

A.4 LABORATORY ACTIVITY: USING UP A METAL (pages 94-95)

This activity demonstrates copper being "used up" by the action of nitric acid. Post-lab questions, demonstration, and discussion emphasize the fallacy of the phrase "used up."

Time

15-20 minutes

Materials (for a class of 24 working pairs)

12 test tubes
6 dropping bottles for dispensing 8 M HNO_3
150 mL 8 M nitric acid solution (76 mL conc. HNO_3/150 mL)
30 cm copper (Cu) wire (24-gauge, non-lacquered); 2.5-cm length/team

Advance Preparation

Advance preparation includes making up the nitric acid solution and cutting the metal samples into pieces. *Caution: if you prepare the 8 M HNO_3 from concentrated nitric acid, remember to slowly add the concentrated acid (with stirring) to water.*

Pre-Lab Discussion

Students should be cautioned regarding the hazards of nitric acid and reminded of the proper procedure for mixing the contents of a test tube.

Lab Tips

Rolling the wire into a ball reduces the amount of nitric acid needed in this lab.

The reaction of copper with nitric acid will liberate a small amount of nitrogen dioxide (NO_2) gas. We have purposely specified a rather small sample of thin copper wire to minimize the generation of this poisonous gas. The quantity generated during this activity is well below the danger level, but *do not allow students to use larger samples of copper! Also, be sure your laboratory room is well ventilated.*

Post-Lab Discussion

The post-lab discussion should focus on the law of conservation of matter. Emphasize that although the copper seems to have been "used up," each and every copper atom is still present, although now in the form of Cu^{2+} ions. In fact, if allowed to crystallize, copper(II) nitrate is itself a valuable product that finds use as a ceramic color; as a mordant and oxidant in textile dyeing and printing; as a pyrotechnical oxidizer; and as a fungicide and herbicide. Or, if desired, that same copper(II) nitrate could be a source of copper (see Demonstration Idea below).

out in recycling, "you can't get something for nothing." That is, to get back a given resource requires use of another resource and/or energy. This demonstration is particularly effective as a transition to the next section (A.5, Resources and Wastes).

Also discuss the physical and chemical changes that occurred during the procedure, and the use of symbols and equations to represent the chemical reaction.

You may also wish to point out that nitrogen dioxide is one of the pollutants formed by any high temperature combustion process (such as in an automobile engine) and contributes to the acid rain problem (more on this in the Air unit). Its acidic properties are readily observed by dissolving a small quantity in water and noting its effect on acid-base indicators that change color slightly below pH 7. It is also possible to expose moistened pH paper to some nitrogen dioxide collected in a bottle.

Answers to Questions:

1. The color is caused by the formation of copper(II) ions when copper metal reacts with nitric acid.
2. The copper metal is converted to copper(II) ions, which are soluble and form a solution with the water.
3. a. $Cu(s) + 4 HNO_3(aq) \rightarrow Cu(NO_3)_2(aq) + 2 NO_2(g) + H_2O(l)$
      copper colorless   blue-green   orange-brown colorless
   b. The copper from the wire dissolved into the water as Cu^{2+} ions.
   c. The colored gas (nitrogen dioxide) contains nitrogen and oxygen.
   d. These atoms came from the nitric acid.
   e. The water molecules also came from the nitric acid.

MICROSCALE PROCEDURE

Procedure

1. Obtain a piece of copper wire and place it into an empty well of a microplate.
2. Carefully add enough nitric acid (HNO_3) to cover the wire. *Caution: if any acid accidentally spills on you, wash the affected area with tap water for several minutes, and notify your teacher.*
3. Observe the reaction for five minutes. Record all observed changes in the wire, the solution, and in the well contents. A piece of white paper behind the microplate will make it easier to observe the reaction.
4. Discard the contents of the well according to your teacher's directions.

A.5 RESOURCES AND WASTE (pages 95-97)

The terms renewable resources and nonrenewable resources are explained in this section. Point out that using and producing resources creates waste, along with vast quantities of discards. Recycling is introduced as a partial solution to part of this problem.

A.6 DISPOSING OF THINGS (pages 97-99)

The quantity of waste produced is examined with graphic data and statistical information. This section points out that increased attention to waste disposal is important for us as individuals, and for communities, nations, and the planet as a whole. New research on uses for waste have resulted in materials for roads, construction, and even a source of oil. Note that both recycling and incineration attempt to regain value from wastes whereas sanitary landfills merely bury them at a cost. You might remind students that just as water had both direct and indirect usages, so does waste: buying and using products produces extensive "indirect" waste.

If time allows, you might consider asking students to keep a "garbage" log, similar to the water usage log kept in the Water unit. Ask students to monitor the types and amount of waste disposed of by their families for a 1-3 day period. It is often instructive to empty your classroom waste containers onto your demonstration table, to illustrate typical waste in the classroom.

CHEMQUANDARY

Wondering about Resources and Waste (page 99)

Facts that may be helpful to students in this question of "paper or plastic": paper is renewable, plastic is not. Paper is easier to recycle. Plastic requires 20-40% less energy to manufacture. There are fewer atmospheric emissions and less waterborne wastes for polyethylene sacks. Polyethylene sacks occupy 70-80% less volume in landfills than paper sacks. Polyethylene sacks are lighter than paper sacks and produce less ash upon incineration.

For Sections A.5-A.7, C.4-C.5, and Part E

1. Sections A.5 and A.6 are focused on municipal waste. The collection and disposal of municipal solid waste costs taxpayers about $5 billion/yr and requires about 10% of the energy consumed in the United States. However, municipal solid waste only represents a small (4%), visible part of the overall solid waste picture. Total solid waste production in the United States is about 18,000 kg/person/yr or 49 kg (108 lbs)/person/day. Most of this is "hidden" waste generated by agriculture (56%), mining (34%), or industrial (6%) processes which produce consumer products.

2. Estimates for 1990 indicate that 80% of the garbage in the United States is sent to landfills, while 10% is recycled and 10% is incinerated. In Japan, 43-53% is incinerated, and 26-39% is recycled. U.S. legislation concerning open dumps and the creation of new dumps has not made a clear distinction between sanitary landfills and open dumps. A sanitary landfill should be a site where solid waste is spread in thin layers, compacted, and covered with a fresh layer of soil each day; this minimizes problems with pests, disease, and air and water pollution. These landfills are located to minimize pollution from runoff and leaching. However, many of today's sanitary landfills are just slightly improved versions of yesterday's open dumps. Recent studies have demonstrated also that some organic materials, such as garden wastes, foodstuffs, and newspapers, do not biodegrade while in the landfill because they are covered, making these materials a permanent part of our waste disposal problem.

3. Many U.S. cities face economic, political, environmental, and physical space constraints in trying to find suitable sites for sanitary landfills. For instance, the cost of burying a ton of trash in a landfill for the city of Minneapolis has risen from $11 in 1980 to $50 in 1991. To counter these costs, this city has become the largest American community to start a source separation recycling program. (See Andrew H. Malcolm, "For Minneapolis, Trash Means Cash," *New York Times*, March 7, 1984). Many other cities and states have instituted very successful recycling programs, and Connecticut has passed a law requiring 20% recycled paper to be used in newspapers and magazines sold within the state; the percentage rises to 90% in 1998.

4. As of 1989, twelve U.S. states (Oregon, Florida, California, Vermont, Maine, Michigan, Iowa, Connecticut, Massachusetts, Delaware, Wisconsin and New York), as well as Sweden, Denmark, Norway, The Netherlands, and several provinces in Canada had mandatory deposit laws on all beverage containers. The results, a 75-86% reduction in beverage container litter, seem quite positive. Given these results, and the over $500 million spent each year to pick up litter, many environmentalists are lobbying for a *national* container deposit law.

5. As of 1991, some 155 energy recovery units for burning municipal trash were in operation in the United States with an additional 29 under construction. Presently few of these units are equipped to separate iron, aluminum, and glass for recycling, although they could be so adapted. Estimates of municipal solid wastes that could be economically recycled range from a low of 25% to a high of 66%. States have spent 39 times more money on incinerators than on recycling programs.

For Sections A.5-A.7, C.4-C.5, and Part E

Many homes are equipped with either home trash compactors or garbage disposals. While they may seem to reduce garbage, both pose their own set of ecological problems. Discuss this paradox.

Answer: (a) Trash compactors reduce the volume of waste, not the total weight, and biodegradation of the compacted waste is slower. Also, compacted waste is harder to recycle. (b) Disposals put organic, oxygen-demanding food waste into drains that enter waterways, increasing biological oxygen demand and creating a water pollution problem.

A.7 YOU DECIDE: CONSUMING RESOURCES (pages 99-100)

This activity enables students to focus on their own resource usage and the strong reliance most of our society has on resources. Your major task is helping students figure out what resources are used in the manufacture of items. By focusing on common materials such as TV sets, pencils, etc., encountered in everyday life, students are sensitized to the variety of things we use and the resources used to make them.

This activity is designed to be completed in one class period. Each student will first answer Questions 1 through 3. Instruct them to list the first items that come to mind. Do not expect correct or complete answers for the resources used to manufacture the items.

After completion of Questions 1-3, divide the class into groups. Each group should work on answers to Questions 4 through 7, and prepare a short group report. A full class discussion will then highlight major points developed by the groups. Encourage students to move through the activity at a fairly brisk, productive pace.

Answers to Questions:

1-5. Answers will vary depending on the students. Check their understanding of renewable and nonrenewable resources in Question 3.

6. Sampling of possible answers:

Item	Nonrenewable component	Renewable substitute
Aluminum beverage can	Aluminum metal	Paper/cardboard
Aluminum foil	Aluminum metal	Cellophane (if not for cooking)
Automobile tire	Synthetic rubber	Natural rubber
Batteries	Zinc or other metal	Rechargeable batteries, solar cells
Clothing	Synthetic fibers	Cotton, wool, silk
Fast food container	Petroleum	Paper
Gasoline	Petroleum	Grain alcohol fuel
Soap/perfumes	Petroleum	Natural fats, oils, plant and animal products
Styrofoam cup	Petroleum	Paper
Tungsten lightbulb wire	Tungsten metal	Electroluminescent panels

7. Commonly, waste disposal is by area or private collection agencies, by personal "hauling to the dump", and so forth. Help students focus on the crisis facing waste management with the closing of landfills, the problems with dumping waste in the oceans, and so on.

PART A: SUMMARY QUESTIONS (page 101)

1. a. The law of conservation of matter: In a chemical reaction, matter is neither created nor destroyed. There is no known instance in which matter has not been conserved in a chemical reaction.

   b. Governmental laws, created by people, are subject to social and economic pressures; scientific laws are part of a natural system immune to human intervention. Scientific laws are not broken—there is no instance where the law of conservation of matter has not been followed in a chemical reaction. Governmental laws are regularly broken by people—observe the speeders caught by state troopers on the highways!

2. a. $Sn + HF \rightarrow SnF_2 + H_2$

   Reactants:
   1 Sn, 1 H and 1 F atom.
   Not balanced.

   Products:
   1 Sn, 2 H, and 2 F atoms.

   b. $SiO_2 + 3C \rightarrow SiC + 2CO$

   Reactants:
   1 Si, 2 O, and 3 C atoms.
   Balanced.

   Products:
   1 Si, 2 O, and 3 C atoms.

   c. $Al(OH)_3 + 3HCl \rightarrow AlCl_3 + 3H_2O$

   Reactants:
   1 Al, 6 H, 3 O, and
   3 Cl atoms.
   Balanced.

   Products:
   1 Al, 6 H, 3 O, and
   3 Cl atoms.

3. According to the law of conservation of matter, atoms cannot be created or destroyed (used up). Hence materials are merely changed in form, but the atoms are never destroyed. It is impossible to throw anything away, because the Earth is self-contained and does not lose materials to space. "Throwing it in the trashcan" is a more proper statement than "throwing it away."

4. a. The Earth may be viewed as a spaceship since Earth and its atmosphere represents (for matter) a closed system. The amount of resources is limited because matter can neither be lost nor gained. The analogy is useful in helping students realize that there is a finite amount of resources—water, fuel, raw materials—on board, and no place to go to get more (although students love to employ science fiction space explorations and mining).

   b. The analogy is a little misleading since Earth continually gains radiant energy from the sun (which is partially returned to space) while a spaceship more closely resembles an isolated (self-contained) system. A truly isolated system exchanges neither mass nor energy with its surroundings.

5. Discarding less waste decreases the waste-storage problem, and recycling waste may lower the demand for resources resulting in a decrease in the price. Other proven benefits include pollution reduction, energy savings, the creation of new jobs, and, in the case of some resources, such as oil, more political stability.

6. Reuse is more advantageous than recycling due to lower costs and energy use.

7. a. Renewable: water, air (nitrogen, oxygen, etc.), paper (trees), biomass.
   b. Nonrenewable: metals (aluminum, iron, copper, etc.), fossil fuels (coal, petroleum, natural gas).

B: WHY WE USE WHAT WE DO

When we choose a resource for a specific purpose we must match the specific requirements with the properties of the resource. Such matching involves knowledge of chemical and physical properties. Part B demonstrates, among other things, how such a matching process was used both in the construction *and* in the restoration of the Statue of Liberty.

Part B emphasizes the organization of chemical and physical properties of elements in the periodic table. Such a table can assist in selecting materials to be used for specific purposes. With the aid of the periodic table, chemists can also modify the structure of molecules, or physically blend substances to produce materials with intended properties. In short, chemical knowledge can be used to modify properties of materials that improve the quality of our everyday lives.

Students have the opportunity to simulate the discovery of some typical patterns in the properties of elements by constructing their own "periodic table." Part B ends with students applying their chemical knowledge to an analysis of the restoration of the Statue of Liberty.

B.1 PROPERTIES MAKE THE DIFFERENCE (pages 102-104)

This section describes some properties of materials used for building the Statue of Liberty. Begin to introduce chemical and physical properties by focusing on the requirements for building the statue. Also remind students of the earlier laboratory activity on changing properties of pennies.

DEMONSTRATION IDEAS

To clarify the differences between physical and chemical properties and changes, some of the following demonstrations may be helpful:

Physical changes: (1) Gently heat a few crystals of iodine in a large, lightly sealed flask. (2) Place a small piece of dry ice (CO_2) in a balloon and seal it for a "self-inflating" effect. (3) Pour a small quantity of liquid nitrogen into an empty 2-L plastic soft drink bottle and seal it with either a balloon or a stopper. (If you use a stopper, be sure to secure the bottle and point the stopper away from students. Doing this outdoors allows arranging the bottle for a vertical takeoff!) (4) Heat a bimetallic strip (20 cm or more in length). (5) Boil water under reduced pressure conditions. (6) Shake and open a warm can of soft drink. (7) Set up a crystal growing "garden", using a diluted solution of sodium silicate ("water glass") and any ionic solid. Even though there appears to be a change here, that requires the interaction of solvent and solute, remind students that dissolving sugar in water merely changes the form, *not* the formula of the sugar or the water.

Chemical changes: (1) Shake a small quantity of powdered antimony in small gas-collecting bottles of bromine and iodine gas. (The slow reaction causes color to disappear; for bromine, it takes about 5-15 minutes, for iodine about one class period.) *Caution: Avoid exposure to fumes by doing this in a fume hood.* (2) Place calcium turnings (calcium metal) in lukewarm water—one only notices the slight evolution of hydrogen gas, yet the addition of an indicator such as phenolphthalein suggests something more is happening. *Caution: Use small amounts of calcium turnings.* (3) Burn a candle. (4) Fry an egg. (5) Demonstrate spontaneous endothermic reactions such as that between 15 g $Ba(OH)_2 \cdot 8H_2O$ and 15 g NH_4SCN. Place the reactants in a 250 mL flask; stopper lightly, shake, wet the outside at the bottom, and set on a small piece of wood. $Ba(SCN)_2 + NH_3 + H_2O$ are formed with a temperature drop sufficient to freeze the flask to the block. The presence of a gaseous base can be detected with moistened red litmus paper.

Properties: Physical or Chemical? (page 103)

In this activity, students practice recognizing the difference between physical and chemical properties. Chemical properties become evident when one substance reacts with another to form new substances, whereas physical properties (such as melting point, boiling point, thermal and electrical conductivity, density, color, hardness, etc.) do not depend on the presence of another substance. In a physical change, the identity of the substance is not altered.

Answers to Questions:

1. Physical — no change in material, depends on nothing else
2. Chemical — material changes, depends on presence of another substance (air)
3. Physical — no change in material, depends on nothing else
4. Chemical — material changes (taste), depends on bacteria in milk
5. Physical — no change
6. Physical — no change in identity, even though shape is changed
7. Chemical — bread rising is due to the production of carbon dioxide by yeast: this involves a change of material
8. Physical — melting does not change a material's identity: when ice melts, the formula of the liquid is still H_2O
9. Physical — no change in the metals

A short follow–up activity, ask students to think of a material or object, then make a list of things to do with that material, or of uses for it. Then describe each use or action as a chemical or physical property. For example, a pencil: used for writing, can be sharpened, heated in a Bunsen burner (still a favorite activity of some students!), etc. The first two properties are physical changes, while the latter is a chemical change.

B.2 THE CHEMICAL ELEMENTS (pages 104-105)

All substances that make up our world are composed of a limited number of chemical elements. Some general properties of elements are given in Table 2 (page 105), and the properties and symbols of common elements (those listed in Table 1) are reviewed through the following *YOUR TURN* activity.

YOUR TURN

Chemical Elements Crossword Puzzle (page 105)

This activity acquaints students with some properties and uses of selected elements. The clues to the crossword puzzle are located on pages 147–149. A blank copy of the crossword puzzle matrix is included in this Teacher's Guide (page 49). Make copies of this puzzle for your students.

Key:

Down		Across	
1. Helium	22. Lead	3. Fluorine	21. Silver
2. Barium	25. Tungsten	5. Calcium	23. Oxygen
4. Lithium	27. Manganese	7. Platinum	24. Silicon
6. Phosphorus	28. Zinc	8. Iodine	26. Cadmium
9. Krypton	29. Chromium	10. Boron	30. Bromine
10. Beryllium	30. Bismuth	11. Nickel	32. Argon
12. Magnesium	31. Cesium	13. Aluminum	35. Antimony
14. Carbon	33. Copper	14. Cobalt	36. Sodium
18. Mercury	34. Iron	15. Uranium	37. Nitrogen
19. Hydrogen	38. Gold	16. Neon	40. Chlorine
20. Potassium	39. Tin	17. Sulfur	

You may wish to highlight the connection between physical/chemical properties and practical use for several of the elements: (1) low density, and nonflammability of helium vs. lower density but highly flammable hydrogen; (2) low density, high strength of magnesium and aluminum used in airplanes and cars vs. lead; (3) high melting point of tungsten used in incandescent bulbs vs. low melting point of bismuth used in fire detection/extinguishing devices; (4) low melting point, high density of mercury (used in electrical switches and barometers); (5) high electrical

CHEMICAL ELEMENTS CROSSWORD PUZZLE

Directions: See Resources unit pages 147-149 for puzzle clues.

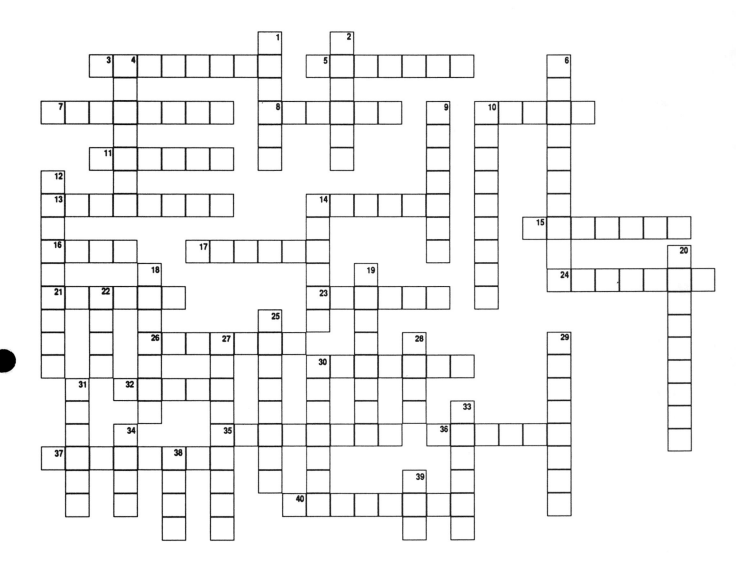

resistance of chromium (used in toasters) vs. high electrical conductivity of copper (used in wiring); and (6) low chemical reactivity of gold (used in jewelry).

OPTIONAL ACTIVITY

Getting to Know the Language of Chemistry

Challenge your students to discover the patterns of symbols for the elements. There are basically four: elements with one–letter symbols which match the first letter of the element name (C, O, F, I); elements with two–letter symbols that match the first two letters of the element (Be, He, Li, Br); elements with two–letter symbols in which the second letter is a prominent sound in the element name (Mg, Mn, Cr, Cd); and elements with symbols from the Latin or Greek names (Fe, Cu, W, Na).

Some popular activities for making elements more familiar are "Element of the Week", in which each day has a small activity pertaining to that week's element; and having each student "adopt" an element, preparing a report and becoming the class expert on the assigned (or chosen) element.

B.3 LABORATORY ACTIVITY: METAL, NONMETAL? (pages 105–107)

In this laboratory activity, students will learn how to distinguish metallic from nonmetallic elements after examining their properties. Seven elements are tested for appearance, malleability, reaction with hydrochloric acid, reaction with copper(II) chloride solution, and electrical conductivity.

Time

One class period

Materials (for 24 students working in pairs)

84 test tubes (13 × 100 mm)
12 test tube racks
12 test tube brushes
12 10-mL graduated cylinders
12 forceps or tweezers
84 vials for elements
12 grease marking pencils
2 burets for solution dispensing (or use dropper bottles)
1 conductivity apparatus (optional)
several small hammers
100 mL 0.5 M hydrochloric acid solution (4.2 mL conc. HCl/100 mL)
100 mL 0.1 M copper(II) chloride solution (1.7 g $CuCl_2 \cdot 2H_2O$/100 mL)

The seven elements to be tested (code on student samples):
25 g carbon (C), graphite sticks, pencil lead or charcoal briquets (a)
35 cm magnesium (Mg), ribbon, cut to 2-cm lengths (b)
25 g silicon (Si), lump (c)
25 g tin (Sn), mossy or foil (d)
25 g sulfur (S), lump roll (e)
25 g iron (Fe), filings or thin wire (better for conductivity) (f)
50 g zinc (Zn), mossy or wire (better for conductivity) (g)

Pre-Lab Discussion

The pre-lab discussion should include a review of the differences between physical and chemical properties. Caution students not to handle any element samples with their hands. Review the proper procedures for using acids safely.

Lab Tips

The materials list specifies seven test tubes per group. If tubes are in short supply, the laboratory activity can be completed with a minimum of 24 test tubes per class, if they are washed and re-used during the activity.

The electrical conductivity portion of this activity is optional, but could be done as a teacher demonstration using larger samples. The microconductivity setups, using 9-volt batteries and light-emitting diodes (LED), could be used here.

Iron sheet or ungalvanized nails could be used for the iron, if wire is unavailable. The nails should be cut to camouflage their identity.

The burets were included in the materials list as an option to dispense the hydrochloric acid and copper(II) chloride solutions. Dropping bottles could also be used.

The edge of a ring can be used as a hammer. To minimize potential damage to countertops, etc., it may be helpful to have students help demonstrate this part of the testing at the front of the room.

Post-Lab Discussion

The post-lab discussion should include a short demonstration of the conductivity of the elements if you chose not to have students complete this portion of the activity. Identify the names, symbols, and (if desired) the connection between observed properties and practical uses of the seven elements featured in this activity.

Expected Results

Code	Element	Appearance	Result of crushing	Conductivity (optional)	Reaction with acid	Reaction with $CuCl_2(aq)$
a	C	gray, dull	brittle	yes	no	no
b	Mg	shiny, silver	malleable	yes	yes, bubbles	yes, darkens
c	Si	gray, shiny	brittle	yes	no	no
d	Sn	shiny, silver	malleable	yes	yes, bubbles	yes, darkens
e	S	yellow, dull	brittle	no	no	no
f	Fe	gray, shiny	malleable	yes	yes, bubbles	yes, darkens
g	Zn	shiny, silver	malleable	yes	yes, bubbles	yes, darkens

Answers to Questions:

1. Group 1: magnesium, zinc, iron, and tin. Group 2: carbon, silicon, and sulfur.
2. Silicon
3. Metals: magnesium, zinc, iron, and tin. Nonmetals: carbon and sulfur. Metalloids: silicon. Students may also classify carbon as a metalloid based on its conductivity—for now, simply point out that since it does not have the characteristic luster, it is not classified as a metal.

MICROSCALE PROCEDURE

Materials needed per student

12-well plates
Beral pipets containing 0.5 M HCl (labeled HCl) and 0.1 M $CuCl_2$ (labeled $CuCl_2$)
7 vials containing small samples of the seven elements.

Procedure

1. Obtain the 12-well plate, Beral pipets, and three pieces of each element.
2. Observe and record the appearance of each element, including color, shininess, etc.
3. Gently tap one piece of the element with a hammer or a ring. A material is **malleable** if it flattens without shattering when struck, and **brittle** if it shatters into smaller pieces. Decide whether the samples are malleable or brittle.
4. Test the conductivity of one of the pieces of each element by touching both electrodes to the piece (make sure the electrodes do not touch each other). Record your observations.
5. Add one piece of an element to one well of the plate, and add 10 drops of HCl. Observe carefully and record your results in the data table.
6. Add one piece of an element to another well of the plate, and add 10 drops of $CuCl_2$ solution. Records the results in the data table.
7. Discard your solids as directed by your teacher. Rinse out the plate four times with water, and return equipment to the designated area.

Everyday uses of the elements:

Carbon, amorphous– also known as activated carbon—used chiefly for clarifying, deodorizing, decolorizing, and filtering (remind students of the Foul Water laboratory activity) due to its large ratio of surface area to volume. Also used as a black pigment for rubber tires, printing inks, and phonograph records.

Magnesium– used as a component of high strength/low weight alloys, flares, pyrotechnics.

Silicon– used for silicone "rubber", in transistors, silicon diodes, and other semiconductors, in photovoltaic cells.

Tin– used as tin plating (such as "tin" cans) and soldering alloys.

Sulfur– used in manufacturing sulfuric acid (#1 industrial chemical), matches, gunpowder, and vulcanized rubber.

Iron– used as a major structural metal (steel).

Zinc– used for galvanizing sheet iron, as ingredient of alloys such as brass, dry cell batteries.

B.4 THE PERIODIC TABLE (pages 107-108)

Scientists search for patterns and regularities in nature which are used to organize chemical information in more under-standable and useful ways. This section describes this process in the development of the periodic table. In the following *You Decide*, students are involved in activities that simulate the historical paths leading to the periodic table as a way to organize knowledge of elements.

A pictorial periodic table, which shows actual samples of each of the elements, is useful here.

B.5 YOU DECIDE: GROUPING THE ELEMENTS (page 108)

In this activity, students develop alternative groupings of 20 elements based on atomic masses, melting points, boiling points, numbers of oxygen atoms in their oxide compounds, and the numbers of chlorine atoms in their chloride com-pounds. Data cards containing selected chemical and physical properties of 20 elements are required for this activity. These cards are found on page 53 and should be duplicated: make a copy for each student. These cards are used to classify the elements according to similarities in their properties.

A similar approach was actually used by Mendeleev who worked with 63 elements and similar data. In discussing the activity, do not emphasize the single "right" pattern, but rather discuss which of the student-generated patterns might be more useful to chemists as an organizing and predictive device.

This works well as a group project in class. It also enables you to monitor each group's work and "prime the pump" with suggestions about grouping the elements. You will need to provide scissors for cutting up the card sheets, and tape for producing the "periodic table."

B.6 THE PATTERN OF ATOMIC NUMBERS (pages 108-112)

After introducing the concept of atomic number, this section expands on the earlier element–group ideas and leads to graphs relating atomic numbers and chemical formulas.

Periodic Variation in Properties (pages 109-110)

This activity demonstrates the usefulness of graphing to discover regularities among observations. The objective is to produce two graphs reflecting the chemical and physical properties of the 20 elements represented on the cards. Students can discover the cyclic (periodic) trends among the data and relate their interpretations to the modern periodic table. Such an approach was used in 1864 by Lothar Meyer, the second developer of the periodic table, when he plotted atomic volumes vs. atomic weights. Unfortunately, Meyer was reluctant to use his scheme to predict undiscovered elements and did not publish his findings until 1870—one year after Mendeleev had published the pattern evident in the formulas of compounds.

Graphing Ranges (for y axis)

Graph 1: Trends in a chemical property
   Total oxygen atoms in the oxide: 0 to 2.5
   Total chlorine atoms in the chloride: 0 to 4
Graph 2: Trends in a physical property
   Melting point: –272°C (helium) to 2037°C (boron) Note: do not use the data for carbon: 3470°C
   Boiling point: –269°C (helium) to 2677°C (silicon) Note: do not use the data for carbon: 4347°C

TWENTY ELEMENTS AND THEIR PROPERTIES

Atomic mass	Melting point (°C)	Boiling point (°C)	O in oxide	Cl in chloride
1	-259	-252	0.5	1
9	1287	2507	1	2
7	179	1327	0.5	1
39	64	757	0.5	1
40	851	1487	1	2
19	-218	-188	0.5	1
20	-248	-246	–	–
32	119	445	3	2
12	3470	4347	2	4
31	44	280	2.5	3
36	-101	-34	0.5	1
27	659	2327	1.5	3
23	97	889	0.5	1
4	-272	-269	–	–
14	-210	-196	2.5	3
16	-219	-183	–	2
40	-189	-186	–	–
10	2037	2527	1.5	3
28	1407	2677	2	4
24	650	1117	1	2

Provide students with the following information, which will permit them to assign atomic numbers and symbols to the 20 elements on their cards. Make sure students distinguish between the two elements with mass 40: argon has no chlorine compound number.

If this assignment is done in groups, it is useful to have groups compare graphs.

Element	Symbol	Approximate atomic mass	Atomic number
Hydrogen	H	1	1
Helium	He	4	2
Lithium	Li	7	3
Beryllium	Be	9	4
Boron	B	10	5
Carbon	C	12	6
Nitrogen	N	14	7
Oxygen	O	16	8
Fluorine	F	19	9
Neon	Ne	20	10
Sodium	Na	23	11
Magnesium	Mg	24	12
Aluminum	Al	27	13
Silicon	Si	28	14
Phosphorus	P	31	15
Sulfur	S	32	16
Chlorine	Cl	36	17
Argon	Ar	40	18
Potassium	K	39	19
Calcium	Ca	40	20

Answers to Questions:

1. a. The melting point and boiling point graphs show repeating patterns. On both graphs, the valleys appear at hydrogen, helium, nitrogen, oxygen, fluorine, neon, and argon; the peaks are at carbon, silicon, and calcium. Of course, the melting-point trend is lower on the temperature scale, but maintains the same overall shape as the boiling-point trend. The chemical properties of the oxides and chlorides show similar peaks and valleys.
   b. The student grouping in the *You Decide* on page 108 may have produced the family groupings which are part of the periodic table. Students may have matched this grouping if they relied on chemical properties for their organizing of the elements, rather than physical properties.
2. The elements between two successive valleys correspond to a period. A table that orders the elements according to such repeating patterns is called a "periodic table."

The text material reinforces the idea of a periodic table by comparison with a given table. Reinforce periodic properties by comparing the graphing results with the arrangement of elements on the table. The definitions of periods and groups or families is given, and the similar chemical behavior of elements in a family is introduced and reinforced by the following *YOUR TURN*.

YOUR TURN

Periodic Table (pages 110-112)

This activity demonstrates the utility of the periodic table in estimating properties of substances. Knowing major properties of a chemical family allows one to predict (via interpolation and extrapolation) the likely behavior of any element in that family.

Answers to Questions:

1. Germanium has a predicted melting point, between 1410°C and 232°C, of 821°C (actual value is 937°C).
2. a. Rubidium has a predicted melting point, between 64°C and 29°C, of 46.5°C (actual melting point is 38.9°C).
   b. Sodium has a higher melting point than rubidium and potassium, at 97.8°C.
3. a. CF_4; b. Al_2S_3; c. KCl; d. $CaBr_2$; e. BaO.
4. The noble ("inert") gases helium, neon, argon, krypton, xenon, and radon.

You may wish to point out that Mendeleev used his periodic table to predict the existence (as a "hole" in the periodic table) and the chemical and physical properties of several undiscovered elements (discovery dates: gallium/1875, scandium/1879, germanium/1886, polonium/1898, rhenium/1925, francium/1939, technetium/1939, and astatine/1940). Invite your students to suggest how such predictions were possible. Emphasize that scientific theories and models need to have both explanatory and predictive power.

To dispel the myth of the standard form of the periodic table being the only "right" and final form, you may wish to display some alternatives such as the periodic spiral, the periodic pyramid, etc. Also point out the controversy surrounding the International Union of Pure and Applied Chemistry (IUPAC) decision to renumber the family designations, and the naming of new elements.

Demonstrations can help students grasp the concept of similarity of properties within a given family. Among the first 20 elements which the students have been considering, the following demonstrations are suitable:

1. Sodium and potassium can be compared in terms of their rapid oxidation or "rusting" ($4 M + O_2 \rightarrow 2 M_2O$) when exposed to air, or their reactivity in water containing an indicator such as phenolphthalein ($2 M + 2 H_2O \rightarrow 2 MOH + H_2$). For the latter, use only a very small, rice-grain piece of metal and proper safety precautions such as a shield. Contrast the behavior of these alkali metals to other metals such as iron, copper, tin, etc.

2. The reaction of magnesium or calcium with water containing phenolphthalein provides a safer (though less dramatic) alternative to (1). To effect a reaction with magnesium, first clean its surface with steel wool, and use boiling water. This reaction also points out the fact that metallic reactivity increases as one proceeds down a given family.

3. Helium, neon, and argon can be shown to be noncombustible vs. hydrogen (pop balloons with a burning candle fastened on the end of a meter stick), and nonsupportive of combustion vs. oxygen (place a burning candle in containers of the various gases).

   Additionally, you may point out that similar properties often lead to similar commercial uses, with abundance/cost often determining which element to use. For instance, consider the antiseptic properties of chlorine, bromine, and iodine or the "inert" property of helium, neon, argon, and krypton.

B.7 CHEMICAL REACTIVITY (pages 112-113)

We can compare the reactivities of elements by observing the relative ease with which different metals combine with oxygen. Gold, iron, and magnesium are discussed as examples of how a metal's reactivity helps determine its value for a specific purpose. Contrast the use of gold for "timeless" items, iron for construction, and magnesium for fireworks.

B.8 LABORATORY ACTIVITY: METAL REACTIVITIES (pages 113-115)

This activity underscores the fact that different metals have varying degrees of reactivity. Students are to rank copper, magnesium, zinc, and silver according to their observed chemical reactivities.

Time

One class period

Materials (for 24 students working in pairs)

48 test tubes (13 × 100 mm)
12 10-mL graduated cylinders
12 test tube racks
12 test tube brushes
12 grease pencils
4 burets or 32 dropping bottles for dispensing
3 sheets of fine sandpaper or emery cloth
24 cm magnesium ribbon, cut to 0.5-cm lengths, 4/team
144 cm copper metal strips (0.5 cm wide), cut to 3–cm lengths, 4/team
144 cm zinc metal strips (0.5 cm wide), cut to 3–cm lengths, 4/team
   Mossy zinc and copper wire can be used as replacements for the copper and zinc metal strips, if desired.
300 mL 0.2 M magnesium nitrate solution [15.4 g $Mg(NO_3)_2 \cdot 6H_2O$/300 mL)]
300 mL 0.2 M copper(II) nitrate solution [14.0 g $Cu(NO_3)_2 \cdot 5H_2O$/300 mL]
300 mL 0.2 M zinc nitrate solution [17.9 g $Zn(NO_3)_2 \cdot 6H_2O$/300 mL]
300 mL 0.01 M silver nitrate solution [store in brown glass bottle if possible] [0.5 g $AgNO_3$/300 mL]

Pre-Lab Discussion

Point out that you can compare the reactivities of metals in various ways (such as observing their relative ease of combustion), but that this laboratory procedure allows one to do so systematically.

Lab Tips

Caution students that silver nitrate solution stains skin (and clothing)—students should avoid any direct contact with it.

If metallic silver is available, you might wish to have students verify (or demonstrate yourself) that all the NRs (No visible Reaction) in the sample data table (page 113) are, in fact, correct observations.

Remind students to record initial appearance of the metal samples (after step 2) in their data table.

It is possible to complete this activity with spot plates, thereby saving chemicals. The spots can be set up just as in the data chart, and all tests completed at one time.

The magnesium/zinc nitrate reaction is difficult to detect; encourage patience for all observations.

Steel wool may be used in place of sandpaper or emery cloth.

Expected Results

Metals	Appearance	Solutions			
		$Cu(NO_3)_2$	$Mg(NO_3)_2$	$Zn(NO_3)_2$	$AgNO_3$
Cu	shiny, copper	—	NR	NR	becomes silvery
Mg	shiny, silver	darkens	—	flaky coating	becomes silvery
Zn	shiny, silver	darkens	NR	—	becomes silvery
Ag	shiny, silver	NR	NR	NR	—

Answers to Questions:

1. Magnesium
2. Silver
3. Magnesium; zinc; copper; silver.
4. Copper was chosen because it is lower than zinc in the activity series. Zinc reacts more readily with oxygen in the presence of water to form oxides.
5 a. Silver would have been a better choice than copper, because of its lower chemical reactivity.
  b. Silver is more expensive than copper. Also, air pollution could have led to the formation of silver sulfide vs. the more pleasing green-colored copper carbonates.
6. a. Silver is most likely to be found in the uncombined state, because of its low reactivity.
  b. Magnesium is least likely to be found in the uncombined state, because of its high reactivity.

Post-Lab Discussion

Point out that a metal's reactivity determines how it is found in nature (free or combined as a sulfide, oxide, carbonate, etc.), how it can be processed into a pure substance, and what its value is for specific commercial uses (more on this in Part D). As examples, you may wish to contrast magnesium, found mostly in the form of compounds—in pure form its high reactivity led to its use in photo flashbulbs–to gold, which is found uncombined and is used for expensive jewelry, where its low reactivity results in permanent beauty. *Both* active and inactive metals have practical uses.

As part of their lab writeup, you may wish to ask students to write balanced equations for all of the reactions that occurred.

MICROSCALE PROCEDURE

Materials needed per student

12-well microplate
4 samples of magnesium, zinc and copper
labelled Beral pipets containing 0.2 M solutions of $Mg(NO_3)_2$, $Cu(NO_3)_2$, $Zn(NO_3)_2$, and $AgNO_3$.

Procedure

1. Obtain four pieces of zinc, and place each in a separate microplate well.
2. To each well add 5 drops of one of the solutions, and observe. Write your observations in the data table.
3. Repeat Steps 1 and 2, using magnesium, then copper. Record your observations.
4. Dispose of your solid samples as directed by your teacher. Rinse out your microplate four times with water, and return all equipment to the designated place.

B.9 WHAT DETERMINES PROPERTIES? (page 115)

Developing a chemical activity series in *Laboratory Activity* B.8 should prompt students to ponder what causes differences in reactivity. The properties of elements are determined by the internal structure of their atoms—particularly by their electronic arrangements. This structure influences reactions, attractions between atoms, and physical properties such as conductivity and melting point.

B.10 MODIFYING PROPERTIES (pages 115-116)

Although the *atomic* structures of elements are fixed in chemistry, the *molecular* structure of substances (that is, the arrangements of atoms) can be modified to produce materials with properties suitable for a given use. Science and technology have expanded the variety of substances available for modern uses. Chemists modify properties of matter by physically blending (alloying metals, for example) or chemically combining substances.

CHEMQUANDARY

Frozen Smoke? (page 116)

So far the only commercial application for aerogels has been in radiation detectors for high-energy physics experiments. Future applications may include using aerogels as insulation between double-pane windows, refrigerators, and buildings. They may also be used for collecting micrometeoroids in space. Perhaps your students will be able to "discover" some other possible uses.

B.11 YOU DECIDE: RESTORING MS. LIBERTY (pages 116-117)

In this activity your students apply their growing knowledge of chemistry to the problem of restoring the Statue of Liberty. They are asked to generate practical solutions to address the three major causes of the Statue's deterioration. Proposed solutions should be based on the chemical and physical properties of the materials that compose the Statue, as well as the costs of the proposed changes. All proposals should be tempered with common sense.

You may have students begin this activity as homework. The assignment could be as follows:

1. In a few sentences, argue for and against the importance of restoring the Statue of Liberty.

2. Write a short paragraph identifying the problems to be considered in restoring the Statue of Liberty.

3. Consider the three major causes of the Statue's deterioration listed on page 117. List chemical concepts
from your previous study that might help explain how each problem arose.

Have students work in groups for 15-20 minutes, pooling their homework answers. Each group will provide a more in-depth chemical explanation to support proposals for preserving the original design of the Statue. Each group should select a spokesperson to give a one-minute report to the class. Each report should contain:

1. Reasons for restoring the Statue.

2. Proposed solutions to the deterioration problems.

3. Economic, chemical, and practical reasons supporting the solutions.

As an alternative make the assignment a group activity in class, including the above suggestions. This will approximately double the class time needed.

Consider ending this activity with a brief presentation of the actual solutions proposed by the French-American Committee for Restoration of the Statue of Liberty. These are summarized below:

1. The Committee decided to preserve the original artistic and structural design as it first looked in October, 1886.

2. The interior was originally covered with many layers of metal-based paint which were used to protect the copper skin. All interior paint was removed and the copper restored to its original state. The copper interior was completely sealed to prevent the entry of water and to keep the stairs dry. The lighting was improved, with proper ventilation and air conditioning installed.

3. The Statue is supported by a central pylon and primary trusswork system composed of iron. The corroding vertical and horizontal iron ribs supporting the armature structure were replaced with new stainless steel ribs.

4. Although actual damage to the external copper was limited, the solution to this corrosion problem (thought to be caused at least in part by acid rain) is unclear. Scientists and engineers need more data. The copper skin will be monitored periodically, comparing exposed and unexposed patina.

OPTIONAL BACKGROUND INFORMATION

An outstanding video on the construction and restoration of the statue is "The Making of Liberty." Copies are available for $29.95 from the National Park Service, Liberty Island, New York, N.Y. 10020.

For the chemistry involved in the restoration of the Statue of Liberty, see the following: Burroughs, Tom. "Statue of Liberty". *CHEM MATTERS*, April 1985, 8-12. This article explains the chemistry of corrosion of the statue. The "Experimenter's Notebook" in the same issue (page 13) contains the following at-home experiment that simulates the galvanic corrosion that weakened the Statue of Liberty.

Statue of Liberty Corrosion

1. Place a folded paper towel in the bottom of a shallow bowl. Dampen the towel with tap water, and add a little table salt and vinegar to it—this imitates Liberty's environment, moist with salt spray and affected by acid from urban and industrial air pollution.

2. Scrape the edges of two pennies until they are clean and shiny. Strip the paper covering from two twist ties, and scrape the wires with a knife or sharp edge until they are shiny. The wires are iron, like Liberty's ribs. Wind one piece of iron wire into a small coil, and wrap the other wire tightly around a penny.

3. Place the wire-wrapped penny, the coil of wire, and a bare penny on the moist towel. The copper in the penny will act like Liberty's copper exterior. Cover the bowl (clear plastic wrap is best) to duplicate seaside humidity.

4. Look in on the experiment every few hours. Which metal sample corrodes faster? Be patient: for the Statue of Liberty, this process took a long time. With these severe conditions, you should see changes within a day.

5. You may try your own variations of this experiment, but be sure to use a control for comparision. Are the acid, salt and water all necessary? What happens if salt is replaced with baking soda? Will pure distilled water produce corrosion? Can you devise ways to stop the corrosion?

(Contributed and tested by Joseph Ciparick, Manhattan Center for Science and Mathematics, New York, N.Y.)

Jones, S. R. and Palmer, D. "Chemicals Help Rejuvenate Miss Liberty." *Chemical Week*, July 2, 1986. A chemical trade magazine provides a brief rationale for the substances actually used in cleaning and restoring the Statue. See the following table:

Some of the Chemical Products Used in the Restoration

Material	Quantity and cost	Uses
Aluminum oxide	240 tons/$90,000	Abrasive for blasting corroded layer of iron from frame
Liquid nitrogen (at 150 psi)	4,000 gal/$75,000	Removed seven layers of paint from the inside of the copper walls without damaging the penny-thin skin. Works by temperature differential
Baking soda (NaHCO$_3$)	40 tons/$12,000	Used to strip/sandblast the dried-out, coal-tar weatherseal from the interior copper—followup to liquid nitrogen step
Polytetrafluorethylene (powdered PTFE)	400 sq yd	Solid Teflon tape, replaced the asbestos as a barrier between the stainless steel saddles and the copper--thus preventing the recurrence of galvanic corrosion

Water-based potassium silicate primer containing zinc dust	525 gal	Quick-drying paint with a high SiO_2/Zn ratio for hardness and good adhesion; generated no volatile organics
Waterborne epoxy polyamide	1000 gal	High gloss, good adhesion topcoat paint that resists graffiti, chemicals and abrasion
Glass/polyvinyl butyral resin	16,000 lb/300 sq ft	Windows of two layers of PBR sandwiched between two layers of laminated glass

Note: (1) The cost of the chemical products used in the restoration constituted only a fraction of the overall cost of approximately $31 million. (2) The July 4, 1986 celebration at the Statue of Liberty was billed as the largest fireworks display ever produced in the United States. During the 30 minute, $1.5 million show, 40,000 fireworks weighing a total of 20 tons were used. The three key chemical components of fireworks include an oxidizer (usually potassium chlorate or potassium perchlorate); a fuel (organics such as dextran, charcoal, shellac, and various epoxies and resins; or powdered aluminum and magnesium); and a coloring agent (strontium compounds for red, copper compounds for blue, sodium compounds for yellow, and barium compounds for green). The point to convey is that chemical technology has practical and aesthetic value.

PART B: SUMMARY QUESTIONS (page 119)

1. a. Physical properties: color, density, melting point. Chemical properties: gold much less reactive, won't tarnish.
   b. Major difference is the unreactivity of helium: contrast the Hindenburg dirigible vs. Goodyear blimp.
   c. Physical properties: color, density, melting point (tungsten's melting point is much higher). Chemical properties: iron is more reactive.
   d. Physical: elements in the lithium family have lower melting points and are much softer than silver. Chemical: silver is much less reactive than lithium, sodium, potassium, rubidium, cesium, or francium.
2. Many nonmetals are gases or liquids at normal temperatures and pressure. Nonmetals that are solid at normal conditions tend to be brittle, and are nonconductors of electricity.
3. a. MgF_2 b. B_2S_3
4. a. KCl
   b. Potassium chloride has properties similar to sodium chloride because sodium and potassium are members of the same chemical family.
5. a. Magnesium; b. Same family; c. Arsenic; d. Carbon.
6. Any material chosen to fit an intended use has disadvantages. The high reactivity of some elements is a drawback in construction, but using silver and gold would be prohibitively expensive, and the two elements are quite dense.
7. If two dissimilar metals are placed in contact, there is a good chance they will react through an electron–transfer reaction (oxidation–reduction). To prevent corrosion of the metals, a nonconducting material may be installed at the metal–metal contact points. The original designers of the Statue of Liberty used shellacked asbestos to perform this function. However, this insulation became damp and allowed an electrical current to flow between the copper plates and iron support structures.

C: CONSERVATION IN NATURE AND THE COMMUNITY

This part explores the chemical supplies of the Earth—where they are located, how society uses them, how nature conserves them, and what conservation means in human terms. The resources are inventoried by kind and location. The real question is not whether we will completely run out, but whether we will wisely manage the resources we have. Although nature conserves atoms at the atomic level, humans need to conserve at the molecular level. The goal of human conservation is to preserve useful molecular arrangements for reuse.

Part C ends with students considering the economic, political, and environmental effects of recycling or reusing three important chemical resources.

C.1 SOURCES OF RESOURCES (pages 120-121)

We rely mainly on the thin crust of the Earth as our source of chemical resources. Unequal distribution of resources requires us to become a part of the world market of imports and exports. Some information about aqua-mining resources is included in this section.

Student reports on the importing of various resources, such as aluminum, copper, platinum, diamonds, mercury, and chromium, might provide further background for this section.

C.2 CONSERVATION IS NATURE'S WAY (pages 122-124)

This section continues with the law of conservation of matter and with developing balanced equations. The fundamentals of equation balancing are discussed, using the example of the synthesis of water from hydrogen and oxygen gas. Some do's and don'ts are given, as well as some helpful hints regarding polyatomic ions and water. The following *YOUR TURN* gives practice in this skill.

YOUR TURN

Balancing Equations (pages 123-124)

Here students practice the mechanics of balancing equations. You may need to reemphasize that the coefficient "one" is implied, but seldom written. Note that these equations refer to particular aspects of actual resource processing or chemical manufacturing. Point out the practical and economic significance of chemical manufacturers being able to rely on balanced equations. For example, properly scaling up a reaction is analogous to multiplying a recipe to feed more people.

Answers to Questions:

1. a. $2\,C(s) + O_2(g) \rightarrow 2\,CO(g)$
   b. $Fe_2O_3(s) + 3\,CO(g) \rightarrow 2\,Fe(l) + 3\,CO_2(g)$
2. a. $Cu_2S(s) + 2\,O_2(g) \rightarrow 2\,CuO(s) + SO_2(g)$
   b. $2\,CuO(s) + C(s) \rightarrow 2\,Cu(s) + CO_2(g)$
3. $4\,NH_3(g) + 7\,O_2(g) \rightarrow 4\,NO_2(g) + 6\,H_2O(l)$
4. $2\,O_3(g) \rightarrow 3\,O_2(g)$
5. $Cu(s) + 2\,AgNO_3(aq) \rightarrow Cu(NO_3)_2(aq) + 2\,Ag(s)$
6. $2\,C_8H_{18}(l) + 25\,O_2(g) \rightarrow 16\,CO_2(g) + 18\,H_2O(g)$
   (octane)

Other sample problems not found in student text:	Answers:
• Two reactions involved in the processing of lead ore:	
$PbS(s) + O_2(g) \rightarrow PbO(s) + SO_2(g)$	$2\,PbS(s) + 3\,O_2(g) \rightarrow 2\,PbO(s) + 2\,SO_2(g)$
$PbO(s) + C(s) \rightarrow Pb(l) + CO_2(g)$	$2\,PbO(s) + C(s) \rightarrow 2\,Pb(l) + CO_2(g)$
• The production of nitric acid from ammonia:	
$NH_3(g) + O_2(g) \rightarrow NO(g) + H_2O(g)$	$4\,NH_3(g) + 5\,O_2(g) \rightarrow 4\,NO(g) + 6\,H_2O(g)$
$NO(g) + O_2(g) \rightarrow NO_2(g)$	$2\,NO(g) + O_2(g) \rightarrow 2\,NO_2(g)$
$NO_2(g) + H_2O(l) \rightarrow HNO_3(aq) + NO(g)$	$3\,NO_2(g) + H_2O(l) \rightarrow 2\,HNO_3(aq) + NO(g)$

C.3 ATOM, MOLECULE, AND ION INVENTORY (pages 124-127)

This section introduces the concepts of the mole and molar mass. Begin by emphasizing the analogy of the "chemist's dozen" and then point out that the reason for such a large number as Avogadro's number is that the units being counted (atoms and molecules) are so much smaller than eggs. Analogies that help emphasize how enormous the number 6.02×10^{23} is include: It is approximately equal to the number of grains of sand on all the beaches of the world, or the number of glasses of water contained in all the oceans.

For other analogies that involve calculations see: Kolb, Doris, "The Mole." *Journal of Chemical Education*, **55**, 728 (1978).

After students have an intuitive sense of the mole (a unit for counting atoms and molecules), build the connection between the concept of the mole and molar mass: chemists typically weigh macroscopic samples of matter rather than actually counting the number of submicroscopic atoms and molecules. It will help if you to perform the following demonstration.

DEMONSTRATION IDEA

David Tanis (from *Chemunications*, the journal of the California Association of Chemistry Teachers, 6(1): 12, 1988) collected several analogies to the mole, which students find fascinating. You may want to share any or all of the following, in addition to the examples used in the text on page 125.

1. Would a mole of rice grains fill up the room? One mole of rice grains would cover all of the land area of planet Earth to a depth of 75 meters! (1)
2. One mole of rice grains is more grains then the number of grains of all grain grown since the beginning of time! (1)
3. A mole of rice would occupy a cube about 120 miles on each edge! (1)
4. Computers can count at the rate of about 80 million counts per second. At this rate it would take a computer a quarter of a billion years to count to 6.02×10^{23}. (2)
5. In order to put a mole of rain drops in a 30 meter diameter tank (100 ft), the sides of the tank would have to be 280 times higher than the distance from the Earth to the sun! (4)
6. A mole of hockey pucks would be equal to the mass of the moon!
7. Assuming that each human being has 60 trillion (6×10^{13} cells) and the world population is 4 billion (4×10^9), the total number of living human body cells on the Earth at the present time is 2.4×10^{23}, or less than half a mole!
8. If you had a mole of pennies and you wanted to buy kite string, you would get your money's worth, even if you had to pay a million dollars an inch for your string! Imagine you were to stretch this string around the Earth one million times and to the moon and back twenty-five times. You would still have enough string left over to sell it back at a penny an inch (a decided loss). With the funds this would bring you, you could buy every single person in the U.S. a $5000 automobile and provide them with enough gasoline to run it non-stop at 55 m.p.h. for a year. Even after this, you would still have enough money to give every person in the world $9,024!! (5) (The basis for calculations: Earth's circumference = 25,000 miles; distance to the moon 240,000 miles; cost of gasoline $1.20/gallon; U.S. population = 220 million; world population 4,020 billion.)
9. If one mole of pennies was divided up and given to every person on Earth, each person would receive 1.5×10^{14} pennies. Personal spending at the rate of one million dollars per day would use up each person's wealth in just over four thousand years. Life would not be comfortable, though— the surface of the planet would be buried in copper coins to a depth of about 420 meters!

After sharing these analogies, ask your students if they would like to see a mole of water. It often happens that someone makes the connection that we use such a large number, because atoms and molecules are so small.

References:
(1) Schrader, C., Dreyfus/Wilson Institute, August, 1982
(2) Editors, *Popular Science*, June, 1979
(3) Kolb, D., *Journal of Chemical Education*, **55**, 728 (1978)
(4) Heikkinen, H. and Atkinson, G., *Reactions and Reasons*, Harper and Row, New York, 1978 p. 69.
(5) Teacher's Guide, *Reactions and Reasons*, Harper and Row, New York, 1973, p.92

Show the relationship between the mole and molar mass via a display showing what one mole of various elements (Al, Cu, Fe, C, Si, S, etc.,) and compounds (H_2O, NaCl, $C_{12}H_{22}O_{11}$/sucrose, C_2H_5OH/ethanol, CH_3COOH/acetic acid, etc.) looks like. Next to the physical display, place cards containing the name, formula, molar mass, and number of particles (6.02×10^{23}). If identical holders are used, a double-pan balance can be used to compare relative masses of the various samples; have students do the weighing. If sealed, labelled, identical glass bottles are used, the samples can be passed around the classroom.

YOUR TURN

Molar Masses (page 126)

This activity provides practice in calculating molar masses for compounds. Answers are based on values on the periodic table, inside back cover or on page 111.

Answers to Questions:

1. 14.01 g/mol
2. 28.02 g/mol
3. 58.44 g/mol
4. 342.30 g/mol
5. 183.54 g/mol
6. 221.13 g/mol

C.4 CONSERVATION MUST BE OUR WAY (page 127)

Since nature does not conserve at the molecular level, "running out" means changing naturally-occurring molecular arrangements faster than they can be replenished, as in the case of petroleum. In reference to metals, "running out" means the dispersion of metal resources originally extracted from high-concentration ores, coupled with the increased cost of extracting metals from low-concentration ores.

Conservation at the human level means proper management of resources. The gathering, processing and using of resources generates unwanted materials. Conservation options that reduce resource demand and/or waste production include replacement, reuse, recycling, and rethinking. This leads to the next decision-making activity on planning a recycling program.

C.5 YOU DECIDE: RECYCLING DRIVE (pages 128-130)

This activity examines resources and their recycling. Students decide which of three resources (paper, aluminum, glass) are the better candidates for recycling efforts by applying their current knowledge of chemistry and other concepts. This activity reinforces the idea that trash can often assume greater economic worth as a resource when it is gathered and organized. The idea is to avoid dispersion of atoms throughout the environment. You may want to reemphasize that waste and dispersion of nonrenewable resources will eventually threaten our present lifestyles. The ideal situation is reuse, which involves the minimum reprocessing; increasing reuse should probably be a long-term goal.

The unit's discussion of glass, paper, and aluminum should be regarded as reference material for students to use in considering the assigned questions. The detailed specifics regarding these three materials are not intended to be memorized or learned—they should be used as needed. As homework, you might suggest that students organize their thoughts through written notes in preparation for their group discussion. These points can guide their thinking:

1. Time required to replace wood or produce the other resources.
2. Abundance and availability (import–export) of each resource.
3. Energy required to process the raw material vs. the energy involved in recycling each resource.
4. Reduction in pollution and associated economic and environmental benefits.
5. Summary chart of information from the student text:

Resource	Renewable?	World-wide abundance	Domestic supplies
Aluminum	No	Great, #1 metal	Not of high grade
Glass	No	Great (sand)	Yes
Paper	Yes	Great, but local shortages exist	Yes

OPTIONAL BACKGROUND INFORMATION

Environmental Benefits of Recycling in the United States

Benefit (as a % reduction)	Paper	Aluminum	Iron and steel
Energy use	30-55	90-95	60-70
Spoil and solid waste	130*	100	95
Air pollution	95	95	30

*Note: (1) % Reduction in BTU's, tons of wastes per ton material recycled, (2) 1.3 lbs of waste paper is required to produce 1 lb of recycled paper.

Source: Brown, L. R. et al. *State of the World 1987: A Worldwatch Institute Report on Progress toward a Sustainable Society*, Chapter 6: "Recycling Materials," page 97 (W. W. Norton & Co., 1987). This chapter also contains information comparing the recovery rates for paper, aluminum, and steel across various nations. See future annual editions of this valuable reference for up-to-date data.

The constraints on recycling and its benefits vary from resource to resource. Each material requires different recycling technologies, policies, and markets. Aluminum recycling is particularly advantageous.

Data Supportive of Recycling of Aluminum Beverage Containers

(You may wish to make this data available to students for this activity and/or the final Putting It All Together activity.)

1. **Volume of waste/resource:** 25% of all U.S. aluminum production goes into packaging, half of which is beverage containers.

2. **Ease of recycling:** "Tin" cans (mostly steel with a thin coating of tin to prevent corrosion) are not easily recyclable because the tin fuses with the steel. Most recycled tin cans are used as a catalyst in copper production where it is actually consumed and cannot be further recovered. Conversely, aluminum cans are readily recyclable.

3. **Economics of recycling:** Scrap aluminum is worth $500/ton or $0.25/lb (end of 1991) vs. scrap glass which is worth about $20/ton or $0.01/lb (for comparison, newspaper is worth about $0.005/lb and cardboard about $0.01/lb).

4. **Energy savings involved:** Energy is used to separate the aluminum ore (bauxite) from unwanted minerals and dirt. A large quantity of electrical energy is needed to break the bonds in Al_2O_3 and to reduce Al^{3+} to aluminum metal. The cost of electrical energy per mole of aluminum produced is very high. Recycling aluminum requires minimal energy (for melting and recasting), compared to the energy needed to process aluminum from its ore. It has been estimated that throwing away an aluminum beverage container wastes as much energy as pouring out such a can half-filled with gasoline (or failing to recycle a weekday edition of the *Washington Post* or *Los Angeles Times*). Glass production and recycling are both energy-intensive processes. Unless a returnable glass container is re-used at least ten times, it offers no energy savings over recycled aluminum cans.

5. **New technologies for collection:** Reverse vending machines accept aluminum cans (rejecting ferrous cans, glass, or other unwanted objects), weigh the aluminum deposited, and dispense money or coupons in payment. Sweden, the world leader in this technology, recycles an estimated 97% of all aluminum cans. At the end of 1988, there were some 12,000 such units in operation in 15 countries throughout the world.

Many concepts relating to the questions were covered in previous sections of this unit. Encourage students to return to concepts such as resource "use," "using thing up," renewable vs. nonrenewable resources, the law of conservation of matter, the Earth's finite resources, metal abundance in ores, and the relation between energy required to extract metals from ores and the reactivity of the metals.

CHEMQUANDARY

Recycling Limits (page 129)

Recycled paper is not as widely available and is rarely used for textbooks, due to the lack of glossiness, which impairs the reading of textual material, and the reproduction of pictures.

Students should generate group answers to the questions in the student edition and report to the class. The following suggested answers may be used as a guide to discussion.

Answers to Questions:

1. a. Aluminum is probably the most important of these resources to recycle for economic reasons, given that nearly 90% of our needed aluminum is imported, and considering the high energy costs in extracting aluminum from bauxite ore. All three resources require energy (either electrical or thermal) for processing. However, the energy required to produce aluminum is by far the greatest.
   b. Aluminum is important to recycle for environmental reasons, due to the large quantity of energy required. Some students, concerned about trees and forests, may argue for paper as the most important environmentally. Encourage students to think of the "hidden" costs of producing aluminum—energy sources, whether nuclear power plants or hydroelectric dams, also have impact on the environment.

2. a. Student answers on this question will vary.
   b. Prime candidates would include all three of these resources. The answer to this question probably goes with the answers to question 1.
   c. One way to enforce significant recycling of aluminum and glass would be laws requiring deposits on all cans and bottles, like the laws in many states—as of 1989, Oregon, Wisconsin, Florida, California, Vermont, Maine, Michigan, Iowa, Connecticut, Massachusetts, Delaware, and New York. The manufacturers of glass containers could be required to produce only returnable or reusable bottles.
      Recycling of paper is more difficult to legislate. Most efforts have focused on the purchasing side. For instance, the U.S. government is required by law to purchase paper made from recycled fibers. Some states have passed similar laws: for instance, 25% of the paper used by the state government of Maryland is from recycled stock. Students may wish to discuss the response of the industrial and private sectors to such measures.

3. Individuals can bundle old newspapers, participate in curbside recycling programs, and return non-returnable bottles and aluminum cans to community collection centers for reprocessing. Paper can be conserved by buying goods in bulk quantities and avoiding excess packaging of goods. Glass containers may be reused in the home for other liquid storage.

4. a. Newspapers, books, magazines, letter–writing, etc.
   b. Printing jobs, delivery people, textbook publishers and sales people, bookstores, reduction in forest product collection, pulp and paper mills.
   c. Computer manufacturers, repair personnel, communications services (wiring, installation, etc.).
   d. A television screen.
   e. More flexible schedule possible, less travel (reduced school busing programs), more leisure time.

OPTIONAL CHEMQUANDRIES

1. Explain the statement: Necessity is the mother of collection (and recycling).
   Answer: Materials will be recycled when it is economically or strategically compelling to do so—witness the recycling efforts during World War II, the present day recycling of aluminum, and the decision by the U.S. government to stockpile certain strategic metals (such as those imported from South Africa). Recycling has reduced U. S. imports of chromium by 25%, and of cobalt 16%. (Source: United States Bureau of Mines Mineral Commodity Summary, 1989.)
2. Provide support for the statement: Waste is a human concept; in nature, nothing is wasted.
   Answer: Matter is continually recycled in nature—even the excretions and dead remains of organisms are broken down and reused.

PART C: SUMMARY QUESTIONS (page 130)

1. a. The three regions of the Earth: atmosphere (100 km thick, composed mainly of gaseous nitrogen and oxygen); hydrosphere (5 km thick, composed mainly of water and ionic salts such as sodium chloride); and the lithosphere—crust, mantle and ore (6400 km thick, composed of silicates, fossil fuels, carbonates, oxides, and sulfides).
   b. The crust of the Earth is the main storehouse of chemical resources used for manufacturing consumer goods.
2. a. $Ca_3(PO_4)_2(s) + 3\ H_2SO_4(aq) \rightarrow 2\ H_3PO_4(aq) + 3\ CaSO_4(s)$
   b. $WO_3(s) + 3\ H_2(g) \rightarrow W(s) + 3\ H_2O(l)$
   c. $2\ PbS(s) + 3\ O_2(g) \rightarrow 2\ PbO(s) + 2\ SO_2(g)$
3.   Possible answers include recycling cans and paper at school; recycling papers, bottles and cans at home; buying products in recyclable/reusable containers; buying products with less packaging.
4. a. 16.00 g/mol   d. 120.38 g/mol
   b. 32.00 g/mol   e. 40.00 g/mol
   c. 48.00 g/mol   f. 180.15 g/mol
5. a. 2 moles Cr_2O_3, 3 moles Si, 4 moles Cr, 3 moles SiO_2
   b. 2 Cr_2O_3: 304.00 g, 3 Si: 84.27 g, 4 Cr: 208.00 g, 3 SiO_2: 180.27 g
   c. mass of reactants = 388.31 g; mass of products = 388.31 g

D: METALS: SOURCES AND REPLACEMENTS

This part begins by examining the sources and uses of copper, and introduces percent composition as a tool for the evaluation of ore quality. Three main techniques for refining a metal from its ore are explained, along with oxidation-reduction terminology. Part D concludes with a look to the future, and to materials such as ceramics, plastics, optical fibers and polymer composites. As usual, an understanding of chemistry is essential to help us understand the use of resources, how to obtain them, and how to solve future problems associated with resource supply, demand, and alternatives.

D.1 COPPER: SOURCES AND USES (pages 131-133)

Copper's long history of use and availability make it a starting point for considering metals and their uses. Students will see that copper is used for much more than the Statue of Liberty. The *YOUR TURN* activity clarifies the many uses of copper, and challenges students to think of possible alternatives to copper in four areas of use.

Uses of Copper (pages 132-133)

Students will connect copper's properties with its uses, and propose a technological change which would alter copper usage.

Answers to Questions:

1. Electrical uses: depend on copper's low reactivity, high electrical conductivity, malleability and ductility. Non-electrical uses: depend on copper's attractive color (jewelry), malleability (anything requiring shaping), low reactivity and high thermal conductivity (pots and pans).

2. a. Increased recycling would decrease our need for imports (the U.S. imports 13% of its copper from Canada, Chile, Peru, and Zambia) and decrease the need for mining and refining lower quality ores.
   b. Recycling has a limit.
   c. Not all copper can be recycled—house wiring must stay in place for the lifetime of the home; the same is true for plumbing, roofing, etc. Also, the benefits of recycling usually reach a maximum in terms of economic profit *before* maximum recycling is reached, thus reducing the impetus to improve recycling.

3. a. Communications: fiber optics, satellites, and cellular phones decrease the need for copper in communication wiring.
   b. Coins: removing pennies from circulation, decreasing the copper in other coins, and using credit cards would lower copper usage.
   c. Power generation: solar power could reduce power generation by turbines, reducing copper used in the wiring of the turbines.
   d. New conducting alloys and aluminum may decrease the demand for copper.

D. 2 EVALUATING AN ORE (pages 133-134)

In considering an ore as a source for a metal, its percent composition is an important factor to be considered. The *YOUR TURN* introduces calculations of percent composition for compounds, and extends these calculations to the evaluation of ore sample quality.

Percent Composition (page 134)

Answers to Questions:

1. a. $CuFeS_2$ = 34.6% Cu $\quad$ $Cu_2CO_3(OH)_2$ = 57.5% Cu $\quad$ $Cu_3(CO_3)_2(OH)_2$ = 55.3% Cu
   b. Chalcocite (done as the example) has the highest percent of copper and would be most profitable. It is interesting to note that its sulfur would have to be controlled to minimize air pollution.

2. Magnetite has the greater amount of iron.
    Fe_2O_3 = 69.9% Fe; Fe_3O_4 = 72.4% Fe
    1 kg $\times$ 0.699 = 0.699 kg Fe in one kilogram Fe_2O_3
    1 kg $\times$ 0.724 = 0.724 kg Fe in one kilogram Fe_3O_4

3. a. Site A: 20 g Fe_2O_3 $\times$ 0.699 = 14.0 g Fe
       Site B: 15 g Fe_3O_4 $\times$ 0.724 = 10.9 g Fe
   b. You may need to help students distinguish between the percent metal in a mineral and the percent metal in an ore. In this calculation, the higher percent of mineral in the ore (20% Fe_2O_3 versus 15% Fe_3O_4) is more important than the higher percent in the mineral (69.9% Fe in Fe_2O_3 versus 72.4% Fe in Fe_3O_4).

Once the ore has been assayed for percent composition and mined, the metal must be processed to remove pure metal from the mineral. This leads to a discussion of reactivities of metals and metallurgy in the next two sections.

D.3 METAL REACTIVITY REVISITED (pages 135-136)

This section presents a brief history of metal usage, leading to the notion that metal reactivity influenced which elements were developed first. A more extensive metal activity series is used to reinforce the organization of the periodic table, and to predict chemical reactions.

CHEMQUANDARY

Discovery of Metals (page 135)

Owing to their low reactivity, copper, silver and gold were less likely to form compounds than the more common and reactive elements of iron, aluminum and calcium. Therefore natural samples of gold, silver, and sometimes copper were easy to find, and their shiny colorful appearance simplified their discovery. Also, copper was easy to separate from its compounds by low-energy roasting processes.

YOUR TURN

Metal Reactivity (page 136)

This activity familiarizes students with the relative reactivity of metals, which will influence chemical reactions. It also explores the energy needed to free metals from their ores.

Answers to Questions:

1. Metallic reactivity generally decreases from left to right across a period of the periodic table. The most reactive metals are located on the left side. The least reactive metals, such as copper, silver and gold, are in the transition metal section, found in the middle.

2. a. Yes
   b. No
   c. Iron is more reactive than lead, according to the metal activity series, and will release lead from a solution. Platinum is less reactive than lead and will not.

3. a. Electrolysis will require the most energy, due to the fact that this process is reserved for the most reactive elements, which require the most energy. Also, the process of heating with coke or roasting produces some energy in the formation of carbon dioxide or gaseous products.
   b. Least expensive is roasting in air, because no chemicals are needed for the decomposition. Electrolysis is most expensive due to the high costs of electricity, and coke and carbon monoxide have costs associated with their production.

4. a. The least reactive metals are not the cheapest, despite having the lowest production costs.
   b. Other factors include demand for the metal (low reactivity is one of the attractive features of gold, silver and copper—all three are long-lasting) and overall availability—gold, silver and copper are much less abundant than aluminum and iron on Earth.

D.4 METALS FROM ORES (pages 136-138)

Oxidation and reduction are defined, and the main reduction techniques for separating metals from their compounds in ores are presented. Students can learn the three techniques more easily by referring to the reduction processes listed with the metal activity series on page 135, and the following laboratory activity. The "LEO the LION says GER" mnenomic works nicely here.

D.5 LABORATORY ACTIVITY: PRODUCING COPPER (pages 138-139)

This activity introduces one technique used to obtain metals from ores. Emphasize the fact that this process is endothermic, contributing to the higher cost of electrolysis for producing metals. The procedure given in the text is microscale.

Time

30 minutes

Materials (for 24 students working in pairs)

12 U-tubes
24 carbon rods (for electrodes)
12 buret clamps
12 ring stands
100 mL 0.2 M $CuCl_2$ (3.4 g $CuCl_2 \cdot 2H_2O$/100 mL)
12 9-V battery connectors with alligator clips attached to the leads
12 9-V batteries

Pre-Lab Discussion

Emphasize that this is an oxidation-reduction reaction: the two **always** go together. Stress the connection between this activity and previous experiences, the table of metal reactivity, and the concepts of oxidation and reduction.

Advance Preparation

U-tubes may be purchased or hand-made from glass tubing. A laboratory assistant might do this, or your students could try it. (Students are often enthralled with glass-working, but the time factor might make this suggestion undesirable). If small glass tubing is used, the pencil leads may need to be adjusted in size.

It is also possible to use 10-cm length of clear plastic tubing (Tygon tubing is excellent). This can be inserted in a empty plastic film canister, and the solution and electrodes added as above.

Pencils sharpened at both ends will serve as electrodes. The batteries, connectors, and small alligator clips can all be purchased from your local Radio Shack store or equivalent.

Lab Tips

Make sure the batteries are fresh (using rechargable batteries, though expensive, reinforces reducing waste).

Caution the students about inhaling chlorine gas—the odor is familiar enough from most swimming pools. Good ventilation in the lab is important to avoid unnecessary irritation and student exposure. To avoid chlorine completely, you could substitute copper(II) bromide, either as a pure solution or a mixture of 0.20 M copper(II) sulfate and 0.40 M potassium bromide. The bromine color at the electrode dissolves in the water, and is easily observed by holding a white piece of paper behind the U-tube.

It is possible to reuse the copper(II) chloride solution; the concentration is not critical to the success of this activity.

Post-Lab Discussion

Review the terminology of oxidation/reduction, and clarify the charges of the cathode and anode in the questions. Remind students of the energy cost of electrolysis. You may wish to point out that the more active elements of groups I and II on the chart, as well as aluminum, are commercially produced by electrolysis. The most active nonmetals, the halogens (group VII), are also produced by electrolysis—chlorine comes from the electrolysis of brine, a concentrated sodium chloride solution.

Students should be able to explain the results of reversing the leads in step 8 of the lab procedure.

Answers to Questions:

1. The negative connection should show a color change on the electrode, hopefully recognized as copper-colored! The odor of chlorine is present at the positive connection.
2. a. Negative electrode is the cathode.
   b. Copper is formed at the cathode.
3. a. Positive electrode is the anode.
   b. Chlorine was formed at the anode.
4. $CuCl_2(aq) \rightarrow Cu(s) + Cl_2(g)$

D.6 FUTURE MATERIALS (pages 139-143)

In 1956 Bob Richards won the pole vault event in the Olympics, clearing a height of 14 feet 11 inches using a bamboo pole. In 1964, using a more flexible fiberglass pole, Fred Hansen cleared 18 feet 8 3/4 inches. This example points out the tremendous changes produced when technology provides substitute materials for present resources. You can supplement the text material by describing changes in sports equipment (tennis racquets, golf clubs, football helmets -- no longer made of leather!, bicycles) to automobiles (extensive use of plastics, aluminum, and some ceramics) and medicine (heart valves, artificial knee and hip joints, skin-like materials to protect burn victims).

The increased demand for resources will also require us to consider substitutes, which hopefully will also improve the function of the application.

YOUR TURN

Alternatives to Metals (page 143)

Answers to Questions:

1. wiring, motors, power distribution–aluminum
   communication equipment–satellite transmissions, fiber optics for phones
   plumbing–plastic pipe
   roofing–aluminum, fiberglass
   coins–other alloys, more credit cards
   jewelry–silver, gold, chromium plated nickel
   pots and pans–aluminum, ceramics
   shell casings–less copper, more zinc
   food preparation machinery–stainless steel
   auto radiators–aluminum, plastics, ceramics

2. Car bumpers, automobile engines, plumbing materials (glass in place of copper), baseball bats, aircraft parts.

3. Silver could replace copper in wiring, due to its higher conductivity. This could also improve efficiency of computers, with lower resistance to electron flow increasing the speed of signal transmission.

PART D: SUMMARY QUESTIONS (page 143)

1. Estimates of future resource supplies depend on future use or demand. Technology might increase or decrease the need, and improvement in mining techniques might alter the supply.

2. a. Recycling is important because it reduces waste disposal problems, and extends the supply of the resource.
   b. Recycling does not guarantee a future supply of the resource, if population growth and rising usage overtake the availability of the resource.

3.

a. Use of Copper	b. Property of Copper
wiring	high electrical conductivity, malleable
plumbing	malleable yet strong
cooking utensils	high thermal conductivity
roofing	low reactivity, malleable and strong
jewelry	low reactivity, pleasing, unique appearance
car radiators	high thermal conductivity, malleable
coins	appearance, low reactivity

4. a. Silver sulfide   87.1% silver
   b. Aluminum oxide   52.9% aluminum
   c. Calcium carbonate 40.0% calcium

5. a. 10% $PbSO_4$ in the ore.
   b. % Pb in $PbSO_4$ = 207.2/303.3 × 100% = 68.3%
   c. 10 g × 0.683 = 6.83 g Pb in 100 g = 6.83%

6. Reaction (b) would occur because zinc is more reactive than silver, and will replace silver from its compound.

7. a. No, it is not a good idea to stir lead(II) nitrate solution with a iron spoon. Iron is more active than lead and will displace it from its compound.
   b. $Fe(s) + Pb(NO_3)_2(aq) \rightarrow Pb(s) + Fe(NO_3)_2(aq)$

8. Group I, the lithium family, also known as the alkali metals; and group II, the beryllium family, also known as the alkaline earths.

9. The three reduction processes all require some sort of electron source, either electricity or a reducing agent, which will lose electrons easily. All of the processes are also endothermic reactions, requiring energy.

E: PUTTING IT ALL TOGETHER: HOW LONG WILL THE SUPPLY LAST?

E.1 METAL RESERVES: THREE PROJECTIONS (pages 144-145)

Students are introduced to a graph (Figure 14) containing three future scenarios related to world reserves of metal ore. Be sure students understand that curved plots represent a change (either increase or decrease) in the rate (a changing slope on the graph) of depletion. You may wish to point out that it has been estimated that by the year 2000, the world will be near the end of its easily mined reserves of lead, copper, zinc, uranium, tungsten, tin, gold, silver, and platinum. Thus, it becomes important for students to understand the assumptions behind such predictions, alternative scenarios, and the possibilities of technology and societal decisions to produce optimal scenarios.

E.2 OPTIONS AND OPPORTUNITIES (pages 145-146)

Students are given information concerning the variety of uses of aluminum, and asked to apply the graph in the previous section to the question of aluminum availability in the future. The questions can either be a homework assignment, leading to group work during class, or group work entirely in class. The first question requires more than a "Yes/No" answer: ask students and groups to give reasons for their answer. Encourage students to consider the topics covered in this unit, which describe recycling's impact on reserves and use; substitution of other materials; possible economic effects of higher prices for resources; and so on. The impact of higher population, plus more affluent societies, should be considered as well.

Answers to Questions:

1. Plot I is probably not a good projection, since a straight line means that the ore will be used up at a constant rate. With higher population and increased industrialization, the rate of depletion will likely increase, as in plot II. This is particularly true for aluminum, due to its increasing role in building and transportation and electrical applications.
2. Plot II, with increased use of reserves, implies that conservation efforts alone will not meet the demands of rising population and/or per capita consumption. This is possible, despite increased conservation in industrialized nations, since developing nations are striking to achieve a better standard of living for their growing populations.
   In aluminum's case, its high abundance (8.1% of the Earth's crust) and improved techniques for extracting it from other ores, make it possible that the known reserves will be increased.
3. Plot III would result if new techniques were developed for extracting aluminum from lower quality ores. Energy considerations may make this unlikely, and increased world-wide use would overtake the beneficial effect. There may be substitutions and technological advances which decrease consumer demand for aluminum. Increased recycling, reuse, and other conservation methods would help make plot III a reality.

E.3 LOOKING BACK AND LOOKING AHEAD (page 146)

Students have encountered working chemical language (symbols, formulas, and equations), laboratory techniques, and basic chemical concepts that are relative to some social problems associated with chemical resource use. Many social problems are too complex to have a simple technological remedy. The solutions usually involve balances between positive and negative aspects of the issues. In these decision-making tasks, chemistry remains an important ingredient. Chemical insight can assist in the definition and rational resolution of many complex problems facing this nation and the world today.

This unit has gone beyond the Water unit in considering the issues associated with general chemical resource use. In contrast to water, which is definitely a renewable resource, resources can be used up or so widely scattered that it is economically impractical to retrieve them. In the next unit students will examine one nonrenewable resource (petroleum) in detail, exploring the problems involved in its wise use—now and in the future. Since it is impossible to recycle petroleum after it has been burned, students will explore alternative supplies to petroleum consistent with the approach of the Resource unit.

SUPPLY LIST

Expendable Items

	Section	Quantity (per class)
Carbon, graphite sticks or charcoal briquets	B.3	50 g
Copper wire, 24 gauge (0.5 mm dia)	A.4	50 cm
Copper strips	B.6	25 g
Copper(II) chloride dihydrate	B.3	14 g
Copper(II) nitrate (optional for A.4)	B.6	40 g
Hydrochloric acid	B.3	33 mL
Iron, filings or thin wire	B.3	50 g
Magnesium ribbon	B.3, B.6	120 cm
Magnesium nitrate hexahydrate	B.6	25 g
Nitric acid, conc.	A.4	130 mL
Silicon, lump	B.3	50 g
Silver nitrate	B.6	2 g
Sodium hydroxide	A.1	36 g
Steel wool pads or sandpaper	A.4, B.6	6 pads
Sulfur, lump roll	B.3	50 g
Tin, mossy or foil	B.3	50 g
Tygon tubing (optional)	D.5	1m
Zinc, mossy or foil	A.1, A.4, B.6	300 g
Zinc nitrate hexahydrate	B.6	25 g

Nonexpendable Items

	Section	Quantity (per class)
Conductivity apparatus (optional)	B.3	12
Grease pencil	B.3	12
Graphite rod	D.5	24
Hammer	B.3	12
Sand paper	B.8	12 pieces
U-tube	D.5	12
9-volt battery and connectors	D.5	12

SUPPLEMENTAL READINGS

"The Fascinating World of Trash." *National Geographic*, April 1983, 424-457. Waste disposal and recycling.

"Element X" and "Family Resemblance," *CHEM MATTERS*, December 1987.

"The Wild World of Compost." *National Geographic*, August 1980, 273-284. Recycling.

"Statue of Liberty." *CHEM MATTERS*, April 1985, 8-12.

"Chemicals Helps Rejuvenate Miss Liberty." *Chemical Week*, July 2, 1986, v. 139 #1. A chemical trade magazine provides a brief rationale on the actual chemicals used in cleaning and restoring the Statue.

"Liberty Lifts Her Head Once More." *National Geographic*, July 1986, 2-20. Historical perspective—little chemistry.

"Brooklyn Bridge, the Structure of Metals," *CHEM MATTERS*, October 1983.

"The Disposal of Wastes in the Ocean." *Scientific American*, August 1974, 10-25.

"Cross–coin Puzzle," "Super–student Conductors," and "Superconductivity", *CHEM MATTERS*, October 1987.

"Storing Up Trouble: Hazardous Wastes." *National Geographic*, March 1985, 318-351.

"Elements and Their Organization." *CHEM MATTERS*, April 1984.

"Aluminum, the Magic Metal." *National Geographic*, August 1978, 186-211.

"Going Against the Flow." *CHEM MATTERS*, December 1986.

"The Miracle Metal—Platinum." *National Geographic*, November 1983, 686-706.

"Element Search." *CHEM MATTERS*, October 1984.

"Gold, the Eternal Treasure." *National Geographic*, January 1974, 1-51.

"Love Always, Francium", "Periodically Puzzling" and "The Wrong Knife," *CHEM MATTERS*, December 1985.

"Silver—A Mineral of Excellent Nature." *National Geographic*, September 1981, 280-313.

"Comic." *CHEM MATTERS*, February 1986.

"Beyond the Era of Materials." *Scientific American*, June 1986, 34-41.

"Tainted Water," *CHEM MATTERS*, February 1988.

"High-Tech Ceramics: a *C&EN* Special Report." *Chemical & Engineering News*, July 9, 1984. Reprints are available for $3 per copy, or $1.75 for 10 or more.

PETROLEUM: TO BUILD? TO BURN?

PLANNING GUIDE

Section	Laboratory Activity	You Decide
A. Petroleum in Our Lives		
Introduction A.2 Petroleum in Our Future?		A.1 It's a Petroleum World A.3 Who's Got the Oil?
B. Petroleum: What Is It? What Do We Do with It?		
B.1 Working with Black Gold B.2 Petroleum Refining B.5 A Look at Petroleum Molecules B.6 Chemical Bonding	B.3 Viscosity B.7 Metal Reactivities B.8 Alkanes Revisited	B.4 Crude Oil to Products
C. Petroleum as an Energy Source		
C.2 Energy: Past, Present, and Future C.3 Energy and Fossil Fuel C.4 The Chemistry of Burning C.6 Using Heats of Combustion C.7 Altering Fuels	C.5 Combustion	C.1 The Good Old Days?
D. Useful Materials from Petroleum		
D.1 Beyond Alkanes D.3 More Builder Molecules D.4 Builder Molecules Containing Oxygen D.5 Creating New Options: Petrochemicals	D.2 The Builders D.6 Petrochemicals	
E. Alternatives To Petroleum		
E.1 Alternative Energy Sources E.2 Builder Molecule Sources		
F. Putting It All Together: Choosing Petroleum Futures		
F.1 Confronting the Issues F.2 Looking Back and Looking Ahead		

TEACHING SCHEDULE

	DAY 1	DAY 2	DAY 3	DAY 4	DAY 5
Class Work	Discuss pp. 152-153 YD pp. 154-155	YD pp. 156-157	Discuss pp. 158-160		YD pp. 164-165 Discuss pp. 166-167 YT pp. 166-167
Laboratory			Pre-lab LA pp. 160-162	LA pp. 160-162	
Homework	Read pp. 152-164	SQ p. 157		CQ p. 164 Read pp. 164-167	Read pp. 167-172

	DAY 6	DAY 7	DAY 8	DAY 9	DAY 10
Class Work			Review Parts A & B	Quiz Parts A & B Discuss pp. 176-177	YD pp.178-179 YT pp.179-180
Laboratory	LA pp. 169-172	LA pp. 172-173			
Homework	YT pp. 171-172 Read pp. 172-175	YT pp. 173-174	SQ pp. 174-175 Read pp. 176-178	Read pp. 178-181	CQ pp. 180-181 Read pp. 181-189

	DAY 11	DAY 12	DAY 13	DAY 14	DAY 15
Class Work	Discuss pp.181-183 YT pp. 183-184		Discuss pp. 189-190 CQ p. 190 YT p. 191	YT p. 194	Review Part C Quiz Part C Discuss pp. 196-197
Laboratory		LA pp. 186-189			LA pp. 197-198
Homework	YT pp. 184-185	Read pp. 189-192	Read pp. 192-195	SQ p. 195 Read pp. 196-199	Read pp. 199-206

	DAY 16	DAY 17	DAY 18	DAY 19	DAY 20
Class Work	Discuss pp. 199-204		SQ p. 206	Review Part D Quiz Part D Discuss pp. 207-212 YT pp. 207-209	PIAT pp. 213-215
Laboratory		LA pp. 204-206			
Homework			Read pp. 207-212	SQ p. 212	

	DAY 21	DAY 22	DAY 23		
Class Work	PIAT pp. 213-215	Review unit	Exam		
Laboratory					
Homework					

LA = Laboratory Activity; **CQ** = ChemQuandary; **YT** = Your Turn; **YD** = You Decide; **PIAT** = Putting It All Together.

Our society is highly dependent on petroleum as an energy source for transportation, heating, electricity, and production of material goods. Petroleum-based products are not as closely associated with material goods in the minds of many citizens. However, petroleum's role as the source of key petrochemicals is vitally important in maintaining our current quality of life. The Earth contains a finite supply of petroleum for nations to share. As primary consumers of this precious resource, we have a responsibility to ourselves and to other nations to make every drop count. Our extensive use of petroleum also has an impact on our foreign policy, particularly towards the nations of the oil-rich Middle East.

The Petroleum unit continues the concepts of resources and conservation, which started in the Water unit and continued through the Resources unit, by introducing students both to the properties and chemistry of petroleum and to some of the central concepts of organic chemistry: naming, properties, covalent bonding, hydrocarbon isomers, and petrochemical synthesis. Although students are exposed daily to many organic-based materials ranging from sandwich bags to cassette tapes to polyester clothing, most chemistry courses postpone any treatment of organic chemistry until the end of the course—if it is covered at all. In *ChemCom* students gain an early view of organic chemistry through its introduction in this unit, and through its extension to important biomolecules in "Understanding Foods."

Knowledge of chemistry can help communities, voters, and lawmakers become wise managers of this important resource. Choices involved in using petroleum for building (petrochemicals) and burning (petroleum-based fuels) serve as the central theme of this unit.

OBJECTIVES

Upon completion of this unit the student will be able to:

1. Compare the usage of petroleum for "building" and "burning", and the benefits and burdens of each usage. [A.2]
2. Identify regions of high petroleum usage and regions of petroleum reserves, and discuss the economic and political implications of petroleum supply and demand. [A.3]
3. Describe the chemical makeup of petroleum and its differences from other resources. [B.1]
4. Identify differences in density and viscosity among common petroleum products, and explain the relationship between the differences and the number of carbon atoms in their molecules. [B.2, B.3]
5. Describe the process of fractional distillation, and list the five major fractions of petroleum distillation and typical products manufactured from each fraction. [B.2, B.4]
6. Name the first ten alkanes and draw structural and electron-dot formulas for each. [B.5, B.7]
7. Describe the processes involved in ionic and covalent bonding. [B.6]
8. State and explain the effect of carbon length and side groups on the boiling point of a hydrocarbon. [B.7, B.8]
9. Define the term isomer and draw structural formulas for the isomers of a given hydrocarbon. [B.8]
10. Trace the history of energy sources and consumption patterns in the United States, and account for major changes. [C.2]
11. Explain endothermic and exothermic reactions in terms of bond breaking and forming, and give examples of each type of reaction. [C.3]
12. Identify energy conversions in the automobile, and calculate the savings resulting from increased automobile efficiency. [C.3]
13. Define the terms heat of combustion and specific heat, and calculate energies of various combustion reactions. [C.4–C.6]
14. Write balanced equations for the combustion of hydrocarbon fuels, including energy changes. [C.6]
15. Define the term octane number, state its relationship to grades of gasoline, and identify two ways of increasing octane number. [C.7]
16. Compare saturated and unsaturated hydrocarbons in terms of molecular models, formulas, structures, and physical and chemical properties. [D.2, D.3]
17. Identify the functional groups for common alcohols, ethers, carboxylic acids, and esters. [D.4-D.6]
18. Describe polymerization and give one example of addition and condensation reactions. [D.5]
19. Describe major sources of energy for the United States of today and alternative sources of fuels for the future. [E.1, E.e2]

A: PETROLEUM IN OUR LIVES

Part A begins by highlighting the wide array of everyday items we rely upon without recognizing their source—petroleum. Since petroleum is a nonrenewable resource that is being rapidly consumed, we must consider the prospect of running out altogether. Petroleum geologists report that world oil production is expected to peak in the near future. Part A considers how much petroleum various nations consume, how it is used (for building and burning), and the geography of oil importing and exporting on a worldwide scale.

A.1 YOU DECIDE: IT'S A PETROLEUM WORLD (pages 154-155)

This activity illustrates the variety of products made from petroleum which students may take for granted. Students examine a scene (Figure 1, page 154) of a household bathroom containing many petroleum-based products. They are asked to identify as many of these products as possible. Items include: bathrobes and drapes (synthetic fibers), plastic shower cap, plastic containers of various cosmetics and medications, plastic cup, toothbrush, hair curlers and box, hair dryer, glasses, electric razor, comb, counter top, and rubber tires on the motorcycle. After completing their lists, students will be given a version of Figure 1 from which the petroleum–based products have been removed. (See Black Line Masters.) They then decide which items would be of least and greatest importance if a severe petroleum shortage occurred.

This opening activity should sensitize students to the fact that we live in a petroleum-based society. It's clear that issues of supply and abundance of this important nonrenewable resource touch all of our lives.

Figure 1. Life without petroleum. The items in Figure 1 (p. 154) of the text book which consume or are made from petroleum have been removed. See Black Line Masters at end of Teacher's Guide for reproducible image.

A.2 PETROLEUM IN OUR FUTURE? (pages 155-156)

This unit presents statistics of petroleum usage, with a look to the past and the future. Remind students of the three graphs predicting resource usage presented in the last unit (page 144) as you discuss Figure 3. The ideas of conservation and technological developments are relevant here although, in the case of petroleum, new technology has increased petroleum usage for plastics, synthetic fibers, and other polymers.

A.3 YOU DECIDE: WHO'S GOT THE OIL? (pages 156-157)

The goal of this student activity is to compare world petroleum consumption and supplies in different geographic regions by examining graphical presentations of data concerning petroleum reserves, population and consumption of petroleum. Once again students see that unequal distribution of a natural resource (and varying demand) lead nations to import-export relationships. Make sure students note the definition of known reserves and understand how to read the various figures.

Answers to Questions:

1. Middle East (57% world's oil/3% world's population)
2. Asia and Pacific (5% world's oil/57% world's population)
3. North America (32% vs. 12%), Asia and Pacific (17% vs. 5%), Eastern Europe (18% vs. 9%), Western Europe (21% vs. 4%), Central and South America (6% vs. 5%)
4. Africa (3% vs. 8%), Middle East (3% vs. 57%)
5. Africa, Middle East
6. Same regions as in Question 3
7. Middle East and North America, Middle East and Western Europe, Middle East and Eastern Europe, Middle East and Asia
8. Differences in political and economic systems
9. a. North America (32%), Western Europe (21%)
   b. North America and Western Europe are highly dependent upon petroleum as fuel for transportation (have more private cars than other nations) and as raw materials for industry.
10. a. (1) North America has a larger population than the Middle East.
       (2) The Middle East produces much more petroleum than it consumes, and it has a much smaller population than North America.
       (3) North America consumes much more petroleum than it produces.
    b. It seems imperative that North America maintain good relationships with Middle Eastern countries, in order to ensure that petroleum supplies will be continuous.

OPTIONAL BACKGROUND INFORMATION

If you want more detailed information on United States imports of oil over time, see the Department of Energy publication *Monthly Energy Review* (often available in large university or public libraries).

Also, as always, encourage students to bring in relevant newspaper and magazine articles, cartoons, and advertisements for a bulletin board and class discussion.

PART A: SUMMARY QUESTIONS (page 157)

1. a. Many answers possible, ranging from drugs, films, clothing, soaps, detergents, vinyl records, synthetic-fiber rugs, etc.
   b. 18 million barrels × 0.87 = 15.66, or 16, million barrels burned.
   c. 18 million barrels × 0.13 = 2.34, or 2.3, million barrels used to build. (The text does not belabor precision and accuracy in calculations. To simplify this and other calculations, encourage your students to have the same number of digits in their answer as their data have—this is a rough approximation of significant figures.)
2. a. If the cost of oil increased 10% and the supply decreased, the short- and long-term costs to consumers would be greater than the initial 10% increase in cost.
   b. Attempts to make up the short-fall in supply would increase expenditures, and consumer costs to support more exploration for oil reserves, as well as for higher-cost alternatives such as oil shale and tar sands.
3. a. A very good trade relationship is needed between North America and parts of the Middle East. As has happened in the past, the Middle East, particularly OPEC countries, have been free to raise their prices and oil-importing countries have been forced to pay the increases. If it becomes too expensive to import oil from the Middle East, North America may turn to more locally available resources or search for petroleum substitutes.
   b. As other nations increase their use of Middle East oil, the North America-Middle East trade relationship will depend on price increases, as mentioned in Question 3. The developing technology in other world regions might lead new nations to become petroleum exporters, providing another North American source for petroleum, and increasing trade relations with those countries.
4. (10,000 mi) × (1 gal gasoline/25 mi) × (42 gal petrol/21 gal gasoline) × (1 barrel/42 gal petrol) = 19 barrels

B: PETROLEUM: WHAT IS IT? WHAT DO WE DO WITH IT?

Part B examines the differences between petroleum and metallic ores, and describes and the major components of petroleum and their separation and their uses. Students explore not only the properties of petroleum fractions, but also the relationship between structure, bonding, and hydrocarbon boiling points through a series of discovery-oriented activities. The rationale behind bonding is introduced, and students use models to explore the structures of common alkanes, their naming, and the concept of structural isomers.

B.1 WORKING WITH BLACK GOLD (pages 158-159)

This section points out that while both petroleum and metallic ores are nonrenewable, petroleum is unique in being a complex mixture that permits both building and burning applications. Crude oil is pumped from the ground as a mixture of hydrocarbons and is transported to a refinery for separation into fractions. Help students recall the *Foul Water Laboratory Activity* distillation, perhaps by setting up the equipment again as a demonstration.

B.2 PETROLEUM REFINING (pages 159-160)

The text and diagrams clarify several aspects of fractional distillation, which separates the crude oil into its components. The contrast between the teacher demonstration of water distillation in the Water unit and fractional distillation may help clarify this process for students. In the distillation of water, the separation is between the volatile water and non-volatile solutes such as table salt and coffee. In a fractional distillation a mixture is separated according to boiling points, a process which necessitates more complicated equipment. The basic difference is a tall tower, which allows lower boiling hydrocarbons to remain gases until they reach the top; higher boiling substances, which condense more easily and require more energy to vaporize, stay at the bottom. There is a difference in temperature throughout the column, promoting the evaporation and condensation of the various fractions at appropriate levels, and their subsequent separation. The "Refinery" film (available from Modern Picture Lending Service), a humorous yet factual look at this process, includes cracking, a topic covered later in this unit.

B.3 LABORATORY ACTIVITY: VISCOSITY (pages 160-162)

In this activity students observe selected physical properties of a number of petroleum fractions, and determine the relative densities and viscosities of several petroleum components. Students find the relationship between the viscosity of a sample and the average number of carbon atoms in its molecules.

Pre-Lab Discussion

The pre-lab discussion can focus on the distillation column used in a petroleum refinery and the relative heights from which each sample used in this activity might have been drawn.

Time

One class period

Materials (for 24 students working in pairs)

72 culture test tubes with caps (20 × 150 mm); 36 mL capacity (or if these are not available, use test tubes with corks)
72 7-mm metal beads
12 25-mL graduated cylinders
several digital timers or stopwatches (or a loud metronome)
450 mL mineral oil (light)
450 mL household lubricating oil (light oil)
450 mL kerosene
450 mL automotive oil, 10W-40
150 g paraffin wax (bars)
150 g asphalt (pieces) from road surface
450 mL distilled water
balances

Advance Preparation

Note: The preparation time for this activity is longer than usual; aides would be helpful if available.

Add a 7-mm metal bead to each of 72 culture tubes. Prepare a set of 5 tubes for each team by completely filling each tube with either mineral oil, kerosene, motor oil, household lubricating oil, or water. Cap all the tubes.

Determine the average mass of an empty culture tube plus cap and the average mass of a metal bead. (Weigh several tubes with caps and several beads to obtain a representative sampling.)

Measure the total volume of water that is held by one culture tube. Then find the volume occupied by one metal bead, using a water-displacement method and based on averaging the volumes displaced by several beads. Finally, calculate the actual volume of liquid held by each prepared vial (total vial volume − volume of one bead).

Place the values for mass of vial + cap, mass of bead, and volume of liquid in vial on the chalkboard. (Students will use these values to complete the density portion of this activity.)

Pre-Lab Discussion

Briefly review the logic of the liquid density determination outlined on page 160. Also review the water-displacement method for determining the volume of irregular solids (such as asphalt and paraffin) as part of this discussion.

Clarify the definition of viscosity as resistance to flow by having students estimate the relative viscosities of common materials such as honey and vinegar. Remind students to leave the sample tubes sealed at all times.

OPTIONAL CHEMQUANDARY

Ask students under what conditions the water-displacement method will *not* work without modification. Have them consider solids such as sugar, sodium metal, wood. (If the solid dissolves in or reacts with water it is necessary to use another liquid, such as oil or mercury. If it is less dense than water it is necessary to use another liquid of lower density than water, or sink it by force with a solid of known volume). What about an object too big to fit into a graduate cylinder? (Use the overflow can approach.) If the materials are on hand this might be done as a inquiry demonstration.

Lab Tips

Some suggestions for the beads: (a) force a BB into the opening of a pop bead and cut off the projection. (b) Use popcorn instead. You may need to change the popcorn each year.

If available, eudiometers allow longer drop times and increased accuracy. You may wish to do this as a Post-Lab inquiry demonstration.

Asphalt can be found at the edge of a newly paved road; or, if you have a refinery close by, the testing facility may provide you with a sample.

Student samples of solids should weigh about 10-20 g each.

It is possible to focus on different oil samples by purchasing motor oils of different ASTM ratings: 10W, 30W, etc. This might be an interesting follow-up demonstration to this activity as well.

Expected Results

Sample	State at room temperature (s, l)	Density	Relative viscosity
Mineral oil	l	0.90	4 (1.91)
Asphalt	s	1.60	—
Kerosene	l	0.74	1 (0.35)
Paraffin wax	s	0.70	—
Motor oil	l	0.88	5 (2.76)
Household oil	l	0.91	3 (0.72)
Water	l	1.01	2 (0.47)

Note: The motor oil used for this data was 10W-40 and the household oil was 3-in-1 oil.

Answers to Questions:

1. Liquid petroleum products are less dense than liquid water and will float on its surface. Oil floating on water can catch fire and burn. This physical property also suggests ways to contain and clean up an oil spill. If oil were more dense than water, it would sink and be out of sight, and the recovery of the oil would be even more difficult.

2. As the number of carbon atoms per hydrocarbon molecule increases, the viscosity increases.

3. a. To date no practical procedure based on differences in viscosity has been developed.
   b. If separation could be performed on the basis of viscosity at room temperature, the energy needed to operate a fractional distillation column could be saved.

CHEMQUANDARY

Gasoline and Geography (page 164)

Since the ease of evaporation of a liquid (its volatility) depends on its temperature, the properties of gasoline must be carefully controlled when it is used in regions with extreme temperatures. When used in car engines in hot southern regions in summer, gasoline composition must contain more higher molecular-weight molecules to prevent excessive vaporization. By contrast, gasoline of a lower average molecular weight is needed when shipped to the colder northern regions of the United States in winter to insure adequate vaporization, and consequent ignition of the fuel at low temperatures.

B.4 YOU DECIDE: CRUDE OIL TO PRODUCTS (pages 164-165)

Students are introduced to the variety of products obtained from petroleum fractions. The objective is to classify various branches shown in Figure 6 (page 165) as petroleum for "burning" and/or feedstocks for "building". This activity establishes important connections between the refined fractions from the original crude oil mixture and the petroleum-based products we use in daily life.

Answers to Questions:

1. and 2. Answers depend on which fraction is used. Petrochemicals are usually "building" uses of petroleum. Fertilizers and ammonia are used in agriculture to "build" food. Residues produce materials used for building (paving, shingles, etc.). Lubricating oils are neither burned nor used for building, and in many cases they can be recycled. Almost every other source falls into the "burning for energy" category.

3. (This sample answer is based on the choice of intermediate distillates; other answers are possible for other branches.) If intermediate distillates were eliminated, homes could not be heated with oil furnaces; there would be no diesel fuel for operating heavy construction and farming equipment; there would be a decrease in the feedstocks available for producing petrochemicals such as alcohols, medicines, plastics, and dyes.

4. Students might choose fuels for heating homes, petrochemicals used in the production of medicines, or fertilizers to grow food, among others. Probe for the reasons supporting specific student choices.

5. Some students might choose the building of new substances as more important, but they will probably still realize that an economy such as ours cannot maintain itself without fuels for transportation. (Other answers are possible, based on the students' own priorities.)

B.5 A LOOK AT PETROLEUM MOLECULES (pages 166-167)

Distillation separates petroleum into fractions because of differences in the boiling points of the substances from which it is composed. This leads to an examination of the boiling points of various hydrocarbons, and the intermolecular forces responsible for the differences in boiling points.

YOUR TURN

Hydrocarbon Boiling Points (pages 166-167)

Students have previously observed differences in physical properties of various hydrocarbon molecules in the laboratory. Chemists determine physical and chemical properties of substances and organize the data in ways that lead to the identification of useful patterns. These patterns often stimulate searches for explanations of such regularities.

In order to encourage students to recognize patterns, one chemistry text suggests changing your seating pattern regularly and asking students to identify the system for assigning seats: alphabetical, by age, sex, height, etc. This prior activity would provide an easy introduction to this YOUR TURN.

Surprisingly, some students prefer to organize chemical information alphabetically. By having them do a few problems with tables arranged according to alphabetical or numerical rank, you can convince them that organizing according to boiling points provides a simpler method for classifying the properties of these compounds. Later the naming system and its relationship to number of carbons and boiling points will reinforce this concept.

Answers to Questions:

1. a. The hydrocarbons in Table 1 are alphabetized by name.
    b. Such an alphabetic arrangement might be useful for locating a specific hydrocarbon in a handbook or index.
2. a. The boiling-point data would be better organized if the hydrocarbons were listed from high to low (or low to high) boiling-point values.
    b.

Hydrocarbon	Boiling Point (°C)
Methane	-161.7
Ethane	-88.6
Propane	-42.1
Butane	-0.5
Pentane	36.1
Hexane	68.7
Heptane	98.4
Octane	125.7
Nonane	150.8
Decane	174.0

3. a. Methane, ethane, propane, butane. You may want to emphasize the point that substances that are gases at room temperature have already boiled at a lower temperature. Students often associate boiling with a high temperature—this mistaken notion can be countered by the following demonstration idea.
    b. Pentane.
4. Decane has a higher boiling point than butane, indicating that the intermolecular forces between decane molecules are greater.
5. a. The greater the intermolecular attraction, the greater the viscosity and the higher the freezing point. Pentane = 1, octane = 2, decane = 3.

DEMONSTRATION IDEA

Either dry ice (–78°C) or liquid nitrogen (–196°C) can be used to liquefy one or more of the hydrocarbons listed in Table 1 (page 166). The latter is capable of liquefying even the lowest boiling hydrocarbon (methane bp = –161.7°C). A convenient setup is to attach a piece of copper tubing, with one end coiled or looped, to a piece of rubber tubing attached to the gas source (either a propane tank or a laboratory gas jet). If the coil is placed in a container of dry ice/acetone or liquid nitrogen, with the other end outside, liquid hydrocarbon can be collected in a beaker and observed to boil as it warms to room temperature. Alternately, a butane lighter can be opened and the evaporation of the liquid observed.

As a flashy alternative that does not involve a hydrocarbon, set a copper tea kettle of liquid nitrogen on a block of ice (or dry ice) and watch it boil as evidenced by the escaping gas.

OPTIONAL CHEMQUANDARIES

1. Why would butane be a poor substitute for propane in the tanks of portable stoves used by campers?
    Answer: Butane has a higher boiling point (–0.5°C vs. –42.1°C) than propane. At winter temperatures (and increased pressure) butane might liquify. Since the stove works on the basis of pressurized gas, this would be a poor fuel choice. Better substitutes might include either ethane or methane since their boiling points are lower than propane.
2. If one looks at a see-through butane lighter, one sees that at least some of the butane is in the liquid state. Explain how this is possible if the lighter is at room temperature. (See also the second *Extending Your Knowledge* item, page 175.)
    Answer: The boiling point of a liquid is dependent upon pressure. As the external pressure increases, so does the boiling point. Apparently, the butane must be under pressure greater that one atmosphere (where it boils at -0.5C) its boiling point at this increase pressure is greater than room temperature.

B.6 CHEMICAL BONDING (pages 167-169)

The bonding responsible for petroleum molecules is explained. Ionic bonding, involving the gain (reduction) and loss (oxidation) of electrons is contrasted with covalent bonding, which is introduced by examining hydrogen molecules and methane. Electron-dot and structural formulas enable students to draw pictures of the compounds and lead to the model building in the next activity.

You can contrast ionic and covalent bonding by referring back to the *Polar/Nonpolar Laboratory Activity* in the Water unit.

B.7 LABORATORY ACTIVITY: MODELING ALKANES (pages 169-172)

In this activity students build three-dimensional models of methane, ethane, and propane. They derive the general formula for alkanes, practice writing electron-dot representations, and begin to view the two-dimensional structural formulas of molecules from a more realistic three-dimensional perspective. A longer-range goal is to give students the tools for understanding the complexity of organic chemistry. In ionic compounds the formulas are different because different elements are used. In organic chemistry the difference seems less because fewer elements are present and the numbers of atoms are the main difference. Constructing models and diagrams enables students to see more clearly the differences in the formulas.

Materials (for 24 students working in pairs)

12 sets of space-filling or ball-and-stick models

Pre-Lab Discussion

One reason organic chemistry is so extensive, in terms of the number of known compounds, is the unique bonding capability of carbon. Carbon forms four covalent bonds and is the only element capable of such repeated bonding to itself. Remind students that carbon needs to add four electrons, so it *always* forms four bonds.

Time

One class period

Answers to Questions:

1. No question
2. a. 6 hydrocarbons
   b. 8 hydrocarbons
   c.

3. a. Ethane   Propane

   b. C_3H_8

4. CH_4, C_2H_6, C_3H_8, C_4H_{10}, C_5H_{12}, C_6H_{14}, C_7H_{16}, C_8H_{18}, C_9H_{20}, $C_{10}H_{22}$

5. Butane   Pentane

$$\begin{matrix} & H & H & H & H & & & & H & H & H & H & H \\ & | & | & | & | & & & & | & | & | & | & | \\ H- & C- & C- & C- & C- & H & & H- & C- & C- & C- & C- & C- & H \\ & | & | & | & | & & & & | & | & | & | & | \\ & H & H & H & H & & & & H & H & H & H & H \end{matrix}$$

6. a. (1) Heptane; (2) Nonane.
   b. C_7H_{16}; C_9H_{20}.
7. a. $C_{25}H_{52}$.
   b. The short version takes much less time and space; the long version won't fit on this page!
8. a. Ethane: C_2H_6: 2(12) + 6(1) = 30 g/mol
   b. Butane: C_4H_{10}: 4(12) + 10(1) = 58g/mol

Post-Lab Discussion

The enormous number and variety of hydrocarbons demand an orderly system of nomenclature, which is introduced here. Students should be familiar with the Greek prefixes used in polygons in geometry (penta-, hexa-, etc.). Using the prefixes enables students to learn formulas for C_5-C_{10} compounds easily. Repeated practice, with as many practical information tidbits as possible (cows "burp" methane, propane is in house trailer fuel, butane in lighters) should help students be able to name hydrocarbons if given the formula, or else write a formula from a name.

Point out the general formula for alkanes, and the alternate ways of writing structures shown in Table 2, page 172. If your students prefer the lines of structural formulas, don't force them to move to condensed formulas. Eventually the ease of writing this style of formula will lead to their change.

YOUR TURN

Alkane Boiling Points: Trends (page 171)

The boiling points of the first ten members of the alkane family are reexamined in terms of the number of carbon atoms involved in each structure. Students are challenged to apply their chemical knowledge to explain the observed boiling-point trend in the alkane series, and predict boiling points of larger hydrocarbons.

After going over their responses, discuss with your students how the chemical explanation for differences in boiling points is based on the notion of intermolecular forces. Longer (and less spherical) molecules have more surface area (and points of contact) for intermolecular attractions to occur.

Answers to Questions:

1. Approximately 30°C; mathematically, this is the slope (change in y/change in x = 30°C/C atom)
2. a. Data can be read from the graph. Or, using $C_{10}H_{22}$ (bp = 174°C) as the base and the figure of 30°C/C atom, the projected boiling points are:

   $C_{11}H_{24}$: 174 + 1(30) = 204°C   b.   (actual: 196°C, 4% error)
   $C_{12}H_{26}$: 174 + 2(30) = 234°C   (actual: 216°C, 8% error)
   $C_{13}H_{28}$: 174 + 3(30) = 264°C   (actual: 235°C, 12% error)

   This type of calculation shows the potential problem with extrapolation: the further removed from the actual known data, the more likely the prediction is to be wrong. This could also be shown by basing the above calculations on methane (bp = –161.7°C)—in this case the percentage errors range from 16-29%.
3. Increasing the number of carbons allows greater intermolecular attractions (due to increased surface area) and thus increased boiling points.

OPTIONAL CHEMQUANDARY

You are given three bottles. One contains a viscous liquid, the other a waxlike solid, and the third a liquid so volatile that it begins to boil when the stopper is removed from the bottle. You are told that the bottles contain one of the following compounds: CH_4, C_2H_6, C_4H_{10}, C_7H_{16}, $C_{14}H_{30}$, or $C_{22}H_{46}$. Identify the three substances.
Answer: Viscous liquid, $C_{14}H_{30}$; waxlike, $C_{22}H_{46}$; volatile liquid, C_4H_{10}.

B.8 LABORATORY ACTIVITY: ALKANES REVISITED (pages 172-174)

Students build models of C_4H_{10} and C_6H_{14} and find that in each case the atoms can be arranged in different (two and five respectively) ways. The notion of isomers is thus introduced. Lead your students to deduce that isomers should exhibit different physical properties because they are different geometric arrangements of atoms—that is, different structures.

Time

One class period

Materials (for 24 students working in pairs)

12 space-filling or ball-and-stick model kits

Note: Since building all five isomers of hexane would require more atoms than found in a typical kit you will need to have students join into larger groups. Since the typical kit contains 12-14 carbon atoms, if each team (with one kit) joins with two others (for a total of three kits/six students), each team can be asked to build as many unique isomers as they can in a given time period.

Run this as a contest if desired—the first to get the maximum number, and be sure that this is the maximum number possible, wins.

Answers to Questions:

1. Two isomers of butane are possible.
2. a. Butane

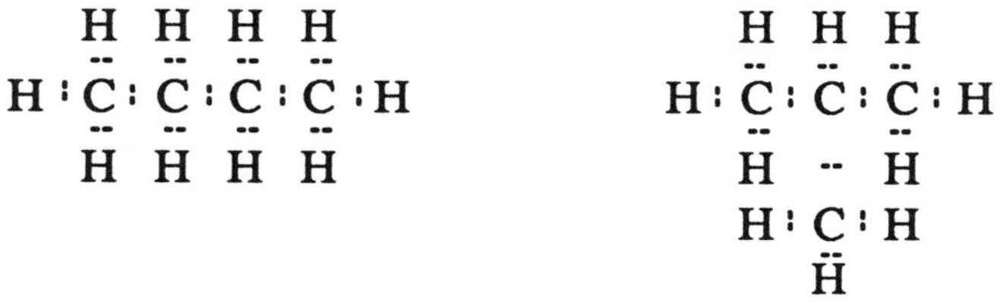

   b. Butane

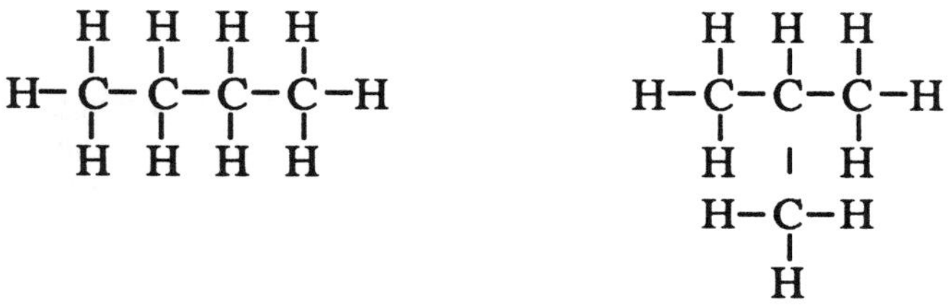

3. No question.
4. 5

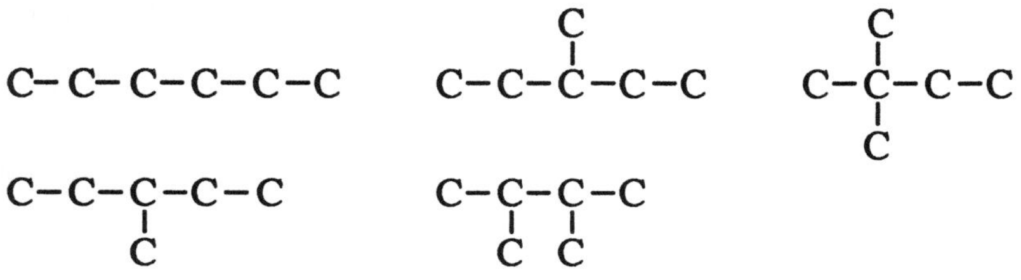

   (The hydrogen atoms are assumed here.)

5 b. 5 isomers of hexane

Alkane Boiling PointY s: Isomers *(pages 173-174)*

The purpose of this activity is to apply and test the idea that increased branching causes hydrocarbon boiling points to decrease. The complications introduced by isomers of hydrocarbons are minimized here, as no nomenclature is required of the students. Instead, students see how isomer structure influences boiling points. Knowledge of this is an important introduction which pertains to altering fuels and octane numbers, which are covered in the next part.

Answers to Questions:

1. The boiling points decrease with increasing number of carbon side chains within an isomer series. This can be explained in light of the fact that intermolecular attractions decrease as the molecular structure approaches a spherical shape (increased branching) which minimizes surface area contact between molecules.

2. a. 3-Methylhexane
    bp = 92.0°C
    b. 2,2-Dimethylpentane
    bp = 79.2°C
    c. Heptane
    bp = 98.4°C

3. a. One would expect its boiling point to be between 99.2 and 117.7°C since it is intermediate in the number of side chains (or branching). Its actual boiling point is 112°C, lower than octane and 2-methylheptane, higher than 2,2,4-trimethylpentane.
    b. One would expect it to have a higher boiling point than all the C_5H_{12} isomers because of its larger size. Some students may think that, since it is branched, it might have a lower boiling point than unbranched pentane. If you have them, carefully compare the two models. Students should be able to see that the C_8H_{18} isomer has the greater surface area. In fact, its boiling point is 112°C (compared to 36.1°C for pentane). Note: (1) Systematic names are provided for you, the teacher only— it is not intended for the students unless you care to support it (the text does not). (2) Again, be sure to emphasize the presence of hydrogen atoms if you write structural formulas in this abbreviated manner.

PART B: SUMMARY QUESTIONS (pages 174-175)

1. a. Petroleum is similar to metallic resources in that it is nonrenewable and world demand is increasing.
    b. Petroleum is different in that it cannot generally be recycled, although used motor oil can be recycled. Petroleum is also a liquid and provides a much greater variety of substances than does any particular metallic resource, since it is a complex mixture of many compounds which can be easily chemically altered.

2. a. 0.702 g/mL
    b. 70.2 g

3. a. Building and burning.
    b. Building—petrochemical products (ethene or ethylene for polyethylene–alcohol solvents–drugs–etc); Burning— fuels (methane/ethane in natural gas used to heat homes–octane and other liquid hydrocarbons in gasoline).

4. $C_{25}H_{52}$

5. a. The number of carbons (which is related to total molar mass) and the degree of branching within a hydrocarbon molecule help determine the relative boiling points.
    b. Boiling points increase as the number of carbons per molecule increase: there is more surface area for intermolecular contact. Boiling points decrease with increasing branching in the molecules because of less surface area for intermolecular contact.

6. a. Isomers are molecules having the same molecular formula but different structural arrangements (and thus different properties).
    b. Any structures from the list (page 85) of nine possible C_7H_{16} isomers will do. (Some students may mistakenly draw two versions of the same isomer and count it twice—to convince them the two represent the same structure, have them build both chains—minus hydrogens—with the model kit and compare them.)

```
C–C–C–C–C–C        C–C–C–C–C–C        C–C–C–C–C–C
                          |                   |
                          C                   C


    C  C                  C   C              C                    C
    |  |                  |   |              |                    |
C–C–C–C–C–C          C–C–C–C–C          C–C–C–C–C          C–C–C–C–C–C
                                          |
                                          C


C–C–C–C–C               C  C
    |                   |  |
    C                C–C–C–C
    |                   |
    C                   C
```

(Hydrogen atoms are assumed.)

In response to a common student question, the number of possible isomers increases rapidly as one moves to higher alkanes. There are 18 isomers for C_8H_{18}; the number rises to 1858 for $C_{14}H_{30}$. How about $C_{20}H_{42}$? The claim is 366,319 isomer possibilities!

EXTENDING YOUR KNOWLEDGE (page 175)

- Students investigating this problem will learn that the actual hydrocarbon blend used in gasoline varies according to seasonal temperatures and climate. In cold winter conditions, more higher-volatility hydrocarbon molecules are used in gasoline to promote proper fuel vaporization in the engine. Under warmer conditions, the gasoline blend contains more lower-volatility hydrocarbon fractions to avoid premature or too rapid vaporization.

  A related effect is experienced with diesel fuels; summer-grade diesel fuel actually thickens to a gel if the temperature drops below about –15°C (not unusually cold for northern climates). "Winterized" diesel fuel is required to prevent automobile stalling under these conditions—this fuel contains a higher fraction of low-viscosity, low molar mass hydrocarbons with higher volatility.

- Butane, normally a gas at room temperature, is a liquid when held under pressure in a closed volume. This increased pressure raises butane's boiling point. When the valve is opened to the lower pressure of the atmosphere, the boiling point of butane becomes lower than room temperature, and the liquid vaporizes. When the pressure is increased, the boiling point is elevated. If the pressure is decreased, the boiling point is decreased.

- Salt water promotes corrosion. Thus although salt will lower the freezing point of radiator fluid, it will contribute to the cooling system's early demise. Ethylene glycol, an organic molecular substance, is virtually unreactive in the radiator environment.

Knowing the differences among properties of petroleum hydrocarbons enables chemists to isolate useful fractions. With a better understanding of petroleum chemistry students are ready to tackle a burning issue—petroleum as a source of energy (Part C).

C: PETROLEUM AS AN ENERGY SOURCE

Students learn some history of how petroleum has affected our lives. Historical perspective is provided when students interview senior citizens about energy and resource use. A look at the future is provided by a science fiction article and a student writing assignment in which students consider their own energy future.

Energy released from burning petroleum is described in terms of bond breaking and bond making. The energy released in hydrocarbon burning is measured by students in a simple calorimetry laboratory activity. The unit concludes by examining improvements in automobile fuels and suggesting some alternate sources.

C.1 YOU DECIDE: THE GOOD OLD DAYS? (pages 177-178)

The assignment of interviewing senior citizens about energy and resource use has challenges for young people, but the insights into previous times are fascinating and revealing. The simplest proposal is for students to contact older family members. The groundwork for involving families in *ChemCom* was already laid in the water use diary in the Water unit, so most families will support a long-distance phone interview. However, many families don't have grandparents nearby, so encourage students to interview an older person in the neighborhood. If your community has a retirement home, it is often possible to bring senior citizens to the school for the class to interview. You may wish to enlist the interest/help of a social studies teacher or run the study as a joint effort. If a large enough number of interviews are conducted it might be possible to break the study down into decades (by those born in the 1900-1910, 1910-1920, 1920-1930, 1930-1940 time periods). In any case, make sure that students record the specific time period(s) being discussed by the person being interviewed.

The class should decide as a group on 10 to 12 questions that will help clarify the connection between lifestyles and energy use. The class should then be divided into several teams. Students first practice their interview plan by questioning other team members so that comparisons will be available to lifestyles in earlier times.

Devise some convenient way for the total class interview data to be summarized and displayed in your classroom. A bulletin board, poster, or even a length of brown wrapping paper taped to the wall can serve this purpose.

Conduct a class discussion focusing on the three questions found on page 178. Be prepared to address other concerns or findings students have identified in this information-gathering activity.

Point out that to make the best use of petroleum today, it is prudent to examine the energy sources that we have used over the past century. This leads to a closer look at past, present, and future energy sources.

Answers to Questions:

1.

Early 1900s	Today
a. Home heating: wood, coal	Gas, electric
b. Lighting: oil lamps, gas	Electric
c. Public transportation: trains, street cars	Buses, subways, planes
d. Private transportation: horse, buggy	Cars
e. Cooking: wood, coal stoves	Gas, electric, microwave
f. Food packaging: little, brown paper	Much, plastic
g. Clothes washing: by hand, dried on clothesline	Electric washing machines and dryers
h. Clothing: cotton, wool	Variety of synthetics and natural fabrics

2. Positive features: slower paced lifestyle, less pollution, more leisure time, greater mobility, etc. Negative features: higher death rate, less adequate medical treatment, smaller variety of consumer items and foods, etc.

3. If alternative energy sources (including conservation) are not adequately developed in time, we could potentially arrive at a situation far worse than the "good old days." That is, our urban and suburban areas have come to rely on centrally distributed power for heating and lighting, on automobiles for travelling to and from work, on trucks to bring in food from outside rural areas, etc. Most of the skills of self reliance (including living off the land, making one's own clothing, etc.,) have been lost by most of the general public—without our energy "slaves" our very survival would be threatened.

C.2 ENERGY: PAST, PRESENT, AND FUTURE (pages 178-181)

This section identifies the sun as the primary source of energy and speculates how the discovery of fire probably influenced the development of civilization. This narrative provides an introduction to the *YOUR TURN* activity which draws attention to a graph representing types of energy use for the past 140 years.

YOUR TURN

Fuel Consumption Over the Years (pages 179-180)

In this activity, based on the graph in Figure 9 (page 179), students trace the growth of energy use and associate it with economic factors and the availability of particular types of energy sources. Note that basic graph-reading skills are practiced and supported in this activity. Make students aware that the joule (J) is the SI unit of energy.

Answers to Questions:

1. a. Our overall energy use has accelerated since 1850.
   b. The rise was due to rising population, increased mobility, and greater consumption of processed or manufactured products per person.
2. a. Up to about 1880, wood supplied more than 50% of our energy needs.
   b. Trains were the chief form of long-distance transportation during this period.
3. a. The decline in local supplies of wood and development of the internal combustion engine (a petroleum-consuming device) contributed to the demise of wood as a primary energy source.
   b. Coal was the next energy source to develop in importance.
4.   Oil burned in lamps was a popular use of petroleum prior to 1910.
5. a. Oil became an important energy source in the 1920s.
   b. Petroleum use increased as the number of automobiles increased, spurred by the Ford assembly line with its mass production approach.
6. a. The most recent energy source to enter the picture is nuclear energy.
   b. To generate electrical power.

CHEMQUANDARY

Life Without Gasoline (pages 180-181)

The Isaac Asimov prediction is very dated, and yet well written. If your school is rural, you may need to explain what a policeman's "beat" is (an area of the city that one patrolled on foot), and why parks are dangerous unless full. Encourage your students to be creative, but realistic, in describing changes in a petroleum-deficient society. Remind them of "The Good Old Days?" interviews for a realistic look backwards.

Answers: 1. Different student responses. 2. a. Purchase more energy-efficient devices (cars, home furnaces, insulation, etc.); conserve by using less energy (lower thermostats, use public transportation and/or bicycle more, etc.); recycle (paper, aluminum, and glass), etc. b. Mandate higher efficiency cars; support public transportation and freight trains; lower thermostats in public buildings; stiffer building codes on insulation, etc.

C.3 ENERGY AND FOSSIL FUELS (pages 181-186)

This section introduces the chemistry of energy storage and release in petroleum molecules. Chemical reactions are discussed in terms of exothermic and endothermic processes, using the synthesis and decomposition of water as an example. You may need to clarify the information about reactions occurring in stages, as shown in Figure 10. Notice that both reactions require an input of energy to start the reaction. You may use the term "activation energy" as a label for this energy— activation energy is covered again in the Air unit. The two *YOUR TURN* activities focus on conversions of energy types, and savings which result from more fuel-efficient cars.

DEMONSTRATION IDEA

The traditional electrolysis of water, producing hydrogen and oxygen, would demonstrate that the decomposition of water is endothermic. Mixing small quantities of hydrogen and oxygen in a test tube, and comparing the sound of its combustion reaction to the combustion reaction in a tube of "pure" hydrogen, should demonstrate adequately the exothermic nature of the formation of water.

The excess thermal energy of exothermic reactions can be harnessed in various ways to do useful work. Students are invited to examine the efficiency of such energy conversions in the following *YOUR TURN*.

YOUR TURN

Energy Conversion (pages 183-184)

Emphasize that the automobile is a system consisting of a series of interconnected and coordinated energy conversion devices.

YOUR TURN

Energy Conversion Efficiency (pages 184-185)

In Figure 14, students consider energy losses associated with various energy conversions in the automobile. Representative calculations illustrate the dollar and resource losses due to inefficient energy conversion.

Answers to Questions:
1. a. (200 mi/wk) × (52 wk/yr) = 10,400 mi/yr
   b. (10,400 mi/yr) × (1 gal/19.0 mi) = 547 gal/yr.
2.   (547 gal/yr) × ($1.35/gal) = $738/yr (rounded from $738.45)
3. a. 75% is wasted, so 547 gal/yr × (0.75) = 410 gal wasted/yr
   b. ($738/yr) × (0.75) = $554 lost/yr = ($1.35) × (410 gal)
4. a. (200 mi/wk) × (1 gal/40.0 mi) × (52 wk/yr) = 260 gal/yr
   b. 547 gal – 260 gal = 287 gal.
   c. (287 gal) × $1.35/gal = $387
5. a. (10,400 mi/yr) × (1 gal/50 mi) = 208 gal/yr; 547 gal/yr – 208 gal/yr = 339 gal/yr saved
   b. (339 gal/yr) × (1.35 gal) = $458/yr; $738/yr – $280/yr = $458/yr.

C.4 THE CHEMISTRY OF BURNING (page 186)

The chemical reaction for combustion is revisited, a good time to review balancing equations, with the addition of the all-important energy component. To make decisions about the wise use of fossil fuels such as petroleum, one needs to know how much energy is released by different fuels. This leads to a method of measuring thermal energy, illustrated by *Laboratory Activity* C.5.

C.5 LABORATORY ACTIVITY: COMBUSTION (pages 186-189)

Students will use a simple calorimeter to trap the energy from a burning candle and perform calculations.

Time

One class period

Materials (for a class of 24 working in pairs)

    12 12-oz, empty soft drink cans (with flip top tab)
    12 ring stands
    24 ring clamps
    12 thermometers
    12 stirring rods
    12 index cards
    Balances
    12 paraffin ($C_{25}H_{52}$) candles
    5 lbs ice
    Matches

Advance Preparation

Consider asking students to bring empty aluminum soft drink cans from home a day or two before the activity is scheduled. Flip up tab on can so glass stirring rod can pass through. Figure 15 (page 187) clarifies what is intended. This minimizes heat loss during heating, but makes stirring and adding the water a little more restricted.

You will need to provide each class with chilled water. It should be at least 5°C below room temperature. Avoid allowing ice to enter the students' cans.

Pre-Lab Discussion

The Pre-Lab discussion should include mention of the related but different concepts of heat and temperature. You may also wish to review the bond-breaking and bond-making involved in burning hydrocarbons (via a generic energy diagram modeled after the combustion of hydrogen), and relate the energy effects to the net exothermic reactions observed. Introduce students to the energy unit, the joule (J). A 100-watt light bulb requires 6000 J (6 kJ) of energy to operate for one minute.

A properly-assembled apparatus (similar to Figure 15) on the front bench makes a good Pre-Lab discussion prop. It will also be useful in the Post-Lab discussion, particularly when sources of experimental error—particularly heat loss—are considered.

Lab Tips

As usual, it's essential that students wear protective eyewear.

Advise students to take particular care in igniting the candle.

It is possible to avoid using ice by using fresh tap water; to improve precision, the temperature change should be at least 10°C

Depending on the calibrations on your thermometers, students may only be able to read to the nearest 0.5°C— you decide what should be required.

Remind students that they should stir the water with a stirring rod, not a thermometer, to avoid breakage problems.

Caution students not to inhale or get carbon soot on skin (it may be an irritant); wash off immediately.

Post-Lab Discussion

As part of the Post-Lab discussion, plan to summarize class data on the chalkboard—heat of combustion (kJ/g) and molar heat of combustion (kJ/mol) for paraffin ($C_{25}H_{52}$). The pooled class values can be used to answer the questions.

Use a pre-assembled calorimeter as a "prop" in discussing Question 5 (page 189).

If alcohol burners are available from the biology department, you might do a quick demonstration of the combustion of a liquid fuel, such as hexane, in the burner. This would reinforce the comparison between liquid and solid fuels explored in the questions.

Answers to Questions:

1. For perspective, the actual heat of combustion for butane (about 49.3 kJ/g) is higher than that for paraffin (approximately 42.0 kJ/g). It's possible, due to the experimental errors associated with this relatively crude calorimeter, that some students will find a higher value for paraffin. In any event, the experimental values should be lower than the accepted values given here (for reasons explored in Question 5).

2. The actual molar heat of combustion is greater for paraffin (approximately 14,800 kJ/mol) than for butane (2859 kJ/mol). Student results should reflect a similar ordering of these two values, although their results will be lower than these accepted values due to heat losses.

3. Since each paraffin molecule on average (paraffin is actually a mixture, not a pure compound) contains 24 carbon-carbon bonds, while a butane molecule contains three carbon-carbon bonds, it might be expected that more energy would result from burning one mole of paraffin than from burning one mole of butane. (Each additional –CH_2 unit produces another molecule of CO_2 and H_2O and additional heat energy.)

4. On a gram-for-gram basis, butane is a slightly better fuel. However, mole-for-mole (or molecule-for-molecule) paraffin is a much better fuel.

5. a. This is not a good assumption. Some thermal energy released during hydrocarbon burning will be lost to the surroundings directly from the flame, through the aluminum can, and from the water surface. This energy will not be accounted for in the calculations—thus a lower experimental heat of combustion should be found than the true value.

    b. Other experimental errors include inaccuracy in weighing, the burning wick's small contribution of energy to the total heat of combustion (its heat of combustion is probably not equal to that of paraffin), and incomplete combustion of paraffin. These factors will tend to lower the total quantity of heat measured.

C.6 USING HEATS OF COMBUSTION (pages 189-192)

Heats of combustion allow students to quantify the energy released in chemical reactions.
This is another opportunity to review balancing equations, enhanced by the following *ChemQuandary*.

CHEMQUANDARY

Splitting Molecules? (page 190)

No. A "half-molecule" of oxygen is possible only at high energy; this is an unstable, and very reactive, oxygen atom. In the case of balanced equations, fractions refer to parts of a mole, which is a very large number of molecules.

YOUR TURN

Heats of Combustion (page 191)

Students establish quantitative and qualitative relationships between the heat of combustion and the first eight members of the alkane series. Point out that since the major use of petroleum is for burning as a fuel, the heat produced per gram is a primary factor in choosing appropriate types of hydrocarbons for burning.

Answers to Questions:

1. Worked in student text as an example.
2. a. $CH_4(g) + 2 O_2(g) \rightarrow CO_2(g) + 2 H_2O(g) + 890$ kJ
    b. $2 C_4H_{10}(g) + 13 O_2(g) \rightarrow 8 CO_2(g) + 10 H_2O(g) + 5718$ kJ
3. a. The molar heats of combustion increase as the number of carbons increases (this is a linear relationship), while the heat of combustion per gram of hydrocarbon decreases slightly.
    b. The molar heat of combustion increases about 660 kJ per carbon atom. Decane's molar heat of combustion should be 5450 + 2(660) = 6770 kJ/mol. The heat of combustion per gram should be slightly less than octane. The exact number can not be determined, due to the non-linear relationship.
    c. The heat of combustion in KJ/mol was easier to predict due to the linear relationship.
4. a. (2 mol C_8H_{18})(5450 kJ/mol C_8H_{18}) = 10,900 kJ
    b. (2,660 g C_8H_{18}/1 gal)(47.8 kJ/g C_8H_{18}) = 127,000 kJ/gal
    c. (20 gal)(2,660 g/gal)(47.8kJ/g)(0.16) = 410,000 kJ
5. a. $C(s) + O_2(g) \rightarrow CO_2(g) + 394$ kJ
    b. Octane (47.8 kJ/g) releases more energy per gram than does carbon:
       (394 kJ/mol C)(1 mol C/12.01 g C) = 32.8 kJ/gC.
    c. Carbon (in the form of coal) could be used in combustion reactions which do not require the vapor phase, such as removing oxygen from metals in ore refining.
    d. Carbon would not work in the automobile engine, which requires the fuel to be in the vapor phase for rapid combustion.

OPTIONAL CHEMQUANDARY

Imagine that you are planning a backpacking trip and wish to take along one of the first eight alkanes as a hydrocarbon fuel. a. Which form of the data (kJ/mol or kJ/g) would be more useful in helping you pick the best fuel? Explain. b. Ignoring the mass of the fuel container, which fuel would be most appropriate for your trip? Why? c. In reality, what other factors would you need to consider if you selected one of the gaseous fuels?

Answers: a. Energy per mass (kJ/g) is the most important consideration. b. Methane is the best choice because it releases the most energy per gram of substance. c. Gaseous fuels such as methane must be kept in pressurized containers to carry a practical quantities. Safety is an important factor to consider.

C.7 ALTERING FUELS (pages 192-195)

The process of cracking, breaking large hydrocarbons into smaller molecules, is presented. Fuel fractions can also be isomerized to produce branched chained hydrocarbons that burn more satisfactorily in car engines, according to a system of octane ratings. This section documents ways in which chemists have been able to engineer hydrocarbon molecules to fulfill particular needs. The older method of adding tetraethyl lead to boost octane ratings is contrasted with oxygen-containing fuels such as methanol and methyl-*t*-butyl ether.

A Burning Issue (page 194)

Answers to Questions:

1. Methanol and ethanol can be considered to be partially combusted molecules since oxygen is already included in their structure. Therefore, less energy should be produced when burned.
2. The gasohol mixture should lower miles per gallon because there is less energy per gallon of gasoline to propel the car.

DEMONSTRATION IDEAS

1. Consider using large molecular models (and/or providing student with their own kits) to demonstrate very directly the meaning of the term **cracking**. Let students see how hydrocarbons with odd number of carbon atoms crack.

2. A catalyst is a substance that can be used to alter (typically speed up) the rate of a chemical reaction (by lowering the activation energy) without becoming consumed in the process. Suitable demonstrations of catalytic behavior:

    a. Attempt to burn a sugar cube with a match—try again after dabbing the cube with some cigarette ashes. (Note: Be sure students understand that sugar is an excellent fuel—their own bodies run on it.)

    b. To show the presence of an activated complex and the reemergence of the original catalyst: Place 60 mL of sodium potassium tartrate (Rochelle salt: $KNaC_4H_4O_6 \cdot 4H_2O$: 25 g per 300 mL solution) in 250- or 300-mL beaker and add 20 mL of 6% H_2O_2 (can be prepared by adding 20 mL of 30% H_2O_2 to 80 mL of water, or some hair-bleaching products are exactly 6%). Heat the solution on a hot plate to about 70°C. The slow evolution of CO_2 indicates a chemical reaction (oxidation of a salt) is occurring at a slow rate without a catalyst. Draw the students' attention to the pink color in a separate beaker containing an aqueous solution of $CoCl_2$ (1 g per 25 mL solution = 0.31 M). Add about 5 mL of the $CoCl_2(aq)$ to the beaker on the hot plate. The reaction instantly speeds up (rapid CO_2 evolution), the pink color disappears and a green color (an activated complex) forms. As the reaction is completed, the green color disappears and the pink color of the regenerated catalyst reappears. (References: "A Versatile Kinetics Demonstration," by Paul T. Ruda, *Journal of Chemical Education*, 1978, **55**, 652 and the ACS book: *Chemical Demonstrations* by Summerlin & Ealy, p. 72).

    In either case, you may want to sketch a general exothermic reaction energy diagram and show that a catalyst lowers the activation energy, but does not change the products produced or the total energy evolved. Also, you may want to mention that catalysts tend to be fairly reaction specific and that the search for cheaper and better catalysts for use in petroleum refining, automobile pollution control, and a host of other processes is a major research project of many companies. The topic of catalysts will be discussed in several of the remaining *ChemCom* units.

PART C: SUMMARY QUESTIONS (page 195)

1. a. Development of the internal combustion engine (automobiles and airplanes)
    b. Increasing industrial technology (refining, processing of crude oil and manufacturing)
    c. Increasing population and demand for electricity (heating and cooling)

2. Wood is a renewable resource, not easily transported (a solid). Petroleum is a nonrenewable resource, easily transported (a liquid or gas, can be piped), burns more cleanly and efficiently than wood.

3. The energy stored in petroleum hydrocarbons was originally captured via photosynthesis in plants. Animals consume the plants. Both the plants and the animals died, eventually forming layers of the underground petroleum deposits we tap today.

4. chemical → thermal → mechanical → electrical → chemical → electrical → mechanical
    (gasoline) (engine) (piston) (generator) (battery) (sun-roof motor) (sun-roof retracting)

5. Energy efficiency represents the quantity of available energy that is converted into useful work by a machine. Two ways to increase energy efficiency are to decrease the number of conversions between the original fuel source and the end use, and to increase the efficiency of each step in the conversion scheme chosen.

6. a. 132,000 kJ × 0.75 = 99,000 kJ wasted.
    b. This energy is lost as heat out the exhaust pipe and radiator, as friction in moving parts, in compressing air in the cylinders, and in running accessories (radio, wipers, heater motor, lights, etc.).

7. $C_3H_8(g) + 5 O_2(g) \rightarrow 3 CO_2(g) + 4 H_2O(g) + 2200 \text{ kJ}$

8. a. The molar heat of combustion of methane (890 kJ/mol) is higher than that of water gas (525 kJ/mol)
   b. (10 mol water gas)(525 kJ/mol) = 5250 kJ

9. a. 2512 kJ/2 mol C_2H_2 = 1,256 kJ/mol
   b. 1,256 kJ/mol × 12 mol = 15072 kJ or 15070 kJ

10. Heat of combustion (kJ/g), molar heat of combustion (kJ/mol), boiling point, density, as well as others

11. a. Octane rating is a scale for measuring the antiknock characteristics of gasoline. Heptane is chosen as the zero point (poor antiknock behavior) and isooctane (2,2,4-trimethylpentane) is rated as 100 (good antiknock behavior).
    b. The octane rating of gasoline can be improved by adding tetraethyl lead or by increasing the percent of branched-chain alkanes (which burn more smoothly) via isomerization processes.

EXTENDING YOUR KNOWLEDGE (page 195)

• Aluminum and magnesium are poor substitutes for petroleum. They are both more expensive per gram than petroleum, requiring large quantities of energy in their reduction from ores. And, quite simply, automobile engines are not designed for the use of such solid fuels.

D: USEFUL MATERIALS FROM PETROLEUM

Part D comes back to the "building" theme for petroleum. It describes alkenes and alkynes and introduces the concept of functional groups and aromatic compounds. This leads to considering how polymers such as plastics are produced. Students are also given an opportunity to be synthetic chemists when they produce a flavoring agent.

D.1 BEYOND ALKANES (pages 196-197)

This section introduces alkenes, hydrocarbons which contain at least one double bond. It is important for students to see the changes in naming and structure that go with unsaturation. They are also introduced to the possibility of including other elements in carbon compounds, which is demonstrated in the next model-building activity.

D.2 LABORATORY ACTIVITY: THE BUILDERS (pages 197-198)

Students work with models to explore the bonding and characteristics of alkenes. Your model systems must be able to represent double bonds—if they can't, the old gum-drop/toothpick standby is preferable. You should clarify the planar nature of the double bond by pre-constructing samples of alkenes. Students are led to construct cyclic compounds, and compounds containing elements in addition to carbon and hydrogen.

Do not dwell excessively on nomenclature during this activity. Its purpose is for students to recognize structural details of common builder molecules.

Time

One class period

Materials (for 24 students working in pairs)

12 sets of space-filling or ball-and-stick models

Part 1 Alkenes

Electron dot structures, structural formulas, condensed formulas, and systematic nomenclature are used to describe molecules with multiple bonds.

Answers to Questions:

1. See molecular structures in text.
2. C_nH_{2n}
3. Students should note that alkene molecules resist rotation about the double bond.

4. a.   Four arrangements of C_4H_8 are possible, as shown by the structural formulas on page 197, as well as cyclobutane:

$$CH_2 — CH_2$$
$$|\qquad\quad|$$
$$CH_2 — CH_2$$
cyclobutane

Point out that this is an isomer, but it doesn't have a double bond in the molecule.

   b.   Yes, Students should recognize the butene isomers; methylpropene is less likely.

5. a.,b.   Each pair alone represents the one substance (not isomers). Some students may correctly point out, however, that the substance represented by the first pair (1–butene) is an isomer of the substance represented by the second pair (2-methylpropene).

6.   Only one isomer is possible for propene.

7.   Isomers: same formula (C_3H_{10}) but different structures.

8. a. $H—C \equiv C—H$ (Ethyne, or acetylene)

   b. $CH_3—C \equiv C — CH_3$ (2-Butyne)

Part 2 Compounds of Carbon, Hydrogen, and Singly Bonded Oxygen

By this time students are familiar enough with isomers that they should find that the two isomers of C_2H_6O are easy to make and identify. Hopefully, students will recognize that the –OH group will be soluble in H_2O (remind them that they have already tested this ethanol in the *Polar/Non-polar Laboratory Activity* in the Water unit) and have more intermolecular bonding possibilities, which gives ethanol a higher boiling point (80°C) than dimethyl ether (–24°C).

Answers to Questions

1. a. Two distinct structures, isomers of C_2H_6O, are possible:

   b. $CH_3—CH_2—O—H$ and $CH_3—O—CH_3$
      Ethanol   Dimethyl ether

   c. Yes, they have the same formula.

2. a. Ethanol is more soluble in water than dimethyl ether.

   b. The boiling points of alcohols are higher than those of isomeric ethers. For example, the boiling point of ethyl alcohol is 78.3°C, while dimethyl ether boils at –24°C. Both phenomena can be explained in terms of the greater hydrogen bonding possible in alcohols.

OPTIONAL QUESTION OR DEMONSTRATION

Repeat the procedure used in Question 1 for molecules containing 4 C atoms, 10 H atoms, 1 O atom (single bonds only):

Four alcohols are possible: (Note: Hydrogen atoms are left off the following structures for convenience only—it is probably best if you do not do this with your students).

C—C—C—C—OH
1-butanol

C—C—C—C
        |
        OH
      2-butanol

                        C
                        |
C— C— C—OH
2-methyl-1-propanol

              C
              |
C—C—C
      |
      OH
2-methyl-2-propanol
(t-butyl alcohol)

Three ethers:

C—C—O—C—C
diethyl ether

C—C—C—O—C
methyl-propyl ether

C—C—O—C
        |
        C
methyl-isopropyl ether

D.3 MORE BUILDER MOLECULES (pages 198-199)

Cycloalkanes, which students may have asked about when making isomers in Part B, and aromatic compounds are presented. The special nature and reactivity of the benzene ring structure are given. Point out that even though cycloalkanes are in a ring, like aromatics, they have more in common (relative unreactivity) with the linear alkanes. You may wish to allow some time for students to build models. Also, the story of how Kekule "discovered" the structure of benzene makes a particularly interesting student research study (or you may wish to recount the story of his dream about the snake swallowing his tail, which made him think of a cyclic structure for benzene, another example of "chance favors the prepared mind.")

D.4 BUILDER MOLECULES CONTAINING OXYGEN (pages 199-200)

The functional groups responsible for alcohols, ethers, acids and esters are diagrammed. It might be helpful for students to see that both alcohols and acids have the –OH functional group, but in ethers and esters the hydrogen is replaced by another carbon group. Point out that the specific functional group(s) found in a given organic compound determine many of its physical (solubility, boiling point, etc) and chemical properties. (Remind them that with the exception of combustion, unsubstituted alkanes are fairly unreactive.) Molecules that contain the same functional group(s) often share similar family properties.

If you have samples of some of these compounds, a safe way to test odors is to add a few drops of the liquid to a piece of cotton, then place the cotton in a stoppered test tube. The tubes can be circulated around the room with little risk of spilling, and odors are minimal. A reminder: ethers are interesting compounds (students have often heard of them as anaesthetics) but *very* flammable, and form explosive peroxides. Automobile starting fluid and Preparation W wart remover are ether-based commercial products which could be purchased to show ether volatility and odor, while minimizing the risks of pure ether.

If you have previously introduced IUPAC nomenclature, you may wish to continue that with the alcohol and ether compounds.

D.5 CREATING NEW OPTIONS: PETROCHEMICALS (pages 200-204)

The extensive variety of petrochemicals presented should give students some sense of the importance of the chemical industry. Whereas alkanes are relatively unreactive (beyond their use as fuels), alkenes which contain unsaturated bonds are quite reactive and therefore quite useful as builder molecules. Two reactions of ethene provide examples. Both alkenes and aromatics (another important class of builder molecules) occur naturally in petroleum, but can also be obtained via cracking and reforming reactions. Once obtained, builder molecules can often be polymerized into macromolecules via addition or condensation reactions. Some of the products and their corresponding properties are discussed.

There is little sense in having students memorize names, formulas, and reactions, though they should have some sense of the need for, and logic of, systematic nomenclature. All students need to understand is that many diverse substances can be created through petrochemical-based, carbon chemistry and that chemists are molecular architects.

Some Important Commercial Polymers and Their Uses

Polymer	Repeating unit(s) in molecule	Monomer(s)	Uses
Polyethylene	$-CH_2-CH_2-$	$CH_2=CH_2$	Films for packaging, molded toys and housewares, plastic bottles, pipe fittings.

Polyvinyl chloride	$-CH_2-CH-$
	         |
	        Cl
		$CH_2=CH-Cl$	Pipe fittings, film and sheeting (vinyl car tops), wire insulations, auto parts, flooring, adhesives, coatings.

Polystyrene	$-CH_2-CH-$ (benzene ring)	$CH_2=CH$ (benzene ring)	Packaging and containers, toys and recreational equipment, disposable food containers and utensils, brush handles.

| Polyester | –O–C– ⬡ C–O–CH₂–CH₂– | | Textile fibers for clothing sheets, curtains, towels, carpets |

Polyester: $-O-C-$ ⬡ $C-O-CH_2-CH_2-$

HO_2C- ⬡ $-CO_2H$

Textile fibers for clothing sheets, curtains, towels, carpets

Polypropylene: $-CH_2-CH-$ | CH_3 ; $CH_2=CH-CH_3$

Fibers for carpeting, wool substitutes, auto parts, rope, nets.

Polyacrylonitrile: $-CH_2-CH-$ | CN ; $CH_2=CH-CN$

Acrylic fiber, carpets, sweaters, single-and double-knit clothing.

Polyamide (Nylon 66): $-C(CH_2)_4-C-NH(CH_2)_6-NH-$ ‖O ‖O ; $HO_2C(CH_2)_4CO_2H$ and $H_2N(CH_2)_6NH_2$

Textile fibers, bristles, carpets upholstery fabrics, yarn, tire cord.

Source: "Plastics: Utilizing the Properties of String-like Molecules," by JCE staff, *Journal of Chemical Education*, 1979, **56**, 42.

DEMONSTRATION IDEAS

1. Students are always fascinated by the breakdown of polystyrene in acetone. Fill a visible container such as a large beaker or graduated cylinder 1/3 full of acetone, and begin adding styrofoam pellets. You will certainly have no trouble getting volunteers to help. The vast quantity which can be broken down, and the bubbling that ensues, makes this the most popular "grabber" in this unit. A simple extension of this activity is to have students think of things to do with the left-over polystyrene.

2. Synthesis of nylon: *Caution: The diamine, the dichloride, and sodium hydroxide are all irritating to the skin, eyes, and respiratory system; using a hood is advised.* It is also advisable to wear plastic or rubber gloves—the demonstration is quite safe if properly performed, but do not directly handle the product until it has been washed with ethanol or water.
   a. Solution 1: Gradually heat a reagent bottle of 1,6-diaminohexane (also called hexamethylenediamine, M.P. = 42°C) in a warm water bath until the solid begins to melt. Decant 2.9 g directly into a 100 mL beaker and dilute to 50 mL with 0.5 M NaOH solution (20 g NaOH per 1 L). The resulting solution will be 0.5 M diamine.
   b. Solution 2: Prepare a 0.25 M diacid chloride solution by diluting 2.3 g of adipoyl dichloride to 50 mL with hexane (density of hexane = 0.66g/cc).
   c. Using equal volumes of the two solutions (as little as 5 mL of each can be effective), slowly pour Solution 2 down the side of the beaker, on the top of Solution 1, so as to minimize agitation between the immiscible solutions.
   d. Use forceps or a bent paperclip to hook the polymer film which forms at the interface, and slowly pull it from the center of the beaker. Continue pulling out the nylon "thread" until the solutions are exhausted. A thread 10-15 feet long is possible. If it breaks first, you may be able to hook a new thread. Longer threads are possible if you avoid contact with the sides of the beaker.
   e. Any excess solution remaining should be placed in a solvent waste container or stirred together, the resulting solid washed and disposed of in a solid waste container.

Reaction:

diamine + diacid chloride → polyamide + 2 HCl

$$H_2N(CH_2)_6NH_2 + Cl-\underset{O}{\overset{O}{C}}-(CH_2)_8-\overset{O}{C}-Cl \overset{NaOH}{\rightarrow} -[\underset{H}{N}-(CH_2)_6-\underset{H}{N}-\overset{O}{C}-(CH_2)_8-\overset{O}{C}]- + 2 HCl$$

Nylon-610

Note: The reaction does not depend on the exact stoichiometry of the reactants, but the concentrations should be close to those indicated for optimal results.

D.6 LABORATORY ACTIVITY: PETROCHEMICALS (pages 204-206)

The synthesis of methyl salicylate illustrates the dramatic changes produced when petrochemicals are formed. The familiar odor of the product of this reaction makes this activity a pleasing experience.

Time

One-half class period

Materials (for 24 students working in pairs)

12 test tubes (15 × 125 mm)
12 10-mL graduated cylinders
12-150 mL beakers
12 test tube holders
12 test tube brushes
12 Bunsen burners, rings, ring stands, and wire screens
12 dropping bottles for dispensing of organic reactants
12 mL concentrated sulfuric acid (18 M H_2SO_4)
1 reagent bottle for dispensing 18 M H_2SO_4
24 mL methanol
6 g salicylic acid
Boiling chips

Pre-Lab Discussion

Discuss the general condensation reaction for forming esters, and once again point out the role of chemists as molecular architects in designing molecules. The flavor and perfume industry is heavily dependent on petrochemical synthesized esters which are either unavailable in nature or available only by expensive extraction from dilute sources. (Many low molecular weight esters have pleasant fruity odors/tastes and are often responsible for the natural odors/flavors of ripe fruit, flowers, and other plant substances.) The continued search for new odors/flavors and for a theory capable of correlating structure with odor/flavor is a potential area for future chemical careers. See Post-Lab discussion as well.

Lab Tips

The students will probably recognize the odor from its use in flavoring candy, perfumes, etc., but caution them not to ingest any of the product.

Remind students to avoid the direct inhalation of any fumes—they should be gently fanned toward the nose.

The boiling chip in the test tube minimizes erratic boiling which can often erupt out of the test tube. The hot water bath needs only to be above the boiling point of methanol (56°C) and does *not* need to boil vigorously.

There are many other combinations which can be used, if you have the available chemicals. Smaller quantities may also be used. Sample procedure: Combine the specified drops of acid and alcohol in a test tube. Add 2 drops of concentrated sulfuric acid. Then heat 5-10 minutes in the hot water bath. Cool, and note the odor.

Acid	Alcohol	Odor
20 drops acetic acid	28 drops n-butyl alcohol	pear
10 drops acetic acid	20 drops octyl alcohol	orange
20 drops acetic acid	30 drops pentanol (amyl alcohol)	banana
20 drops propanoic acid	15 drops ethanol	rum
0.4 g benzoic acid	20 drops ethanol	"fruity"

Post-Lab Discussion

The Post-Lab discussion can include some discussion of the reaction of the functional groups in the organic acids and alcohols that produced esters. Emphasize that a seemingly small difference in molecular structure can make a very noticeable difference in properties. Have students note the contrasting odors of reactants and products. To reinforce this point, you may want to synthesize several other esters (see suggestion in Lab Tips). Be sure to pass samples around the classroom and write out the specific reaction(s):

$$\text{Alcohol} \ + \ \text{Organic acid} \ + \ \text{Acid catalyst} \ \rightarrow \ \text{Ester.}$$

Also, be sure to distinguish organic acids (RCOOH) from common laboratory inorganic (or mineral) acids such as HCl, H_2SO_4, and HNO_3. Organic acids are typically weak acids and were originally obtained by isolation from plant and animal ("organic") sources. For example, formic acid (HCOOH/methanoic acid) was first obtained from the destructive distillation of ants, and acetic acid (CH_3COOH/ethanoic acid) from the aerobic fermentation of a mixture of cider and honey. Students have probably smelled butyric acid ($CH_3CH_2CH_2COOH$/butanoic acid) in rancid butter or body odor. Today commercially used organic acids are typically synthesized from petroleum or coal tar, as are the alcohols with which they are reacted.

Answers to Questions:

1. a. Methanol: CH_3OH.   Salicylic acid (2-hydroxybenzoic acid): $C_7H_6O_3$
   b. Methanol + Salicylic acid → C_8H_8O + H_2O
                                          Methylsalicylate
             (2-Hydroxybenzoic acid)   (methyl-2-hydroxybenzoate)
   (common names used, IUPAC names in parentheses)
2. Amyl acetate: $CH_3COOC_5H_{11}$ (pentyl ethanoate)
3. a. Alcohol (1-butanol)
   b. Ester (methylpropanoate)
   c. Acid (3-methylpentanoic acid)
   d. Acid (ethanoic acid)

OPTIONAL BACKGROUND INFORMATION

The nomenclature of organic compounds is something of a mess. Historically speaking organic compounds were named for the material from which they were created. So formic acid, with one carbon, derived its name from the Latin word for ants, in which the acid was first detected. Acetic acid came from the word for vinegar. Salicylic acid was named after the Latin word for willow tree, in which compounds containing this acid were found.

As organic chemistry became more organized and extensive, a more uniform naming system was required, and this has been established by the International Union of Pure and Applied Chemistry (IUPAC). This system establishes root words for the number of carbons (meth- = 1, eth- = 2, etc.) and names the functional groups by adding suffixes to the roots: alcohols add "-ol" to the root, acids add "-oic", etc. This system would require that acetic acid, with two carbons, be named ethanoic acid, and formic acid, with one carbon, be named methanoic acid. Salicylic acid, because of the benzene ring, is named as a derivative of benzoic acid, with a hydroxy group (-OH) located on carbon number 2 of the benzene ring. Alkenes are named the same as alkanes, except for the "-ene" ending, and alkynes have a "-yne" ending.

However, many chemical companies still sell their materials using the old naming system, and polymers in particular still retain the common names (polyvinyl chloride doesn't sound the same as polychloroethene, or polyphenylethene instead of polystyrene!). So we continue to use common names—acetic acid; acetylene, which is properly called ethyne; and amyl alcohol for pentanol—and teach the IUPAC system, planning for the day when the systematic system is adopted by everyone. That day may be long off: after all, we can't even adopt one world-wide system of units for measuring distance and mass!

PART D: SUMMARY QUESTIONS (page 206)

1. a.   Ethene, C_2H_4. Common name "ethylene"; this colorless, odorless gas is used as an agent to speed ripening of fruits, and in the synthesis of polyethylene, polystyrene, and other plastics.

    b.   Benzene C_6H_6. A clear, colorless, highly flammable liquid used as a solvent and in the synthesis of a wide variety of products including medicines, dyes, plastics, etc.

2. $C_6H_{14} \rightarrow C_3H_8 + C_3H_6$ or C_4H_8 and C_2H_6 (several possibilities)

3. a.,b.,c. (1) Acetic acid:

$$CH_3 - \overset{\overset{\textstyle O}{\|}}{C} - OH$$

        (2)   Dimethyl ether: $CH_3 - \boxed{O} - CH_3$

        (3)   Methyl formate:

$$H - \overset{\overset{\textstyle O}{\|}}{C} - O - CH_3$$

        (4)   ethanol: CH_3CH_2OH

4.   The rings in benzene (and other aromatic compounds) exhibit unusual chemical stability.

5.   (1) Carbon can bond with itself or other atoms to form single, double, or triple bonds.
         (2) It can form straight chain or cyclic compounds and a variety of isomers.

6.   Polymerization is the chemical process of joining together many small molecules to make a very large molecule. Examples include ethene (or ethylene) $\rightarrow$ polyethylene

7. a. Cl_2
    b. H_2O (in presence of an acid catalyst)
    c.   NaOH

OPTIONAL QUESTION

$C_6H_6 + \underline{\ \ ?\ \ } \rightarrow C_6H_5NO_2 + H_2O$
The addition of nitric acid ($HONO_2$ or HNO_3) to benzene produces nitrobenzene, which is an important intermediate in the manufacture of dyes.

E: ALTERNATIVES TO PETROLEUM

We are confronted with a future of dwindling supplies of petroleum (and other nonrenewable fossil fuels). Part E briefly surveys various petroleum substitutes presently under consideration. The promise of chemical technology is that economically and environmentally suitable alternatives to petroleum for both building and burning will be developed. Conservation not only provides scientists and engineers with time to develop such alternatives, but is a wise course of action environmentally. Insights gained here lead to the culminating activity in Part F—the examination of several views regarding how best to promote the conservation of petroleum.

E.1 ALTERNATIVE ENERGY SOURCES (pages 207-211)

After calculations concerning the present usage and future availability of petroleum and other fossil fuels, the text presents several suggestions to improve our energy future: offshore oil drilling, the extraction of petroleum from oil shale, coal liquefaction, and petroleum plantations. Biomass conversion, various forms of solar energy, and nuclear fission and fusion are options for replacing petroleum's role as an energy source at fixed site locations. Several topics raised here, such as nuclear energy and the greenhouse effect, will be covered in future units.

YOUR TURN

Energy Dependency (pages 207-209)

This activity reestablishes the need to search for petroleum substitutes as fuel for our transportation needs. Students are asked to interpret the pie chart in Figure 18 (page 209), as well as data contained in Tables 5 (page 208) and 6 (page 209). It is not important for students to remember the actual numbers from the problems, but they should understand the logic (the idea of using data on known reserves and extrapolating the present use) of such calculations. Remind students that these are estimates, and do not include unexplored resources such as the Arctic Natural Wildlife Reserve.

Answers to Questions:

1. Petroleum

2. 88.6% (petroleum, natural gas, and coal)

3. 4.2% (hydro/solar/wind/geothermal, and wood)

4. Petroleum: $(1000 \times 10^{18}\,J)(1\text{ quad}/1.05 \times 10^{18}\,J) = 1619$ quad;

        1619 quad - 462 quad = 1157 quad left by the year 2000.

5. Assuming that the % use by fuel type in Figure 18 remains constant:
   (1) Natural gas: (0.231)(2000 quads) = 462 quads of natural gas used by the year 2000.
           Available reserves $(1700 \times 10^{18}\,J)(1\text{ quad }/1.05 \times 10^{18}\,J) = 1619$ quads;
           1619 quads - 462 quads = 1157 quads left by the year 2000.
   (2) Coal: (0.232)(2000 quads) = 464 quads of coal used by the year 2000.
           Available reserves $(40,300 \times 10^{18}\,J)(1\text{ quad}/1.05 \times 10^{18}\,J) = 38,381$ quads;
           38,381 quads - 464 quads = 37,917 quads left by the year 2000.

6. 100% − 63% = 37%

7. Coal will more readily substitute for petroleum at stationary sites. Transportation vehicles are engineered to use liquid fuels.

OPTIONAL BACKGROUND INFORMATION

Synthetic fuels: To obtain any source of energy requires a certain amount of energy for building the necessary equipment and for mining, pumping, or processing the fuel. Given the present technology and oil prices, all the synthetic fuels have lower net energy values than the 50 to 1 ratio for domestic oil.

Coal-derived fuels: Ordinary coal is composed primarily of carbon, with only about one hydrogen atom for every 16 carbons. Since liquid and gaseous hydrocarbon fuels contain a much higher percent hydrogen (methane: 4H/1C; octane: 2.25H/1C; heavy fuel oil: 1H/6C), temperature and pressure are used to break down the chemical structure of coal in order to admit more hydrogen atoms.

In World War II, when it was unable to import oil, Germany used coal-derived gasoline to fuel its planes and tanks. More recently, South Africa's Sasol I plant has been producing oil from coal since 1960. The net energy value for coal-derived fuels is about 17 to 1.

Shale oil: Of the 1.8 trillion barrels of shale oil in the states of Colorado, Wyoming, and Utah, about 600 billion are considered recoverable. Unfortunately, there are a number of problems with its extraction. Problems include: (1) the energy needed to extract the kerogen (temperatures of approximately 500°C result in a net energy value of only about 6.5 to 1); (2) the volume of water needed (1-4 barrels per barrel of oil produced); and (3) the waste generated (1.5 tons of shale per barrel of oil produced). Thus, at the present time, shale oil is expensive from both an economic and environmental perspectives.

Alcohols: As partially oxidized hydrocarbons, alcohols have a lower energy content than an equal volume of gasoline. The chief advantage of using alcohols is that they can be "grown" and then mixed with regular gasoline up to concentrations of 10-15% without the need for engine redesign. Methanol is made from wood and, given its high affinity for water, is commonly used as a fuel drying additive. Ethanol is made from grain (wheat or corn) and is less toxic and corrosive than methanol. However, its use in gasohol has led some to question the ethics of using food to run machinery.

E.2 BUILDER MOLECULE SOURCES (pages 211-212)

Even though some petroleum currently used as an energy source can be diverted to producing petrochemicals, the world's supply of petroleum will eventually be used up. Industrialized nations are turning to coal as a source for needed raw materials. Biomass conversion is second. Finally, since all of the mentioned alternatives for building and burning uses of petroleum have disadvantages, it is wise to consider conservation goals for extending present supplies through efficiency measures, recycling/reuse, and gradual changes in consumer behavior.

This factual information could be enhanced by asking students to make a chart listing each alternative source, its advantages, drawbacks, and renewability.

PART E: SUMMARY QUESTIONS (page 212)

1. (1) Petroleum, a nonrenewable resource, supplies nearly half our energy needs. Our high rate of consumption, compared with estimated future supplies, poses a serious threat to our way of life. Transportation accounts for over half of our use of petroleum for burning; industry uses 10% of the total oil per year for building products that support our lifestyles.

    (2) As petroleum supplies dwindle, we are faced with the choice to build or to burn—or with making decisions regarding the relative amounts of building and burning to be supported.

2. a. Research and development regarding renewable sources of energy give some hope that new technologies will be in place to serve as substitutes for dwindling supplies of petroleum and other nonrenewable resources. This policy will help to avert drastic and sudden change in lifestyles.

    b. It is least likely that hydroelectric, solar, or wind power will be used for future transportation purposes, if by "transportation" we mean what the term signifies today (emphasis on privately owned motor vehicles).

    c. The community may have large towers for wind turbines, large fields of solar cell arrays, and/or individual residences with solar collectors and wind-powered generators.

3. (1) Coal is in plentiful supply, but environmental problems are associated with its widespread use. The cost of converting coal to liquid fuel is currently greater than the cost of producing the same quantity of fuel from petroleum.

    (2) Petroleum can be extracted from **oil shale** and tar sand, the crude oil condensed from kerogen vapor. These shales contain a large supply of pre-oil. However, the technology is not adequately developed, access to federally owned land containing the shale has not yet been secured, and conversion processes require large amounts of water–already scarce in the western states where the oil shale and tar sands are found.

    (3) Petroleum can be extracted from some **plants**, implying the development of "petroleum plantations," yet larger-scale technology supporting this idea has not yet been demonstrated. Also, there would be the ethical question of whether it is justifiable to devote large tracts of land to "growing petroleum" when much of the world's population is starving.

4. The burning of petroleum is most likely to be curtailed before building. The petrochemicals needed to produce detergents, drugs, plastics, and fibers will likely be given higher priority than certain uses of petroleum as fuel.

5. Desirable compounds would be obtained quickly, with low expense in terms of energy to process the compounds, and with low environmental impact -- no air or water pollution. The compounds would also not require the use of resources used to meet other societal needs, such as food.

OPTIONAL QUESTION

Grain alcohol (ethanol) has been proposed as a substitute for imported oil. A typical U.S. car traveling 10,000 miles per year at 15 mpg would require an amount of grain–produced alcohol equal to 14,700 lb of grain. For comparison, a subsistence diet for a typical adult male is the equivalent of about 400 lb of grain. Assuming that it takes approximately one acre of land to produce 1890 lb of grain a. How many acres of land would need to be set aside to grow the grain needed for one car? b. How many people could be fed on the yearly diet of a alcohol-driven car?

Answers: a. (14,700 lb)(1 acre/1890 lb) = 7.78 acres b. (14,700 lb) (1 person/400 lb) = 37 persons

F: PUTTING IT ALL TOGETHER: CHOOSING PETROLEUM FUTURES

F.1 CONFRONTING THE ISSUES (pages 213-215)

This activity seeks to involve students in a mock debate concerning policy options for dealing with an energy crisis. Instead of assigning team members, it is possible for each student to select which option to support—but encourage a reasonable number balance among the teams. The group should focus on three arguments in favor of the chosen option and develop questions for the other groups. Unless there is a particularly strong personality in the class, it is probably best for you to serve as moderator. Conducting the class in a "Donahue-type" setting, with commercial breaks to allow rebuttal, helps provide some positive stimulus to the ensuing discussion. If you have not videotaped your Town Meeting in the Water unit, the videorecorder camera may have a positive impact here.

It may be advisable to make the team assignments at least one week prior to the end of the unit to allow time for library research. The class time allotment for this activity is somewhat open-ended. Try to schedule debate and letter-writing to complete one class period—the letter may be a homework assignment if the debate is particularly lively. A possible specialization of team effort might be as follows: Assume three teams of 10 students each. Each team could consist of two students who will serve as the panel representatives (they will assimilate and present the findings of their fellow team members and be prepared to answer questions from the class), four students who will serve as support staff/ researchers to gather evidence in support of the assigned plan (if one or more is artistic, charts, graphs, etc., might be prepared), and four who will split off to research the problems associated with their opponents' plans (two for each of the other plans).

As homework, consider instructing students in each group to select and read a magazine or newspaper article related to their topic. This will help focus the assignment and ensure that each student contributes to the group effort. If you choose to have students find their own articles (rather than providing references for them), be sure students understand that some popular newspapers and periodicals specialize in reliable, authoritative science writing. Some good journals for the nonspecialists include *Popular Science*, *Discovery*, *National Geographic*, and *Nature*. The more detailed and scientific journals include *Scientific American*, *Science*, *Bioscience*, *CHEMTECH*, *Journal of Chemical Education*, and *Chemical & Engineering News*. The high school student chemistry magazine, *CHEM MATTERS*, published by the American Chemical Society, also includes useful student-centered readings. The definitive sources of statistics and predictions on energy use are the U.S. government DOE Energy Information Administration publications: *Monthly Energy Review* and the *Annual Energy Outlook*. These are likely to be available only at a large university or public library, but DOE/EIA can supply free brochures and pamphlets (like *Energy Outlook 199X*) that summarize some of this data. Also, the most recent information found in popular publications can be found by using the *Reader's Guide to Periodic Literature* and other indexing services found in school or community libraries. You may need to help students identify key descriptors for their assigned plan.

If your students have difficulty analyzing their assigned issue, you can assist their preparations by giving them the following questions:

Plan #1: Federal Restraints on the Fuel Uses of Petroleum

a. How have the relative costs of gasoline in the United States and Europe changed since World War II?
b. How has fuel consumption "voluntarily" shifted in favor of higher mileage automobiles and/or less reliance on the private automobile since OPEC nations have changed oil prices?
c. How have similar restraining actions by other governments, like Brazil, stimulated the development of domestic alternatives such as ethanol?
d. What are the costs (or risks) and benefits of government intervention (who and/or what section of the country or economy stands to "pay" the highest costs or receive the most benefits)?
e. If in favor of restraints, what types and/or degree of restraint would be most effective and fair?

Plan #2: Government Subsidized Mass Transit with Restrictions on the Use of Automobiles

a. What have been the successes and failure of Japan and European nations in this area of national policy?
b. How successful have mass transit rail systems such as those used in San Francisco (BART) and Washington, D.C. (Metro) been in enticing riders to give up their cars? What conditions encourage use of such systems?
c. What alternatives exist to laws restricting the use of automobiles if one wants to "encourage" people to use mass transit?
d. What fuels power such mass transit systems and what are their environmental and economic costs and benefits?
e. Does the government have the right to enact such a policy and, if so, how would such a policy be instituted? How would people of various age groups respond to this plan? This question lends itself to a brief survey by students.
f. How would the costs and benefits of such a national policy be distributed across the various sections of our country (which sections would be least/most likely to benefit in the short run)?

Plan #3: Student-generated policy option

Questions will vary.

The activity concludes with each student writing a letter to the editor, supporting one of these options. One suggestion for improving the validity of this assignment is to have the students choose the best letter and submit it to a local newspaper, or at least the school newspaper. Present the class with three or four choices. If you have more than one *ChemCom* class, you could have each class choose the best letter from the other class, to preserve anonymity of the author during the selection process.

F.2 LOOKING BACK AND LOOKING AHEAD (page 215)

This brief transitional section to the next unit suggests an unexpected connection between petroleum use by society and food use by individuals—in both cases we confront building/burning issues. Using a resource for energy and for building is an elegant lead-in to the Food unit, which will develop this insight in detail.

SUPPLY LIST
Expendable Items

	Section	Quantity (per class)
Acetic acid, conc. (optional)	D.6	10 mL
Amyl alcohol, pentanol (optional)	D.6	10 mL
Asphalt, pieces	B.3	15
Benzoic acid (optional)	D.6	10 g
Candles, paraffin	C.5	20
Ethanol (optional)	D.6	25 mL
Hexane (optional)	C.5	100 mL
Ice, crushed	C.5	10 lbs
Kerosene	B.3	750 mL
Methanol	D.6	40 mL
Octanol, octyl alcohol (optional)	D.6	10 mL
Oil, automotive	B.3	750 mL
Oil, household lubricating	B.3	750 mL
Oil, mineral (light)	B.3	750 mL
Paraffin wax, flakes or shavings	B.3	150 g
Propanoic acid (optional)	D.6	10 g
Salicylic acid	D.6	20 g
Sulfuric acid, conc.	D.6	150 mL

Nonexpendable Items

Space–filling or ball-and-stick model kits
Digital timers or stop watches
7-mm plastic or metal beads
Culture test tubes with caps (20 × 150 mm); 36 mL capacity (or test tubes with corks)
Aluminum pop cans
Index cards

SUPPLEMENTAL READINGS

"Special Report on Energy." *National Geographic*, February 1981. Whole issue.

"Methane." *CHEM MATTERS*, February, 1983.

"Liquid Crystals." *CHEM MATTERS*, December, 1983.

"Across Australia by Sun Power." *National Geographic*, November 1983, 600-607. Solar powered car.

"Polymers." *CHEM MATTERS*, April 1986. Several articles: "Polymers" (4-6), "Tyvek" (8-10), "Polysaccharides" (12-14), and "Silly Putty" (15-17).

Polymer Chemistry. National Science Teachers Association, 1986. Book available from the American Chemical Society, Education Division. Developed by teachers in cooperation with Arco, DuPont, and Shell, the first half of the book provides background information on natural polymers; chapters on polymer structure, synthesis, economic value, and a minitext for students. The second half of the book is demonstrations, lab exercises, games, and other activities. The disk is a student tutorial for use on the Apple II computer.

"The Interrupted Party." *CHEM MATTERS*, October, 1984.

"The Blackened Bucket." *CHEM MATTERS*, December, 1984.

"Polymer Samples for College Classrooms." *Journal of Chemical Education*, February 1984, **61**,#2, 161-3. Presents a list of companies, contact persons, and the type and form of polymer that can be provided gratis.

"The Missing Warning." *CHEM MATTERS*, October, 1985.

"Flash Point" and "Natural Dyes", *CHEM MATTERS*, December, 1986.

Playing with Energy. National Science Teachers Association, 1981, 106 pp. Games and simulations for grades 9-12 selected from the Project for Energy-Enriched Curriculum learning packets.

"Dissolving Plastic", Exploring the Solubility of Polyvinyl Alcohol", and "Non-Safety Glass", *CHEM MATTERS*, October, 1987.

"Dissolving Plastic" and "The Exploding Tire", *CHEM MATTERS*, April, 1988.

"What's Best for Your Car." *Consumer Report*, February 1987, 88-94. Discusses and compares the viscosity, additives, etc., found in major brands of motor oil.

"The Smell of Danger" and "Killing for Oil", *CHEM MATTERS*, October, 1988.

"Fireside Dreams." *CHEM MATTERS*, December, 1988.

"Geochemistry in the Search for Oil." *Chemical & Engineering News*, February 10, 1986, 28-43. Special report reprint: fairly technical, more for teachers than students.

"The Fruits of Ethylene." *CHEM MATTERS*, April, 1989.

"Nylon" and "Rudolph Diesel's Engine", *CHEM MATTERS*, December, 1990.

Energy Facts, 1984 (First Edition). U.S. Government Printing Office, 1985, 56 p, (S/N 061-003-00453-9). Published by the Energy Information Administration, provides a quick reference to a broad range of domestic and international energy data over the time period 1973-1984 through the use of graphs and charts. Also lists other sources of information.

UNDERSTANDING FOOD

PLANNING GUIDE

Section	Laboratory Activity	You Decide
Introduction		Keeping a Food Diary

A. Foods: To Build or to Burn?

A.1 Nutritional Imbalances A.3 Why Hunger?		A.2 Dimensions of Hunger

B. Food as Energy

B.1 Food for Thought–and for Energy B.3 Carbohydrates: The Energizers B.4 Fats: Stored Energy with a Bad Name		B.2 Energy In–Energy Out

C. Foods: The Builder Molecules

C.1 Foods as Chemical Reactants C.2 Limiting Reactants C.3 Proteins	C.4 Milk Analysis	

D. Other Substances in Foods

D.1 Vitamins D.3 Minerals: Part of Our Diet D.5 Food Additives	D.2 Vitamin C D.4 Iron in Foods	D.6 Food Additive Survey

E. Putting It All Together: Nutrition around the World

E.3 Looking Back and Looking Ahead		E.1 Diet Analysis E.2 Meals Around the World

TEACHING SCHEDULE

	DAY 1	DAY 2	DAY 3	DAY 4	DAY 5
Class Work	Discuss pp. 218-219	Discuss pp. 220-223 YD pp. 221-222	Discuss pp. 224-226 YT pp. 225-226	Discuss pp. 228-232 YT pp. 231-232	Review Parts A & B
Laboratory					
Homework	YD p. 219 Read pp. 218-223	SQ p. 223 Read pp. 224-228	YD pp. 226-228 Read pp. 228-232	Read pp. 232-235	SQ pp. 234-235 Read pp. 236-241

	DAY 6	DAY 7	DAY 8	DAY 9	DAY 10
Class Work	Quiz Parts A & B Discuss pp. 236-241 YT pp. 237-238	Discuss pp. 241-246 YT pp. 243-244			Review Part C
Laboratory			LA pp. 246-247	LA pp. 247-250	
Homework	YTs pp. 239-241 Read pp. 241-250	YT p. 246			SQ p. 250 Read pp. 251-256

	DAY 11	DAY 12	DAY 13	DAY 14	DAY 15
Class Work	Quiz Part C Discuss pp. 251-254		Discuss pp. 256-259 YT p. 258		CQ p. 264 Discuss pp. 260-265
Laboratory		LA pp. 254-256	Pre-lab LA pp. 259-260	LA pp. 259-260	
Homework	YT p. 253	Read pp. 256-260		Read pp. 260-266	YD pp. 265-266

	DAY 16	DAY 17	DAY 18	DAY 19	DAY 20
Class Work	Review Part D	Quiz Part D YD pp. 267-268	YD pp. 268-269	Review unit	Exam
Laboratory					
Homework	SQ p. 266 Read pp. 267-269				

LA = Laboratory Activity; **CQ** = ChemQuandary; **YT** = Your Turn; **YD** = You Decide; **PIAT** = Putting It All Together.

The study of food, our most important renewable resource, ties in with the Petroleum unit's theme of burning and building, by examining food for energy and food for structure. Underlying the information concerning food chemistry are two themes: global hunger and personal nutrition.

In using this unit students examine chemical structures and behavior of carbohydrates and fats (as energy sources), and proteins (as builder molecules). Vitamins and minerals are included to round out the nutrition story, and students are introduced to the pervasive role of additives. The highlight of the unit is an activity involving a personal food diary and its analysis, which leads to a comparison with meals eaten from different countries around the world.

OBJECTIVES

Upon completion of this unit the student will be able to:

1. Compare the uses of food in terms of "building and "burning." [Introduction, B.1]
2. Distinguish malnutrition from undernutrition, and identify parts of the world where these problems are most acute. [Introduction-A.3]
3. Define calorie and joule, and calculate energy changes from calorimetry data. [B.1]
4. Correlate weight gain or loss with caloric intake and human activity. [B.2]
5. Compare and contrast mono-, di-, and polysaccharides in terms of structural formulas and properties. [B.3]
6. Identify key functional groups in carbohydrates and fats, and write an equation for the formation of a typical fat. [B.3, B.4]
7. Distinguish between saturated and unsaturated fats, and relate the consumption of each to health. [B.4]
8. Define and illustrate the concept of limiting reactant using biochemical examples and calculations. [C.2]
9. Describe how functional groups in amino acids interact in protein formation. [C.3]
10. Describe five functions of proteins in the body. [C.3]
11. Discuss the concepts of essential amino acids, complete protein, and complementary protein with respect to a balanced diet. [C.3]
12. Separate and measure protein and carbohydrates in nonfat milk, and calculate a sample's caloric value. [C.4]
13. Distinguish water-soluble from fat-soluble vitamins (with specific examples of each), and discuss the implications of these differences in terms of dietary needs. [D.1]
14. Analyze the vitamin C content of foods by performing titrations. [D.2]
15. Identify minerals used in the body, and distinguish between macrominerals and trace minerals. [D.3]
16. Determine the iron content of foods by colorimetry. [D.4]
17. Discuss the relative risks and benefits of various food additives in terms of their purposes, and provide specific examples. [D.5]
18. Discuss the role of the Food and Drug Administration and federal regulations in ensuring food safety. [D.5]
19. Compare and contrast menus from several cultures in terms of calories and nutritional balance, and analyze the nutritional quality of food recorded in a personal food diary. [E.2]

SOURCES ON NUTRITION RELATED TO WORLD HUNGER

In keeping with the "need-to-know" approach, this unit does not provide extensive information concerning nutrition or world hunger. Students who have special interests in these areas may be interested in the following sources:

1. Center for Science in the Public Interest, 1501 16th St., NW, Washington, DC 20036. 202-332-9110. Offers a wide variety of low cost books, posters, and software on nutrition. Ask about reprints of past issues of the *Nutrition Action Health* (news) *Letter*.

2. Chemical Manufacturers Association, 2501 M St., NW, Washington, DC 20037. 202-887-1223. Free pamphlets on chemicals naturally found in typical meals (Your Breakfast—Lunch—Dinner Chemicals) and on chemical additives.

3. Council for Agricultural Science Technology (CAST), 250 Memorial Union, Ames, IA 50011. Inquire about a free subscription to the magazine *Science of Food and Agriculture*.

4. Food & Drug Administration, 5600 Fishers Lane, Rockville, MD 20857. 301-443-1544. Publishes the journal *FDA Consumer* and a variety of free nutrition pamphlets.

5. National Dairy Council, 6300 North River Road, Rosemont, IL 60018-4233. 708-696-1020. Offers a variety óf free and low cost nutrition materials including an inexpensive set of colorful Comparison Cards (#Bo43) highlighting the nutritional quality of common foods.

6. Office of Disease Prevention and Health Promotion (ODPHP) Health Information Center, P.O. Box 1133 Washington, DC 20013-1133. 800-336-4797 (toll free). A variety of useful low-cost publications including: *Nutrition Activities of the Department of Health and Human Services* (1984/422pp/No. U0005), *Nutrition and Your Health: Dietary Guidelines for Americans* (1985/24pp/No. U0003), *Behavior Patterns and Health* (1985/36pp/No. E0001), *Staying Healthy: A Bibliography of Health Promotion Materials* (1984/42pp/No. E0002). Also helpful as an information referral service.

7. Population Reference Bureau, Inc., 1875 Connecticut Ave., N.W., Suite 520, Washington, DC 20009-5728. (202) 483-1100. Publishes up-to-date fact sheets on world demographics and nutrition/health parameters.

8. Society for Nutrition Education. 1700 Broadway St., Suite 300, Oakland, CA 94612. 510-444-7133. Publishes the *Journal of Nutrition Education* (4 issues/yr) and other useful materials.

9. United Nations agencies such as the World Food Council, Food and Agriculture Organization (FAO of the UN) New York, NY 10017), UNICEF, etc.

10. United States Department of Agriculture, Independence Ave., SW, Washington, DC 20250. 202-447-2791 (info). National Agriculture Library, Baltimore Blvd., Beltsville, MD. 301-344-3355.

11. World Health Organization, 525 23rd St., NW, Washington, DC 20037. 202-861-3200. Publishes the magazine *World Health* and other useful materials on the topic of world hunger.

12. World Hunger Education Service, 1317 G St., NW, Washington, DC 20005. 202-347-4441. Books and audiovisuals.

13. Worldwatch Institute, 1776 Massachusetts Ave., NW, Washington, DC 20036. 202-452-1999. Variety of books and pamphlets including the annual *State of the World* report (paperback 1987: covers most of the topics in the various *ChemCom* units)

INTRODUCTION

The necessity of food for life is discussed, along with its dual role of burning and building. These topics lead to a comparison of our diets with those of other cultures. Terminology introduced includes developed and developing nations, and undernourishment and malnourishment.

You may wish to expand on the analogy between petroleum and food. Almost any fuel (wood, coal, etc.) can be used for burning, but specific types of hydrocarbons (especially petrochemicals) are most useful for building. Similarly, meeting body fuel needs (which means avoiding undernourishment) is easier than meeting body-building requirements (avoiding malnourishment). This point will be brought home to students as they are asked to begin collecting data on their own typical diet. These data will be analyzed later (pages 267-268) in terms of the level of several key nutrients.

YOU DECIDE: KEEPING A FOOD DIARY (page 219)

Once again *ChemCom* ventures into the private lives of students—this time into their eating habits. This important and often revealing activity needs to be handled carefully. Make the assignment early in the unit, reminding students that one diary day should be over a weekend.

Encourage precise reporting about quantities: you may want to bring a typical meal (or typical food samples) into the classroom to help students practice the estimates they will need to make in collecting their data. Some teachers allow class time for writing the diaries and collecting the sheets daily. It is also helpful to distribute data sheets with blanks for different meals and categories of food: proteins, vegetables, beverages, etc. To encourage information-gathering, ask students what they had for breakfast, dinner last night, or after-school snacks. It is amazing how willing kids are to talk about what they eat!

You should collect the completed diaries and hold them for redistribution and analysis for the *You Decide* E.1 on pages 267-268.

A: FOODS: TO BUILD OR TO BURN?

The introduction provides a starting point for considering global nutrition and food supply. This part revisits malnourishment and undernourishment, then considers world hunger and its effect on adults and children. It concludes with a sobering look at the connection between infant mortality rates and world hunger.

Be sure that students understand the two ways in which food shortages can be viewed as an energy crisis.

A.1 NUTRITIONAL IMBALANCES (pages 220-221)

Two types of hunger—malnutrition and undernutrition—are developed here. Stress the fact that the body's energy needs take priority over building needs, just as it might be necessary to burn some of a wooden house to heat the rest of it. Note that even overweight individuals can be malnourished, and both malnutrition and undernutrition ultimately have observable (and sometimes tragic) consequences. This section provides background for the following *You Decide* activity.

A.2 YOU DECIDE: DIMENSIONS OF HUNGER (pages 221-222)

This activity highlights two dimensions of hunger: too little energy in the diet (undernutrition) and the lack of certain molecular building blocks (malnutrition). The purpose of this activity is to explore the magnitude of the problem of world hunger by identifying regions of high infant mortality rates (IMR) as indicators of the world's nutritional condition.

A world map is included in your teaching support materials. Make a copy of the map for each student. They will shade in (with colored pencils or markers) all countries having infant mortality rates higher than 50. Next to the name of each country students record the appropriate infant mortality rate. The current-year *World Population Data Sheet* with IMR data may be obtained from: Population Reference Bureau, Inc., 1875 Connecticut Ave, N.W., Suite 520, Washington, DC 20009-5728. 202-483-1100. You must write for their brochure. There is a small charge for the IMR data. Your librarian may be able to help with ordering this data sheet, or have another source which will provide similar information. An almanac for the current year will provide much help on this and other *ChemCom* topics.

You will find that students need help in locating some of the smaller, developing countries (information you can share with your social studies department.). The followup questions for this activity give some insights into the influence of climate and geography on world hunger.

OPTIONAL INFORMATION

Factors	Developed nations	Developing nations
Population (billions)	1.18	3.76
Per capita GNP	$9510	$700
Infant mortality rate	17	92
Life expectancy at birth	73 yr	58 yr
Population increase (annual %)	0.6%	2.0%
Population doubling time in years (at current rate)	117 yr	35 yr

WORLD MAP

Directions: See Food unit, page 222.

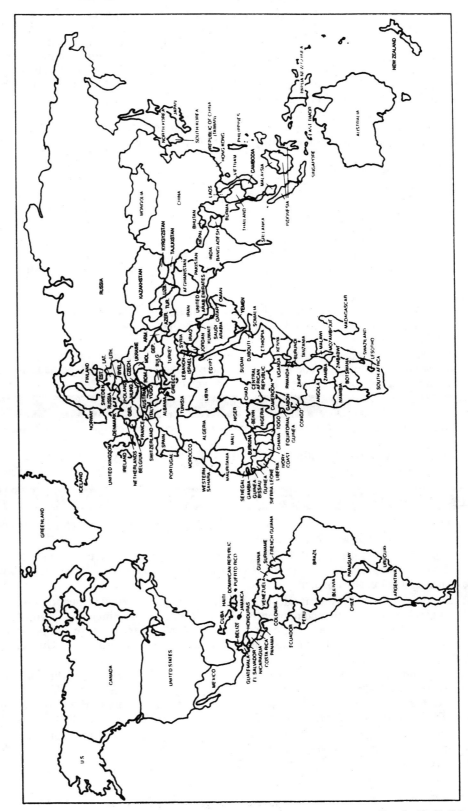

Answers to Questions:

1. The body tissues of children are constantly growing and require the full spectrum of daily nutrients to support physical development. Since adults are already physically developed, they can tolerate malnutrition conditions better than children can.

2. Undernutrition means too few calories in the diet to support bodily work and energy-requiring processes. If this condition continues, the body will consume its stored fat and turn to structural tissue for fuel. The kidneys, liver, and endocrine system often begin to function improperly. A shortage of carbohydrates, which play a vital role in brain chemistry, affects the mind. Lethargy and confusion set in. Starvation victims often seem unaware of their plight. Because body defenses are weakened, disease kills most famine victims before they have time to starve to death. An individual begins to starve when he has lost about a third of his normal body weight. Once this loss exceeds 40%, death is almost inevitable.

3. The totals would be larger. An individual may consume sufficient calories (to avoid being classified as under-nourished) and yet be malnourished (lacking in certain dietary substances).

4. a., b. A number of nations have an infant mortality rate (IMR) considerably lower than other nations in their sector of the world:
    Eastern Africa (Mauritius, Reunion, Seychelles)
    Western Asia (Cyprus, Israel, Kuwait)
    Southern Asia (Sri Lanka, Brunei, Malaysia, Singapore)
    East Asia (Hong Kong, Japan, Macao, Taiwan)
    Central America (Costa Rica, Panama)
    A number of nations (including all of Northern Europe, Austria, Belgium, France, Germany, the Netherlands, and Switzerland) have IMRs lower than that of the United States.
    It is interesting to note that IMR has racial overtones as well. For example, in South Africa, the IMR for white children is 10; for colored 30; for blacks 90.

   c. Individual student responses.

5. Yes, with a few significant exceptions (including Venezuela, Costa Rica, Panama, Chile, Argentina, Uruguay, Sri Lanka, China, Japan, Singapore, Australia, and New Zealand) world hunger tends to divide along a rather distinct line separating North from South.

For human society as a whole, world hunger: (1) weakens family structure and cultural unity; (2) lowers economic productivity; (3) destabilizes governments and threatens world peace and security; and, perhaps the worst of all, (4) represents the waste of our most precious resource—human spirit and creativity.

OPTIONAL QUESTION

The textbook cites the figure of 10-14 million people dying each year from hunger. Calculate the number of people who die each minute.

Answer: If we take the low limit, 10 million, the answer is $(10 \times 10^6 \text{ people/yr}) \times (1 \text{ yr/365 days}) \times (1 \text{ day/24 hours}) \times (1 \text{ hour/60 min}) = 19 \text{ people/min}$

OPTIONAL QUOTES ON WORLD HUNGER FOR USE THROUGHOUT UNIT

"No society can surely be flourishing and happy, of which the far greater part of the members are poor and miserable." From: *The Wealth of Nations* by Adam Smith (Scottish economist), mid-1700s.

"We hold the moral obligation of providing for old age, helpless infancy and poverty, is far superior to supplying the invented wants of courtly extravagance." From: *The Declaration of the Friends of Human Peace and Liberty* by Thomas Paine (American political writer), mid-1700s.

"The essential component of social justice is adequate food for all mankind. Food is the moral right of all who are born into this world...without it, all other components of social justice are meaningless.... If you desire peace, cultivate justice, but at the same time cultivate the fields to produce more bread; otherwise there will be no peace." From: *The Nobel Peace Prize Acceptance Speech* by Dr. Norman E. Borlaug (American agronomist, Green Revolution creator), December 10, 1970.

"Every man, woman, and child has the inalienable right to be free from hunger and malnutrition in order to develop fully and maintain their physical and mental facilities. Society today already possesses sufficient resources, organizational ability, and technology, and hence the competence to achieve this objective." From: *Universal Declaration on the Eradication of Hunger and Malnutrition* of the United Nations Food Conference, Rome, 1974.

If there is the political will in this country and abroad...it should be possible to overcome the worst aspects of widespread hunger and malnutrition within one generation." From: *World Food and Nutrition Study* of the National Academy of Sciences, National Research Council, 1977.

A.3 WHY HUNGER? (pages 222-223)

This section raises the question of adequate food for the world, in light of information about the availability of food. The text material encourages students to incorporate the ideas of food supply (which is apparently adequate, if shared equally), demand, and distribution into addressing world hunger problems. The basic factors presented in Section A.3 can be expanded to include undersupply–the limitations on plant and animal growth imposed by climate and soil fertility (remember the IMR world map); demand–the much higher populations in developing countries such as India and China; and distribution–the losses due to spoilage and to profiteering among shippers who divert the supplies for their own needs and distribution. The stage is set for the study of food and how the body uses it.

PART A: SUMMARY QUESTIONS (page 223)

1. a. Food is used both for burning for energy and for building bodily structures—much as we found for petroleum.
   b. Burning takes priority when insufficient food is provided. If the body runs out of fuel, it dies. The building functions of food may be given secondary priority when insufficient fuel is available to maintain body temperature and activity.
2. a. World hunger is an energy crisis in at least two senses: (1) large numbers of people are undernourished (inadequate caloric intake for fueling bodily activity); and (2) the energy needed to produce fertilizers and pesticides and to carry out large-scale crop planting/harvesting is not usually available in developing countries where food shortages exist.
   b. World hunger is a national resource crisis in that: (1) food is used for building, much like other chemical resources; and (2) developing countries often lack chemical resources (phosphates, natural gas, petroleum, and others) just as they lack adequate protein, carbohydrate, fat, vitamin, and mineral resources from food. They also lack the money to buy these resources from other nations.
3. a. Undernourishment is associated with an insufficient number of calories for burning, and implies malnourishment as well. Malnourishment means inadequate supplies of needed nutrients.
   b. Yes, one can be malnourished even if sufficient numbers of calories are ingested.
   c. No, being undernourished means insufficient calories, which means that the body will not have sufficient materials for building either.
4. a. Eating a variety of foods increases the likelihood of an adequate balance of all needed nutrients.
   b. Calorie-counting alone is inadequate, since the body uses food for building as well as burning.

B: FOOD AS ENERGY

Part B highlights the energy aspect of foods by introducing students to carbohydrates and fats.

The energy units of the dietitian (Calorie) and the chemist (calorie and joule) are defined and related quantitatively. Students are reintroduced to calorimetry, the measurement of quantities of heat energy, and make conversions between dietary units. This leads to decision-making on activity as a factor in human weight loss and weight gain.

Students examine the basic structures of fats and carbohydrates in terms of organic functional groups. Some controversy surrounds the eating of saturated and unsaturated fats, and knowledge of the structural properties of these substances helps students understand the issues.

B.1 FOOD FOR THOUGHT—AND FOR ENERGY (pages 224-226)

Calories, and the determination of caloric content, lead to calculations which convert food energy into heat energy. Highlight that one Calorie (Cal), the food unit, is different than a chemist's calorie (with a small "c"). The food Calorie is the same as a kilocalorie (kcal), 1000 calories. Students are introduced to the joule, the SI unit, which is more widely used on other parts of the world. In some European countries the energy values of common packaged foods are currently declared in units of kilojoules (kJ). Because more public data is available in Calories, that unit will be used throughout this unit.

DEMONSTRATION IDEA

To demonstrate the connection between the energy content of fuels (such as hydrocarbons like paraffin) and food: place an almond or peanut sliver (Brazil nuts do nicely as well) wick in the end of a peeled and trimmed banana "candle." You could use a cork borer to make a straighter candle, or use a potato instead. Light the candle as you discuss how the caloric value of foods can be released quickly by burning or more slowly by cellular metabolism.

YOUR TURN

Calorimetry (pages 225-226)

This activity illustrates how the energy content of foods can be determined from calorimetry data. The calculations continue to move students away from formula usage and toward quantitative reasoning. Beginning with the definition of a calorie in terms of mass of water and temperature change, help students grasp the algorithm of multiplication by asking several questions: How many calories to raise 1 g of water 5°C? 1 g of water 10°C? 2 g of water 5°C? (Twice as much water or change in temperature, twice as much energy) 5 g of water 10°C? etc.

Answers to Questions:

1. $(1 \text{ cal/g°C}) \times (160 \text{ g}) \times (60 - 22)°C = 6080 \text{ cal}$ (6.08 Cal = 6.08 kcal)
2. a. $(1 \text{ cal/g°C}) \times (6 \text{ glasses}) \times (250 \text{ g/glass}) \times (37 - 0)°C = 55,500 \text{ cal}$
   b. 55.5 Cal
   c. 220 Cal = 220,000 cal
      220,000 cal/55,500 cal = x glasses/6 glasses
      $x \sim 24$ glasses.
   d. No, this is not an efficient way to diet. One would need to consume large quantities of water, although health food advocates suggest this technique to assist in a diet program.

OPTIONAL PROBLEMS

1. How much energy is released by a burning peanut, if a 60 mL sample of water is heated from 15°C to 38°C?
   $$[(1 \text{ cal/g°C}) \times (60 \text{ mL}) \times (1 \text{ g/mL}) \times (38 - 15)°C = 1380 \text{ cal} = 1.4 \text{ kcal} = 1.4 \text{ Cal}]$$

2. How much energy is released by burning 0.2 g of candle wax, if the temperature of a 80 g sample of water at 20°C is heated to 45°C?
   $$[(1 \text{ cal/g°C}) \times (80 \text{ g}) \times (45 - 20)°C = 2000 \text{ cal} = 2.0 \text{ kcal} = 2.0 \text{ Cal}]$$

3. How much water could be heated by burning 1 g of paraffin (heat of combustion is about 10,000 cal/g) from 0°C to body temperature (37°C)?
   $$[(10,000 \text{ cal/g}) \times (1/37°C) \times (1 \text{ cal/g°C}) = 270 \text{ g}]$$

4. What would be the final temperature of a 250 g sample of water, beginning at 20°C, if 1.5 grams of propane were burned to heat the water? Heat of combustion of propane is 12.0 kcal/g.
   $$\text{Energy} = 1.5 \text{ g} \times 12.0 \text{ kcal/g} = 18 \text{ kcal} = 18,000 \text{ cal}$$
   $$18,000 \text{ cal} = (250 \text{ g}) \times (\text{temp. change}) \times (1 \text{ cal/g°C}) \quad \text{temp. change} = 72°C$$
   $$72°C = T_f - 20°C \quad T_f = 92°C$$

B.2 YOU DECIDE: ENERGY IN—ENERGY OUT (pages 226-228)

This activity establishes links among the caloric values of foods, human activity, and weight gain or weight loss. Students keep a diary of personal activity for a 24-hour period and calculate the total energy expended. Since all that we do uses energy, every minute counts. Encourage students to consider broad categories, such as 5 or 6 hours sitting at school, 1/2 hour (or more) driving, 15 minutes of standing doing dishes or housework, etc. The total time should equal 24 hours, which will require fairly accurate accounting of activities.

The calculations of energy use are similar to the "Water Use Diary": basically the time multiplied by the energy use/hour, given in Table 2. You may have some students who are more active in sports such as wrestling or basketball. It is safe to project a higher caloric output, 1100 to 1200 kcal/hour, for those activities.

Students compare the total energy consumed in food with the total energy expended by physical activity. One of three conditions results: maintenance of body weight, weight gain, or weight loss.

This activity can be completed in one class period. If you give the assignment as homework, less class time will be required, and the diary probably will be more accurate.

Answers to Questions:

1. a., b. Individual student answers
   c. Individual answers; generally students have been below average in energy use, except for those involved in sports.

2. a. Calories = 250 + 125 = 375 Cal.
   (1) 375 Cal/ 420 Cal (one hour of tennis) = 0.89 hours
   (2) 375 Cal/210 Cal (for walking one hour) = 1.79 hours. Distance = (1.79 hr) × (2.5 mi/hr) = 4.46 mi
   (3) 375 Cal/430 Cal (for one hour of swimming breast stroke) = 0.87 hours
   b. (375 Cal) × (1 lb/4000 Cal) = 0.094 lb
   c. (3 × 16) sundae × (0.094 lb/sundae) = 4.5 lb
   d. Students will express their own opinions here.
3. a., b. One could, for example, lower the intake of some other high-energy foods in their regular diet to compensate for the energy in the ice cream. There have been students who would skip meals, drink lots of water, or regurgitate the food.

OPTIONAL CHEMQUANDARY

In general, why is a weight-loss plan based only on exercise (without controlling caloric intake) not likely to be adequate for anyone that is more than a little overweight?

Answer: Physical activity typically accounts for only about 30% of the body's energy output; the other 70% is to maintain basal metabolism. Even the most vigorous physical activity burns off only about 1000 Cal/hour, and one pound of excess body fat is equivalent to about 4000 Cal. Thus, if one wanted to lose 1-2 lbs/week (the maximum recommended without medical supervision), one would need to set aside 4-8 hours per week for vigorous physical exercise. Such a plan would probably be unhealthful or even impossible for a very overweight person. A balanced combination of low-calorie diet and exercise would be more effective in both the short and long term.

B.3 CARBOHYDRATES: THE ENERGIZERS (pages 228-230)

Carbohydrate structure brings students back to photosynthesis as our energy-storing process in foods. Different types of saccharides, including polysaccharides, are presented, along with some nutritional advice.

The chain form of sugars such as glucose closes to the ring form, which is more stable. The closure involves a condensation reaction between the –OH group of carbon 5 and the aldehyde group on carbon 1 of the sugar (additional functional groups are listed on page 231), forming a compound know as a "hemiacetal." The key to ring formation in sugars is the C=O (carbonyl) group. Fructose (also shown in the margin on page 232), having a carbonyl group one carbon from the end of the sugar chain, forms a smaller, 5-membered ring system. The 6-membered sugar ring systems are called pyranoses, and the 5-membered ring systems are called furanoses.

The difference in bonding between starch and cellulose (Figure 4, page 229) is small but critical. In digestible starch, both rings are attached to the connecting oxygen (this is similar to an ether functional group) so that the oxygen is below both rings (this is called the alpha-linkage). In cellulose, the oxygen is above one ring and below the other (the beta linkage). Human enzymes can break down the more exposed bonding in the starch molecule, but only certain animals have enzymes capable of digesting cellulose.

OPTIONAL BACKGROUND INFORMATION

The body is able to metabolize various classes of biomolecules to meet its caloric needs:

1. **Carbohydrates**—4 Cal/g—the body's preferred fuel—are ultimately converted into glucose which circulates at all times in the blood, providing an ever present source of energy to cells. Excess glucose is converted into glycogen which is stored in the liver and muscle cells or, if these "banks" are full, as fat stored in fat cells.

2. **Fats**—9 Cal/g. Of the fats in food and the human body, 95% are triglycerides (the other 5% are phospholipids), which consist of a glycerol ($C_3H_8O_3$) backbone with three attached fatty acids. Under conditions of low to moderate activity, both components can be burned along with glucose to fuel voluntary muscles. Unfortunately, the red blood cells, brain and other nerve cells are unable to utilize fatty acids for fuel. Also, under conditions of strenuous activity, even the muscles shift to using only glucose (from stored glycogen) since oxygen can not be supplied fast enough to support the larger oxygen demands of fat oxidation. Still, by serving as a fuel under normal conditions, and serving as a means of storing glucose that is in excess of glycogen storage capacity, fats serve a vital energy role.

3. **Protein**—4 Cal/g—is not normally used as a fuel (provided the body is supplied with adequate calories from carbohydrates and fats) since it has more important "building" functions. Under starvation conditions the body will break down protein to fuel the brain, nerve cells, red blood cells, and maintain vital organ functions.

4. **Alcohol (ethanol)**—7 Cal/g—provides empty calories, meaning lacking any of the 44 nutrients needed to sustain life. It can displace more nutrient-dense foods and contribute to malnutrition and obesity, as well as acting as a central nervous system depressant.

1. Use of refined sugar in the United States has risen from about 76.4 lb/person/year around the turn of the century to about 102 lb/person today. Only about 25% of this consumption is due to direct consumer use—most comes "hidden" in soft drinks and sugared "fruit" drinks (for individuals between the ages of 15-34 these account for the single largest portion). Many nutritionists argue that such empty calorie/low nutrient-density "food" items can cause malnutrition by displacing nutrient-dense alternatives in the diet. This issue makes a fascinating library research project or in-school survey study (comparing students' use of, and nutritional understanding about, real fruit juice vs. soft drinks and fruit drinks).

2. Coinciding with a rise in sugar consumption, the amount of fiber in a typical American's diet has decreased from 6 g/day at the turn of the century to about 4 g/day today. What is dietary fiber, and what role(s) does it play in a healthy diet?

B.4 FATS: STORED ENERGY WITH A BAD NAME (pages 230-234)

The section begins with a review of the formation of esters, which was covered in the Petroleum unit. The terms lipid, saturated and unsaturated, and glyceride are explained, and students taught to identify more functional groups.

The IUPAC terminology, referred to in the Petroleum unit, takes a beating in biochemistry. Traditional names (glycerol, from the Greek word *glykeros*, meaning sweet, for 1,2,3-propanetriol or palmitic acid, from palm oil, instead of hexadecanoic acid [16 carbons]) dominate and are used by the text. One can simplify much of the chemistry of the different types of biomolecules by focusing on the functional groups. Notice that unsaturated carbons, meaning carbon-carbon double bonding (alkenes), play a significant role in the body's ability to handle and store fats.

OPTIONAL CHEMQUANDARY

A hunter/hiker wishes to minimize the quantity of food that he needs to carry with him on a short trip in a cold climate. If he also wishes to maximize his caloric intake, would he be better advised to take along high carbohydrate or high fat foods?

Answer: Fats— since they deliver more Cal/g than do carbohydrates.

YOUR TURN

Functional Groups in Biomolecules (pages 231-232)

This activity is designed to review the concept of functional groups, first introduced in the Petroleum unit. Students are invited to identify structural features in some of the food-related molecules considered thus far.

Answers to Questions:

1. a.

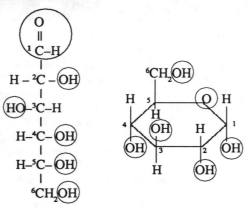

   b. Aldehyde (–C=O), alcohol (–OH), and ether (–O–).
          |
          H

   c. (1) The aldehyde and the alcohol.
      (2) The aldehyde is on C-1, the alcohol on C-5.

2.   Fructose has a ketone group, located on carbon-2. Glucose has an aldehyde group on carbon-1. Fructose has an alcohol on C-1. Glucose has an alcohol on C-2.

3.

$$\begin{array}{c}
\text{H} \\
| \\
\text{H} - \text{C} - \text{OH} \\
| \\
\text{H} - \text{C} - \text{OH} \\
| \\
\text{H} - \text{C} - \text{OH} \\
| \\
\text{H}
\end{array}
\quad + \quad 3\ \text{HO} - \overset{\displaystyle \overset{O}{\|}}{\text{C}} - (\text{CH}_2)_{16} - \text{CH}_3 \quad \rightarrow$$

$$\begin{array}{c}
\text{H} \quad\quad O \\
| \quad\quad \| \\
\text{H} - \text{C} - \text{O} - \text{C} - (\text{CH}_2)_{16} - \text{CH}_3 \\
| \quad\quad O \\
| \quad\quad \| \\
\text{H} - \text{C} - \text{O} - \text{C} - (\text{CH}_2)_{16} - \text{CH}_3 \\
| \quad\quad O \\
| \quad\quad \| \\
\text{H} - \text{C} - \text{O} - \text{C} - (\text{CH}_2)_{16} - \text{CH}_3 \\
| \\
\text{H}
\end{array}$$

Glycerol   Stearic acid (octadecanoic acid)   glyceryl tristearate

4. a. It contains one carboxylic acid group and one C = C double bond.

$$\text{H} - \text{O} - \overset{\displaystyle \overset{O}{\|}}{\text{C}} - (\text{CH}_2)_7 - \overset{\displaystyle \overset{\text{H}\ \ \text{H}}{|\ \ |}}{\text{C} = \text{C}} - (\text{CH}_2)_7 \text{CH}_3$$

   b. It is a fatty acid, indicated by carboxylic acid functional group and long carbon chain.
   c. It is unsaturated, because of the carbon-carbon double bond.
   d.

$$\text{HO} - \overset{\displaystyle \overset{O}{\|}}{\text{C}} - \text{CH}_2\ \text{CH}_2\ \text{CH}_2\ \text{CH}_2\ \text{CH}_2\ \text{CH}_2\ \text{CH}_2 - \text{CH} = \text{CH} - \text{CH}_2\ \text{CH}_2\ \text{CH}_2\ \text{CH}_2\ \text{CH}_2\ \text{CH}_2\ \text{CH}_2\ \text{CH}_3$$

OPTIONAL BACKGROUND INFORMATION

Fats in the Diet: An optimal diet should contain (as % of total calories):
10-15% protein (which yield 4 Cal/g)
20-30% fats (which yield 9 Cal/g)
55-70% carbohydrates (which yield 4 Cal/g).

Note: If an adult male obtained 25% of his daily 2700 Calories from fat, he would consume the equivalent of 17 teaspoons (or 75 g) of fat— an adult female on a 2000 Calorie/day diet would consume 13 teaspoons (or 56 g). This may seem like a lot until one considers the high percent of fats in a number of the foods we consume (see the following table).

In moderate amounts, fats in the diet serve several useful functions: as a concentrated source of energy (9 Cal/g); as a carrier and aid to absorption of fat-soluble vitamins (A, D, E, and K); as a source of linoleic acid (an essential fatty acid); as a heat insulator and padding for the body; and as a source of flavor and aroma in food. However, nutrition experts believe that the typical American diet (40% or more of the Calories from fat) contains too much fat. In fact, as many as 25% of all U.S. adults (or 34 million Americans) are overweight and thus at increased risk for heart disease, high blood pressure, diabetes, and cancer. Some studies have shown a relationship between high fat diets and cancer of the colon and breast. Obesity can be attributed to a combination of overeating, lack of exercise, psychological factors, and heredity. Regardless of the specific cause(s), given the high caloric value of fats, monitoring fat intake is important in controlling weight.

Percent fat (by mass)	Food
90-100	Salad and cooking oils, fats, lard
80-90	Butter, margarine
70-80	Mayonnaise, pecans, macadamia nuts
50-70	Walnuts, almonds, baking chocolate, bacon
30-50	Broiled choice T-bone, spareribs, cheddar and cream cheese, potato chips, french dressing, chocolate candy
20-30	Choice beef pot roast, broiled choice lamb chops, hotdogs, ground beef, chocolate chip cookies
10-20	Broiled choice round steak, broiled veal chop, roast turkey, eggs, ice cream, french fried potatoes, apple pie
1-10	Fish, crab meat, cottage cheese, milk, creamed soups, sherbet, most breakfast cereals
< 1	Baked potato, most fruits and vegetables, chicken

PART B: SUMMARY QUESTIONS (pages 234-235)

1. $(1 \text{ cal/g°C}) \times (1000 \text{ g}) \times (62 - 22)°C = 40,000 \text{ cal} = 40 \text{ Cal}$

2. a. Hydrocarbons contain only carbon and hydrogen atoms, while carbohydrates contain carbon, hydrogen, and oxygen atoms.
   b. Both represent organic molecules such as methane (hydrocarbon) and glucose (carbohydrate). A variety of molecules of both occur due to the many ways in which carbon atoms can link.
   c. Hydrocarbons are typically nonpolar and insoluble in water. Carbohydrates may be polar or nonpolar.

3. a. 6:00 – 11:00 P.M. —light housework: (180 kcal/hr) × (5 hrs) = 900 Cal
   11:00 P.M. – 7:00 A.M. —sleeping: (80 kcal/hr) × (8 hrs) = 640 Cal
   Total = 1540 Cal. Your body needs to replace this much energy in order to begin functioning.
   b. Lack of sufficient food to run the body would prevent the body from supporting high-energy activity needed in many jobs.

4. a. Fats deliver the most calories per gram.
   b. (3000 Cal) × (1 g/9 Cal) = 333 g
   c. No; a diet of pure fat would neglect the "building" needs of the body.

5. a. $(2.5 \times 10^9 \text{ lb}) \times 4000 \text{ Cal/lb} = 1.0 \times 10^{13} \text{ Cal.}$
   b. $1.0 \times 10^{13} \text{ Cal}/(2600 \text{ Cal/day} \times 365 \text{ days/yr}) = 1.05 \times 10^7 \text{ people}$ (slightly more than 10 million.)

6 a. (1) HC – CH – CH – CH – CH – CH$_2$ – OH   (2) $CH_3 \; CH_2 \; CH_2 \; CH = CHCH_3$
   $\quad\quad\;$ ‖ | | | |
   $\quad\quad$ O   OH   OH   OH   OH

   b. The carbohydrate is more soluble in water due to the presence of the polar –OH groups, which will be attracted to water.

7. Carbohydrate ($C_3H_6O_3$)

EXTENDING YOUR KNOWLEDGE (page 235)

- a. (1 mo/30 days) × (30 lbs/2 mo) × (4000 Cal/lb) = 2000 Cal/day must be removed from the diet.
  b. 3000 Cal – 2000 Cal = 1000 Cal/day can be consumed.
  c. This is quite unwise; a 67% reduction in caloric intake is not likely to provide a balanced diet unless carried out under a doctor's close supervision.

C: FOODS: THE BUILDER MOLECULES

Part C focuses on food nutrients as builder molecules. The role of protein is primarily to build (or rebuild) body structures, but protein fragments can also be used as fuel.

The concept of a limiting reactant is introduced and interpreted in relation to the basic nutritional concern of eating the right amounts and kinds of foods. Students are introduced to amino acids and protein structure, as well as the essential amino acids and the concept of complementary proteins. Milk's nutritional value is tested in the laboratory activity found at the end of Part C.

C.1 FOODS AS CHEMICAL REACTANTS (pages 236-237)

This section seeks to prepare the student for the concept of food for building by examining the breakdown of glucose in the body. This process, called cellular respiration, is clarified by explaining the function of enzymes, which will be studied more completely in the Risk unit.

DEMONSTRATION IDEA

Students might enjoy seeing a sugar cube (sucrose) burn in a brief demonstration—a simple way to highlight the energy stored in carbohydrates. As you may already know, the trick to getting the cube to burn is to sprinkle a few ashes on its surface before lighting. The ashes act as a catalyst.

For a dramatic "model" of how the caloric content of sugar can be released without a visible flame: Cover the bottom of an old 250-mL Pyrex beaker (to facilitate cleanup you can coat the inside of the beaker with vegetable oil spray) with white, granular table sugar (about 50 g of $C_{12}H_{22}O_{11}$). Place the beaker on a noncombustible surface under a fume hood and pour approximately 30 mL of concentrated sulfuric acid on top of the sugar. Stir with a glass rod (and, if desired, place a thermometer near the surface). The sugar will turn from yellow to brown to black, carbon residue will raise above the beaker, and steam will be visible. The highly exothermic reaction involves the dehydration of sugar:

$$C_{12}H_{22}O_{11}(s) + 11\ H_2SO_4(l) \rightarrow 12\ C(s) + 11\ H_2SO_4 \cdot H_2O(g)$$

Avoid touching the carbon residue unless it is first washed with an aqueous solution of baking soda to neutralize excess acid. Note: In the human body, sugar is burned in a series of steps to form carbon dioxide and water.

C.2 LIMITING REACTANTS (pages 237-241)

Students are usually familiar with the idea of following a recipe in cooking, and make the connection between recipes and balanced equations. Emphasize the limiting reactant in each *YOUR TURN* question. Limiting reactants lead to some revealing information concerning vitamins and elements in the body. In Question 4 of the Plants and Humans *YOUR TURN*, refer students to the margin note on page 241.

YOUR TURN

Limiting Reactants (pages 237-238)

This activity gives students practice in predicting the amounts of products obtained when one reactant is limited. As you will note, these calculations correspond closely to those routinely required in chemical systems. We have purposely limited this early practice to concrete, macroscopic systems in the hope that the central ideas will be understood by all students.

To clarify the concept of limiting reactant, work through the possible number of "products" possible with each "reactant." It is easy to see which material is the "limiter."

Answers to Questions:

1. a. Seven cakes
   b. Flour is the limiting reactant.
2. a. Napkins used up first.
   b. Napkins
   c. 19 lunch boxes possible.
3. a. 20 booklets
   b. The pages are the limiting reactant.
   c. 60 staples, 20 covers left over.

YOUR TURN

Limiting Reactants: Chemical Reactions (page 239)

This activity is an extension of the previous one. Students should understand that the concept of a limiting reactant operates at both an atomic and macroscopic level and is of crucial importance in the industrial production of fertilizers.

Answers to Questions:

1. a. One mole nitrogen, three moles hydrogen, 2 moles ammonia
   b. 28 g nitrogen, 6 g hydrogen, 34 g ammonia.
2. a. Nitrogen is the limiting reactant.
   b. Hydrogen is found in excess.
   c. 3 kg
   d. 34 kg of ammonia

DEMONSTRATION IDEA

This demonstration provides a chemical example of the limiting reactant concept, involving easily-observed chemical changes.

Part A Constant amount of HCl, variable amount of Mg

Tube 1: 5 mL 3 M HCl + 2.5 cm Mg ribbon (3 M HCl: 12.5 mL conc. HCl/50 mL solution)
Tube 2: 5 mL 3 M HCl + 5.0 cm Mg ribbon
Tube 3: 5 mL 3 M HCl + 10.0 cm Mg ribbon
Tube 4: 5 mL 3 M HCl + 15.0 cm Mg ribbon

Part B Constant amount of Mg, variable amount of HCl
Tube 5: 2 mL 3 M HCl + 5.0 cm Mg ribbon
Tube 6: 4 mL 3 M HCl + 5.0 cm Mg ribbon
Tube 7: 6 mL 3 M HCl + 5.0 cm Mg ribbon
Tube 8: 10 mL 3 M HCl + 5.0 cm Mg ribbon

Materials

8 culture tubes
8 4-inch round balloons
50 mL 3 M hydrochloric acid
55 cm magnesium ribbon

Demo Tips

Fill the tubes with the proper volumes of hydrochloric acid, clearly label them by number, and place them in test-tube racks. Fold each piece of magnesium metal so it will fit inside an uninflated balloon.

Now work with one tube at a time. (The procedure that follows is one you should practice before going public with your students. The manipulations are not difficult—just a bit awkward.) Hold the tube at an angle to prevent the ribbon from prematurely falling into the acid. Carefully stretch the neck of the balloon containing the magnesium strip over the lip of the culture tube. Then stand the tube upright, so that the magnesium ribbon falls into the acid. Note the relative volume of hydrogen gas generated (as indicated by the size of the balloon) and how much of the original sample of magnesium ribbon is consumed. Repeat for each tube in the set.

Finally, display and compare all the tube-balloon systems. Invite students to interpret the results in terms of the limiting reactant concept. Ask students—after they have observed all tubes carefully—which reactant apparently limited the production of hydrogen in the Part A tubes (1-4). Then invite them to analyze the Part B tubes in a similar fashion. Also help students sense how this demonstration helps to clarify the limiting nutrient ideas developed in the unit.

YOUR TURN

Limiting Reactants: Plants and Humans (pages 240-241)

Answers to Questions:

1.   Boron is not an essential nutrient for humans.
2.   Sodium, chromium, cobalt, fluorine, iodine, nickel, and selenium are not essential for plants.
3. a.,b. Answers will vary depending on product examined.
4. a.   Nitrogen, phosphorus, potassium
   b.   It would be difficult to grow adequate quantities of food.

DEMONSTRATION IDEA

Limiting Reactants at Home

You may wish to challenge your students to try some at-home experiments on the concept of limiting reactant. Some options to explore include: (a) Varying the amount of baking powder used in making biscuits, and noting differences in the rising of the final product due to varying amounts of CO_2 produced; and (b) Adding varying amounts of baking soda to a constant vinegar/soap solution and noting the varying amounts of bubbles produced by the CO_2 given off—or keeping the amount of baking soda constant and varying the concentration of vinegar used.

C.3 PROTEINS (pages 241-246)

This section surveys the major functions of proteins, and then highlights the structures of amino acids and the reactions which produce proteins. Students also learn about complementary proteins (see Table 7 on page 245) and essential amino acids. The structures of amino acids and their conversion to peptide units may need some clarification. Consider constructing some crude "models" of amino acids, consisting of a ruler with a red tail (for carboxyl group) and a blue tail (for amino group). Point out that amino acids always combine so that red (carboxyl) reacts with blue (amino): the product still has an amino group and a carboxyl group free to react. Also point out that there are two ways to unite the amino acids, producing two different peptides.

Unique Characteristics of Proteins:

Carbohydrates and fats	Proteins
Contain C, H, and O	Also contain N (and sometimes S or P)
Excess is stored	Excess rebuilt into glucose or fat and nitrogen excreted
Built from a few basic monomers	20 different repeating units
Primary use: Burning	Building

Protein Structure: An Analogy

From a base of 26 letters thousands of unique words can be formed by varying the number, type, and sequence of letters. However, some combinations lack meaning and are gibberish (zxq or hjt). A similar argument can be made for proteins. (For instance, sickle cell anemia is caused by one mistake in a sequence of approximately 300 amino-acid units.) Given that proteins (words) contain from 50 -10,000 amino-acid units (letters), the usual accuracy of protein synthesis is remarkable—relatively few "misspellings" occur.

Just as different countries have different languages, different species have their own characteristic protein languages. In fact, individual humans each have our own tailor-made set of more than 30,000 different proteins. (Of course, similarities in protein language between humans is greater than between individual humans and other mammals, which is greater than between mammals and nonmammals). This biochemical uniqueness is most evident in the action of antibodies and the immune system.

Protein Use: An Analogy

The increased structural complexity and variety of proteins, as compared to carbohydrates and fats, is similar to the difference between petroleum and petrochemicals (remember their relative usefulnesses for building purposes). A similar argument can be made that using protein as a fuel source is like heating your house by burning dollar bills (remember Mendeleev's quote about petroleum). This is why the human body only uses protein for burning as a last resort. Only when reserves of carbohydrates and readily available fat have been exhausted does cellular metabolism turn to burning proteins.

Complete and Complementary Proteins

Most animal sources of protein (such as meat, fish, poultry, milk, and eggs) are considered complete in that they closely match the needs of the human body for all the essential amino acids. (Note: Gelatin is one of the few animal sources that is not complete—it contains almost no tryptophan and only small amounts of threonine, methionine, and isoleucine). Due to the concept of a limiting reactant, efficient use of protein intakes requires that all the essential amino acids be eaten within about a four-hour period. This can be accomplished without eating meat if one chooses nonmeat options that complement each other. In general, legumes, such as nitrogen-fixing beans and nuts—deficient in methionine, and grain (deficient in tryptophan and lysine) combinations are effective. Examples include: peanut butter and whole wheat bread, black beans or soybeans and rice, beans and corn, etc. Vegetarian diets can be quite healthy for adults, especially if dairy products are included (only concern is for adequate vitamin B_{12}). In any case, given the overconsumption of protein in the typical American diet, the presence of fat in most meat, and the negative health impact of saturated fats and cholesterol, eating less meat and more vegetables and fruits is a good idea.

Molecular Structure of Proteins (pages 243-244)

This activity provides students practice in writing structural formulas for amino acids and helps them realize how a staggering number of different protein structures can be based on a limited number of amino acids.

Answers to Questions:

1 a.

Alanine   Glycine

b. Functional groups: both contain the amino group (NH_2) and the carboxyl or carboxylic acid group (COOH).
c. Alanine has a methyl (CH_3) group in place of the hydrogen atom on glycine.

2. a.

$$H-\underset{\underset{NH_2}{|}}{\overset{\overset{H}{|}}{C}}-COOH + H-\underset{\underset{NH_2}{|}}{\overset{\overset{H}{|}}{C}}-COOH \rightarrow H_2N-CH_2-\overset{\overset{O}{\|}}{C}-NH-CH_2-COOH$$

Peptide bond

b.

$$H_2N-CH_2COOH + H_2N-\underset{\underset{CH_3}{|}}{CH}-COOH \rightarrow H_2N-CH_2-\overset{\overset{O}{\|}}{C}-NH-\underset{\underset{CH_3}{|}}{CH}-NH_2$$

Glycine   Alanine   Glycyl-alanine

$$H_2N-\underset{\underset{CH_3}{|}}{CH_2}COOH + H_2N-CH_2-COOH \rightarrow H_2N-\underset{\underset{CH_3}{|}}{CH}-\overset{\overset{O}{\|}}{C}-NH-CH_2-NH_2$$

Alanine   Glycine   Alanyl-glycine

3. a. abc, bac, cab, acb, bca, cba — 6 ($3 \times 2 \times 1$)
   b. 27 ($3 \times 3 \times 3$)
   c. 24 ($4 \times 3 \times 2 \times 1$)
   d. Millions-plus
   e. With such a vast number of arrangements, unique "genetic codings" would be possible at the molecular level.

YOUR TURN

Protein in the Diet (page 246)

Since the body cannot store excess protein, it is important that we provide the body a continual supply of these biomolecules for body maintenance "building" and "rebuilding" purposes. Here students examine and apply the Recommended Daily Dietary Allowances (RDAs) for protein, as summarized in Table 8 on page 245.

Answers to Questions:

1. 45 g of protein
2. Student answers will depend on their age and gender: for the 15-18 year old female students, 44 g; for the 15-18 year old males, 59 g.
3. a. Probably lower than 50 g
   b. With lower body weight, protein needs are less.
4. Infants develop rapidly and need protein for essential "building" purposes. Adults have already developed most of the needed body structures, and therefore need lower quantities of protein (which are devoted mainly to rebuilding and maintenance uses.)
5. a. Mother's milk
   b. Nursing (lactating) women require higher supplies of protein to produce breast milk.
6. a. Infant needs for protein were already mentioned in Question 4 above.
   b. If infants fail to receive adequate amounts of protein, early death is more likely than for adults. The old, the young, and pregnant or nursing women suffer most from inadequate protein supplies.
7. Milk formulation requires clean, safe drinking water, knowledge of proper proportions, and adequate money. Individuals in developing nations are more likely to encounter and use low-quality water supplies. Some may also elect to dilute the formulation—lowering the quantity of nutrients delivered per milliliter—to make it last longer, in an attempt to economize. With no refrigeration, more will be wasted as well.

1. Some athletes load up on protein for "extra energy" by eating steak before a competition. Is this a wise choice?

   Answer: No, the body's preferred fuels in order of decreasing use are carbohydrates, fats, and proteins. The average American consumes almost twice as much protein as is needed. The athletes would be wiser to consume pre-game meals high in carbohydrates.

2. What is the logic and myth behind high protein diet formulations?

   Answer: The logic behind trying to lose weight by eating only protein is that the body does not store excess protein (or even excess amino acids). However, after excreting the nitrogen, the body does convert excess protein by-products into fat. Another problem is that the metabolism of excess protein puts a strain on the liver and kidneys. For the most part, the $70-million-a-year protein supplement industry is supported by mistaken ideas about protein requirements and utilization.

3. Since most animal sources of protein are "complete" one could build a healthy diet by excluding other food types. True or false?

   Answer: False, an adequate diet must also contain carbohydrates, fats, vitamins, and minerals.

C.4 LABORATORY ACTIVITY: MILK ANALYSIS (pages 246-250)

This activity allows students to conduct a relatively complete analysis of the main components of skim milk. The procedure allows students to determine the percent protein, carbohydrate, and water in a milk sample, and to calculate its energy content.

    The use of nonfat milk simplifies this procedure considerably. Attempting to separate fat from whole milk is time–consuming and not very effective. Sharing class data and finding a class average will simplify the comparisons with standard values for milk at the end of this activity.

Time

Two class periods

Materials (for a class of 24 working in pairs)

Day 1
12 50-mL beakers
12 150-mL beakers
12 25-mL graduated cylinders
12 short-stemmed funnels
12 ring clamps
12 10-cm dropping pipets (medicine droppers)
12 ring clamps
12 ring stands
12 spatulas
12 dropping bottles for dispensing acetic acid
12 wash bottles
Paper towels
48 filter papers (Whatman #1, 11 cm diameter)
180 mL nonfat milk
50 mL concentrated acetic acid (approximately 17.4 M)

Day 2

Group One	Group Two
Protein Determination	**Carbohydrate/Water determination**
24 test tubes	6 evaporating dishes
6 test tube racks	6 glass stirring rods
6 100-mL graduated cylinders	6 10-mL graduated cylinders
6 250-mL beakers	6 250-mL beakers
6 ring clamps	6 ring clamps
6 ring stands	6 ring stands
6 wire gauze	6 wire gauze

6 Bunsen burners
6 dropping bottles
25 mL Molisch reagent (1.25 g 1-naphthol/100 mL 95% ethanol)
25 mL concentrated sulfuric acid H_2SO_4, 18 M
50 mL Biuret reagent (see Advance Preparation below)

6 Bunsen burners
boiling chips

Advance Preparation

Both the Molisch and Biuret reagents may be ordered as prepared solutions. The Biuret reagent may also be available from your neighboring biology teacher.

Molisch reagent should be freshly prepared. Dissolve 1.25 g 1-naphthol (alpha-naphthol) in 100 mL of 95% ethanol (ethyl alcohol).

To prepare the Biuret solution, dissolve 0.25 g copper(II) sulfate pentahydrate ($CuSO_4 \cdot 5H_2O$, fw = 249.68) in 100 mL solution, producing a 0.01 M $CuSO_4$. Also prepare 250 mL of 10.0 M sodium hydroxide (100 g NaOH/250 mL). (Store in a plastic bottle to avoid glass-etching.) Prepare the Biuret solution when needed by adding 6-7 mL of the copper sulfate solution (with stirring) to the 250 mL sodium hydroxide solution.

Pre-Lab Discussion

Caution students about using concentrated acetic acid and sulfuric acid. Encourage precision in measuring and care in following procedure steps.

Lab Tips

Remember to remove the filter paper containing the protein from all students' funnels for overnight drying at the close of the first day. Remove the filter papers from the funnels after about two hours; set them on a paper towel to dry overnight. Try not to leave the protein sample in one large clump (low surface area retards drying). Spread the protein thinly on the filter paper with a spatula (as you would butter bread). Leave the appropriate beaker and filtrate next to each filter paper.

Air drying admittedly will not remove all the water from the protein. However, the mass of water remaining is not large enough to affect the student results in any significant way. If desired, the protein could be given a final wash with ethanol to hasten drying.

Only the casein fraction of the protein is removed during the protein extraction. The whey proteins (beta-lactogobulin and alpha-lactalbumin) are not precipitated. Thus it is incorrect to claim that this procedure measures "all" the milk protein. However, since casein is the major protein source in milk, student results will still be quite acceptable.

Caution students regarding the hazards of concentrated sulfuric acid (used in conjunction with the Molisch test). Place concentrated sulfuric acid in a suitably labelled dropping bottle or reagent bottle. Instruct students to keep the dropper in a vertical position so the acid cannot run back into the rubber bulb, and to add the acid slowly to avoid splattering. The test tube should be held at a slight angle, not pointing at anyone. It is possible to use Benedict's reagent in place of the Molisch reagent for the carbohydrate tests: see your school's biology instructor for help.

It is helpful to keep the two parts of the Biuret reagent separate and add as follows: mix the protein with the NaOH solution first, and allow to stand 10-15 seconds. Then add the copper sulfate solution: often the purple color appears without any heating.

Have students add a boiling chip to their water bath (step 9, page 248).

The primary carbohydrate in milk is lactose.

Post-Lab Discussion

The Post-Lab discussion should reexamine the statement that milk is a nearly perfect food in light of student results in this activity.

Answers to Questions:

1. Students should find that their results compare favorably with the accepted values.
2. The milk sample contains other substances such as minerals and vitamins.
3. The calculations in this activity assume that the milk sample is composed of only three components. Thus, the sum should equal 100%.
4. Protein, formed in the same manner as the protein was separated from the carboyhydrates in this activity.
5. a. (8.0 g) × (9 Cal/g) = 72 Cal in fat; (8.0 g) × (4 Cal/g) = 32 Cal in protein; (11.0 g) × (4 Cal/g) = 44 Cal in carbohydrate. Total = 72 + 32 + 44 = 148 Cal.
   b. 80 Cal for nonfat milk , which is 68 Cal less than the whole milk sample analyzed. Note that this is almost entirely due to the fat content of whole milk.

PART C: SUMMARY QUESTIONS (page 250)

1. a. The proper amounts and kinds of foods are important both for building and for "burning". Certain nutrients are needed for the efficient energy-use of food. Any nutrient lacking or in short supply can become a limiting reactant.
   b. The same is true for plant nutrients.

2. a. Essential amino acids are needed by the body, but cannot be synthesized.
   b. Deficiencies in these amino acids can adversely affect body chemistry by inhibiting protein synthesis and rebuilding of body structures.

3. a. Nitric acid
   b. 80 g
   c. Ammonia
   d. 17 g left over

4. The body requires other nutrients in addition to the amino acids found in proteins. Fats, carbohydrates, minerals, and vitamins all have essential roles to play. In fact, excessive protein intake can cause health problems.

5.

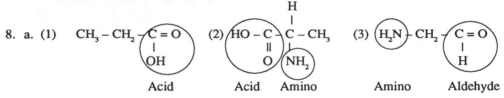

   Amino   Acid   Amino   Acid   Amino   Peptide   Acid

6. a. 44 g (Students answers may be slightly higher)
   b. 23.5 g/16 oz = 44 g/x = 30 oz (May vary depending on answer a)
   c. 8.03 g/1 C = 44.g/x = 5.5 C (May vary depending on answer a)

7. a. Complete protein contains all of the essential amino acids, incomplete does not.
   b. One needs all the essential amino acids to continue protein synthesis in the body. Complementary proteins, such as nuts and grains, should be included in the diet.

8. a. (1) $CH_3 - CH_2 - C = O$ with OH (2) $HO - C - C - CH_3$ with O, NH_2, and H (3) $H_2N - CH_2 - C = O$ with H

       Acid   Acid   Amino   Amino   Aldehyde

   b. (2) is an amino acid.

D: OTHER SUBSTANCES IN FOODS

Part D discusses the roles of vitamins, minerals, and intentional food additives—substances whose combined daily intake is less than several grams. Students learn why vitamins and minerals are essential to the diet. They learn that vitamins function to catalyze biochemical reactions at body temperatures; about the importance of the RDAs; and how to experimentally determine the amount of vitamin C in certain beverages. Similarly, through text and laboratory investigation, the functions of minerals in biochemical catalysis and in body structure are investigated. Finally, students briefly examine the benefits and risks of some common intentional food additives.

D.1 VITAMINS (pages 251–254)

This section surveys sources of vitamins, their contributions to human well-being (see Table 9 on page 252) and the effects of too many and too few. The notion of water-soluble and fat-soluble vitamins helps to explain health problems resulting from "too many". This section also discusses the role of vitamins in the catalyzing important chemical reactions in the body and examines the RDAs for selected vitamins.

YOUR TURN

Vitamins in the Diet (page 253)

This activity allows students to apply their knowledge of vitamins to several dietary concerns. It also provides practice in interpreting tabulated data (Table 10 on page 252), and in using such data to solve practical problems.

Answers to Questions:

1. a. B_{12} (cobalamin) deficiency results in pernicious anemia and exhaustion; D (calciferol) deficiency results in rickets.

    b. By taking vitamin supplements

2.

	RDA		Servings	
	B_1	C	B_1	C
Green Peas	1.5	60 (male)	3.88	1.03 (male)
	1.1	60 (female)	2.84	1.03 (female)
Broccoli	1.5	60 (male)	25.9	0.73 (male)
	1.1	60 (female)	19.0	0.73 (female)

   a. Vitamin B_1

   b. Since there is no perfect food, variety in the diet assures that the proper kinds of vitamins are consumed. Green peas are a better source of B_1 than broccoli, but not as good a source of vitamin C.

   c. Even if the caloric intake is adequate, vitamin deficiencies from the lack of variety in foods will limit body building and burning processes.

3. Fat-soluble vitamins: they can be stored and thus accumulate in the body, while excess water-soluble vitamins are more likely to be excreted.

4. a. Water-soluble vitamins.

   b. Long cooking can dissolve and often destroy these vitamins from vegetables. Raw or lightly steamed vegetables retain maximum concentrations of vitamins.

OPTIONAL BACKGROUND INFORMATION

Vitamins are indispensable, noncaloric, organic compounds needed in small amounts (less than 100 mg/day) in the diet. They serve as coenzymes in catalyzing a wide variety of biochemical reactions. For a substance to qualify as a vitamin, its absence must cause a specific deficiency disease that is cured when the substance is resupplied. There are 13 known vitamins required by humans: four are fat soluble (A, D, K, and E) and nine are water soluble (C and eight B-complex). A variety of mistaken notions surround the characteristics and use of these vitamins:

1. Despite their name, many are not amines, and they do not supply energy (they are non-caloric), meaning they provide no "vitality."

2. There is no reliable scientific evidence to support the notion that vitamin supplements have beneficial effects for the average person eating a balanced diet. Vitamin deficiency is rare unless a person's diet is extremely unbalanced. Yet, a $3 billion a year industry has as many as 40% of all adult Americans buying vitamin or mineral supplements. Many advertisements are designed to inspire insecurity and are often based on the deception that "if some is good for you, more is better."

    Actually, since they act as catalysts, vitamins are reused by the body many times before being broken down or excreted. The small amounts needed to replace such lost vitamins are more than accounted for in RDAs that have been set high intentionally to account for individual differences (beyond age and sex)—many persons actually do not need the full 100% of the RDAs.

3. Many people believe that vitamin supplements are needed when they are under emotional stress and "can't hurt" at other times. Studies that have indicated an increased need for vitamins have identified stress as physical— burns, extreme heat/cold, infection, injury/bone fractures, radiation, and other traumas—not psychological. Most nutritionists prefer the balanced nutrient intake provided by a varied diet to popping pills. If people choose to take supplements as an added "insurance" along with a balanced diet, nutritionists recommend products that contain approximately 100% of the RDAs for a wide range of vitamins and minerals. Taking excessive amounts of one or more nutrients can adversely affect the body's absorption, utilization, and excretion of others. (As an example, high zinc intakes seem to impair copper's function, and many Americans get too little copper anyway.) Also, especially for the fat-soluble vitamins (but also true for megadoses on the order of $100 \times$ RDA of the water soluble ones), high doses can have serious toxic effects.

4. Advertisements sometimes promote the idea that natural vitamins (such as the vitamin C in rose hips) are better than synthetic ones. Actually, individual cells cannot tell the difference between the two. From a chemical point of view, vitamin C is vitamin C. Yet, people will pay much more for the "natural" pill when the only natural source that nutritionists prefer is a balanced diet over pills.

5. Tables of nutrient values do not tell the whole story on food quality. The actual vitamin content of a given food depends to a large extent on how the food was processed, stored, and made ready for eating. In general, the fresher and less processed the food, the higher its vitamin content. In particular, fresh (from the garden) fruits and vegetables are slightly better than frozen ones, which are significantly better than canned ones. In cooking, the less time and/or water used the better. Steaming, pressure cooking, or microwaving are preferred over boiling in water which can destroy some fat-soluble vitamins and leach out many of the water-soluble ones. If the latter method is used, nutritionists recommend using the drained-off water in making soups and gravies.

Note: The text contains no structural formulas for the vitamins. You may wish to display some of these on an overhead transparency and ask the students to identify differences between the water-soluble and fat-soluble ones. They should be able to see that the fat-soluble vitamins (A, D, E, and K) contain a larger nonpolar hydrocarbon component, with only one or two oxygen atoms. In contrast, the water-soluble vitamins (B complex and C) contain more electronegative oxygen and nitrogen atoms (as hydroxy or amine groups), which make the molecule polar and allow for hydrogen bonding (and solubility) with water. The ability of the body to store fat-soluble vitamins can be related back to the earlier "like dissolves like" principle.

OPTIONAL STUDENT REPORTS

1. In developing nations, vitamin A deficiency (xerophthalmia/night blindness) is second only to general malnutrition in its negative effects. Investigate this deficiency (its cause, symptoms, groups and locations most affected, and efforts being made to combat it) and/or other vitamin/mineral deficiencies (such as: beriberi and pellagra/vitamin B complex, scurvy/vitamin C, rickets/vitamin D, anemia/iron, and endemic goiter/iodine).
2. Examine the controversy surrounding health claims made about vitamin C and colds (or cancer), or vitamin E and breast lumps, or vitamin B_6 and premenstrual syndrome, etc.

D.2 LABORATORY ACTIVITY: VITAMIN C (pages 254–256)

This activity demonstrates how small amounts of vitamins in foods can be detected and analyzed. Students titrate a known amount of vitamin C to obtain a conversion factor used to calculate the amount of vitamin C contained in several beverages.

Time

One class period

Materials (for a class of 24 working in pairs)

12 150-mL beakers
12 25-mL graduated cylinders
12 125-mL Erlenmeyer flasks
12 ring stands
12 buret clamps
12 burets
12 dropping bottles for dispensing 1% starch solution
450 mg vitamin C (ascorbic acid)
200 mL 1% starch solution (2 g soluble starch/200 mL distilled water)
1 L iodine (I_2) solution (see Advance Preparation)
5 beverage samples

Advance Preparation

To prepare the iodine solution, dissolve 10.0 g potassium iodide (KI) in 250 mL of 0.01 M potassium iodate (KIO_3) (fw = 214.0). Transfer the solution to a 1-L graduated cylinder. Add 60 mL of 3 M sulfuric acid. Then add enough distilled water for a final volume of one liter. Mix thoroughly. (Since students will base their "conversion factor" on the iodine solution as actually prepared, the iodine solution need not be prepared with high precision.)

The reaction in the iodine solution:

$$IO_3^- + 5 I^- + 6 H^+ \rightarrow 3 I_2 + 3 H_2O$$

The 1% starch indicator may be ordered directly from a chemical supplier or easily prepared from solid soluble starch. Prepare a thin paste composed of 2 g of soluble starch and enough room-temperature distilled water to make a "pourable" suspension. Bring 200 mL of distilled water to boiling. With stirring, pour the starch suspension into the boiling water. Allow to cool, then place in dropping bottles.

The vitamin C standard for the conversion factor can be obtained in one of two ways: (1) prepare a fresh stock solution of 0.500 g ascorbic acid per 500 mL solution, and dispense 25.0 mL aliquots to the teams as suggested in the student text; or (2) pre-weigh 25.0 mg solid samples of ascorbic acid needed by students (to avoid student errors).

Pre-Lab Discussion

The Pre-Lab discussion may center around the small quantities of vitamins needed daily by the body (RDAs) and the common function of vitamins as catalysts (or catalyst-helpers) in biochemical reactions. You may also decide to review the health hazards of vitamin C deficiencies. This is also a good place to associate the water-solubility of some vitamins such as vitamin C with the actual solvents selected for use in this activity.

If you elect to dispense the student samples of ascorbic acid via 25.0 mL aliquots from a stock solution (see Advance Preparation), then display the original ascorbic acid solute as part of the Pre-Lab discussion, so they associate the pure substance with solid rather than with liquid.

Demonstrate the technique of titration, with emphasis on the actual end-point work: one drop of iodine solution should cause the color change.

Lab Tips

The activity is more interesting if the beverages selected vary considerably in vitamin C content. Group beverages tested into five major categories. Select any five beverages for the activity, one from each category, if possible.

Vitamin C content	Beverage
Very high	Pineapple or orange juice
High	White grape juice, grapefruit juice, or High C
Medium	V-8 juice
Low	Apple juice, milk, or Tang
Very low or none	Gatorade, Sprite

Beakers with stirring rods may be substituted for the Erlenmeyers if there is a shortage.

The theoretical conversion factor expected in Part 1 (for your information only; let students determine their own values) is 1.32 mg vitamin C per mL of iodine solution. You can verify this for yourself by working through the equation given for the solution preparation, and noting (from the equation provided) that vitamin C reacts with iodine in a 1:1 mole ratio. If you actually complete these calculations you will appreciate the much more direct student calculations specified in the procedure.

You may wish to do one sample calculation of the type students are asked to complete in their data table.

Even if some students are not too adept at titrating to the starch endpoint, their data will still allow them to rank-order the beverages quite reliably in terms of vitamin C levels.

To simplify endpoint determination it is helpful to choose beverages of the same color. For example, you may use Tang, orange Kool Aid, orange Gatorade, orange juice, and orange soda.

Post-Lab Discussion

The Post-Lab discussion could center around the choice of drinks needed to acquire the recommended RDA for vitamin C. You can initiate a discussion of possible sources of error by displaying and comparing the variability of individual teams results regarding the same beverage.

Answers to Questions:

1.a.,b. See Lab Tips for the general levels of vitamin C found in recommended beverages. (What surprises one student may not surprise another—various responses are possible here.)

2. Among others: pineapple juice, white grape juice, Hi-C, and Hawaiian Punch.

Materials (for a class of 24 working in pairs)

12 13 × 100 mm test tubes
12 Beral pipets containing vitamin C solution (0.100 g in 100 mL water. *Prepare Fresh*)
12 Beral pipets containing iodine solution (dissolve 1.0 g of potassium iodide in 25 mL of 0.01 M potassium iodate solution -- 0.5 g KIO_3 in 250 mL water). Add 6 mL of 3 M sulfuric acid, then add water to make total volume 250 mL. The iodine pipets cannot be cleaned, so use the same pipets for every class.
12 Beral pipets containing 1% starch solution (1 g of starch in 100 mL water: see Advance Preparation section for activity on preparing the solution).
Beral pipets containing juice samples.

Procedure

1.   Place 25 mL of vitamin C solution into the test tube. Hold the dropper vertically to obtain drops of uniform size.
2.   Add one drop of starch solution.
3.   Add one drop of iodine solution and tap to mix. Continue adding iodine solution, one drop at a time with agitation, until a permanent blue to blue-black color appears. Record the number of drops of iodine solution used.
4.   Clean test tube, and add 25 drops of juice sample in place of vitamin C. Add one drop of starch solution, and titrate with iodine as in step 3. Record the number of drops of iodine solution used.
5.   Perform the following calculations:
     a. Conversion factor = 25/number of drops of iodine used for vitamin C titration.
     b. Relative mg of vitamin C = (Conversion factor) × (number of drops of iodine used for juice titration)
        Sample: 25/11 drops = 2.27 (conversion factor)
        3 drops × 2.27 = 6.81 mg equivalent per 25 drops juice.
6.   Calculate the mg vitamin C for each beverage tested.
7.   Rank your samples by number, from 1 for highest to 5 for lowest.

Consider using the nutritional scorecards, films, etc. available from such groups as: Center for Science in the Public Interest, Chemical Manufacturers Association, Food and Drug Administration, National Dairy Council, Society for Nutrition Education (see beginning of Food unit, page 107 for their addresses).

D.3 MINERALS: PART OF OUR DIET (pages 256–259)

This brief section emphasizes the role of minerals as cofactors in biochemical catalysis and as component parts of the body's structure. Minerals are classified as macrominerals and trace minerals. Health problems occur with either too much or too little of any needed mineral—once again an issue of "what kind" and "how much." Students examine recommended amounts of several different minerals in the following *YOUR TURN*.

Minerals in the Diet (page 258)

Students are asked to interpret the RDAs for various minerals in terms of their availability in various foods, and in terms of preventing mineral-deficiency health problems.

Answers to Questions:

1.   19 slices of bread for females, 15 for males.
2.   4.2 cups of milk
3. a.   Fe; Ca: 2.3% for males and females, Fe: 3.3% for males 2.7% for females
   b.   Divide the amount given by RDA × 100.
4. a.   438 g of each.
   b.   Bones and teeth contain large amounts.
   c.   Ca: milk, dairy products; P: animal protein.
   d.   No.

e. Ca: rickets, osteomalacia, and osteoporosis.
   P: no deficiency
f. Yes; infants and women.
5. a. Prevention of goiter.
   b. Seafoods.

Minerals, like vitamins, are present in foods in relatively small amounts (compared to carbohydrates, fats, and proteins), and their absence in a diet can lead to specific deficiency diseases. However, they are different from vitamins in a variety of ways:

1. They are found in food as relatively simple inorganic compounds (vs. the more complex organic vitamins).

2. The mineral content of a given food is dependent more upon the kind of soil from which it was derived (the vitamin content is based more on the genetics of the plant).

3. Most minerals (especially the trace elements) have smaller safety ranges than do vitamins. Essential minerals are more likely to become toxic if taken in abnormally large doses. As with vitamins, one is mistaken in following the "if some is good, more is better" philosophy.

4. Minerals may or may not serve a catalytic role. Several of the macrominerals serve a structural function in the body.

5. Whereas vitamin deficiencies are rare among individuals who eat a reasonably balanced diet, a fair number of Americans may need to supplement their diet to obtain sufficient quantities of several minerals. Specific examples include menstruating, lactating, or pregnant women and growing, adolescent boys who need to be concerned about their intake of iron (among college-age women as many as 25% may have depleted iron stores and some symptoms of anemia) and calcium (osteoporosis, a thinning of the bones, is found in 25% of all postmenopausal women due to earlier deficiencies).

6. In some areas, food provides unreliable sources of some minerals; specifically, fluorine (if water has not been fluoridated) and iodine (which is why salt is iodized).

7. The typical American's intake of one mineral, sodium, is unhealthfully high. Health specialists recommend that intake be less than 5 g (1 teaspoon) sodium chloride equivalent per day, but the typical American consumes between 6-18 g/day. Such high salt intakes have been linked to high blood pressure (in the United States about 1 in 6 persons have this), hypertension, heart disease, and strokes. A variety of strategies can be used to lower sodium intake: (a) eat fewer foods with salt on the surface (potato chips, french fries, pretzels, salted nuts, crackers); and (b) reduce use of processed foods which often contain relatively large quantities of "hidden salt." For example, 100 g serving of fresh peas contain 0.9 mg sodium, frozen peas contain 100.0 mg, and canned peas contain 230.0 mg. In the typical American diet, the top "offenders" include: white bread, rolls and crackers (12.1% of dietary sodium); hot dogs, ham and lunch meats (9.8%); canned or powdered soups (6.6%); cheeses (5.4%); and potatoes (5.1% as french fries, hash browns, etc.).

1. Beyond the pleasure of eating/cooking, why is it unlikely that a pill will ever replace a balanced diet?
   Answer: Even discounting the need for sizable quantities of protein, carbohydrates/fiber, and fats, the RDA for all the essential vitamin and minerals will not fit into a pill that can be swallowed (for instance, the RDAs for Ca, P, and Mg range from 400-1200 mg). Also, some elements known to be essential in trace amounts do not have RDAs at the present time (examples: Se, Cr, Mo, and Mn). Additionally, food may be found to contain yet further essential (though trace) elements.

2. If an adult who had previously had a healthy diet was forced to subsist on a diet too low in calories, which vitamins (fat or water soluble) would most likely be the first to show up as a deficiency?
   Answer: Water-soluble vitamins. Fat-soluble vitamins can be stored up in the body and these reserves drawn on at a later time, preventing the onset of deficiency symptoms.

D.4 LABORATORY ACTIVITY: IRON IN FOODS (pages 259–260)

This activity demonstrates how the mineral content of foods can be determined. Students determine the relative levels of iron in several common foods.

Time

One class period

Materials (for a class of 24 working in pairs)

24 crucibles, porcelain
24 clay triangles
12 ring stands
12 ring clamps
12 Bunsen burners
12 tongs
12 test tubes
12 funnels
12 50 mL beakers or larger
12 rubber stoppers
2 burets for dispensing HCl and KSCN
30 filter papers (Whatman #1, 11 cm diameter)
200 mL 2 M hydrochloric acid (HCl) (33 mL conc. HCl/200 mL solution)
200 mL 0.10 M potassium thiocyanate (KSCN) (fw=97.18) (1.9 g/200 mL)
10 mL 1.0% Fe^{3+} ($FeCl_3$) stock solution for preparing Fe^{3+} color standards
Several 2.5-mg samples of foods (see Lab Tips)

Advance Preparation

Sets of six color standards for a range of Fe^{3+} levels should be prepared for this activity. This procedure will provide you with a set of color standards for each bench: Dissolve a 0.50-g sample of iron(III) chloride hexahydrate in 9.5 mL of distilled water, producing a 1.0% Fe^{3+} standard. Continue following the serial dilution flow chart provided below. Be sure to mix the solution in each beaker thoroughly before continuing on to the next.

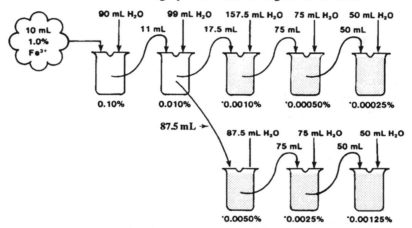

The shaded solutions shown in the flow chart above become your six color standards, providing $(FeSCN)^{2+}$ color intensities equivalent to original Fe^{3+} levels of 0.0050%, 0.0025%, 0.00125%, 0.0010%, 0.00050%, and 0.00025%. Pour 5 mL of each shaded solution into a test tube and add 5 mL of 0.10 M KSCN. Invert each tube once to mix. Note that you have enough of each color standard to prepare up to 20 sets of these test-tube-size standards. Label and place a rack of these standard tubes at each bench.

The standards should include the same volume of solution in the same size/type test tube as the students will be using in order to make optimal comparisons. If a light table is available, you may wish to set up a set of standards there.

Pre-Lab Discussion

The Pre-Lab discussion should include the use of color-intensity methods in quantitative analysis, the reaction that produces the iron(III) complex in this activity, the small amounts of iron in foods, and associated health problems accompanying iron deficiency.

OPTIONAL DEMONSTRATION

To highlight the presence of iron in foods, you may try the extraction of iron from breakfast cereals such as Total: having a magnetic stirrer makes this demonstration a breeze. Fill a 1000 mL beaker with dry cereal, add water and a magnetic stirring bar, and stir. After the mixture is suitably mixed, remove the magnet: iron filings should be present on the stirring bar. You might wish to discuss with your students the ability of the body to absorb elemental iron, perhaps by adding the stirring bar to a beaker of dilute HCl.

Lab Tips

We suggest that food samples be assigned as follows, to insure good variety in the iron levels determined. Half the teams should work with parsley and raisins. The other teams should be assigned spinach and cauliflower. Use fresh-produce samples (the raisins, of course, will be boxed). Either broccoli or canned kidney beans could be substituted for the raisins or cauliflower.

Food samples should be chopped into small pieces; whole beans and raisins produce a hard product (ash) which is difficult to dissolve.

Some of the foods may produce a fair amount of smoke. This can be controlled by leaving the lid over the crucible until the food is charred black (about two minutes). Then, remove the lid until the food has completely turned to ash.

Dry breakfast cereals provide a simple, and cleaner, alternative to the food samples listed in the textbook. You may consider having the students use the following (iron content is for one ounce of cereal:

Brand of Cereal	milligrams iron
Total	18.0
Product 19	18.0
Bulgar	9.5
Kellogg's 40% Bran Flakes	8.1
Cap'n Crunch	7.5
Cheerios	4.5
Post 40% Bran Flakes	4.5
Fruit Loops	4.5
Kellogg's Raisin Bran	4.5
Kellogg's Corn Flakes	1.8
Rice Krispies	1.8
Kellogg's Frosted Flakes	1.8

Color standards may fade somewhat overnight (especially if not placed in a dark area).

A more elegant (and precise) analysis procedure can be followed if you have access to a spectrophotometer. With such equipment an absorbance-Fe^{3+} working curve can be prepared from the color standards. Students then determine the absorbance of their sample and can read the % Fe^{3+} directly from the standard curve.

You may want to remind students of the Bunsen burner adjustments that produce a very hot "bluish" flame. Demonstrate the proper technique for handling crucibles, and caution students about their fragility.

Post-Lab Discussion

The Post-Lab discussion may include a summary of class results and the ordering of the foods according to their iron concentrations.

Answers to Questions:

1. No, the iron(III) concentration values found by color-comparison in the solutions do not represent the percent iron in the original food samples—they just represent the percent iron in the water solution.
2. See student values from Post-Lab summary. For a comparison with accepted values, see the Appendix beginning on page 523.
3. Macrominerals such as calcium and sodium, and trace minerals such as chromium, zinc, and fluorine.

D.5 Food Additives (pages 260-265)

This section briefly surveys the benefits (and some associated risks) involved in modern use of food additives. Legislation has been passed to protect food quality in the United States, and students are introduced to the Delaney clause, as well as to case studies of two relatively familiar additives, nitrites and aspartame. The optional information given is valuable when debating the overall concept of additives in foods.

Nitrite Additives (page 264)

This *ChemQuandary* probes the benefits and risks of nitrites as food additives. Here are some possible questions to give some sense of what is expected of students.

Answers:
1. How conclusive is the evidence that nitrosamines cause cancer in laboratory animals and in humans?
2. What is the evidence suggesting that the reactions of nitrites with stomach acid, and nitrites and nitrous acid with secondary amines, occur to any measurable extent?
3. What are some possible substitutes for sodium nitrite as a preservative?
4. Have substitutes been used? What are the risks involved with them?

OPTIONAL BACKGROUND INFORMATION

Labeling, Additives, and "Chemicals" in Food

1. The student text briefly outlines the requirements of the Fair Packaging and Labeling Act. Beyond legislating what must be on a food label, this act also specifies what *cannot* be on a label. Specifically, a label may **not** claim that: (a) The food is an effective treatment for disease; (b) A normal, balanced diet is nutritionally insufficient; (c) Soil or other factors may be responsible for lowered nutritional quality; (d) The product delivers unsubstantiated dietary benefits; and, (e) Natural vitamins are superior to synthetic. Unfortunately, advertising of health foods and articles and books are not so well monitored for nutritional "lies", and the public is often misled in matters of nutrition.

2. The use of food additives raises once again the concept of risk vs. benefit. To set the risk in perspective, consider that in 1975 the Commissioner of the Food and Drug Administration listed food additives **last** on his "worry list" about food quality. In descending order of importance the factors were: food-borne diseases (which kill some 9000 people/yr), nutritional quality, environmental contaminants, naturally occurring toxins, pesticide residues, and intentional food additives. Even additives that have been banned (such as the cyclamate sweeteners, and FD&C Orange dye #1, Yellow dyes #3 and #4, and Red dye #2) are believed to have presented an extremely low risk. However, the policy of eliminating specific additives that have little or no health benefit and likely health risks, such as certain coloring agents or sweeteners, is probably wise and certainly more responsible than calling for a ban on all "chemicals" in food.

3. To help students overcome a fear of "chemicals" in food and an exaggerated sense of confidence in the "natural," consider sharing the following information:
   a. About 700 "natural" plants found in the Western hemisphere have been reported to cause death or serious illness. Some of the more infamous examples include: hemlock, mistletoe berries, wild *Amanital* mushrooms, apricot pits, and rhubarb leaves.
   b. A freshly picked strawberry contains (besides many other things) acetone, acetaldehyde, methyl butanoate, ethyl caproate, hexyl acetate, methanol, acrolein, and crotonaldehyde—all "natural" substances and yet every one is poisonous. The point is that when considering risk, one must consider dose—it is perfectly safe to eat strawberries in any reasonable quantity. Other examples include: carcinogenic safrole, which occurs naturally in cinnamon and nutmeg; aflatoxins, which are potent natural carcinogens (as much as 10 million times more potent than saccharin) produced by molds on nuts and grains. The FDA has set a limit of 20 ppb for aflatoxins in these foods.
   c. The Chemical Manufacturers Association publishes three free menus (breakfast, lunch, and dinner) that list a sampling of the natural chemicals found in typical meals. The point is that all foods are a mixture of chemicals, many of which would be poisons in larger amounts, and that the addition of certain chemicals to improve quality or safety is not necessarily "unnatural."
   d. Although about 2800 different intentional additives are presently being used, "natural" sugar, salt, corn syrup, citric acid, baking soda, vegetable colors, mustard, and pepper make up *over 98%* (by weight) of all additives used.
   e. A truly natural, organic food product is one that is fertilized with manure and/or compost, grown without pesticides, and processed without additives. Typically such products cost at least 50% more than those grown and processed with modern chemical technology, yet have essentially the same nutritional quality. The movement of the population away from farms, the increased desire for convenience foods, and competition for sales have all contributed to the increased use of additives and make "organic" foods an option for only a limited segment of the population.

D.6 YOU DECIDE: FOOD ADDITIVE SURVEY (pages 265-266)

The purpose of this activity is to make students more aware of food additives found in common products that they consume everyday. Students survey 10 packaged foods, identifying and describing the purpose of 10 of the additives from information found in Table 14 on page 261. This activity will take about 30 minutes to complete. You may want students to collect the raw data as homework. In discussing the activity, you may wish to develop these points: (1) the place of additives on the label; (2) which additives show up most frequently, and in what types of items; and (3) which additives could be eliminated without associated health risks.

OPTIONAL STUDENT REPORTS

1. Research the benefits and risks associated with various sugar substitutes: cyclamates, saccharin, aspartame, and polyhydroxy alcohols (sorbitol and xylitol).
2. Research the controversy surrounding the use of antibiotics in animal feed and their subsequent occurrence in foods as an incidental additive/contaminant. As much as 50% of the 9 million kg of antibiotics produced annually in the United States are used in animal feed. Note that humans consume incidental additives involuntarily without knowing it.

OPTIONAL CHEMQUANDARIES

1. Maraschino cherries are bleached with sulfur dioxide and then dyed with an organic food coloring. What are the benefits and risks of maraschino cherries relative to the "natural" product? What problems would a manufacturer face in switching to the natural?

   Answer: The benefits are cosmetic and esthetic—people have come to expect such cherries to be bright red. The risks include the possibility of a reaction in people sensitive to sulfur dioxide and the possibility that, at some future date, the particular red dye used may come under suspicion as a health threat. A manufacturer would need to change people's image of what a good cherry is supposed to look like.

2. Would you expect to find golden raisins in an organic food store? Why or why not?

   Answer: Probably not, since they have been bleached with sulfur dioxide.

3. Why does it make sense to ban additives that have been shown to possibly cause cancer when carcinogens occur naturally in food?

   Answer: Although risks from either may be quite small, the former is an avoidable risk, while the latter is not. In general, the greater the exposure to carcinogens, the more likely one will develop cancer. The human body seems to have mechanisms for dealing with low levels of natural carcinogens, but it does not make sense to risk overloading these mechanisms by intentionally adding known, synthetic carcinogens to foods.

OPTIONAL DEMONSTRATION

Prepare 100 mL of approximately 0.2 M sodium sulfite (2.5 g Na_2SO_3/100 mL). Dip a fresh slice of apple into the sulfite bath for one minute. Remove it and place it on a towel near an untreated "control" apple slice. Continue with your lesson and view the slices later in the class period. The treated slice will continue to look fresh long after the control slice begins to darken. Ask students which apple slice they would be more likely to choose at a salad bar (assuming they were unaware of the sulfite treatment).

OPTIONAL LABORATORY ACTIVITY: SULFITING AGENTS

Many raw foods rapidly discolor or otherwise lose their natural appearance when exposed to air. Sulfiting agents have been found to be quite effective in slowing these oxidative processes. However, recent medical findings have linked sulfiting agents to adverse reactions in sensitive individuals, primarily asthmatics. Such individuals have had trouble breathing, and a few have gone into shock after eating sulfite-treated foods. Also, the FDA is concerned about possible long-term effects of sulfite exposure on all consumers.

In this activity you will determine the relative concentration of sulfiting agents in a number of food products. The key reactions involve the conversion of sulfite into sulfate by reaction with hydrogen peroxide:

$$SO_3^{2-} + H_2O_2 \rightarrow SO_4^{2-} + H_2O$$

The sulfate is precipitated by reaction with barium chloride:

$$Ba^{2+} + SO_4^{2-} \rightarrow BaSO_4(s)$$

The relative amounts of sediment and turbidity thus give an indication of the quantity of sulfiting agent that was originally present in the food.

Time

One class period

Materials (for a class of 24 working in pairs)

12 glass stirring rods
12 funnels, short stemmed
12 test tubes
12 test tube racks
12 rubber stoppers
12 10-mL graduated cylinders
24 150-mL beakers
2 burets
12 dropping bottles for dispensing $BaCl_2$ solution
20 filter papers (Whatman #1, 11 cm diameter)
1 L 3% hydrogen peroxide (H_2O_2) solution
400 mL 3 M barium chloride (saturated—see Advance Preperation) (250 g $BaCl_2 \cdot 2H_2O$/400 mL)
Assorted fruit (see Lab Tips)
Distilled water

Advance Preparation

Prepare the saturated barium chloride solution by placing the barium chloride in a large (one-liter) beaker and adding distilled water to the 400 mL mark. Stir vigorously and allow the remaining solid to settle. Pour the clear supernate into dropping bottles.

The 3% hydrogen peroxide solution can be ordered directly or prepared from the available 30% hydrogen peroxide reagent. Place 40 mL of the 30% hydrogen peroxide reagent in a 600-mL (or larger) beaker. *Caution—handle this reagent with care; avoid spilling on hands or clothing.* Then, with stirring, add distilled water to the 400-mL mark.

Pre-Lab Discussion

The Pre-Lab discussion can center on the use of sulfiting agents. Primary uses of this additive are in bakery dough (as a conditioner), in preserving fresh fruits and vegetables, in some salad bars, in dried fruits and vegetables (prunes, dried potatoes, raisins), in peeled or cut fresh potatoes, and in seafood for display.

Lab Tips

Buy a bag of mixed, dried fruit (apricots, pears, prunes, apples, and apricots), a small jar of red cherries, a box of dates, and a box of dried potato slices (found in au gratin and scalloped potato products). The dried apricots, dried pears, and dried apples have higher sulfite values than cherries, dried prunes, and scalloped potatoes. You may wish to use five different samples.

Since it is possible that some of the food products may also contain sulfate, this may also contribute to the cloudiness observed in the procedure, due to $BaSO_4(s)$. It may be worthwhile to extend this activity by running controls to judge the amount of the turbidity which actually can be attributed to sulfite in this activity. The only change needed in the procedure would be the addition of a control test—just substitute an equal volume of water for the 3% hydrogen peroxide. You might like to see if students are able to devise such an extension to the procedure on their own.

The major problem with sulfites in the consumer market has been with fresh salads, where some restaurants "overdo" the sulfiting treatment and fail to wash the leaves adequately before use. The level of sulfite on dried food products is controlled by FDA standards.

Procedure

Follow this procedure for each of the food products used.

1. Place several pieces of the supplied food product in a 150-mL beaker and add distilled water to the 40-mL mark.

2. Allow the food to soak for 15 minutes. As the food soaks, occasionally press the pieces with a stirring rod to force out any trapped water or juice.

3. Place a clean, rinsed short-stemmed funnel with filter paper into the top of a clean test tube supported in a test tube rack.

4. Pour the juice into the funnel. Collect enough filtered juice so that the test tube is about one-quarter full. Replace the test tube with a beaker to collect the remaining liquid, which can be discarded.

5. Add enough 3% hydrogen peroxide solution to the test tube to double the total volume. Place the cork stopper in the test tube and invert once to mix.

6. Add 5 mL 3 M barium chloride solution to the test tube and mix.

7. Estimate the relative level of sulfite by the degree of turbidity observed in the tube. You may wish to store the test tube overnight to better judge the amount of sediment present.
8. Compare your results with other teams and rate the food products from highest to lowest relative to their sulfite levels.

Questions

1. How important is the appearance of a food product to your willingness to buy it or eat it? More specifically, would you be prepared to buy off-color fresh or dried fruit if you knew they did not contain sulfiting agents?
2. Examine the labels of packaged food products at home. Identify any that contain one or more sulfiting agents.
3. Should restaurants be required to list the use of sulfiting agents in salad bars and foods on their menus?

Post-Lab Discussion

The Post-Lab discussion can involve the risks and benefits of sulfiting agents, relating the comments to the ranking of the foods tested.

PART D: SUMMARY QUESTIONS (page 266)

1. a.   Water solubility: you need more water-soluble vitamins on a daily basis.
   b.   Toxicity: too high a dose of fat-soluble vitamins can be toxic.
   c.   Cooking precautions: over cooking can break down fat-soluble vitamins and leach out water-soluble vitamins.
2.   Macrominerals are needed in larger amounts, but microminerals are no less important. If a diet is deficient in a micromineral, it can act as a limiting reactant. Macrominerals: calcium, phosphorus, potassium, chlorine, sulfur, sodium, magnesium. Trace minerals: fluorine, chromium, manganese, iron, cobalt, copper, selenium, zinc, molybdenum, iodine.
3. a.   Calcium: dairy and eggs
   b.   Iron: meats
   c.   Vitamin C: fruits and vegetables
   d.   Vitamin A: fruits, vegetables, and dairy
4. a.   Orange juice—0.5 cup
   b.   Milk—26 cups.
5.   Some food additives are natural substances, such as vitamin C and minerals needed by the body. All foods may be viewed as chemicals.
6. a.   Least essential—coloring agents. Has no impact on nutritive value.
   b.   Most essential—preservatives that prevent bacterial growth. Prevent the spread of disease and illness.
7.   If the benefit is deemed high, and the risk is judged to be low, and no better substitute is available, then the wisest course is to continue to use the additive.
8. a.   Delaney Clause— "No additive shall be deemed safe if it is found to induce cancer when ingested by man or animal."
   b., c.   Some students may try to insist that this clause is based on a weakly established assumption: that if a substance causes cancer in test animals in high doses within a short time period, it will also cause cancer in humans at low doses over a longer period of time. But if an error needs to be made concerning this assumption, we probably should do so on the side of caution—hence the Delaney Clause.

E: PUTTING IT ALL TOGETHER: NUTRITION AROUND THE WORLD

The final part involves an analysis of the food diaries completed earlier, using nutritional data given in the textbook, and a comparison of different diets from around the world. The often-surprising conclusion is that many people in the United States, despite an abundance of food and choices of food stuffs, tend to be undernourished or malnourished.

E.1 YOU DECIDE: DIET ANALYSIS (pages 267-268)

In this activity students apply their newly-acquired knowledge of the chemistry of foods to analyzing their own diets. The objective is for students to evaluate the adequacy of their daily diet in terms of Calories, protein, iron, and vitamin B_1. This activity will take one class period.

    The diet analysis is often quite revealing. Respect the confidential nature of this material, and allow students to discover the sometimes inadequate nature of their own diets.

Sources of More Extensive Nutrient Tables

1. *Food Values of Portions Commonly Used* (14th ed.) by Jean A. T. Pennington & Helen Nichols Church. Harper & Row Publishers, 1985. The complete nutrient content—calories, cholesterol, salt, fat, vitamins, and much more—of all the foods you eat.
2. *Nutritive Values of Foods*. Home & Garden Bulletin No. 72. United States Department of Agriculture, 1981. Available from the United States Government Printing Office.
3. *Nutrition: Concepts and Controversies* (2nd ed.) by E. M. N. Hamilton and Eleanor Noss Whitney. West Publishing Co., 1982. Excellent nutrition textbook with valuable appendices.

Follow-up Discussion

The typical teenager's diet is characterized by: (1) highly variable caloric needs (a rapidly growing, active 15-year old boy may need 4000 Calories/day, while a less active girl of the same age may only need 2000 Calories) and (2) irregular eating habits, with as high as 25% of the caloric intake coming from snacks. Such a diet is likely to be low in calcium, iron, vitamin A, and folacin (folic acid) and high in sodium, fats, and sugar. In general teenagers would be well advised to eat more low (or nonfat) milk, whole grains, and fresh fruits/vegetables and less sugar, fat, and salt.

This activity provides a good opportunity to review the concept of essential nutrients (carbohydrates, lineoleic acid from fat, eight amino acids from protein, 13 vitamins, 15 minerals, and water). "Empty calories—junk food" can be contrasted to foods with a high nutrient density by way of the nutrient tables provided in the text or cards available from the National Dairy Council. For example, you may wish to contrast two drinks that contain 84 Calories:

Flavored sugar drink or soft drink	Nonfat milk
84 calories from sucrose	36 calories from protein
no nutrients	48 calories from lactose
	296 mg calcium
	10 IU vit A
	0.44 mg riboflavin

E.2 YOU DECIDE: MEALS AROUND THE WORLD (pages 267-269)

In this activity, which serves as an extension of *You Decide* E.1, students can compare typical diets of various co relative to their ability to supply the RDAs for several key nutrients.

Encourage students to back their choices of "best meal" by giving reasons. There may be some differences of but ideally students should look for meals that are balanced with adequate supplies of protein, iron and vitamin

Answers to Questions:

1. a., b. Individual student answers.
2. a. Multiply values by three.
   b. Iron: 18 mg; Vitamin B_1: females need 1.1 mg, males need 1.4 mg; protein: 44 g of females, 59 g fo
   c. Students should refer to the tables on pages 252 and 257.
   d. These foods provide additional sources of the needed vitamin or mineral.

OPTIONAL CHEMQUANDARIES

1. Provide specific support for the contention that, given energy-intensive farming methods, America to eat as much oil as food.
   Answer: As an example, corn production in an 10,000 meter squared area (= 1 hectare = 2.5 acre oil equivalent of about 263 gallons (96 gal for fertilizer + 72 gal direct use as fuel + 38 gal for m for power + 26 gal for transportation/misc. + 4 gal for pesticides). However, many would argue input is well spent since each calorie equivalent of input results in as much as three Calories of output (due to the aid of the sun). But, given that over half of all U.S. grain production is fed to picture is not so positive, since every calorie of grain invested in beef cattle results in only abo actual food. On the other hand, without energy-intensive farming a much larger portion of the the developed nations) would go hungry.
2. Although people go hungry in this country, the United States sells food to other countries. C surplus food production be given to the hungry in this country rather than sold abroad? Why
   Answer: This is a complex question. Consider the farmer, the producer of our food supply, economic community. Farmers need to sell grain to continue financing their yearly food pr can not just be given away. On the other hand, some people are going hungry in the Unite

government subsidizes farmers for not producing food—but at the same time subsidizes the production of non-food crops such as tobacco.

E.3 LOOKING BACK AND LOOKING AHEAD (page 269)

As the text suggests, this marks the half-way point in the full *ChemCom* program. Four additional units complete this one-year course. Use this as an opportunity to look back with students at the range of topics and issues already developed in this program, and to do some foreshadowing of the units to come—units of study that deal with *Nuclear Chemistry in Our World; Chemistry, Air, and Climate; Health: Your Risks and Choices;* and *The Chemical Industry: Promise and Challenge.* A reminder: the first four units are designed to be sequential, with each topic following the next. The last four units can be done in any order, depending on teacher or student interest or preference. At least one of the last four units should be covered in order to provide students with enough basic chemical concepts.

SUPPLY LIST
Expendable Items

	Section	Quantity (per class)
Acetic acid, conc.	C.4	50 mL
Ascorbic acid, vitamin C, crystals	D.2	1 g
Balloons, small (4 in. dia.) demo.	C.2	8
Beverage samples (for vit. C analysis)	D.2	400 mL each
Biuret reagent	C.4	50 mL
Boiling chips	C.4	100 g
Ethanol (for Molisch reagent)	C.4	100 mL
Filter paper, Whatman 1, 11-cm dia.	C.4, D.2	80
Food samples	various	
Hydrochloric acid, conc.	D.4	60 mL
Iron(III) chloride hexahydrate	D.4	5 g
Magnesium ribbon	demo/C.2	60 cm
Milk, nonfat	C.4	300 mL
1-Naphthol (for Molisch reagent)	C.4	5 g
Potassium iodate (for iodine solution)	D.2	1 g
Potassium iodide (for iodine solution)	D.2	10 g
Potassium thiocyanate	D.4	2 g
Starch, soluble	D.2	2 g
Sulfuric acid, conc.	C.4	8 mL

Optional Laboratory Activity

	Section	Quantity
Barium chloride	D.5	70 g
Filter paper, Whatman #1, 11-cm dia.	D.5	15
Hydrogen peroxide, 3%	D.5	150 mL
Sodium sulfite	D.5	2 g

SUPPLEMENTAL READINGS

"The Pesticide Dilemma." *National Geographic*, February 1980, 145-183.

"Krill: Untapped Bounty." *National Geographic*, May 1984, 626-43. Future source of protein.

"The World's Urban Explosion." *National Geographic*, August 1984, 179-85. Population growth and the rural to urban shift.

"Egg Cookery." *CHEM MATTERS*, December, 1984.

"Potential New Crops." *Scientific American*, July 1986, 33-37. Plants that show promise as new sources of food and industrial materials.

"What's so Bad About Oxygen?" *CHEM MATTERS*, February, 1986.

"Food Additives ... Who Needs Them." Chemical Manufacturers Association. Free 11-page booklet which answers questions frequently asked about food additives.

"Polysaccharides." *CHEM MATTERS*, April, 1986.

"Your Breakfast (Luncheon, Dinner) Chemicals." Chemical Manufacturers Association. Three, free pamphlets which lists the chemicals which occur naturally in typical meals—the point: All foods are made up of chemicals; the term "chemical" should not be indiscriminately associated with "harmful."

"Chocolate", "Chocolate-covered Cherries", and "The Salmonella Search," *CHEM MATTERS*, April, 1987.

"Using Concepts of Exercise and Weight Control to Illustrate Biochemical Principles." *Journal of Chemical Education*, **61**, 882-85, 1984. Discusses such topics as forms of stored energy, caloric balance, exercise, weight control, physiological effects of endurance training, biochemistry of running a marathon, and exercise programs.

"Vanilla.", *CHEM MATTERS*, February, 1988.

"Nutrition (Diet) and Athletics (Chem I Supplement)." *Journal of Chemical Education*, **61**, 536-39, 1984. Discusses such topics as nutritional requirements, energy use, carbohydrate loading, and myths and fallacies regarding food and athletic performance.

"Future Food.", *CHEM MATTERS*, April, 1989.

"Biochemistry Off the Shelf." *Journal of Chemical Education*, **62**, 796-97, 1985. Provides sources of inexpensive, readily available, nonanimal biochemical materials for use in experiments.

"Oil Changes" and "Making the Grade.", *CHEM MATTERS*, December, 1989.

"The Vitamin Pushers." *Consumer Reports*, March 1986, 170-75. Discusses the science and the hype behind vitamin supplements and associated advertising "myths."

"Fast Food.", *CHEM MATTERS*, February, 1990.

"How Good Is Your Breakfast?" *Consumer Reports*, October 1986, 628-37. Rates (and rank orders) 59 cereals in terms of fiber, protein, sugar, sodium, fat, calories, and cost with some surprising results relative to advertising images. The same issue contains two other articles of interest: "What Has All-Bran Wrought?" pp. 638-39—explores health claims in cereal advertising and "The Fiber Furor" pp. 640-42—state of research on fiber's role in colon cancer prevention.

FDA Consumer. Issued 10 times per year, this periodical contains colorful, plain-language articles and features on consumer-oriented FDA activities.

NUCLEAR CHEMISTRY IN OUR WORLD
PLANNING GUIDE

Section	Laboratory Activity	You Decide
Introduction		Public Understanding

A. Energy and Atoms

A.2 Different Kinds of Radiation A.3 The Great Discovery A.4 Nuclear Radiation A.6 Gold Foil Experiment A.7 Architecture of Atoms A.9 Isotopes in Nature	A.1 Radioactivity A.5 The Black Box A.8 Isotopic Pennies	

B. Radioactive Decay

B.2 Natural Radioactive Decay B.3 Half-life: A Radioactive Clock B. 5 Radiation Detectors B.7 Artificial Radioactivity	B.1 α, β, and γ Rays B.4 Half-life B.6 Cloud Chambers	

C. Nuclear Power: Powering the Universe

C.1 Splitting the Atom C.2 The Strong Force C.3 Chain Reactions C.5 Nuclear Power Plants C.6 Nuclear Fusion		C.4 The Domino Effect

D. Living With Benefits and Risks

D.2 Benefits of Radioisotopes D.4 Measuring Ionizing Radiation D.5 Radiation Damage: Now and Later D.6 Exposure to Radiation D.7 Radon in Homes D.8 Nuclear Waste: Pandora's Box D.9 Castastrophic Risk: A Plant Accident		D.1 The Safest Journey D.3 Putting Atoms to Work

E. Putting It All Together: Separating Fact From Fiction

E.1 Processing the Survey Information E.2 Looking Back		

TEACHING SCHEDULE

	DAY 1	DAY 2	DAY 3	DAY 4	DAY 5
Class Work	Discuss pp. 272-274	Discuss pp. 275-277 CQ p. 277	Discuss pp. 278-280 CQ p. 279	Discuss pp. 282-284 YT pp. 284-285	
Laboratory		LA p. 275	LA p. 281		LA pp. 285-286
Homework	Nuclear Survey Read pp. 272-277	Read pp. 278-281	Read pp. 282-285	Read pp. 285-290	YT pp. 287-288

	DAY 6	DAY 7	DAY 8	DAY 9	DAY 10
Class Work	SQ p. 290	Review Part A Quiz Part A		Discuss pp. 295-297 YT p. 298	Discuss p. 299
Laboratory		LA pp. 291-294	LA pp. 291-294		LA pp. 299-301
Homework	Read pp. 291-294		Read pp. 295-299	Read pp. 299-302	YT pp. 301-302 Read pp. 302-304

	DAY 11	DAY 12	DAY 13	DAY 14	DAY 15
Class Work	Discuss pp. 302-303	Discuss pp. 304-305 YT pp. 305-306	CQ p. 306 SQ pp. 306-307	Review Part B Quiz Part B Discuss pp. 308-310	YD pp. 310-311 Discuss pp. 311-314
Laboratory	LA pp. 303-304				
Homework	Read pp. 304-307		Read pp. 308-310	Read pp. 310-314	Read pp. 314-315

	DAY 16	DAY 17	DAY 18	DAY 19	DAY 20
Class Work	Discuss pp. 314-315	Review Part C Quiz Part C Discuss pp. 316-317 YD pp. 317-318 CQ p. 318	YD p. 322	YD p. 322 Discuss pp. 322-324	Discuss pp. 324-327 CQ. p. 325 YT pp. 326-327
Laboratory					
Homework	SQ p. 315 Read pp. 316-318	Read pp. 318-322	Read pp. 322-324	Read pp. 324-327	Read pp. 328-334

	DAY 21	DAY 22	DAY 23	DAY 24
Class Work	CQ p. 331 SQ p. 334	PIAT pp. 335-337	Review unit	Exam
Laboratory				
Homework				

LA = Laboratory Activity; **CQ** = ChemQuandary; **YT** = Your Turn; **YD** = You Decide; **PIAT** = Putting It All Together.

The Nuclear unit confronts one of the most emotional issues in our society: the use of nuclear energy. The unit traces the history and development of nuclear energy, from the discovery of radioactivity to modern-day reactors and fusion, and emphasizes strongly the benefits and risks of nuclear technologies. The nuclear accidents at Chernobyl and Three Mile Island add an element of urgency to this unit, which probes the structure of the nucleus as well as the question of nuclear waste.

Our goal here is not to end debates on the use of nuclear energy by providing the "right" answers, but rather to provide a firm grounding in basic facts related to nuclear science and technology. In this context, it is important to note that neither the technology nor the ethics of nuclear weapons (the most controversial of nuclear technologies) is addressed in this unit.

OBJECTIVES

Upon completion of this unit the student will be able to:

1. List at least three examples of nuclear technology and/or natural radioactivity that affect daily life. [A.1]
2. Distinguish between ionizing and nonionizing radiation and their biological effects. [A.2]
3. Discuss general properties of electromagnetic radiation, and specific properties of various regions of the electromagnetic spectrum. [A.2]
4. Describe the experiments of Roentgen, Becquerel, the Curies, and Rutherford, and explain how they led to modifications in the atomic model. [A.3, A.4, A.6]
5. Describe the properties and locations of the three major subatomic particles. [A.7]
6. Define the term isotope, and interpret nuclear isotope notation. [A.7, A.8]
7. Use molar masses and isotopic abundance data to calculate average mass and relative abundance of elements. [A.9]
8. Compare and contrast the general properties of alpha, beta, and gamma radiation, including penetrating power, and discuss safety considerations in terms of shielding abilities of cardboard, glass, and lead. [B.2]
9. Balance nuclear equations and use them to describe natural radioactive decay. [B.2]
10. Explain the concept of half-life and discuss the implications of half-life for natural radioactivity and nuclear waste disposal. [B.3, B.4, D.8]
11. Describe radiation detectors and their operating principles. [B.5, B.6]
12. Define nuclear transmutation using a nuclear equation to illustrate the process. [B.7]
13. Distinguish nuclear fission from nuclear fusion. [C.1, C.6]
14. Use the equation $E = mc^2$ to compare the energies produced by nuclear fission and by typical exothermic chemical reactions. [C.2]
15. Explain the energy effects of a chain reaction and compare a controlled and an uncontrolled reaction. [C.3-C.5]
16. Identify the main components of a nuclear power plant and describe their functions. [C.5]
17. Assess relative risks and benefits of various nuclear technologies (such as power generation, medical applications, industrial tracing techniques). [D.2, D.5-D.7]
18. List and briefly explain some factors that determine the amount of biological radiation damage. [D.5]
19. Compare the ionizing radiation produced by various sources, including radon, that are encountered by a typical United States citizen. [D.6, D.7]
20. Discuss the problems and possible solutions associated with nuclear waste generation and disposal. [D.8]

SOURCES ON NUCLEAR SCIENCE/TECHNOLOGY

1. American Nuclear Society (ANS), Public Communications Department, 555 North Kensington Avenue, LaGrange Park, IL 60525. 708-352-6111 or toll free 800-323-3044. ANS is an organization of professional scientists and engineers involved in peaceful uses of the atom. ANS publishes a variety of free materials including: a newsletter (*Re–actions*—includes teaching tips, projects, and information sources); a nuclear science curriculum guide; Energy Chase board game; a resource contact list; and assorted pamphlets, audiovisuals, and computer software (free loan from Audio Visual Lending Library). Local sections typically maintain a speakers' list. ANS also has an associated teacher affiliate group, ANSTA (membership includes monthly *Nuclear News*).

2. Atomic Industrial Forum (AIF), Educational Services, 7101 Wisconsin Avenue, Bethesda, MD 20814-4805. 301-654-9260. Industry-sponsored group involved in the development and utilization of constructive uses of nuclear energy, monitoring of government regulations, and the production of educational materials. Makes available a variety of free slide sets and videotapes/films (on loan), transparency masters, information sheets and pamphlets.

3. Critical Mass (Public Citizen's Critical Mass Project), 215 Pennsylvania Avenue, SE, Washington DC 20003. 202-546-4996. A nonprofit research and advocacy group that opposes nuclear power and promotes "safe" energy alternatives. Publishes approximately 70 titles ranging in price. Request publications list.

4. Edison Electric Institute (EEI), Educational Services Department, 1111 19th Street, NW, Washington, DC 20036-3691. 202-828-7587. A trade association representing investor-owned electric companies. Request: (1) *Directory of Educational Services*—lists names, addresses, and phone numbers of member companies across the country, and (2) *EEI Publications Catalog*–includes the "Experiments You Can Do" series.

5. Fund for Renewable Energy and the Environment (FREE), 1001 Connecticut Avenue, NW, Suite 638, Washington, DC 20036. 202-466-6880. Formerly called Solar Lobby and the Center for Renewable Resources, this nonprofit, membership organization publishes a variety of books and reports on energy alternatives as well as a variety of energy-saving products and toys.

6. National Science Teachers Association (NSTA), 1742 Connecticut Avenue, NW, Washington, DC 20009. 202-328-5800. Publications related to nuclear science include: *Playing with Energy* (#PB34/1981/106 p), *Teaching About Nuclear War* (#PB51/1985/72 p), *Supplement of Science Education Suppliers* (#PB58/annual/104 p), and *Energy & Education* (5 issues/yr newsletter) plus annual *Directory of Energy Education Materials*.

7. Nuclear Information and Resource Service (NIRS), 1616 P Street, NW, Suite 160, Washington, DC 20036. 202-328-0002. A nonprofit, membership organization (includes quarterly issues of *Groundswell*) designed to serve as an information clearinghouse for anti-nuclear individuals and organizations. Publications of interest to teachers include: *Teaching Nuclear Issues—a Kit for Secondary School Teachers* (includes fact sheets, articles, maps, posters, cartoons, annotated resource guide, etc.), *Growing Up in a Nuclear Age—A Resource Guide for Elementary Teachers* (32-page annotated list of educational materials), energy fact sheets on a variety of nuclear issues, and a *Publications Distribution Service List* (annual publication with quarterly updates that annotates recent articles dealing with nuclear issues—also serves as a low-cost source of reprints of the same).

8. Union of Concerned Scientists (UCS), 26 Church Street, Cambridge, MA 02238. 617-547-5552. Nonprofit organization concerned with the societal impact of advanced technologies; specifically, UCS is involved with independent research, public advocacy, and education on issues related to nuclear arms, energy policy alternatives, and nuclear power safety. Publishes a variety of free briefing papers: *Disposal of Radioactive Waste, Energy Strategies, Safety of Nuclear Power Reactors, Decommissioning of Nuclear Power Reactors, etc.*, and low-cost books: *Radioactive Waste: Politics, Technology, and Risks, Nuclear Power Plants in the United States; Safety Second: A Critical Evaluation of the NRC's First Decade; Three Mile Island: Thirty Minutes to Meltdown;* etc. UCS tends to be critical of nuclear energy technologies as they are presently employed.

9. United States Department of Energy (USDOE):

   a. Energy Information Administration (USDOE/EIA), 1000 Independence Avenue, SW, Washington, DC 20585. 202-586-8800. Since its inception in 1977, the EIA has been the nation's primary source of comprehensive national energy statistics and objective energy analysis. Of special interest are these publications: *Annual Energy Review*—typically released in May or June at a cost; *Annual Energy Outlook*—published in February provides energy market projections for the next 10 years; and the *Monthly Energy Review*—available for a nominal charge per issue or a yearly subscription. These publications contain extensive up-to-date statistics, tables, and graphs on the U.S. energy picture and may be ordered over the phone (or found in most university libraries). Your school library may be able to get single, complimentary copies of these and other EIA publications.

   b. National Energy Information Center (NEIC), Energy Information Administration, U.S. Department of Energy, EI-20, Forrestal Bldg., Washington, DC 20585. 202-586-8800. The following publications are available free: *EIA*

Publications Directory 199X: A User's Guide, Annual Report to Congress (a 60-page report based on other EIA publications), EIA New Releases (bimonthly), *Energy Information Directory* (semiannual), and assorted information sheets. NEIC also publishes a yearly booklet, *Energy, Facts 199X* .

c. National Technical Information Service (NTIS), Document Sales, 5825 Port Royal Road, Springfield, VA 22161. 703-487-4650. NTIS makes all EIA publications available to the public in microfiche form.

d. Office of Civilian Radioactive Waste Management, U.S. Department of Energy, Mail Stop RS-40, Washington, DC 20585. 202-252-5722. A series of pamphlets on nuclear waste (DOE/RW-0104—RW-0112): *Overview—Nuclear Waste Policy Act. The Illustrated Mechanics of Nuclear Waste Isolation. What Will a Nuclear Waste Repository Look Like? What Is Nuclear Waste? What Is Spent Nuclear Fuel? Can Nuclear Waste be Transported Safely? Radiation and Nuclear Waste—How Are They Related? How Much High-level Nuclear Waste Is There? What Rock Types Are Being Considered for Nuclear Waste Repositories and Why?*

e. Energy, P.O. Box 62, Oak Ridge, TN 37831. Disseminates a variety of free DOE publications including a series on nuclear energy: *Nuclear Energy Economics, Understanding Radiation, Nuclear Powerplant Safety: Operations—Design and Planning, The Nuclear Fuel Cycle, Electricity from Nuclear Powerplants*, etc..

INTRODUCTION

This section introduces the energy production aspect of nuclear energy and begins raising the question of risk versus benefit. Remind students of the general *ChemCom* theme that scientific understanding alone will not provide the answers to the more complex societal problems surrounding technological issues, but it can provide a more rational basis for public debate and action.

YOU DECIDE: PUBLIC UNDERSTANDING (pages 272-274)

This survey allows students not only to test their own understanding of nuclear chemistry, but more importantly to involve others as well. This pre-test type of activity should not be graded, and should encourage the students to range beyond their normal circle of contacts for the survey.

Time

Allow 15 minutes of class time for students to answer the questions themselves. This activity is meant to introduce the unit. Allow one week for students to complete this activity by surveying the other three individuals.

Materials

In the Teacher's Guide, on pages 143 and 144, you will find a printed version of the survey suitable for duplicating. Each student will need five copies of the survey to complete the task. Encourage students to keep each person's answers separately, so that no one will be influenced by other answers.

Suggestions

Give the survey to the students immediately following the introduction. Be sure to stress that the survey will not be graded. Follow this up with a quick tabulation of the results on the chalkboard or overhead. It is easiest for students to claim no knowledge of topics, and so discourage the "U" response as much as possible. Do not enter into a discussion concerning the correct answers, since the students should discover that many of their questions are answered in subsequent activities.

Next, discuss the general procedure you wish students to use in administering the survey to others. The assignment is due in seven days. Students are to survey one high school student *not* currently taking chemistry, one adult born before 1950, and one adult born after 1950.

Answers to Questions:

1. The survey results are likely to reflect a general lack of understanding of nuclear-related phenomena and issues. Probably most of those interviewed have had little, or no education on nuclear issues.

2. The survey results are likely to reflect that public fear of radiation is high. Students may point to the association of radioactivity with weapons, war, and cancer; a media focus on a possible nuclear holocaust; governmental defense policies; nuclear power plant accidents; and not being able to sense the presence of nuclear phenomena. It is

Nuclear Phenomena Survey

These statements are designed to survey your understanding of nuclear-related phenomena. Indicate whether you agree (A), disagree (D), or are unable to answer because of insufficient knowledge (U).

For Item "0" Fill in your major source of knowledge of nuclear phenomena:

School = 1; Television = 2; Scientific magazines = 3; Magazines/newspapers = 4; Conversations with others = 5.

Yourself	Non-chem Student	Pre-1950 Adult	Post-1950 Adult		
					0. Major source of knowledge of nuclear phenomena. (See choices above)
					1. The atom is the smallest particle in nature.
					2. Home smoke detectors may contain radioactive materials.
					3. Radioactive materials and radiation are unnatural. They did not exist in the world until created by scientists.
					4. All radiation causes cancer.
					5. Most space occupied by an atom is "empty."
					6. Electromagnetic radiation should be avoided at all costs.
					7. Human senses can detect radioactivity.
					8. Nuclear wastes are initially "hot" both in temperature and in radioactivity.
					9. All atoms of a given element are identical.
					10. Radiation can be used to limit the spread of cancer.
					11. Individuals vary widely in their ability to absorb radiation "safely."
					12. Small amounts of matter change to immense quantities of energy in nuclear weapons.
					13. The human body naturally contains a small quantity of radioactive material.
					14. Physicians can distinguish cancer caused by radiation exposure from cancer having other causes.
					15. Television tubes emit radiation.
					16. Most nuclear waste generated to date has come from nuclear power plants.
					17. Radioactive and nonradioactive forms of an element have the same chemical properties.
					18. Cells that divide rapidly are more sensitive to radiation than are cells that divide slowly.
					19. Physicians use injections of radioactive elements to diagnose and treat certain disorders.
					20. Medical X rays involve potential risks as well as benefits.
					21. Nuclear reactors were originally designed to generate electricity.
					22. Nuclear plants are the only electric power plants that create serious hazards to public health and the environment.
					23. To date, no one has died from radiation released by nuclear power plants.
					24. Nuclear power plants do not emit air pollution during normal operation.
					25. Regardless of risks, nuclear power plants are needed to keep the nation functioning and less dependent on foreign oil.
					26. An improperly-managed nuclear power plant can explode like a nuclear weapon.

					27. The main difference between a nuclear power plant and a coal-fired power plant is the fuel used to boil the water.
					28. Some nuclear wastes must be stored for hundreds of years to prevent dangerous radioactivity from escaping into the environment.
					29. Nuclear power presently supplies more than 10% of our country's total energy needs and is increasing in importance each year.
					30. If the half-life of a radioactive substance is six hours, all of it will decay in 12 hours.
					31. More federal funds have been spent on nuclear power development than on all other "alternative" energy sources (wind, solar, etc.) combined.
					32. In the United States, the largest quantity of human-produced radiation comes from nuclear power plants.
					33. Some states have banned construction of new nuclear power plants.
					34. Nuclear wastes can be neutralized or made non-radioactive.
					35. Nuclear power plants produce material that could be converted into nuclear weapons.
					36. A nuclear power plant uses a much smaller mass of fuel than does a coal-fired plant.
					37. A nuclear power plant is less expensive to build than is a coal-fired plant.
					38. A national system for long-term storage of radioactive wastes is now operating.
					39. The rate of radioactive decay can be slowed by extreme cooling.
					40. The United States should increase its reliance on nuclear power to generate electricity.

interesting to note that Marie Curie once said: "Nothing in life is to be feared. It is only to be understood."

3.   Answers will vary.

Remember that most of the survey questions will be addressed in later sections of this unit; avoid providing many answers now. The answers to each survey item are found in Part E of this guide.

A: ENERGY AND ATOMS

This Part introduces students to radiation, its characteristics, and the electromagnetic spectrum. Students learn that human beings daily encounter both naturally occurring radioactivity and the fruits of nuclear technology.

One of the laboratory activities simulates the accidental circumstances that led to Becquerel's observations later associated with radioactivity. This particular example of serendipity, as well as several others, are discussed in further detail in Section A.3 on pages 278-280. Students learn to distinguish among the different kinds of radiation and to describe the damage they do. They trace the discoveries of X rays, as well as alpha, beta, and gamma rays. They learn about the structure of atoms, and about how Ernest Rutherford discovered the atomic nucleus. Isotopes and nuclear notation are introduced. In addition, students learn that certain scientific discoveries are accidental, and that in searching for the answers to one question, scientists often stumble upon another question that is even more interesting.

A.1 LABORATORY ACTIVITY: RADIOACTIVITY (page 275)

Students are made aware that some common objects in their environment are radioactive and, although the radiation is not directly observable with the senses, it can be detected by other means. This activity provides a macroscopic view of nuclear events. It is not necessary to explore the substances or reasons for these events in detail at this time.

The critical element in this activity is the sun-sensitive paper. You should test it out on your radioactive materials before beginning this activity with students; long-term exposure may be necessary.

Time

Allow 24 hours for students to complete this activity. You will need 15 minutes to prepare sun-sensitive paper for observation the next day.

Materials (for a class of 24 working in pairs)

12 sheets of sun-sensitive paper
12 lantern mantles (unused)
12 forceps
Smoke detectors (ionization) [optional]

Note that students may use any assorted objects, but they **must** include the lantern mantles to obtain at least one positive result. Low-sodium salt that contains potassium, (Lite salt) and luminescent watch faces that contain radium are both radioactive, but the quantity of radiation emitted is too low to react with the sun-sensitive paper to produce an image. (You may later wish to use a Geiger counter or other more sensitive radiation detection instrument.)

Advance Preparation

Sun-sensitive paper exposed to radiation forms an image when developed in water. This paper is sold in stores that stock educational games and activities for children under a variety of names such as "solar graph paper" or "nature print paper." Sun-sensitive paper (nature print paper) can be ordered from Chaselle Inc., 9645 Gerwig Lane, Columbia, MD 21046, 1-800-492-7840 (catalog number 240002) or the New England School Supply, Division of Chaselle, Inc., P.O. Box 1581, Springfield, MA 01101 (catalog number 240002), 1-800-628-8608. Also sold by Solar Encounters, Educational Insights, Compton, CA 90220 and Solargraphics Co., P.O. Box 7091 P, Berkeley, CA 94707.

Lab Tips

Although many different objects may be used, the lantern mantles will produce a clear spot on the paper after developing the paper with tap water.

Make sure that you try this activity in advance. Results with sun-sensitive paper have been erratic and it appears that long exposure times are necessary. *ChemCom* seeks to make a low-key introduction to nuclear phenomena by using simple materials: a Geiger counter is not simple, and does not fit with this objective. However, the important point is to see the relative commonness of radioactivity, so it is better to use a Geiger counter than to lose observations for the students.

Be sure that all objects are completely dry. Any moisture will stain the print paper, leading to erroneous results.

If you have access to photographic film and developing processes at school, that gives you a reasonable alternative activity. Ask your photography colleague or students to prepare small samples of film in light-proof containers, and lay your samples on top. You will need to be able to differentiate the film samples after development.

Post-Lab Discussion

The Post-Lab discussion should focus on the idea that radioactivity is all around us and plays an important role in our lives. This activity provides an observational foundation for the historical discussion of the discovery of atomic energy that follows.

Answers to Questions:

1. Lantern mantles will be found to be radioactive.

2. The reason lantern mantles are radioactive is that thorium is intentionally added to the mantle material—thorium oxide serves as a surface catalyst for the combustion reaction. The radioactivity itself serves no useful purpose, but simply accompanies the thorium.

3. Gieger counter (You may want to demonstrate the radioactivity associated with low-sodium Lite salt using a Geiger counter.)

A.2 DIFFERENT KINDS OF RADIATION (pages 275-277)

This section defines radiation in the context of the electromagnetic radiation (EMR) spectrum. There is extensive vocabulary and information here. Help students understand the variety of energies in the EMR, and the differences between ionizing and non-ionizing radiation.

CHEMQUANDARY

Always Harmful? (page 277)

"All radiation" includes the entire electromagnetic spectrum, both ionizing and nonionizing. It also includes visible light, which is responsible for sight and cannot be considered harmful. Avoiding all radiation would mean isolation from radio waves, microwave equipment, all lighting sources, and infrared heaters–not a very practical existence or a possibility. The statement is obviously false.

A.3 THE GREAT DISCOVERY (pages 278-280)

In most cases students are given scientific information with little background as to the hows or whys of the research. This section provides an historical background to the discovery of radioactivity, including the often startling revelation that Becquerel's crucial experiment, the exposure of photographic film by uranium, was something of an accident. The discovery of radioactivity demonstrates how scientific inquiries are often interrelated; in this case Roentgen's X rays led to Becquerel seeking X rays from fluorescent materials such as uranium.

You may want to demonstrate the fluorescence of the common objects, such as certain types of Frisbees and children's toys that glow in the dark. These objects contain compounds that are excited by ultraviolet light and reemit the absorbed energy as visible light. Also point out the role of UV-excited fluorescent materials in such diverse situations as supermarket packaging of products that get clothes "whiter than white," and in certain flowers on cloudy, overcast days.

You may also want to borrow a cathode-ray tube (CRT) from the physics teacher to discuss fluorescence. The use of a CRT reinforces the idea of using a beam of electrons to excite certain substances. As the substances return to a more stable state, energy may be released in the form of X rays. Note that TV picture tubes and X-ray machines are both CRTs that have been modified to decrease and increase the emission of X rays, respectively.

You may wish to focus attention on the medical benefits of X rays, using Figures 2 and 3 and the students' own experiences.

You may also want to explain how Becquerel's plates were exposed by the uranium. Photographic plates are composed of silver halides (chloride, bromide, or iodide), salts that are sensitive to radiant energy. Any radiation more energetic than infrared radiation will cause a photochemical reaction leading to silver metal ions being reduced to the metal during development. This reduction results in the darkening of a photographic plate. Thus, uranium must have given off some type of radiant energy, as hypothesized by Becquerel.

Scientific Discoveries (page 279)

Each discovery was similar to Becquerel's uranium work in that something unexpected happened. In all cases, some of which were accidental, the observed events were discrepant with current scientific explanations. Becquerel did not expect the plates to be exposed; nor Goodyear expect sulfur to harden the rubber; nor Fleming expect a mold to kill bacteria; nor Plunkett expect to find a polymer of tetrafluoroethene. The point to be made is that careful observation is needed to challenge current scientific conceptions. Pasteur's statement, "Chance favors the prepared mind," is good advice for all scientists, present and future.

These cases could be used for student library research reports.

A.4 NUCLEAR RADIATION (page 280)

Rutherford's investigations into the nature and energy of radioactive particles are presented here. The differences among the three types of radiation are important background for subsequent sections, so make sure your students are familiar with the differences highlighted in this section.

A.5 LABORATORY ACTIVITY: THE BLACK BOX (page 281)

The purpose of this activity is to simulate how scientists use indirect evidence to infer properties that cannot be perceived by our senses. Remember that there are no chemical facts in this activity; instead the focus is on the reasoning processes used.

Time

One-half to one class period

Materials (for a class of 24 working in pairs)

24 sealed, coded boxes (cigar or shoe box), each containing three unconstrained objects such as small scissors, small rubber ball, small fork, plastic comb, large paper clip, or small key. Commercially available black boxes could also be used.

Pre-Lab Discussion

You may want to discuss how the study of atomic structure is analogous to a black box. A brief review of how Rutherford discovered alpha, beta, and gamma rays may be useful here.

Lab Tips

Using common materials such as test tube holders, flame spreaders, or rubber stoppers may be advisable. One suggestion is to include cotton balls, since their detection is as difficult as it was to find neutrons in the atom.

Post-Lab Discussion

Focus attention on how the activity simulated the work of atomic scientists rather than on which students got right answers.

Answers to Questions:

1. Hearing
2. The structure of the atom is like a black box, which scientists cannot directly open in order to examine its contents.
3. Indirect evidence is used to formulate explanations (theories) about the natural world whenever the events occur beyond the range of our perceptions. The events may occur over too short (kinetic molecular theory) or too long (theory of evolution) a time period, or the objects of interest may be too small (atomic theory) or too large (theory of the universe).

You may wish to demonstrate an interesting variation on the black box that involves suspending a disposable object (key, paper clip, fork, comb, scissors, rubber ball, etc.) inside a sealed shoe box. Rigid construction paper strips (white) about 56 cm long by 7 cm wide cut to fit through slits in the back of the box serve as an imaging screen. Spray paint through a hole in the front of the box to create a "shadow image" of the concealed object. The demonstration reinforces the notion of using a source and a target, and shooting something at the object in the atomic black box to acquire information about its contents, as Rutherford did in his gold foil experiments.

Another possibility for identifying an unknown object is to imbed the object in modeling clay, and attempt to define its shape or identity by sticking pins into the clay. Commercial Rutherford simulations are also available, which involve a piece of Styrofoam under a board; the investigator rolls marbles under the board, and notes the trajectory of the marble after it collides with the "hidden" object.

A.6 GOLD FOIL EXPERIMENT (pages 282-283)

This section provides a prime example of how experimental observations may be considered "unexpected" based upon a current model. The discovery of the nucleus was not the purpose of the gold foil experiment; Rutherford was investigating the behavior of alpha particles with metals. The unusual deflections prompted him to propose a very heavy, very small particle— the nucleus.

It is important that students acquire a picture of the physical arrangement of the experiment for the resulting deductions to make sense. Focus the discussion on Figures 5 and 6 on page 282. Point out that not only is there a considerable amount of space between atoms, but most of the volume occupied by an individual atom is empty space. (The actual numbers illustrating these ideas appear in Section A.7 on p. 283).

A.7 ARCHITECTURE OF ATOMS (pages 283-285)

The components of the nucleus are identified, and isotopes related to mass numbers. Nuclear notation is used to help identify the number of neutrons in a given isotope.

As mentioned in previous units, analogies can help students gain a sense of atomic scale. Emphasize several points:

1. Atoms are very small. If persons were the size of atoms, the entire population of the world (5.4 billion persons) would fit on the head of a pin.
2. Most of the volume of an atom is empty space. If the nucleus were the size of a billiard ball, the average electron would be located about 1.6 kilometers away.
3. Protons and neutrons account for most of the mass of the atom since they weigh approximately 2000 times as much as an electron. You may wish to encourage your students to come up with an appropriate analogy.

Reemphasize the idea that, in an uncharged atom, the number of electrons must equal the number of protons (or the atomic number). The number of neutrons may vary without perturbing the neutrality of the atom, and the mass of an atom of a given element varies with the number of neutrons it contains.

YOUR TURN

Isotope Notation (pages 284-285)

This activity provides practice in using nuclear chemistry notation and illustrates that atoms with higher atomic numbers have a larger neutron/proton ratio than do the elements with lower atomic numbers.

1.

	Symbol	Name	Atomic number (# protons)	# Neutrons	Mass number	# Electrons
a.	$^{12}_{6}C$	carbon	6	6	12	6
b.	$^{14}_{7}N$	nitrogen	7	14	7	7
c.	$^{16}_{8}O$	oxygen	8	8	16	8
d.	$^{24}_{12}Mg^{2+}$	magnesium ion	12	12	24	10
e.	$^{200}_{80}Hg$	mercury	80	120	200	80
f.	$^{238}_{92}U$	uranium	92	146	238	92

2. a. The lighter elements exhibit roughly a one-to-one ratio between the number of neutrons and protons.
   b. Heavier elements such as mercury (1.5/1 ratio) and uranium (1.6/1 ratio) exhibit higher neutron/proton ratios. Given the students' prior exposure to the law of electrostatics (like charges repel), you may wish to have them hypothesize an explanation for this trend (they might think of neutrons as a type of glue that keeps protons in the nucleus from flying apart).

A.8 LABORATORY ACTIVITY: ISOTOPIC PENNIES (pages 285-286)

This activity simulates the determination of isotopic composition of an element. The simulation involves finding the number of pre- or post-1982 pennies in a sealed container of 10 pennies without opening the container. The calculations are typical of the type found in the *YOUR TURN* section that follows.

Time

One class period.

Materials (for a class of 24 working in pairs)

12 sealed opaque containers (coded) with different 10-coin mixtures of pre-1982 and post-1982 pennies.
12 pre-1982 pennies
12 post-1982 pennies
Balance

Advance Preparation

Weigh one of the empty containers. Provide the mass for students to use in their calculations, as specified in the procedure.
    Make up the containers using various combinations of pennies. Code them.

Pre-Lab Discussion

Emphasize that most (but not all) elements are composed of mixtures of two or more isotopes.

Lab Tips

Consider working through a sample calculation for the students.
    Avoid using 1982 pennies (unless you use **all** 1982 pennies), since some have the pre-1982 composition; others are similar to post-1982 pennies. There were 7 different mintings of pennies in 1982.
    If you want all the pennies to look equally "new," soak them briefly in dilute hydrochloric acid (not nitric acid), then rinse and dry.
    To discourage "cheating" put the pennies in preweighed film containers packed with cotton. Or use a Seal-a-Meal machine to seal the pennies into little bags, after squashing all the pennies together so students cannot see the dates on the individual coins.
    To minimize differences in pennies you may want students to weigh 10 pre-1982 pennies and 10 post-82 pennies, then use the average masses in their calculations.

Post-Lab Discussion

Provide a conceptual bridge between the simulation and "real" isotopes. Point out that atomic-weight values for various elements represent weighted averages of the masses of each element's isotopes.

Answers to Questions:

1. Mass
2. a. The pre- and post-1982 penny mixture represents different isotopes of the fictitious element "coinium." For practical purposes, they look and "behave" the same, yet they have different masses. Similarly, isotopes of the same element behave the same chemically, yet have different masses.
    b. This analogy is limited in that: (1) there are only two "isotopes", while actual elements often have more than two; (2) in the case of real isotopes, one isotope is quite likely to be worth more than another—this is especially true in the case of isotopes used in nuclear applications; and (3) one cannot weigh real individual atoms using a balance.
3. Eggs (small, medium, large), car tires, soft-drink containers, any class of objects that come in different masses, but serve the same function.

A.9 ISOTOPES IN NATURE (pages 286-289)

This section points out that most elements occur naturally as mixtures of isotopes which exhibit the same chemical properties, but physically differ by mass due to differing number of neutrons.

Molar Mass and Isotope Abundance (pages 287-288)

This activity is designed to provide students with practice in (1) calculating the average molar mass of an element from isotopic molar masses, and (2) determining the percent abundances of isotopes from an element's molar mass and isotopic molar masses. The similarity between these calculations and those done in the laboratory activity should facilitate student work on these problems.

Answers to Questions:

1. Student data, students should use the formula given in the text.
2. Student data
3. Check student responses: the answers to #1 and #2 should be the same.
4. a. 238. The final mass should be closest to the most prevalent isotope.
   b. Average mass = $(238.1) \times (0.9928) + (235.0) \times (0.0071) + (234.0) \times (0.000054) = 238.1$

PART A: SUMMARY QUESTIONS (page 290)

1. Characteristics: energy with no mass, travels at the speed of light, travels through a vacuum, emitted by decaying or energized atoms, consists of packets of energy called photons, higher frequencies have higher energy. Examples: infrared, ultraviolet, visible, gamma rays, X rays, microwaves, radiowaves.
2. Roentgen's discoveries of X rays and fluorescence prompted many others to study this phenomenon. Becquerel was studying the fluorescence of other elements, and working with uranium. His work with uranium led to the discovery of radioactivity.
3. Fluorescence depends on a material absorbing ultraviolet light that is subsequently reemitted as visible light. Since the uranium compound was able to cause blackening of the photographic plate, while not being previously exposed to UV-radiation from sunlight, the radiation from the compound had nothing to do with fluorescence. The penetrating power of this new radiation was observed to be much greater than X rays generated by a cathode-ray tube (see Section A.3, pages 278).
4. See Section A.4 (page 280). Simply, two components of the beam emitted by the radioactive source were influenced by a magnetic field. This behavior is typical of charged particles, but not of electromagnetic radiation.
5. See Section A.6 (page 282).
   a. The majority of heavy and charged alpha particles passed straight through gold foil.
   b. A small percent of the alpha particles were reflected back at large angles.
6. Electron (9.1095×10^{-28} g) and, proton (1.67261×10^{-24} g); because a proton is about 1836 times more massive than an electron, it takes approximately 2000 electrons to equal the mass of one proton.
7.

Symbol	Name	Atomic number	Number of protons	Mass number	Number of neutrons
$^{12}_{6}C$	carbon	6	6	12	6
$^{60}_{27}Co$	cobalt	27	27	60	33
$^{207}_{82}Pb$	lead	82	82	207	125

8. a. If the average molar mass of the element potassium is 39.098 g/mol, then the most abundant isotope has a molar mass of 38.964 g/mol.
   b. Its molar mass is closest to the average as compared with the other isotopes. The element name and mass number is potassium-39.
9. a. The average molar mass of the element neon has to be closer to 20 g/mol, since the natural abundance of the Ne-20 isotope with a molar mass of 19.992 g/mol is over 90%.
   b. Average molar mass of Ne = 0.9051(19.992) + 0.0027(20.994) + 0.0922(21.991) = 20.179 g/mol.
10. B-11, since the molar mass is closer to 11 g/mol than it is to 10 g/mol.
11. Atomic weight, and atomic number or chemical symbol.

- Students can use references such as the *Readers' Guide to Periodical Literature* and the *McGraw-Hill Encyclopedia of Science and Technology* to find fairly recent information on these particles. A clearer understanding of these particles might help us to better understand questions ranging from the origin of the universe to ways of dealing with nuclear waste.

B: RADIOACTIVE DECAY

In Part B students learn the nature of the three most common types of radioactivity: alpha, beta and gamma. The concept of half-life is introduced, along with techniques of radiation detection. Man-made radioactivity provides a link to the next part, concerning the development of the nuclear power industry.

B.1 LABORATORY ACTIVITY: α, β, AND γ RAYS (pages 291-294)

This activity is designed to acquaint students with nuclear radiation emissions from radioactive materials and with the ability of materials to serve as radiation shields. In Part 1, students compare the penetrating power of alpha, beta, and gamma radiation through cardboard, glass, and lead. In Part 2, they determine the effect of distance on radiation intensity; and in Part 3, they compare the radiation shielding effects of lead and glass. Students also discover that their everyday environments contain natural radiation as they account for background radiation in their observations.

Time

One to two class periods, with the students working in groups of two to five, depending on equipment availability. If equipment is in short supply, this activity can be presented as a demonstration. Since the activity covers a number of important concepts, students should witness at least a demonstration.

Materials

1-5 ratemeters (such as Sargent-Welch catalog number S-72110-36)
Alpha, beta, gamma radiation source set (such as Sargent-Welch catalog number S-72118-40)
12 meter sticks
12 lead plates (5 cm × 5 cm × 0.1 cm)
24 glass plates (5 cm × 5 cm × 0.1 cm)
12 pieces of cardboard (5 cm × 5 cm × 0.1 cm)
Calipers (optional)
12 sheets graph paper
12 forceps or crucible tongs (for isotope handling)
12 pairs disposable safety gloves
Radiation safety instructions (see summary later in this section)

Advance Preparation

You may want to check the background radiation before students begin the activity. If student readings are questionable, have them take more readings.

Lab Tips

The thickness of the lead plates may be more easily determined by standing the plate on a centimeter scale. Most plates available for this purpose are 1 mm thick. To save time you may wish to give this value to your students.

The alpha, beta, and gamma sources and ratemeters may be ordered from the Sargent-Welch Scientific Company (catalog numbers are given above) and from other sources.

The Sargent-Welch toll-free number is 800-SARGENT. Ask for an educational products salesperson.

Due to the fact that a few centimeters of air will block alpha particles, the thick glass on the probe of the ratemeter will not allow alpha particles to be detected. You may want to order a very thin glass plate or totally remove the glass plate on the end of the ratemeter probe so that students can discover how a few centimeters of air act as a barrier to alpha particles.

If you order the alpha, beta, and gamma sources from Sargent-Welch you will receive three 25-mm diameter disks. The alpha source (Po-210) will be plated on a foil disk and may be stored in any plastic or glass container. The beta source (T1-204) should be stored in a lead container. The gamma source (Co-60) must be stored in a lead-walled container. Students are not to touch any source directly, especially the beta and gamma sources. All of these radiation sources degrade to stable isotopes. It is of the utmost importance that you keep track of all radiation sources. Given the small quantities used, it is easy to misplace them, and other teachers or students may unknowingly handle them directly with the fingers.

Disposable safety gloves may be ordered from American Scientific Products, Division of American Hospital Supply Corporation, McGraw Park, IL 60085, catalog number G-7-238-2.

Check with your local American Nuclear Society for the loan of Geiger counter set-ups if your budget does not allow such purchases. These machines will require some practice on your part and may require modification of the procedure. Another potential source of Geiger counters is the local Civil Defense authority.

Radiation Safety

Great care must be taken to ensure that the radioactive materials which you and your students will be handling do not constitute a danger. Minimize the amount of radiation you are exposed to during the activity by following these basic rules:
1. Always wear rubber or plastic gloves and use tongs when handling radioactive materials.
2. Do not bring food into the room when working with radioactive materials (or at any other time). Food can be easily contaminated during handling and could result in the ingestion of radioactive materials.
3. Be sure that no radioactive material comes in contact with your radiation counter. The counter can become contaminated with radiation and record higher than normal background readings.
4. All materials used in the activity should be collected in an appropriate storage container and checked for radiation before disposal through normal channels.
5. Check students' hands with a radiation monitor before they leave the laboratory.

Part 1 Answers to Questions:
1.   Lead greatly reduces the level of alpha, beta, and gamma radiation. Cardboard greatly reduces the alpha radiation.
2.   Gamma radiation is the most penetrating; alpha radiation is the least penetrating. Gamma radiation is without mass and is classified as electromagnetic radiation. Alpha radiation is the most massive and can be retarded easily by collisions with other particles.
3.   a. Lead
      b. Cardboard
4.   Density and thickness

Part 2 Answers to Questions:
1. Four
2. No
3. The intensity varies inversely as the square of the distance $(I = k/r^2)$.

Part 3 Answers to Questions:
1. a. Glass is not very effective.
   b. Lead is effective.
2.   Lead, because it is most effective at blocking radiation, including electromagnetic radiation such as X rays.
3.   Gamma radiation is the most dangerous to the body because it has the greatest penetrating power and the greatest energy per photon; thus it can cause the most cellular and molecular disruption.

Post-Lab Discussion

Focus on the questions.

B.2 NATURAL RADIOACTIVE DECAY (pages 295-299)

The characteristics of alpha, beta, and gamma radiation are explained, along with the concept of balanced nuclear equations. In the Resource unit students learned that mass is conserved in chemical equations. Emphasize that mass numbers are conserved in nuclear equations, and while elements may change in a nuclear reaction, the atomic numbers (which may include electrons) are also conserved. After working through examples of nuclear equations, draw attention to Table 6 on page 297, which summarizes radioactive decay and the nuclear notation used.

Given that radioactive decay is a dynamic process occurring at a level below human vision, computer and film simulations can be effective aids to enhance student understanding (see the listing provided in this guide).

Nuclear Balancing Act (page 298)

This activity provides the students with practice in balancing nuclear equations. Emphasize that during beta decay, a neutron is transformed into a proton and an electron. Since the new proton remains in the nucleus, the atomic number increases by one while the mass number remains the same. (The mass number remains constant because the number of neutrons is decreased by one and the number of protons is increased by one; the consequence is the formation of another element.)

Answers to Questions:

1. a. $^{14}_{6}\text{C} \rightarrow {}^{14}_{7}\text{N} + {}^{0}_{-1}\text{e}$

   beta decay

   b. $^{241}_{95}\text{Am} \rightarrow {}^{237}_{93}\text{Np} + {}^{4}_{2}\text{He}$

   alpha decay

2. a. $^{4}_{2}\text{He}$

   b. $^{228}_{89}\text{Ac}$

   c. $^{228}_{90}\text{Th}$

   d. $^{224}_{88}\text{Ra}$

   e. $^{224}_{88}\text{Ra}$

B.3 HALF-LIFE: A RADIOACTIVE CLOCK (page 299)

This section briefly introduces the concept of half-life, a term with which most students are familiar. However, check their understanding by giving several examples of half-life, starting conveniently with multiples of two, such as 64 or 128. This avoids fractional or decimal considerations, inevitable when dividing ordinary numbers by 2. Propose a sample half-life, such as 5 days, and ask how many atoms would be left after 10 days. Table 7 (on page 299) helps make the point that different nuclei have different stabilities, a property whose importance is emphasized in subsequent exercises.

B.4 LABORATORY ACTIVITY: HALF-LIFE (pages 299-301)

This activity models the exponential decay curve of radioactive samples through several half-lives. The "decaying" atoms are pennies, and decay is represented when a coin flips from heads to tails. The simulation makes the point that because atomic decay is a statistical matter, we can never be sure which atom will survive and which will decay.

Time

One-half class period

Materials (for a class of 24 working in pairs)

960 pennies
12 cardboard boxes
12 sheets of graph paper

Advance Preparation

Set up 12 containers, each holding 80 pennies.

Pre-Lab Discussion

None; discussion of Section B.3 can follow completion of this activity. Results will approximate a radioactive (exponential) decay curve.

Post-Lab Discussion

Connect the concepts from Sections B.2-B.4 and focus on the questions. Make sure that students understand the random, statistical nature of radioactive decay. One cannot know when a given atom will decay, but can predict the rate of decay for a large number of atoms. Once again, various computer and film simulations can be helpful.

Answers to Questions:

1. a. The graph slopes downward as a curved line.
   b. The pooled class data should smooth out the curved line and be more reliable, representing a larger number of decayed atoms.
2. 75 (600 – 300 – 150 – 75)
3. 4 (2800 –1400 – 700 – 350 – 175)
4. The simulation is similar to the actual radioactive decay in that the graph results model the half-life relationship. However, this simulation involves only 80 pennies, and the half-life of any radioisotope is based upon a **much** larger number of particles. The decay of nuclei is a chance occurrence as observed in this simulation. The averaging of several results from the simulation helps adjust for this fact. Also, identification and separation of real radioactive atoms are not as simple as this simulation might suggest.
5. To model the half-lives of different isotopes, use an object with more than two sides, such as dice or sugar cubes.
6. a. Four half-lives
   b. (1) Yes   (2) Yes
7. a. No
   b. A half-life value only predicts how many of the total number of atoms will decay, not which ones.
8. If you have $1000 and spend half of the amount remaining every 10 days, the half-life of the money is 10 days.

YOUR TURN

Half-lives (pages 301-302)

This activity provides practice in calculating half-lives.

Answers to Questions:

1. a. 5th year; ($500—1st year, $250—2nd year, $125—3rd year, $62.50—4th year, $31.25—5th year.)
   b. $0.98. ($15.62—6th year, $7.81—7th year, $3.91—8th year, $1.95—9th year, $0.98—10th year.)
2. a. 750 mg Cobalt-60. No; the rate of decomposition is independent of usage, and depends only on time.
   b. (1) 4, 50%—1 half-life, 25%—2 half-lives, 12.5%—3 half-lives, 6.25%—4 half-lives.
      (2) 7, 3.125%—5 half-lives, 1.563%—6 half-lives, 0.7813%—7 half-lives.
      (3) 10, 0.3906%—8 half-lives, 0.1953%—9 half-lives, 0.09766%—10 half-lives.
3. a. Theoretically, the person never reaches the telephone booth since one half of any distance still leaves a distance to be traveled to the destination.
   b. See the following table:

Distance (cm)	Time(min) [half-lives]	Percent
51,200	0	100
25,600	1	50
12,800	2	25
6400	3	12.5
3200	4	6.25
1600	5	3.13
800	6	1.56
400	7	0.78
200	8	0.39
100	9	0.20
50	10	0.10
25	11	0.05

4. a. (1) 0—1963; 1—1990.6; 2—2018.2; 3—2045.8; 4—2073.4; 5—2101.
        (2) 1963—100%; 1990.6—50%; 2018.2—25%; 2045.8—12.5%; 2073.4—6.25%; 2101—3.125%.
        (3), (4)

Year versus percentage

    b. 50% in 1991
    c. 3% (values can be read from graph).

B.5 RADIATION DETECTORS (pages 302–303)

This section discusses two types of radiation counters: ionization and scintillation. A Geiger counter is a kind of ionization counter that can detect alpha, beta, and gamma particles. Although gamma radiation is better detected by scintillation counters, the large alpha particles may go undetected by this device.

B.6 LABORATORY ACTIVITY: CLOUD CHAMBERS (pages 303–304)

In this activity students construct a simple device for indirect observation of ionizing radiation. The objective is to use a simple cloud chamber with supersaturated 2-propanol (isopropyl alcohol) vapor and observe the condensation trails generated by the radiation source.

Time

One class period.

Materials (for a class of 24 working in pairs)

    12 cloud chamber kits (see note below)
    100 mL 2-propanol
    1 lb dry ice (thin slabs)
    6-12 high-intensity light sources (20W or greater)
    12 squeeze bottles for 2-propanol
    24 disposable gloves
    12 crucible tongs

Advance Preparation

You will need to obtain cloud chambers and dry ice. (See Lab Tips below).

Pre-Lab Discussion

The Pre-Lab discussion may center on how radiation can be detected indirectly. You may want to emphasize the mechanics of radioactive counters as presented in Section B.5 and the counter used in the previous activity in *Laboratory Activity* B.1.

Lab Tips

Cloud chamber kits, including the radiation source, may be purchased from Sargent-Welch (catalog number 6829). Cloud chambers may also be made from transparent plastic disposable boxes, the kind containing salads in grocery stores or fast-food restaurants.

High-intensity (20-W, spot) light sources may be purchased from Sargent-Welch (catalog number S-58225-06). The light sources used for the Tyndall effect in the Water unit are ideal. A flashlight covered with tape, leaving a pencil-sized hole, may be used.

The Sargent-Welch toll free number is 800-SARGENT. Ask for an educational products salesperson.

It is advisable for you to place the radioactive source in the cloud chamber with forceps. Students need only dampen the felt band (on the top cover) with alcohol, thus avoiding direct contact with the radiation source. A small piece of lantern mantle can be used in the chambers.

The dry ice should be crushed unless you have a smooth block of dry ice. The objective is to get good contact between the dry ice and the bottom of the cloud chamber.

For best observation, the light source should be directed at an oblique angle rather than straight down.

The procedure calls for students to wear gloves while the radioactive source is being handled with forceps.

Experience with this activity indicates that practice is necessary. Rubbing the chamber with a silk cloth to promote a charged atmosphere may also help.

Post-Lab Discussion

Discuss the conditions that support the direct observation of ion vapor trails (supersaturation) and the importance of cloud chambers even in current research.

Answers to Questions:

1. The length, direction, and intensity of the tracks will differ due to variations in temperature and the degree of alcohol saturation.
2. Alpha particles will produce more visible tracks due to their larger mass and size, and to their relatively slow speed.
3. As the dry ice sublimes directly from the solid to the gaseous state, it absorbs heat from the surroundings. Thus, the dry ice forms a heat sink, reducing the temperature of the gases inside the closed chamber. This allows the already vaporized alcohol to supersaturate the air. The supersaturated gaseous mixture is unstable, and will condense to the liquid form with the slightest disturbance, such as the introduction of an ion speeding through the chamber.

B.7 ARTIFICIAL RADIOACTIVITY (pages 304-306)

A brief history of the development of nuclear chemistry is given, beginning with Rutherford's experiments with alpha particles and transmutation, the goal of the alchemists. The phosphorus-30 produced by the Joliet-Curies had a half-life of 3.5 minutes, changing into silicon-30 by the elimination of a positron, an electron with a positive charge:

$$^{30}_{15}P \rightarrow {}^{30}_{14}Si + {}^{0}_{+1}e$$

YOUR TURN

Bombardment Reactions (pages 305-306)

This activity provides more practice in balancing nuclear equations, while also helping to identify the target, the projectile, the product nucleus, and the ejected particle in transmutation equations.

Answers to Questions:

1.

target nucleus		projectile		product nucleus
$^{59}_{27}Co$	$+$	$^{1}_{0}n$	$\rightarrow$	$^{60}_{27}Co$
cobalt–59		neutron		cobalt–60

2.

target nucleus		projectile		product nucleus		ejected particle
$^{96}_{42}Mo$	$+$	$^{2}_{1}H$	$\rightarrow$	$^{97}_{43}Tc$	$+$	$^{1}_{0}n$
molybdenum-96		hydrogen-2		technetium-97		neutron

3.	target nucleus		projectile		product nucleus		ejected particles
	$^{209}_{83}Bi$	+	$^{4}_{2}He$	→	$^{211}_{85}At$	+	$2\,^{1}_{0}n$
	bismuth–209		alpha particle		astatine–211		neutrons

CHEMQUANDARY

Transmutation of Elements (page 306)

Transmutation has made possible the creation of elements not found on our planet, and the conversion of common elements such as lead into less common ones such as gold. In the latter case, the number of steps and the expense involved prohibit it from being a practical process. Yet, considering the many commercial radioisotopes produced to date, it could be argued that we've done better than the alchemists ever dreamed possible.

PART B: SUMMARY QUESTIONS (pages 306-307)

1. a. Beta decay
   b. Gamma decay
   c. Alpha decay

2. a. $^{60}_{27}Co \rightarrow \;^{60}_{28}Ni \;+\; ^{0}_{-1}e$

   b. $^{40}_{19}K \rightarrow \;^{40}_{20}Ca \;+\; ^{0}_{-1}e$

   c. $^{241}_{95}Am \rightarrow \;^{237}_{93}Np \;+\; ^{4}_{2}He$

   An ionization smoke detector uses the radioactive isotope americium-241. This isotope, half-life of 450 years, is located in a sensing chamber. As long as the alpha particles are emitted by the radioactive decay of Am-241, an electric circuit stays open, and the alarm is held in check. But, when smoke blocks the emission of alpha particles, the flow of electricity is reduced, which activates the alarm.

3. $t_{1/2}$ = 15 minutes

4. 6.25 micrograms

5. a.

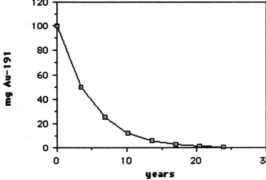

   b. (1) After 10 h = ~12.5 g; (2) after 24 h = ~0.781 g; (3) after 34 h = ~0.098 g
   c. Consumers might not like an expensive product which is almost totally gone after a day and a half. In addition, the radioactivity emitted by the jewelry could be harmful to the wearer.

6. Cloud chamber, scintillation counter, photographic film, electroscope, or rate meter.

7. $^{4}_{2}He$, alpha particle; $^{0}_{-1}e$, beta particle, and $^{0}_{0}\gamma$, gamma ray.

8. a.

		projectile	element formed		ejected particle
	$^{239}_{94}Pu$	+ $^{4}_{2}He$	→ $^{242}_{96}Cm$	+	$^{1}_{0}n$
			curium-242		

b.

	projectile	element formed	ejected particles

$$\underset{92}{\overset{239}{}}U \ + \ \underset{1}{\overset{1}{}}H \ \rightarrow \ \underset{93}{\overset{238}{}}Np \ + \ 2\underset{0}{\overset{1}{}}n$$

neptunium-238

9.   Pu-239 would be more useful for a space probe. One would need a long-lasting fuel supply, and the uranium-239 would be exhausted before the space probe reached a planet of our own solar system.

EXTENDING YOUR KNOWLEDGE (page 307)

- Since neutrinos have such a small mass, you would not expect the supersaturated gaseous particles in a cloud chamber to condense on the neutrino to form an observable vapor trail.
- This is a library research question.
- 7 alpha particles and 4 beta particles.
- The book *Problem Solving in General Chemistry* by Ronald A. DeLorenzo (D. C. Heath, 1981) contains an interesting discussion and problems in Chapter 15 on radioactive dating. This book is a good source for applied chemistry problems and useful analogies in teaching introductory chemistry.

C: NUCLEAR ENERGY: POWERING THE UNIVERSE

Part C introduces students to the prodigious energy that is locked inside the atom. They learn about the strong force (the force that holds together the nucleus) and see how nuclear energy is released via chain reactions in power plants and in weapons. Major components of a nuclear plant, are pictured and discussed. Students see the enormous difference in the magnitude of nuclear energy compared with chemical energy. Fusion, the nuclear reaction that powers the sun and the stars, and a potentially potent source of energy on Earth, is also introduced.

C.1 SPLITTING THE ATOM (page 308)

The brief historical sketch about the first nuclear fission reaction introduces Lise Meitner, an Austrian who worked with Hahn and Strassman, but who fled from Nazi Germany because of her religion. Hahn met with Meitner in Copenhagen and communicated the results to her. The complication was that barium compounds were used to precipitate out the expected radium products, which would then be subsequently separated. Hahn and Strassman were unable to separate any radium. Meitner explained this result by proposing that barium was one of the products, and that the uranium atom had split into two almost equal parts. In addition, she and her nephew, Otto Frisch, predicted that krypton would also be formed, which was verified by Hahn and Strassman.

C.2 THE STRONG FORCE (page 309)

Students are introduced to the major force that holds the nucleus together, often called the binding energy of the nucleus. Energy changes in nuclear reactions and chemical reactions are compared and contrasted. You may want to emphasize that a small quantity of mass, when converted into energy, yields an enormous quantity of energy, according to the equation $E = mc^2$. (To calculate energy in joules with this famous Einstein equation, mass should be expressed in kilograms, and the speed of light in parts of meters per second.)

C.3 CHAIN REACTIONS (pages 309–310)

The terminology of nuclear reactions is introduced and explained. Use Figure 17 (page 310) and the following *You Decide* activity as the focal points for your discussion. You may wish to point out that the term critical mass (or more properly, critical density) refer not only to nuclear bombs and explosions, but also to controlled reactions.

C.4 YOU DECIDE: THE DOMINO EFFECT (pages 310-311)

This activity demonstrates the concepts of expanding versus limited chain reactions. You will need 2-3 sets of dominoes for a demonstration, or as many as 30-45 sets for a student centered activity.

Answers to Questions:

2. b. This is an expanding (out of control) chain reaction, because the number of falling dominoes increases with each row.

4.   This is a limited (controlled) chain reaction, because the number of falling dominoes does not increase as it did above.

5. a. The expanding chain reaction models the explosion of a nuclear weapon, in which a tremendous amount of energy is released in a very short time.

   b. The limited chain reaction models a nuclear power plant. In this case the energy is released over a longer period of time, and is kept under control.

6. a. The limited reaction may be stopped by removing one or two of the dominoes in the sequence.

   b. It is nearly impossible to stop the expanding reaction since you would have to remove a whole row of dominoes at one time, instead of one or two. The farther the expanding reaction proceeds, the lower the probability of stopping.

7. a. This simulation is like a nuclear chain reaction in that the falling dominoes (representing disintegrating atoms) release energy, which triggers other atoms to fission.

   b. The simulation is limited in that it does not produce new substances or produce measurable quantities of energy.

   c. Other modeling alternatives include loaded mouse traps, ping-pong balls in a closed container, or popcorn popping.

DEMONSTRATION IDEA

The September 1986 issue of *Reactions* (free newsletter of the American Nuclear Society) suggests an alternative class participatory demonstration. Each student is given two corks (or other light objects: styrofoam balls, crumpled paper) and instructed to shut his/her eyes. The teacher begins the chain reaction by tossing a cork over his/her head in the direction of the student desks. If hit, a student must toss his/her corks in a random direction. Within the confines of an average classroom, the reaction usually takes a few tries to get started and dies out before all the corks are thrown. Repeat the process by having students take one step closer together. It's amazing how close they have to get to reach "critical mass." This demonstration is an effective model of the concept of chain reaction and critical density.

C.5 NUCLEAR POWER PLANTS (pages 311-314)

This section deals with the basic components and operation of nuclear power plants. The mechanics of plants are explained, with diagrams and text. Several different types of reactors are in use: pressurized water, boiling water, high temperature gas-cooled, and breeder reactors in Europe. The text explains pressurized and boiling water reactors.

OPTIONAL BACKGROUND INFORMATION

State of Nuclear Power Industry in the United Sates

From the days of the post-World War II "Atoms for Peace" program and the opening of the world's first large-scale nuclear power plant (December 18, 1957 in Shippingport, PA), the United States has been an acknowledged leader in nuclear power-plant construction. Today, due to a variety of technological, economic, and political factors, the U.S. nuclear industry is in a state of decline. The last year a nuclear plant was ordered, and not subsequently canceled, was 1974. Between 1975 and the end of 1983, some 87 plant orders were canceled. Construction delays and costs overruns mean that the 20-odd plants originally scheduled for completion between 1986-1990 will likely take an average of about 12 years to build, and cost about twice as much as comparable coal plants. While nuclear power is certainly not dead–the number of operable reactors grew from 100 to 111 between 1986 and 1990–the early promise of "safe, clean electricity too cheap to meter" has certainly proved to be a myth.

Reliance on Nuclear Power on an International Level

On a worldwide basis, nuclear power accounts for a sizable fraction of total electric power generation in several countries. Most notable among these is France where, faced with scarcely any domestic supplies of coal, oil, and natural gas, the government has adopted an aggressive nuclear plant development program. Yet, especially in the wake of the Chernobyl accident, nuclear power is being hotly debated. The following countries have adopted a public policy of not building additional nuclear power plants and phasing out existing ones: Australia, Austria, Denmark, Greece, Ireland, Luxembourg, New Zealand, Philippines, and Sweden.

Worldwide Nuclear Power Commitment, July 1, 1986

Country	Number of plants operating		Number under construction		% of electricity generated
		Megawatts		Megawatts	
United States	110	96,331	4	4,284	19
France	55	52,588	9	12,245	75
Soviet Union	46	34,230	26	22,180	12
Japan	39	29,300	12	10,629	28
West Germany	24	22,716	1	295	50
United Kingdom	39	11,242	1	1,188	22
Canada	18	12,185	4	3,524	16
Sweden	12	9,817			45
All others	103	49,862	39	24,562	
World total	426	318,271	96	78,907	15

Source: International Atomic Energy Agency.

[1]Note: The Energy Information Administration's *Monthly Energy Review* is another reliable source of energy data.
[2]**Alternatives to Reliance on Nuclear Power**

Electric utilities interested in providing adequate power at the lowest feasible price (with an accounting of the environmental costs) are turning to other options to fill the demand for electricity, which is rising at a lower rate than in previous years. During the 1970s electric demand rose at a rate of about 7%/year; in the 1980s, less than 2%/yr.

OPTIONAL CHEMQUANDARY

Proponents of nuclear power claim that it is held to unrealistically strict standards. In particular, they argue that its environmental and health impacts are less than those of coal plants, which are estimated to kill thousands of people each year by polluting the air. In assessing the relative risks of nuclear and coal plants, what considerations might lead to especially strict standards being imposed on nuclear power plants?
Answer: Although the probability of a catastrophic nuclear accident may be small, the potential effects in terms of massive loss of life and property damage far exceed the destruction possible from the worst coal plant accident. One might compare the regulation on construction, maintenance, and traffic control within the airline industry, with that on the automotive industry. Also, the impact on future generations (in terms of genetic disorders and long-term environmental contamination) is greater in general. This is not to say that coal plants are the best answer to our energy needs.

C.6 NUCLEAR FUSION (pages 314-315)

This section explains nuclear fusion, introduces positrons, previously mentioned in the Teacher's Guide on page 156.
   The tremendous energy potential, as well as difficulties with fusion plants, are explained.
   The promise of generating electricity via nuclear fusion is that fusion reactors will produce neither the troublesome air pollutants of fossil fuel plants, nor the long-lived fission fragments and transuranic elements resulting from fission reactors. However, fusion reactors will contain a large inventory of radioactive tritium, and during operation structural components will become radioactive. Mid-1970s forecasts for the development of fusion power, which suggested full-scale power plants by the year 2020, are now considered unrealistically optimistic by most experts. You may wish to encourage some of your students to prepare reports on the history and present state of fusion research.

PART C: SUMMARY QUESTIONS (page 315)

1. a. Uranium-235, uranium-233, plutonium-239, and californium-252.
   b. Radioactive decay is a naturally occurring process resulting in more stable isotopes. Nuclear fission can also be artificially induced by neutron bombardment.
2. $$ _{0}^{1}n \ + \ _{92}^{235}U \ \rightarrow \ _{35}^{87}Br \ + \ _{57}^{146}La \ + \ 3 \ _{0}^{1}n $$
3. a. A chain reaction is a self-propagating reaction where the products of one reaction initiate the next. When fissionable material such as U-235 is bombarded with neutrons, the atom is split into lighter isotopes, producing more neutrons. These split other U-235 particles, thus setting off a chain reaction.
   b. A chain reaction will occur when the fissionable mass is near enough or large enough to capture neutrons creating the chain reaction. The specific mass in this volume is termed the **critical mass.**
   c. An expanding chain reaction cannot be stopped and will lead to an explosion, such as in a nuclear bomb. In a controlled reaction the process is slowed down to a manageable speed, mainly by absorbing neutrons and lowering the percentage of fissionable material.

4. Nuclear reactor fuel pellets are only 3% U-235. Weapon-grade uranium is enriched to 90% U-235, enough to cause a chain reaction leading to a nuclear explosion. The uranium used in nuclear power plants is sufficient to allow the chain reaction to occur, but not enough to cause a nuclear explosion.

5. a. Both nuclear power plants and fossil-fuel-fired plants heat water to drive turbines that generate electricity.
    b. Nuclear power plants generate energy by bombarding fissionable materials with neutrons, while fossil-fuel-fired plants burn materials such as coal to release thermal energy.
    c. The advantage of using fossil fuels is that they are readily available. Two disadvantages are the large amounts of carbon dioxide and other gases such as sulfur dioxide dumped in the atmosphere. One advantage of nuclear fuel is that relatively small volumes are needed; a disadvantage is that large volumes of radioactive waste will accumulate over the years.

6. The control rods limit the chain reaction in nuclear power plants. See Section C.5 on page 313.

7. Nuclear fission is the splitting of radioisotopes by neutron bombardment into lighter isotopes; nuclear fusion is the high-temperature combining of lighter isotopes to form heavier isotopes with the evolution of energy. In effect, fission is the break down of larger nuclei into smaller pieces, and fusion the joining together of small nuclei to make a larger one.

8. a. Nuclear fusion in the laboratory is full of problems and is not yet available for use as an economically sound source of energy.
    b. Since energy from the sun drives the physical and chemical processes of the Earth, and since this energy is derived from nuclear fusion reactions, it may be said that nuclear fusion is already the "number one" energy source on Earth.

EXTENDING YOUR KNOWLEDGE (page 315)

- In a very real sense, we are "star dust." Astrochemists believe that all the heavier elements found in the universe were produced via fusion reactions from the lighter elements (hydrogen and helium) that make up the bulk of the mass of stars.
- Check the *Readers' Guide to Periodical Literature* and the Supplemental Readings list provided on page 172.

D: LIVING WITH BENEFITS AND RISKS

What risks are worth taking to reap the benefits of nuclear technology? Before delving into the intricacies of this question as it relates to nuclear technology, students perform risk-benefit comparisons of several means of transportation. Next, they explore the benefits of radioisotopes used widely in medicine, industry, and biological research. They also learn how ionizing radiation is measured, and what its effects on living systems are. Following that, students evaluate their own exposure to both natural and manmade sources of radioactivity, in an exercise designed to teach them to balance the risks of individual exposure. Finally, they examine two of the most important hazards of nuclear power—the dangers of nuclear waste and the risk of a catastrophic accident.

D.1 YOU DECIDE: THE SAFEST JOURNEY (pages 317-318)

This activity is designed to point out that risks and benefits enter into every part of our lives. How we deal with risks may be a function of information coming from the news media. Controversy arises when the risk is in close balance with the benefits, or when society (government) makes decisions in opposition to the wishes of certain individuals or sub-groups.

Answers to Questions:

1. a. Bicycle: $(500) \times (0.000001/10) = 0.00005 \ (5 \times 10^{-5})$
        Auto: $(500) \times (0.000001/100) = 0.000005 \ (5 \times 10^{-6})$
        Scheduled airline: $(500) \times (0.000001/1000) = 0.0000005 \ (5 \times 10^{-7})$
        Train: $(500) \times (0.000001/1200) = 0.00000042 \ (4.2 \times 10^{-7})$
        Bus: $(500) \times (0.000001/2800) = 0.00000018 \ (1.8 \times 10^{-7})$
    b. Bus travel is the safest.
    c. Bicycling is the least safe.

2. Persons may have the false impression that air travel is the most dangerous form of transportation due to media attention given to air disasters; rarely do we read about nonlocal bicycle fatalities. Persons may have the false impression that automobiles are the safest due to their familiarity with driving.

3. a. Traveling by bicycle is the cheapest, auto/train/bus are faster than the bicycle, and transportation by scheduled airline is the fastest and most worry-free in terms of responsibility for guiding the vehicle. Traveling by train/bus takes longer than scheduled airline but offers the lowest risks.
   b. Increased time, cost, personal responsibility, and risk factor would be seen by most as negative factors. In this case the cost savings of traveling 500 miles by bicycle probably does not seem worth the risk of death and the extended travel time. Other factors (see Question 5), such as the desire for personal mobility and luggage room provided by the auto, may influence decision-making.
4. These statistics would probably not hold true for 25 years given an increase in the population and increased automobile transportation and air travel. On the other hand, new technologies might make transportation safer or even provide totally new options.
5. Cost (including meals and overnight expenses), time, responsibility for vehicle, space, payload (luggage, load), freedom to stop vehicle or detour, exercise, pleasure, lack of worry regarding mechanical breakdown.
6. a. Individual answer
   b. A risk/benefit analysis could never yield the same result for every individual since each person places a different priority and value on each mode of transportation.

CHEMQUANDARY

Risk-Free Travel (page 318)

There is no risk-free way to visit a friend. Even deciding not to visit a friend may pose unknown risks. A conscious decision not to take a risk may put one in a position of greater risk, because of future events unknown at the time of the decision. The idea here is to minimize total risk by taking a more knowledgeable approach towards known risks. Your students may be surprised to learn that traffic fatalities per passenger mile (30 deaths per 10^8 passenger-miles on horses) were approximately 30 times higher in the pre-automobile days than today (car), or roughly equivalent to the fatality rate for motorcycles. There is no such thing as a risk-free ride.

OPTIONAL CHEMQUANDARIES

1. How has the government attempted to legislate certain safety factors into the transportation industries? Should it attempt to outlaw the transportation mode with the highest risk factor? Why or why not?
   Answer: Seat belts, infant car seats, driver's licenses, helmets for motorcyclists, drug testing (for pilots, conductors, and bus drivers), stop lights, maintenance requirements for aircraft, etc. are legislated safety factors. Given our transportation problems, especially in large cities, outlawing a transportation mode because of its high risk factor is not an educated approach to solving either the problems that contribute to the high risk, or our transportation problems.

2. Briefly list some other aspects of modern life that involve both risks and benefits.
   Answer: Risks and benefits in everyday life include diagnostic X rays, jogging, sunbathing, pharmaceuticals, childbirth, power lawn mowers, chlorinated drinking water, smoking, and many, many others.

D.2 BENEFITS OF RADIOISOTOPES (pages 318-322)

The medical benefits of radioisotopes, including diagnostic tracers, high-energy emitters for cancer treatment and sterilization of equipment, are discussed in this section.

OPTIONAL BACKGROUND INFORMATION

Medical Applications: Diagnosing and Treating Diseases
The medical applications of nuclear technology date back to Roentgen's X-ray imaging of his wife's hand on December 22, 1895. Within a few years, X-ray machines were being widely used for medical diagnosis. As understanding of nuclear chemistry increased, so did its medical applications. Today, nuclear medicine techniques are involved in diagnosis, therapy, and research on a wide variety of disorders. Some indicators of the extent of medical applications include: (1) Over 10 million nuclear medicine procedures are performed each year in the United States. (2) One patient in four admitted to a general hospital has some type of nuclear diagnostic procedure performed. (3) Nearly 90% of all new drugs require evaluation using radioisotopes before receiving FDA approval. (4) About 30% of all biomedical research involves radioisotopes. (5) Syringes, bandages, first-aid dressings, gowns, masks, tongue depressors, and sutures are often sterilized by nuclear radiation.

As with all technologies, nuclear medicine involves risks as well as benefits. Early X-ray machines exposed patients to unnecessary and excessive amounts of radiation. Advances in technology have allowed far improved imaging with considerable reduction in the radiation required. However, encourage your students to question their doctors and dentists about shielding (especially of the reproductive organs) and exposure rates. Even today, many nuclear medicine specialists feel that some patients are needlessly exposed to unnecessary amounts of radiation in some medical examinations (scoliosis radiography without the use of breast shields, fast film-screen combinations, and compensating filters; X-ray pelvimetry; some mammography; etc.). In general, however, the benefits of most medical uses of nuclear chemistry far outweigh the risks.

Other Applications of Radioactive Substances

Radiation technologies now play an important role in manufacturing goods and providing services worth $2–3 billion (Atomic Industrial Forum). Specific examples include:

Homes: In ionizing smoke detectors, a small amount of polonium in the tone arm of some turntables reduces static.

Offices: In photocopying machines, to eliminate static

Manufacturing plants: For thickness gauging, product testing, paint curing, coating nonstick (Teflon) frying pans

Police labs: In crime detection

Art: In the process of authenticating works of art

Agriculture: To measure utilization of fertilizers and uptake of pesticides

Food: Radiation can be used to preserve food (and thereby reduce the need for post-harvest fumigants and other chemicals that can leave potentially cancer-causing residues). Gamma rays from Co-60 or Cs-137 have been shown to inhibit sprouting of potatoes and onions, delay ripening of fresh fruits and vegetables, eliminate many of the bacteria responsible for food poisoning, and even sterilize food outright. On April 18, 1986, the FDA authorized broad use of this technology, despite opposition from some consumer groups who contended that the generation of unique radiolytic products in irradiated food had not been adequately studied for human safety.

Health care products: To reduce or eliminate microorganisms in baby powder, baby bottles, infant teething rings, artificial eye lashes, mascara, contact lens solutions, cotton swabs, burn ointments, and talcum powder

City water departments and large oil companies: To detect pipeline leaks or determine flow

Highway construction companies: To measure density and moisture in soil for road building

Laboratories: In chemical analysis, archaeological dating, etc.

OPTIONAL CHEMQUANDARY

Availability and storage of medical radioisotopes pose special problems that are not common to most other chemicals used in medicine. Why?

Answer: Most medical radioisotopes have short half-lives and as a result cannot be stockpiled for any reasonable length of time. This means that supplies must be either generated on site as needed (in cyclotrons or small-scale nuclear reactors) or shipped via a rapid delivery system.

D.3 YOU DECIDE: PUTTING ATOMS TO WORK (page 322)

This section involves students in both individual and group exploration of ideas concerning the use of radioisotopes to solve particular problems. We hope students will see the benefits of nuclear technology in this exercise, providing some balance or perspective often lacking in public reports of nuclear technology, which tend to focus on sensational bad news. Even one who opposes nuclear weaponry and/or nuclear power generation, might rationally argue in favor of other nuclear technologies that have given us numerous benefits with relatively few risks. While students may be aware of some medical applications, they are probably less familiar with the industrial and agricultural uses. Review the examples given in the text, paying attention to the half-lives of isotopes and the role of isotopes in detecting the presence or absence of a particular ion, such as iron in the bloodstream.

Time/Grouping

This activity can be run in a variety of ways: (a) Assign the four items from the text as individual homework for everyone. The next day, after discussing students' ideas, pose a subset of the seven optional items as additional large or small group discussion items. Alternatively, (b) use both the four items from the text and the five optional items listed on page 224 in the Teacher's Guide. On the day before the in-class activity, assign students to groups of three to five. Each group will be responsible for working on a subset of the nine items. (You may wish to build in some overlap by having each group work on two or three items). That night, students should begin the activity as an individual homework assignment—outlining (in written form to be handed in if desired) their own ideas on the problems. On the next class day, allow approximately 20 minutes for individual and group sharing of ideas, and 30 minutes for group presentations and full class discussion.

Support

An excellent teacher/student reference volume for this activity and the entire unit is the *Chemical Rubber Company (CRC) Handbook of Radioactive Nuclides*. Specifically, for this activity, Part VI—"Radionuclides for Medical Application" and Part VII— "Radionuclides for Industrial Applications" are most useful. Also of value (and more readily available, though less detailed) are: (1) *CRC Handbook of Chemistry & Physics*—see the extensive table of isotopes which lists all the known isotopes, their half-lives, mode of decay and natural abundance; (2) *The Merck Index: An Encyclopedia of Chemicals and Drugs*, 11th edition (1991). Misc-1 to Misc-20 contains the above information plus more detailed information on isotopes used in medical therapy and diagnosis; and (3) *McGraw-Hill Encyclopedia of Science and Technology*.

Pre-Activity Discussion

This activity exposes the students to a wide range of applied problems involving radioisotopes and gets them thinking about possibilities. It is not intended that they should conduct extensive library research in order to obtain all the right answers. The specificity of answers (refer to the four problems given in the student text) that a given team will be able to provide may vary somewhat with each of the items; in many cases they should be able to indicate the specific element(s) (or in some instances even the specific isotope) needed, in other cases perhaps only the solubility characteristics of the compound used. Also, students should have some sense of the order of magnitude of half-life needed and any safety precautions. Even if no reference materials are available, students should be able to use the general and nuclear chemistry they have learned to date to list some of the information relevant to the application. In general, most of the nine items involve one-shot, tracer applications of radioisotopes.

Answers to Questions:

1. In the case of oil, either an oil-soluble or insoluble compound could be used; in the case of natural gas, a gaseous form would be needed. Given that the fuel may be traveling across country, isotopes with half-lives on the order of tens of days or years would be needed. If the company placed different tracers at the start of each new shipment, the receiver of the shipments could monitor the arrival of each new shipment. Actual isotopes used for this purpose include Sb-124 (60.3 d), Co-60 (5.24 yr), and Cs-137 (30.23 yr).
2. Use radioactive forms of metals that are a normal part of the piston rings and/or cylinders with half-lives less than the expected life of the engine parts (ten years, though actually these tests are conducted under accelerated time schedules). Possibilities include: Fe-55 (2.6 yr), Zn-65 (245 d).
3. The chief elements of concern are nitrogen, phosphorus, and potassium. The radioisotopes of nitrogen are too short lived to be of practical value. P-32 (half-life: 14.3 d) has been used extensively for this purpose, as have some forms of potassium (K-42/12.4 h and K-43/22.4 h).
4. Unlike the tracer applications, this medical treatment requires the use of a high-energy emitter. Co-60 (5.24 yr), a powerful gamma emitter, is commonly used for this purpose. The source is outside the body, finely focused, and the dose level carefully controlled.

OPTIONAL ITEMS

1. A local conservation group wants to determine how long it takes for underground water to leach from an old, illegal chemical dump site to a lake five kilometers away, and to estimate the amount of water leaching.
   Answer: Use radioisotopes in water-soluble compounds with half-lives on the order of days to years. Actual radioisotopes used for this purpose include: H-3 (12.26 yr), Br-82 (35.5 h), I-131 (8.1 d).
2. An industrial chemist wishes to compare the cleansing effectiveness of a new detergent formulation to that of a well-established product.
   Answer: Prepare special dirt, oil, or grease containing a radioisotope such as P-32 (14.3 d) and measure the radioactivity in the soiled clothes before and after laundering.
3. Underground pipelines sometimes spring leaks that are heard to detect. It is expensive as well as time-consuming to dig up huge sections of pipeline, in hopes of finding the leak. a. What could be used to find the leaks, so that they can be repaired at minimum cost? b. Would the same substance work for both oil and water pipelines? c. If not, how and why should they differ?
   Answers: a. Use a radioisotope which could be detected by a scintillation or Geiger counter. b. No. c. One would a water-soluble isotope for water pipes, and a nonpolar substance for an oil pipeline. The half-lives would have to be on the order of minutes to hours, depending on the length of pipe being searched. Also, in the case of pipes carrying domestic water, public safety would require short-lived isotopes.
4. A chemist develops a new herbicide that is intended to attack broadleaf weeds selectively and not be absorbed at harmful levels by grain crops. How might this be tested?
   Answer: A radioactive form of the herbicide would need to be synthesized. P-32 (14.3 d) and I-131 (8.1 d) have been used.
5. A medical researcher wishes to know how efficient the body is in extracting a key mineral needed by red blood cells from a given diet.
   Answer: The mineral required by red blood cells is iron, so either Fe-55 (2.6 yr) or Fe-59 (45.1 d) would be useful.

Post-Activity Discussion

Be sure students have a general overview of the use of radioisotopes. Uses of radioisotopes fall into one of five broad categories: (1) tracing, (2) gauging, (3) radiography, (4) ionizing radiation (including sterilization, cancer treatment, etc.), and (5) heat and power generation (ranging from electric generating plants, to spacecraft, to pacemakers).

The principal advantage of radioisotopes lies in their ability to get things done better, or faster, or more cheaply than by other methods. Also, in some cases they allow us to do things that cannot be done in any other way.

The disadvantages include safety concerns, and the needs for trained personnel, sophisticated equipment, and licensing regulations. A decision to use radioisotopes in a given situation must be made by weighing the various benefits and risks specific to that application.

D.4 MEASURING IONIZING RADIATION (pages 322-323)

A quantitative scale for measuring radiation exposure is introduced. There are several units to measure radiation, such as the Curie, the Roentgen, the Rad, the Rem, and the Sievert. The text has limited its discussion to the Rem, which concerns the impact of radiation on the human body. You will need to remind students of the prefix milli– and perhaps briefly explain what a Roentgen is, in order to clarify the marginal note.

OPTIONAL INFORMATION

Curie (Ci): 3.7×10^{10} disintegrations per second, which is the measured activity of one g of radium. Radiation "concentration" in air is often expressed in terms of Ci/L.
Roentgen: A measure of the capacity of a source to cause ionization—one Roentgen causes the formation of 2.1×10^9 units of charge in 1 cm^3 of air at 0° C and one atmosphere of pressure; 87 ergs of energy absorbed per gram of air.
Rad (*R*adiation *a*bsorbed *d*ose): Absorption of 100 ergs of ionizing radiation per gram of any material.
Rem (*R*oentgen *e*quivalent *m*an): The dose equivalent for any type of ionizing radiation absorbed by body tissue in terms of its estimated biological effect on exposure to one roentgen of X rays.
Sievert (The SI unit of radiation): equal to the dose of one milligram of radium enclosed in a platinum container with walls 1/2 mm thick, sent to a distance of 1 centimeter in one hour. Is approximately equal to 8.38 Roentgens.

D.5 RADIATION DAMAGE: NOW AND LATER (pages 323-324)

This section considers the biological effects of ionizing radiation on human tissue, as well as the link to cancer. Focus student attention on Tables 10 and 11 on page 324. The text outlines the uncertainties involved in estimating the long-term effects of exposure to low dose levels (below 25 mrems). This is related to the similar discussion concerning the Delaney Clause in the Food unit.

One problem with assessing radiation damage is that radiation does not produce unique effects which can be unequivocally identified. One in five persons will die of cancer from a variety of causes—experts estimate that only about 2% of cancer deaths are due to radiation. Contrast this to 30% of all cancer deaths attributed to voluntary cigarette smoking, 3% attributed to alcohol consumption, and 2% attributed to asbestos exposure.

D.6 EXPOSURE TO RADIATION (pages 324-326)

This section clarifies sources of radiation and presently accepted levels of exposure. This material and Section D.7 provide the background needed for the *YOUR TURN* activity. The point is made that we are all regularly exposed to natural, low level, background radiation (of about 135 mrem/year) and that, through various conscious or unconscious decisions (such as where we choose to live, what medical treatment we receive, etc.), we are exposed to varying amounts of additional radiation generated as a result of human activities. Note that, just because human additions to the natural background radiation seem small (with the exception of medical X rays—see Figure 26 on page 325), it does not mean that they are unimportant in terms of potentially negative health effects. Also, individual risks are determined by individual, not group average exposures.

CHEMQUANDARY

Radiation Exposure Standards (page 325)

Certain individuals, having more rapidly dividing cells, such as children and pregnant women, are more at risk at a given dose level than the general population. On the other hand, for some people needing the benefits of radiation treatment, higher doses may be worth the risks. Also, for practical (including economic) reasons, occupationally exposed individuals (such as medical X-ray technicians and workers handling radiopharmaceuticals, defense program workers, and civilian nuclear power plant workers) are allowed to receive higher, though apparently still safe, doses.

D.7 RADON IN HOMES (pages 326-328)

The radon issue is a particularly interesting one to explore with your students because: (1) It is a new problem that has only received widespread attention since the mid-1980s. (2) It is a case where nature, not man, is the source of the pollution. (3) It illustrates how solving one problem (high home energy bills) can sometimes help create a new one. (4) It provides a good context for discussing risk-benefit analysis and the concept of comparative risks.

The following background information was obtained from: (1) *A Citizen's Guide to Radon: What It Is and What to Do About It*. U.S. Environmental Protection Agency, OPA 86-004, August 1986. (2) *Radon Reduction Methods: A Homeowners Guide*. U.S. Environmental Protection Agency, OPA-86-005, August 1986. (3) Radon Information Council, 1250 Eye Street, NW, Suite 300, Washington, DC 20005. 202-682-0690.

OPTIONAL BACKGROUND INFORMATION

Radon is a naturally occurring, chemically unreactive, dense (five times as dense as air), colorless, odorless, tasteless gas that results from the radioactive decay of uranium. It can be found in high concentrations in soils and rocks containing uranium, granite, shale, phosphate, and pitchblende. In outdoor air, radon is diluted to harmless concentrations. However, if a well-insulated home is built on soil where radon concentrations are naturally high, dangerous levels can occur. In the United States, regions that have been identified as being radon hot spots include the "Reading Prong" region running through Pennsylvania, New Jersey, and New York; the phosphate lands of Florida; certain regions of Colorado; and some mining areas in Montana and other western states. (See the map in "Radon Gas: A Deadly Threat," *Newsweek*, August 18, 1986). It is especially ironic that many energy-conscious homeowners, insulating their homes to keep the cold out, are at the same time sealing radon in.

As with other materials that emit ionizing radiation, the main health risk is that of cancer. Based on studies of miners exposed to radon in their work underground, scientists estimate that from 5,000-20,000 lung cancer deaths a year in the United States may be attributed to radon. For comparison, the Surgeon General and the American Lung Association attribute around 110,500 deaths, or 85% of the total of 130,000 lung cancer deaths per year to smoking. Of course, the risk of developing cancer increases as the level of radon (concentration) and the length of exposure (time) increase. Also, although radon is an inert gas (and therefore might be expected to leave the body readily), its decay products—polonium, bismuth, and lead— are solid particles that can become trapped in the lungs, where they continue to do damage even after the original source of radon is removed.

Solutions to the radon problem depend on being able to assess its concentration and source. Charcoal canisters and alpha track detectors that measure working levels of radon gas in terms of picocuries per liter (pCi/L: a picocurie is 10^{-12} curies) can be obtained from private firms or state radiation protection offices. While radon levels vary with season (higher in the spring after the winter ground thaw) and location within a home (higher in basement), if screening measurements result in concentrations below 4 pCi/L (or 0.02 working levels), there is probably little reason to be concerned if one does not smoke (there is evidence for a synergistic health risk), or sleep in the basement. Higher levels may indicate the need for one or more radon reduction techniques.

If high radon levels are found, a variety of radon reduction methods are potentially of value: (1) Increase natural ventilation—In the average American house, all the interior air is replaced by outside air about once every hour. Newer, tighter houses have air exchange rates as low as one tenth of that rate. The latter run a greater risk of having higher radon concentrations. (2) Provide forced ventilation—Air can be blown into houses and allowed to exit through windows or vents. However, using exhaust fans to pull air out of the house is likely to decrease interior air pressure and draw more radon inside. (3) Utilize heat recovery ventilation. (4) Provide furnaces and clothes dryers with separate sources of external air. (5) Cover exposed earth. (6) Seal cracks and openings. (7) Install drain tile suction, block wall ventilation or sub-slab suction.

In some New England states (especially Maine) radon-contaminated well water is also a concern. Public drinking water supplies from surface sources are not an appreciable source of radon. Other rare sources of radon contamination are homes constructed with building materials made from radioactive shales, mine tailings, or other contaminated materials.

The risk estimates in the chart on page 167 are based on the assumption that 75% of the person's time is spent in the home. If less time is actually spent there, the risk is reduced.

Radon Risk Evaluation Chart

pCi/L[a]	Working Level[b]	Estimate additional lung cancer deaths (per 1000 people)	Comparable exposure levels	Comparable risk
0.2	0.001	1-3	Average outdoor level	20 chest X Rays/yr
1	0.005	3-13	Average indoor level	Non-smoker risk of dying from lung cancer
2	0.01	7-30	10× avg outdoor level	200 chest X rays/yr
4	0.02	13-50		5× nonsmoker risk
10	0.05	30-120	10× avg indoor level	
20	0.1	60-120	100× avg outdoor level	1-2 packs a day smoker
100	0.5	270-630	100× avg indoor level	20,000 chest X rays/yr
200	1.0	440-770	1000× avg outdoor level	14 packs/day smoker or more than 60× nonsmoker risk

[a]One pCi/L represents the decay of about two radon atoms per minute in one liter of air. The EPA has set 4 pCi/L as the level above which it recommends remedial action be taken. The higher the levels, the sooner action should be taken. [b]Working level is a measurement of radon decay products. One working level is about 200 pCi of radon gas.
Source: Adapted from *A Citizen's Guide to Radon: What It Is and What to Do About It.* U.S. Environmental Protection Agency, OPA-86-004, August 1986.

OPTIONAL CHEMQUANDARIES

1. Why would radon concentrations tend to be higher in basements than in upstairs rooms?
   Answer: The source of radon is the ground, and radon is considerably denser than air.
2. Why do radon daughter products pose a greater health risk than radon itself?
   Answer: Radon, being an inert gas, would pass readily out of the body once inhaled. Its daughter products exist as charged particles that become readily attached to dust and can remain lodged in the upper respiratory track and lungs where their harmful effect can continue over long periods of time. They can also deposit themselves in the room and continue to be radioactive.

YOUR TURN

Your Annual Radiation Dosage (pages 326-327)

This activity heightens student awareness of the sources and quantities of radiation in their everyday environment. They are asked to estimate the dosage of radiation received per year, analyze the sources, and consider means of reducing the quantity of their annual radiation exposure. "Personal Radiation Dose Charts" are available in quantity at special prices from the American Nuclear Society, Public Communications Department, 555 North Kensington Avenue, La Grange Park, IL 60525, (708) 579-8265.

Answers to Questions:

1. a. Individual student answers.
   b. Individual student answers, comparing their numbers to 135 mrem.
2. a. Yes: some ways are practical, such as avoiding unnecessary X rays by keeping records of medical and dental X rays, inquiring about the necessity of each, and being sure technicians use lead to shield body parts not being photographed. Other ways to reduce your exposure are harder to implement—flying less (this may involve switching to more accident-prone automobile transportation), moving to lower elevations, or switching to a wood home, etc.
   b. Individual answers. In many cases, such as choice of living or type of home, students have no choice to make.

D.8 NUCLEAR WASTE: PANDORA'S BOX (pages 328-331)

This section introduces students to a highly controversial and pressing issue. The nature of nuclear waste disposal, with the proposed solution, is presented and clarified. Even if all nuclear waste generation were to end tomorrow, this would still be an issue requiring continued attention. You may wish to encourage students to watch newspapers for articles pertaining to this issue. Given the nature of radioactive decay (particularly the concept of half-life), the problem will not go away in the near future.

Nuclear Power Worldwide (page 324)

Most other countries have higher percentages of electricity supplied by nuclear power, and substantial plans for expanding its usage. The United States has been fortunate in having abundant supplies of coal, oil, and natural gas. Although our prodigious consumption of oil makes it necessary to import oil, our coal supplies are more than adequate to meet electrical production needs well into the future. Most other nations are not as fortunate regarding domestic energy resources.

OPTIONAL BACKGROUND INFORMATION

Nuclear waste disposal presents both technological and political problems. While billions were spent in the 1950s and 1960s to develop nuclear power and produce nuclear weapons, the government spent only a few tens of millions for research on ways to solve the waste problem. Additionally, most regions of the country have taken a "not in my backyard" view of the problem. As of 1982, some 22 states had passed laws barring or limiting nuclear waste disposal within their borders. In the meantime, the volume of nuclear waste continues to grow. As of 1985, radioactive waste included: 3 million cubic meters of low-level wastes (approximately 26% from commercial/74% from military); 225 million tons of uranium mine tailings in AZ, CO, NM, ND, OR, PA, SD, TX, UT, WA, WY. Every ton of uranium produced from milled ore results in almost 200 tons of tailings (approximately 83% from commercial/17% military); 12,500 metric tons of spent commercial, nuclear reactor fuel; 368,000 cubic meters high-level radioactive waste from military reprocessing of irradiated fuel and the fabrication of plutonium for weapons—another 5000 cubic meters has been generated by commercial nuclear power plants; 251,000 cubic meters of transuranic waste (mainly from the reprocessing of irradiated fuel and the fabrication of plutonium by the military).

These wastes are broadly classified as:

High-Level Radioactive Wastes (HLRW). These wastes include spent fuel rods from nuclear reactors which contain fission products and transuranic elements. Each year approximately one-third of the fuel rods within a reactor need to be replaced. These are presently being temporarily stored in pools at commercial nuclear reactor sites. Also, such wastes include liquid by-products of the reprocessing of spent fuel for the extraction of plutonium for nuclear weapons. These wastes are presently being stored in tanks at military installations. HLRW may contain radioisotopes with half-lives of 1000 years or more.

Low-Level Radioactive Wastes (LLRW). These wastes contain less than 10 nanocuries of transuranic elements per gram of material; LLRW result from almost all processes involving nuclear applications, from mining to medicine.

Technologically, storage solutions must take into account the fact that radioactive wastes are both thermally hot and extremely radioactive. The common rule of thumb is that radioisotopes must be stored for 10-20 half-lives before they decay to safe levels. Thus, LLRW can be stored for comparatively short periods of time and then diluted and dispersed. Most scientists consider LLRW to be more of a political than a technological problem. HLRW, on the other hand, contain radioisotopes with very long half-lives and must be effectively stored for hundreds or thousands of years. Designing containers and finding geological locations that will remain stable for long periods of time is a difficult technological problem.

Government actions to ensure safe, long-term storage have been slow and have met with considerable public opposition. The Low-Level Radioactive Waste Policy Act of 1980 urged all states to form interstate regional compacts to ensure the existence of sufficient facilities for disposal of each region's low-level waste by January 1, 1986. This deadline was later extended to 1992. The Nuclear Waste Policy Act of 1982 calls for the Department of Energy to establish two permanent repositories for high level radioactive waste—one in the West by the end of the century and one in the East to open after that. As the debate continues, many commercial nuclear power plants are running out of short-term storage and many states have refused licensing for new plants until the problem of wastes is resolved.

OPTIONAL CHEMQUANDARIES

1. One proposal for solving the nuclear waste problem is to use rockets or the space shuttle to shoot the wastes out into space. What are some problems with this idea?

   Answer: High costs, possibility of a launch accident spreading radioactive wastes across a wide area.

2. Even if all nuclear power plants were phased out in the next five years, nuclear waste disposal would remain a controversial issue that still needed attention. Why is this so?

   Answer: We would still need to find an effective solution for the waste accumulated at the power plants and the much larger volumes produced by military weapons production. Additionally, some small, non-electricity generating reactors would need to remain in operation to supply radioisotopes for medical and other essential applications.

D. 9 CATASTROPHIC RISK: A PLANT ACCIDENT (pages 331-334)

The two major accidents in nuclear power plants at Three Mile Island and Chernobyl are discussed, including the reasons for the accidents. The section concludes by posing the central question found in many parts of *ChemCom*: Are the benefits worth the risk?

PART D: SUMMARY QUESTIONS (page 334)

1. a. A tracer is a radioisotope used to locate the site of a specific medical problem.
    b. See Section D.2 and Table 9 on page 321.
2. a. Industrial: Thickness gauging, paint curing, measuring liquid levels. Agricultural: food irradiation and preservation, utilization of fertilizers and uptake of pesticides.
    b. In general, the risks are relatively low and the benefits high due to the small amounts of radioisotopes used, their short half-lives, and the lack of suitable cost-effective alternatives with lower risks.
3. Ionizing radiation such as alpha, beta, or higher-energy electromagnetic radiation (high-UV and above on the spectrum) causes disruption of molecular bonds in cellular materials. Nonionizing radiation is not sufficiently energetic to affect such changes (visible, infrared and the lower end of the spectrum).
4. Rem (roentgen equivalent man)—a measure of the radiation's power to ionize human tissue.
5. Alpha particles have little penetrating power. If the source is outside of the body, the skin blocks their entry. However, if the source is internal, the alpha particles can travel short distances and cause damage to surrounding tissue. If an alpha emitter is ingested or inhaled and if its chemical behavior causes it to be concentrated by the body, more severe damage can result.
6. (1) Length of time of exposure; (2) area of body and type of tissues exposed; (3) density of ionization; (4) the number of ionizations per unit of tissue, the dose, and the quantity of radiation received.
7. a. Benefit: radiation treatment for cancer.
    b. Risk: undesirable radiation exposure for fetuses and infants.
8. a. Problems: lack of good experimental data, complication of long-term effects (somatic cell cancer and germ cell mutations), and the difficulty of isolating radiation-induced effects from other environmental factors.
    b. Without better data, regulations may err on the side of safety (overly cautious) and not assume a threshold between no effect and hazardous effect.
9. a. Minimum dose for immediate observable effects: 25,000 mrem (25 rem);
    b. Maximum allowable dose for radiation workers: 5000 mrem (5 rem);
    c. Maximum allowable dose for individual: 170 mrem (0.17 rem). Both the occupational and individual allowances are substantially less than the minimum short term effect dose, to allow for the possibility of delayed, long-term effects. Radiation workers are allowed higher levels, at least partially because of the practical difficulties of protecting them at the 170 mrem level.
10. Radioactive materials decay at a rate (half-life) that is unalterable by any known means. If one uses the 10 half-lives rule of safety, we must be able to isolate most of our nuclear wastes for thousands of years from both man-made (terrorism and war) and natural (water seepage and geological shifts) intrusions. The shielding material must block beta and gamma radiation and withstand thermal, mechanical, and chemical stress.
11. a. Burial in a geologically stable location.
    b. Problems include concentrating and safely shipping the wastes to the site, and the political reality that nobody seems to want such a site near their home (or even in their state).

E: PUTTING IT ALL TOGETHER: SEPARATING FACT FROM FICTION

When they finish this unit, students should be prepared to make sounder choices about nuclear technology in their personal lives and in the voting booth. As they process the results of the nuclear survey they took earlier, students will have an opportunity to examine their original misconceptions about nuclear issues and gauge how much they have learned since then.

E.1 PROCESSING THE SURVEY INFORMATION (pages 335-337)

By this point most students should be able to answer correctly the 20 survey items which are direct statements (true or false) of scientific facts concerning nuclear issues. The purpose of this activity is to have students see how much their understanding has grown and critically analyze statements for implicit political overtones or values.

Using either the suggested grouping pattern (from student text) or an alternative of your own choosing, have the class focus on approximately 10 items that they feel deserve further discussion. In this final discussion do not strive for unanimity of opinion, but do encourage students to consider the best available scientific information when discussing their positions. At the close of the discussion you may wish to stress the desirability of keeping their new understanding up-to-date by suggesting some reliable sources of information.

Answers to Survey Questions:

Survey question	True/False	Categories	Location in text[a]
1.	False	S	A.6, A.7
2.	True	S	B
3.	False	S, V	A
4.	False	S, V	A.2
5.	True	S	A.6, A.7
6.	False	S, V	A.2
7.	False	S	B.5
8.	True	S	D.8
9.	False	S	A.10
10.	True	S	D.2
11.	True	S	D.5, D.6
12.	True	S,P,V	C
13.	True	S	D.6
14.	False	S	NA
15.	True	S	A.3
16.	False	P,V	D.9
17.	True	S	A.10, D.2, D.3
18.	True	S	D.2, D.3, D.6
19.	True	S	D.2
20.	True	S	D.5, D.6, D.8
21.	False	P, V	C.5
22.	False	S, P, V	D
23.	False	P, V	NA
24.	True	S, V	C
25.	False	P, V	NA
26.	False	S, V	C, D.10
27.	True	S	C.5
28.	True	S, P	C.5
29.	False	P	D.9
30.	False	S	C.5
31.	True	P	B.3, B .4
32.	False	S, P	NA
33.	True	P	D.6-D.8
34.	False	S	NA
35.	True	S, P, V	D.9
36.	True	S	D.9
37.	False	P	C.5
38.	False	P	NA
39.	False	S	D.9
40.	Good question!	V	NA

[a]The locations in the text are provided for teacher reference; students are not expected to search for these.

E.2 LOOKING BACK (page 337)

This closing section underscores a fundamental *ChemCom* viewpoint: decisions that take into account the best available scientific information are better than those based on misconceptions and untruths. Individual choice and decision making are vital to a democratic society. This unit has tried to avoid the emotional issues that distort the true picture about nuclear energy. Hopefully *ChemCom* students can continue to lead the way in seeking knowledge to better understand technological and political issues.

SUPPLY LIST

Expendable Items

	Section	Quantity (per class)
Alcohol, 2-propanol (isopropyl)	B.6	100 mL
Boxes, cigar or shoe	A.5	8
Boxes, cardboard with lid	B.4	12
Cardboard, pieces (5 cm × 5 cm × 0.1 cm)	B.1	12
Cloth, pieces	A.1	12
Dry ice, (CO_2 solid)	B.6	5 lb
Gas lantern mantles	A.1	12
Gloves, disposable	B.1, B.6	24 pair
Graph paper	B.1, B.4	50 sheets
Paper bags, small brown	A.8	24
Salt, low-sodium (optional)	A.1	1 box
Sun-sensitive paper	A.1	12 sheets

Nonexpendable Items

	Section
Assorted objects, (2 pairs of scissors, 2 keys, 2 small rubber balls, 2 plastic forks, 2 plastic combs, and 2 large paper clips)	A.5
Balances	A.8
Calipers, vernier (optional)	B.1
Cloud chamber kits	B.6
Coins, assorted	A.1
Dominoes (for chain reaction activity)	C.4
Forceps	A.1
Glass plate (5 cm × 5 cm × 0.1 cm)	B.1
Lead plates (5 cm × 5 cm × 0.1 cm)	B.1
Light source, intense	B.6
Meter sticks	B.1
Pennies, (100 pre-1982, 100 post-1982)	A.8, B.4
Radiation sources (alpha, beta, and gamma)	B.1
Ratemeter	B.1
Ruler, metric (cm scale)	A.5
Squeeze bottles	B.6
Tongs, crucible	B.1, B.6
Watch faces, luminescent (optional)	A.1

SUPPLEMENTAL READINGS

"Worlds within the Atom." *National Geographic*, May 1985, 634-63. Includes historical milestones and a discussion of accelerators, quarks, etc.

"Nuclear Diagnosis." *CHEM MATTERS*, December 1985, 4-7.

"Radioactive Dating: A Method for Geochronology." *Journal of Chemical Education*, **62**, 580-84, 1985. Gives historical background on the discovery of natural radiation and discusses the use of K-40/Ar-40 in geochronology.

"Starborn—The Origin of the Elements.", *CHEM MATTERS*, October, 1984.

"Nuclear Synthesis and Identification of New Elements." *Journal of Chemical Education*, **62**, 392-95, 1985. Reviews nuclear terms, reactions, and experimental methods.

"Radioactivity in the Service of Man." *Journal of Chemical Education*, **59**, 735-38, 1982.

"Radiation Chemistry: State of the Art Symposium." Special issue of *Journal of Chemical Education*, **58**, Feb. 1981. This contains many articles including: "Principles and Techniques of Radiation Chemistry" (84); "Radiation Chemistry and the Preservation of Food" (162-67); "Radiation Processing: Industrial Applications of Radiation Chemistry" (168-73); "The Effects of Ionizing Radiation on Mammalian Cells" (144-56); and "Radiolytic Damage to Genetic Material" (135-39).

"Biological Effects of Low Level Radiation." *Scientific American*, February 1982, 41-9.

"Radon Gas: A Deadly Threat." *Newsweek*, August 18, 1986, 60-61.

"Carbon-14 Dating", *CHEM MATTERS*, February, 1989.

A Citizen's Guide to Radon: What It Is and What to Do About It (OPA-86-004/August 1986) and *Radon Reduction Methods: A Homeowners Guide* (OPA-86-005/August 1986). Two publications of the Environmental Protection Agency.

"Safety of Fission Reactors", *Scientific American*, 242 (#3), 33, 1980.

"Special Report on Energy." *National Geographic*, February 1981 (whole issue). "Uranium: Too Hot to Handle?" (66-67).

"Fusion Power." *Chemical and Engineering News*, April 2, 1979, 32-47. An overview of theory and research.

"Fission in the Fusion Camp", *Discover* , December, 1989.

"Light Your Candy", *CHEM MATTERS*, October, 1990.

Overview-Nuclear Waste Policy Act. The Illustrated Mechanics of Nuclear Waste Isolation. What Will a Nuclear Waste Repository Look Like? What Is Nuclear Waste? What Is Spent Nuclear Fuel? Can Nuclear Waste be Transported Safely? Radiation and Nuclear Waste—How are They Related? How much High-level Nuclear Waste Is There? What Rock Types Are being Considered for Nuclear Waste Repositories and Why? A series of publications of the Office of Civilian Radioactive Waste Management, U.S. Department of Energy, Mail Stop RW-40, Washington, DC 20585.

"The Promise and Peril of Nuclear Energy." *National Geographic*, April 1979, 459-93.

"The Disposal of Radioactive Wastes from Fission Reactors." *Scientific American*, June 1977, 21-31.

"The Reprocessing of Nuclear Fuels." *Scientific American*, December 1976, 30-41.

"New Era of Inherently Safe Nuclear Reactor Technology Nears." *Chemical and Engineering News*, June 30, 1986, 18-22. Describes work that has occurred since the TMI accident and contrasts U.S. plant designs to Chernobyl.

"Rethinking Nuclear Power." *Scientific American*, March 1986, 31-39. Discusses how a new generation of lower power, centrally fabricated reactors could avoid the safety and political impasse facing nuclear power; figures show the growth of the industry both here and abroad from 1964-present and details the new designs.

The Three Mile Island Accident: Diagnosis and Prognosis. American Chemical Society, 1986. The story of the TMI accident told by the scientists and engineers who operated the reactor or worked in the national laboratories that analyzed the accident.

"The Three Mile Island Story." *Time*, April 9, 1979.

"Decisions, Decisions." *Discover*, June 1985, 22-31. Research shows how irrational considerations play a large part in human decisions.

"Advanced Light-Water Reactors", *Scientific American*, 262 (#4), 82, 1990.

CHEMISTRY, AIR, AND CLIMATE
PLANNING GUIDE

Section	Laboratory Activity	You Decide

A. Living in a Sea of Air

A.3 Air: The Breath of Life	A.2 Lab Demonstration: Gases	A.1 The Fluid We Live In

B. Investigating the Atmosphere

B.2 A Closer Look at the Atmosphere B.4 Air Pressure B.5 Boyle's Law: Putting on the Squeeze B.8 Ideally, Gases Behave Simply	B.1 The Atmosphere B.6 T-V Relationships	B.3 Atmospheric Altitude B.7 A New Temperature Scale

C. Atmosphere and Climate

C.1 The Sunshine Story C.2 Earth's Energy Balance C.3 At the Earth's Surface C.4 Changes on the Earth's Surface C.8 Off in the Ozone	C.6 CO_2 Levels	C.5 Trends in CO_2 Levels C.7 Reversing the Trend

D. Human Impact on Air We Breathe

D.1 To Exist Is to Pollute D.2 Smog: Hazardous to Your Health D.5 Pollution Control D.6 Industrial Emission of Particulates D.8 Photochemical Smog D.10 Controlling Automobile Emissions D.11 Acid Rain D.13 pH	D.7 Lab Demonstration: Cleansing Air D.12 Acid Rain	D.2 What Is Air Pollution? D.3 Major Pollutants D.9 Autos and Smog

E. Putting It All Together: Is Air a Free Resource?

E.1 Air Pollution Control: A Success? E.3 Paying the Price E.4 Looking Back		E.2 Just Another Resource?

TEACHING SCHEDULE

	DAY 1	DAY 2	DAY 3	DAY 4	DAY 5
Class Work	Discuss p. 340 YD pp. 341-342		SQ p. 345	Review Part A Quiz Part A	
Laboratory		LD pp. 342-343		LA pp. 346-351	LA pp. 346-351
Homework	Read pp. 340-345	YT pp. 343-344	Read pp. 346-351		Read pp. 351-356

	DAY 6	DAY 7	DAY 8	DAY 9	DAY 10
Class Work	YTs pp. 353 & 354	Discuss pp. 356-359 CQ pp. 357-358	Discuss pp. 359-361 CQ p. 359 YT pp. 361-362		CQ p. 365 YT pp. 365-366
Laboratory			Pre-lab LA pp. 362-363	LA pp. 362-363	
Homework	YD pp. 354-356 Read pp. 356-359	Read pp. 359-363	Read pp. 364-366	YD pp. 364-365	Read pp. 366-369

	DAY 11	DAY 12	DAY 13	DAY 14	DAY 15
Class Work	Discuss pp. 366-369 CQ p. 366 YT pp. 367-368	Review Part B Quiz Part B YT p. 373 CQ p. 373	Discuss pp. 374-379 CQ p. 374		YD pp. 381-382
Laboratory				LA pp. 380-381	
Homework	SQ pp. 368-369 Read pp. 370-374	Read pp. 374-380	YD pp. 379-380 Read pp. 380-381	Read pp. 381-382	Read pp. 382-384

	DAY 16	DAY 17	DAY 18	DAY 19	DAY 20
Class Work	SQ p.384	Review Part C Quiz Part C Discuss pp. 385-390	Discuss pp. 390-392 CQs pp. 390 & 392	Discuss pp. 392-396 YD pp. 393-394 CQ p. 396	
Laboratory			LD p. 392		LA pp. 399-400
Homework	Read pp. 385-390	YDs pp. 387-388 Read pp. 390-392	Read pp. 392-397	Read pp. 398-400	Read pp. 400-402 YT p.401

	DAY 21	DAY 22	DAY 23	DAY 24	DAY 25
Class Work	SQ pp.401-402	Review Part D Quiz Part D PIAT pp. 403-404	PIAT pp. 405-407	Review unit	Exam
Laboratory					
Homework	Read pp. 403-407	YD pp. 404-405			

LD= Lab Demonstration; **LA** = Laboratory Activity; **CQ** = ChemQuandary; **YT** = Your Turn; **YD** = You Decide; **PIAT** = Putting It All Together.

Returning to the theme of resources, the Air unit examines another vital and little understood resource: our atmosphere. The physical properties of gases, including changes in volume caused by pressure and temperature changes, are explored, and the chemical properties of the air's major constituents are tested in laboratory activities. The primary technological issue explored is global warming caused by carbon dioxide; ozone and smog are also discussed.

OBJECTIVES

Upon completion of this unit the student will be able to:

1. Describe common physical and chemical properties of air. [A.1, A.2]
2. Compare the chemical properties of nitrogen, oxygen, and carbon dioxide. [B.1]
3. Identify the major components of the troposphere and indicate their relative concentrations. [B.2]
4. Show how Avogadro's law and the concept of molar volume clarify the interpretation of chemical equations involving gases. [B.2]
5. Describe with words and mathematical equations the interrelationships among amount, temperature, volume, and pressure of a gas (Avogadro's, Charles' and Boyle's laws), and list one practical application of each law. [B.2, B.4-B.8]
6. Define and apply in appropriate situations the terms molar volume, standard temperature and pressure (STP), Kelvin temperature scale, and absolute zero. [B.2, B.5-B.8]
7. Sketch or graph the relationship between altitude and air pressure. [B.3]
8. Discuss air pressure and explain how to measure it. [B.4]
9. Account for the gas laws in terms of the kinetic molecular theory of gases. [B.8]
10. Compare the various components of solar radiation. [C.1]
11. Describe how reflection, absorption and re-radiation of solar radiation account for the Earth's energy balance. [C.2]
12. Explain how differing heat capacities and reflectivities of various land covers and water can influence local climates. [C.3]
13. Describe the greenhouse effect, its natural incidence and causes, and the significance of industrial contributions. [C.4, C.5]
14. Use graphical extrapolation to predict future CO_2 concentrations, and outline assumptions and problems associated with such predictions. [C.5]
15. Compare the production of CO_2 from combustion with that from respiration. [C.6]
16. Describe the function of the ozone layer and how human activities may be affecting it. [C.8]
17. List the major categories of air pollutants and discuss the relative contributions of various human and natural factors to each category. [D.1-D.3]
18. Describe major general strategies for controlling pollution, and specific strategies for particulates [D.5-D.7]
19. Describe chemical reactions and geographic and meteorological factors which contribute to photochemical smog. [D.8]
20. Interpret graphs and tables related to automotive-induced air pollution. [D.8-D.10]
21. Explain the role of activation energy in a chemical reaction, and give an example of how a catalyst affects it. [D.9]
22. Describe the role of catalytic converters in reducing automotive emissions of unburned hydrocarbons, carbon monoxide, and nitrogen oxides. [D.10]
23. Describe sources and consequences of acid rain. [D.11-D.13]
24. Define the terms acid and base, give examples of each, describe their formation with balanced ionic equations, and relate hydrogen ion concentration to the pH scale. [D.11-D.13]
25. Interpret historical emissions data to assess the success of various pollution control efforts. [E.1]
26. Discuss air pollution in terms of the trade-offs between control costs and damage costs. [E.2-E.3]

SOURCES ON THE CAUSES AND SOLUTIONS OF AIR POLLUTION

1. Acid Rain Foundation, Inc., 1630 Blackhawk Hills, St. Paul, MN 55122. 612-455-7719. Public supported, tax-exempt, nonadvocacy organization. Educational resources on acid rain include: a resource directory, posters, AV rental programs, curriculum packets, t-shirts, and gift items. To receive a free catalog, send a stamped, self-addressed #10 envelope.
2. LaMotte Chemical Products Co., P.O. Box 329, Chestertown, MD 21620. 301-778-3100. Sell a variety of air, water, and soil testing kits. Request a catalog and the free *Test Equipment Digest* newsletter.
3. Library of Congress, Science Reference Section, Science & Technology Division, 10 First Street, SE, Washington, DC 20540. 202-287-5639. Prepares *Tracer Bullets* on a variety of subjects pertaining to science and technology—each tracer is a free bibliographic listing. Ask for LC *Science Tracer Bullet* ISSN 0090-52232 which lists the presently available titles.
4. Office of Technology Assessment (OTA), U.S. Congress, Washington, DC 20510. 202-226-2115 or 224-9241 (Pub. Affairs). OTA is a nonpartisan, analytical agency that serves the Congress by providing objective analysis of major public policy issues related to scientific and technological change. OTA has produced more than 100 reports assessing the present and future impact of various technologies. Report summaries are free from OTA—full texts of reports can be purchased from the Government Printing Office (GPO) or the National Technical Information Service (NTIS). Request free publications catalog. For the Air unit, the title: *Acid Rain and Transported Air Pollutants* (June 1984/ OTA-0-204) may be of interest.
5. U.S. Environmental Protection Agency, 401 M Street, SW, Washington, DC 20460. Publishes a wide variety of both technical and nontechnical materials. The *EPA Journal* is especially useful for the *ChemCom* course. For this particular unit: Sept. 1986, *Special Supplement on Acid Rain* (OPA-86-009) and Dec. 1986, *Greenhouse Effect and Ozone Depletion.*
6. EPA Public Information Center. 202-475-7751. Supplies a free educator's kit containing publications on air, water, and solid waste pollution (including: 115 p booklet: *Environmental Progress and Challenges: An EPA Perspective*). Also serves as a referral service for accessing other EPA offices.
7. EPA Office of Air and Radiation, Office of Air Quality Planning and Standards, Research Triangle Park, North Carolina 27711. Source of *Air Quality and Emissions Trend Report, 1983* cited in Part E.

INTRODUCTION

The unit begins with some information about our uses and abuses of the atmosphere. In the context of raising questions about air quality, topics on the chemistry of air are introduced.

A: LIVING IN A SEA OF AIR

The major issues to be considered in this unit are introduced by having students complete a true-false questionnaire designed to assess their understanding of gases. The questions go further than normal true-false tests by asking students to make corrections in their false answers, and to give applications of true answers. You then perform several demonstrations of physical properties of gases, mainly involving mass and movement.

You may wish to increase your students' awareness of the local/regional aspects of air quality by having them: (1) keep a record of the atmospheric pressure, temperature, forecast, and Air Quality Index (see Table 6 on page 389) as announced on the TV news or local newspaper; and (2) create a bulletin board, using articles clipped from local papers related to air quality. Consider assigning different teams for different weeks.

A.1 YOU DECIDE: THE FLUID WE LIVE IN (pages 341-342)

This activity focuses students' attention on a key resource (air) that they probably take for granted. It is somewhat similar to the nuclear survey, but in this case only the student answers the questions. As before, discourage students from using the "U" answer, which requires no correction or application statement.

Answers to Questions:

1. True—An air supply is needed for scuba diving or flying at high altitudes.
2. True—Since the volume is a function of the temperature and pressure, it is necessary to purchase gases on the basis of mass (or moles). If volume is used, the temperature and pressure must be specified.
3. False—All forms of matter have mass. The different densities of various gases account for sinking and floating of balloons filled with different gases.
4. True—The human body is designed to handle external pressures of approximately 15 lb./in.^{2}. Underwater diving (increased pressures) and high-altitude flying (reduced pressures) require special pressurized suits or compartments for bodily protection.
5. False—If this were true, living organisms would be much more limited in the habitats they could occupy.
6. True—Most of the more harmful, high energy ultraviolet radiation is absorbed by ozone and other triatomic species (CO_2, H_2O).
7. False—Except under conditions of a thermal inversion, temperature decreases with increasing altitude within the troposphere, as experienced by mountain climbers or snow skiers.
8. True—Atmospheric water is a part of the hydrologic cycle. Also, together with carbon dioxide it plays a crucial role in helping to maintain the heat balance on our planet.
9. True—Oxygen is number five (35 billion lb./yr in the United States) on the top 10 chemicals list. This atmospheric gas is used in welding, medicine, propellants, and steel production. Nitrogen is number three (37 billion lb./yr in the United States) on the top 10 chemicals list. This atmospheric gas is used as an inert atmosphere for metals, electronics, as a freezing agent for foods and cryogenics, and in the production of ammonia.
10. True—In the lower atmosphere, ozone is a product of photochemical smog that causes direct damage by corroding metals and plastics, and by oxidizing biological tissue. Ozone also interacts with other smog components such as hydrocarbons to produce such pollutants as peroxyacetyl nitrate (PAN). In the stratosphere, ozone acts to filter out harmful ultraviolet radiation.
11. False—Air is a mixture that contains a variety of substances (N_2, O_2, Ar, CO_2, H_2O, etc.) of both biological and industrial importance.
12. False—Air pollution is at least as old as civilization. Early fires from cooking and smelting operations polluted the air in ancient times. Today the planet supports a much larger number of people. Since the Industrial Revolution the amount of pollution/person has increased along with our increased use of resources.
13. False—Air pollution disasters include: Oct 27-31, 1948 in Donora, PA—18 excess deaths; Nov 26-Dec 1, 1948 in London, England—700-800 excess deaths; Dec 5-9, 1952 in London, England—4000 excess deaths; Dec 7-10, 1962 in Osaka, Japan—60 excess deaths; and Jan 29-Feb 12, 1963 in New York City—200-400 excess deaths. While such extreme cases have not occurred for a number of years, medical authorities attribute the higher rate of deaths due to lung diseases in cities and industrial centers at least partially to polluted air.
14. True—The eruption of the Mt. St. Helens and Mt. Pinatubo volcanoes, and the fires that ravage Western states, vividly demonstrate sources of natural air pollution.
15. True—Estimates of economic loss due to air pollution run as high as $16 billion annually in the United States.
16. False—In Section D.2 the students will discover that in terms of total mass of pollutants emitted, industrial activity falls behind transportation, space heating, and electricity generation. The important implication is that each of us, by our lifestyle and personal habits, contributes to air pollution. It is not just a problem of industry.
17. True—The greenhouse effect is one of the Earth's natural thermal control mechanisms. Part C discusses this in some detail.
18. False—The rain in industrialized nations has become more acidic, causing both environmental and political problems.
19. True—Major combustion products are CO_2 and H_2O, and a variety of other substances such as unburned hydrocarbons, NO_x, SO_x, and particulates.
20. False—In the final activity the students will examine progress in improving the air quality within the last eight years. Air pollution disasters have not occurred since the 1960s.

A.2 LAB DEMONSTRATION: GASES (pages 342-343)

The purpose of these demonstrations is to promote further interest in the sea of gases. Since atmospheric gases are colorless, odorless, and tasteless, students tend to think of them as nothing. Actually, gases, like solids and liquids, have definite physical and chemical properties.

Do as many demonstrations as possible. Encourage your students to predict the results whenever practical, and make sure that the desired conclusion is produced. It is useful to identify each demonstration by name, and provide a brief summary of the procedure for each demonstration.

Guidelines for Effective Demos

Although you should allow your natural personality to come out in the presentation, there are some general rules:

1. Always test the demonstration before showing it to your class.
2. Clear the demonstration table of clutter that presents a safety and/or visibility problem.
3. Provide a general introduction to help focus student attention, but let nature do your talking for you. Show, don't tell—adopt a "Let's see what happens" approach.
4. As you perform the demonstration, check with your audience to be sure that they can see, hear, smell, etc., the phenomenon. Do not attempt to give long-winded explanations at the time of the demonstration—allow sufficient time for the full sensory impact to be felt before eliciting student response.
5. Be honest with your students. If the demonstration does not work out as planned, remember that experiments never fail—you may wish to invite your students to explain what did occur and/or consider this yourself and attempt to repeat the demonstration, altering a perceived flaw. If a second attempt is unsuccessful, avoid desperate attempts to salvage the failure.
6. Ask probing questions and encourage similar questions from your students. If you are unsure of the answers, do not attempt to hide your lack of knowledge but attempt to display scientific reasoning.
7. If the demonstration requires the use of safety precautions, follow these precautions and warn your students not to attempt to duplicate your actions.

Note: Demonstrations should always be presented as a "minds-on" learning activity. Additionally, many demonstrations (including many in this set) allow for some hands-on experiences via volunteers from the audience or even whole-class participation.

Time

One class period

Materials

10 large balloons
Meter stick
1-L beaker
3 50- or 100-mL beaker
250-mL Erlenmeyer flask
Ring stand, clamp, wire gauze, ring
Bunsen burner
Empty soft drink bottle
2 drinking glasses
Piece of cardboard or waxpaper
Clean plastic bottle with lid (detergent bottle)
Thin-walled can with lid or rubber stopper (duplicating fluid can)
Empty aluminum soft drink can
Bottle of perfume
100 mL concentrated ammonia, $NH_3(aq)$
5 mL phenolphthalein
Large plastic or glass container
Long glass tube, open on both end with stoppers
Package of cotton balls
10 mL concentrated hydrochloric acid, $HCl(aq)$
Cylinder of hydrogen gas (optional)
Cylinder of helium (optional)
Cylinder of carbon dioxide (optional)
Cylinder of sulfur hexafluoride (optional)
Natural gas, methane (burner gas)
Small funnel
Soap-glycerin mixture
Dry ice
Package of birthday candles
Iodine crystals

Pre-Demo Discussion

As you do the demonstrations, ask students to:
1. try to predict the outcome of the demonstration;
2. carefully observe the demonstration and, if necessary, account for differences between your prediction and the actual outcome;
3. try to identify the specific gas property that is being demonstrated, and why gases exhibit this property; and
4. think of practical consequences of atmospheric gases' exhibiting this property.

Demo Tips

1. When performing the demonstrations be sure to capitalize on the element of surprise.
2. Avoid forecasting expected outcomes to your students.
3. On several demonstrations you may wish to elicit ideas on what will occur and why. For others, you may wish to do the demonstration before eliciting student response.
4. You do not need to try to "teach" concepts on this day; all will be developed more fully in subsequent text and activities.
5. The main purpose at this time is to raise questions and motivate your students to want to learn more about the "sea" of gases in which we live.

The following demonstrations have been chosen for several reasons:
1. They are effective for illustrating that gases (even colorless, odorless ones) are "real" substances.
2. They draw students' attention to very important properties of our atmosphere in a dramatic fashion.
3. They use readily available materials.
4. If properly performed, they are quite safe—in fact, you may encourage your students to repeat some of the demonstrations for themselves at home.
5. They do not require elaborate setups or much advance setup time—the entire sequence can be performed between the first and second days of the unit.

Answers to Questions:

Is air really matter?

Demonstration 1: Air has weight Balance two filled balloons by hanging them from a meter-stick balance. Pop one of the balloons and note the direction of the balance's tilt. Though "invisible," gases do possess weight, like other forms of matter.

Demonstration 2: Air occupies space Lower an "empty" glass, open end downward, into a larger beaker (or pot) of water. Does water fill up the "empty" glass? Tilt the glass. What happens? Does air occupy space? Lead your students to realize that even colorless gases occupy space.

Demonstration 3: Not all gases are colorless Place several crystals of iodine in a sealed flask. A faint purple color will soon be observed in the flask above the crystals. (This is best displayed with white paper behind the flask.) Contrast this "colored gas"–iodine vapor, $I_2(g)$–to air or other colorless gases.

What is air pressure all about?

Demonstration 4: Air has pressure Select a balloon that is easy to blow up. Put the balloon inside a volumetric flask or an empty soft drink bottle and stretch its neck over flask or bottle mouth. Ask a talkative student to try blowing up the balloon so it fills the container as in the above experiment. Ask the class to explain the difficulty of this task.

Demonstration 5: Temperature changes the air pressure Before class, place 10 mL water in a 250-mL Erlenmeyer flask. Bring the water to boiling. Remove the flask from the source of heat and quickly place an empty balloon over the mouth of the flask. As the flask cools, the pressure differential between the outside and inside of the flask will cause the balloon to invert and line the inside walls of the flask. In class, ask your students to explain how the balloon was placed inside the flask and for suggestions on how to remove the balloon from the flask. Note that heating will cause the liquid water to vaporize and push out the balloon. This demonstration is also effective if performed "live." Students seem to enjoy the demonstration even more if the balloon pops before completely filling the flask.

Demonstration 6: Air exerts pressure upward and in all directions Fill a glass to the rim with water. Cover this with either wax paper or a piece of cardboard (as cut from a cereal box). Press down along the edges to make a tight seal, turn the glass upside down over a sink, and let go of the cover. Air pressure (acting upwards against the covering) will support the weight of the water and prevent it from spilling out. Repeat the process without filling the glass completely.

Ask your students to account for the difference in the two cases.

Demonstration 7: Air exerts pressure sideways and in all directions Punch a small hole in the side of an empty, clean, plastic bottle (such as a liquid-detergent bottle). Holding your finger over the hole, completely fill the bottle with water. Replace the cap. Remove your finger from the hole. Explain your observation. Unscrew the lid and note what happens. Ask students to account for the difference. Why is it difficult to pour juice from a can if only one hole is punched in it?

Demonstration 8: Air exerts pressure in all directions Place about 20 mL water in a clean, empty aluminum soft drink can. Bring the water to a rapid boil and quickly invert the can in a container of water. It is not necessary to submerge the can, but only to have its top touch the water's surface. The boiling process will force out much of the air that is inside the can and replace it with water vapor. Upon cooling, the water vapor will condense, leaving a partial vacuum inside the can. Atmospheric pressure outside the can will crush it. In addition to demonstrating the substantial pressure of our atmosphere, this demonstration contrasts the relative densities of liquid and gaseous states of matter. You may wish to reserve it for that purpose. If duplicating-fluid cans are still available in your school, you may wish to do the demonstration with an empty one. Add about 100 mL of water to the can, heat the water to boiling, and then seal the can with a rubber stopper. The advantage of this method is that the crushing of the can is more prolonged, and partially reversible; you can restore the shape of the can by blowing into it.

Why does air sometimes carry odors?

Demonstration 9: Diffusion of gases Standing at the front of the classroom, open a bottle of perfume or other safe, odoriferous substance with high vapor pressure. Have each student raise a hand as soon as the odor is detected.

Demonstration 10: Diffusion rates of gases Place two small beakers (one containing ammonia, the other, water and phenolphthalein) under a large plastic or glass container. Ask your students to account for any changes they note. (You will see a pink ring forming nearing the water surface.) Placing the setup on an overhead projector will make it more visible and cause the diffusion to occur faster. A similar demonstration using hydrochloric acid and ammonia (to produce ammonium chloride "smoke" particles) can also be performed with this setup or with the traditional long glass tube setup.

Is air heavy? Will it burn?

Demonstration 11: Density Contrast the behavior of equal sized balloons filled with available gases such as hydrogen (0.09 g/L), helium (0.18 g/L), natural gas from a burner jet which is mainly methane (0.55 g/L), "air" (1.0 g/L), carbon dioxide (2.0 g/L), or sulfur hexafluoride (SF_6–6.6 g/L) by attempting to throw them. Alternative and especially effective demonstrations include: (a) "blowing bubbles" using a liquid soap-glycerin mixture and a funnel connected via rubber tubing to a gas source and (b) pouring carbon dioxide down a tilted tray of candles. Or, simply observing a container of dry ice is also effective.

Demonstration 12: Flammability Contrast the flammability of hydrogen (or methane) to that of helium, using either balloons or soap bubbles. If using the soap bubble approach be sure to run enough gas through the tubing to clear out any air. This precaution will insure even, rapid, nonexplosive burning. Use a candle attached to a meter stick to attempt igniting the bubbles after they detach themselves from the bubble generator and begin to rise.

At-Home-Activities

1. A straw enables one to suck fluids "uphill" against gravity by using atmospheric pressure. In sucking, air pressure is decreased at the end of the straw in one's mouth. The greater pressure acting on the surface of the fluid pushes the fluid up. If two straws are used in the manner described, one cannot build up a difference in pressure and therefore the fluid does not rise. Some of your brighter students may use their tongue to close off the end of the second straw and thereby be able to accomplish the task.
2. As above, the action of a straw depends on a pressure difΣferential. If the container is filled, atmospheric pressure cannot act on the surface of the fluid and the fluid will not rise.

Post-Demo Discussion

Additional ideas you may wish to use on the first and second days include:
1. newspaper or magazine clippings relating to the atmosphere, climate and/or air pollution; and
2. slides of nature-generated air pollution (such as Mt. St. Helens) or human created air pollution—in either case, local or regional "news" will probably have the most impact on your students.

A.3 AIR: THE BREATH OF LIFE (pages 343-344)

This section focuses on oxygen's unique role in the atmosphere, and the reversibility of its supply, being consumed by respiration and produced by photosynthesis. Figure 1 on page 344 is central to the *YOUR TURN* exercise. An enjoyable extension of the *YOUR TURN* questions is to have all students measure their own respiration rates, and use those in all calculations.

Breath Composition and Glucose Burning (pages 343-344)

This activity is designed to compare quantitatively the composition of inhaled and exhaled air by humans, and to reemphasize the role of oxygen in the metabolic conversion of glucose to water and carbon dioxide.

Answers to Questions:

1.a. The percent of oxygen decreases, carbon dioxide increases, and nitrogen plus other gases increase.
   b. Animal respiration converts carbohydrates (or fats and proteins) and oxygen into carbon dioxide, water vapor (accounting for the slight increase in percent of nitrogen plus other gases), and energy.
   c. (14 breaths/min)(60 min/h)(24 h/day) = 20,160 breaths/day.
   d. Factors which could change this include level of activity, personal metabolism (age, sex, weight, body build), and food consumption. In general, larger, more active people require more oxygen for respiration.
   e. (1) (500 mL/breath)(1 L/1000 mL)(14 breaths/min) = 7 L/min
      (2) (7 L/min)(60 min/h)(24 h/day) = 10,080 L/day
2. Oxygen requirements for animal respiration are considerably less than the total quantity of oxygen inhaled. If all of the oxygen were used, our metabolism rate and body temperature would be considerably higher.
3. $C_6H_{12}O_6 + 6 O_2 \rightarrow 6 CO_2 + 6 H_2O + 686$ kcal/mol glucose
   a. Given an equal number of molecules, oxygen would be the limiting reactant, since six molecules are needed for every one molecule of glucose.
   b. Given equal masses, oxygen would be the limiting reactant, because for each 180 g (1 mol) of glucose, 192 g (6 mol) of oxygen are needed.
   c. (20 moles O_2/day)(1mole glucose/6 moles O_2) = 3.33 moles glucose/day.
   d. (3.33 moles glucose/day)(180 g glucose/mole glucose) = 600 g glucose/day.
   e. (600 g glucose/day)(17 kJ/g glucose) = 10200 kJ/day.
4.a. Water vapor is a by-product of animal respiration.
   b. Exhale into cold air or onto a cold surface to confirm the presence of water.

Set up the following optional classroom display. Place an *Elodea* plant in a large test tube containing water and an indicator that changes color near pH 6 (such as bromthymol blue). Set up another test tube (without plant) as a control. Breathe into both test tubes and note the color change due to the presence of carbon dioxide in your exhaled breath. Stopper both test tubes and place them in the light. Note the change in color in the first test tube as the carbon dioxide is used by the plant.

PART A: SUMMARY QUESTIONS (page 345)

1. a. Air is similar to other resources in that our planet has fixed quantities of air and other resources, and its behavior can be explained in terms of atoms and molecules.
   b. Air is different in that it is uniformly distributed across all geographic locations and therefore is not usually mined, stored, or exported; it is absolutely essential for life; and it is a gas at room temperature and pressure.
2. a. True—Empty bottles do not usually contain a vacuum, but contain air molecules.
   b. False—The can was crushed from all directions.
   c. True—Recall the demonstrations on diffusion.
   d. False—Recall the demonstration contrasting helium, air, carbon dioxide, and hydrogen.
3. The density of air is less at higher altitudes. Therefore the same volume of air contains less mass at higher altitudes, and consequently less oxygen. In order to maintain metabolism the rate of respiration must increase.
4. a. (0.15 mol O_2)(1 mol glucose/6 mol O_2)(180 g glucose/mol glucose) = 4.5 g glucose.
   b. (4.5 g glucose)(17 kJ/g glucose) = 76.5 kJ.
   c. (76.5 kJ)(1 Cal/4.2 kJ) = 18.2 Cal.

B: INVESTIGATING THE ATMOSPHERE

Part B explores the chemistry of the individual gases contained in the atmosphere, and then turns to the gas laws, including Avogadro's hypothesis, standard temperature and pressure, Boyle's law and Charles' law; the latter is covered in a laboratory activity. Some mathematical manipulation of the gas laws is included, and the concepts of absolute zero and the Kelvin temperature scale are introduced. This Part concludes with applications of the kinetic molecular theory to explain the behavior of gases.

B.1 LABORATORY ACTIVITY: THE ATMOSPHERE (pages 346-351)

This activity provides students with a basic understanding of chemical properties of the gaseous components of air. Specifically, students prepare two of the major components (oxygen or carbon dioxide) of air and discover the flammability, acidity, and basicity characteristics of each gas.

Manipulating the equipment necessary for collecting gases by water displacement may be demanding for many students, but the exciting and spectacular testing results are worth it. The oxygen generation part is last, to make a great climax to this activity.

Time

Two class periods

Materials (for a class of 24 working in pairs)

12 ring stands
12 wire screens (with ceramic center)
24 iron rings
12 500-mL Florence flasks
12 10-mL graduated cylinders
12 1-hole rubber stoppers to fit flask
12 30-cm lengths of glass tubing (right angle bend)
12 15-cm lengths of glass tubing (right angle or hooked bend)
12 60-cm lengths of rubber tubing
24 250-mL gas collecting bottles
12 pneumatic troughs or trays
24 glass plates
36 test tubes (18 × 150 mm)
36 corks or rubber stoppers to fit test tubes
12 spatulas
12 tongs
12 Bunsen burners
Balances
12 grease pencils
12 dropper bottles (for universal indicator solution)
180 cm magnesium ribbon (15 cm/team)
300 mL universal indicator solution (25 mL/bottle)
72 mL limewater (6 mL/team)
12 g steel wool (1 g/team)
12 wood splints (1/team)
1.5 L 3% hydrogen peroxide (H_2O_2) solution (200 mL/team)
12 g manganese dioxide (MnO_2) (1.0 g/team)

Advance Preparation

Cut magnesium strips to length. If your budget is limited, shorter pieces (at least 3 cm) should be long enough.

Bend glass for generator tubes: a bend at each end will prevent the pinching of the tube (see Figure 2 on page 349).

Check the oxygen generation reaction ahead of time. The hydrogen peroxide solution should be freshly prepared; if the rate of generation is too slow, you may need to increase the amount of catalyst or secure fresher hydrogen peroxide. Consider buying 30% hydrogen peroxide from a chemical supplier, and diluting it down. *Caution: 30% hydrogen peroxide is **very** corrosive: contact with skin produces very rapid blistering, and contact with clothing is usually irreversible and damaging.*

Pre-insert the glass bends into the one-hole stoppers before class to prevent the risk of student accidents. Follow conventional precautions, lubricate glass tubing with glycerin and insert the tubing gently with a twisting motion. A handy technique for reducing the chances of breakage involves choosing a brass cork borer just large enough to go around the glass tubing. Insert the cork borer into the stopper, pass the tubing through the borer until it is in the right position, then remove the cork borer, leaving the glass entrapped within the stopper.

You can use sodium hydrogen carbonate (baking soda) and dilute acetic acid in place of the antacid tablets. Pretest your mixtures, to determine which concentration of acid and amount of solid, produces a satisfactory amount of CO_2 and an appropriate rate of gas generation.

Pre-Lab Discussion

Be sure your students understand that the substances being generated in this activity are normal components of the gaseous mixture called air. The gases will be prepared in the activity, since this is easier than trying to separate them directly from air.

Review the general technique of collecting gases by water displacement (you may wish to have your students explore what factors limit the general use of this technique, such as the extent to which a given gas is water soluble or reacts with water).

Lab Tips

Remind students to collect gas as soon as possible after beginning the reaction, or the reagents will be used up.

The pH testing of oxygen and nitrogen gases may give inconsistent results. Suggest that students pool their data or collect extra tubes of gas for testing. It would also be helpful to have students measure the pH of the water in the trough and then see if there is a change in pH with the gas measurement.

You can demonstrate the synthesis of nitrogen gas by heating 5 g NH_4Cl + 7 g $NaNO_2$ + 50 mL H_2O. Generating nitrogen by this reaction is a convenient way to model the setup for student generation of gases. However, remind students that they *don't* need to heat their mixtures. Heating is necessary only to begin the reaction; after bubbling starts, the residual heat is enough to continue gas generation. *Don't overheat!* It is important that you demonstrate the properties of nitrogen by using the gas that you generate or obtain from a cylinder.

The reaction of magnesium with pure oxygen often generates enough heat to crack the gas collecting bottles. Be prepared to replace several every year.

Expected Results

Test	Nitrogen	Oxygen	Carbon dioxide
Mg ribbon	Burns	Burns	Burns
Steel wool	No reaction	Burns	No reaction
Wood splint	Extinguished	Burns	Extinguished
Limewater	Clear	Clear	Cloudy
Indicator	Yellow (neutral)	Yellow (neutral)	Red (acid)

Post-Lab Discussion

Reemphasize the points made during the Pre-Lab discussion. Prepare a class summary table of the results and discuss the questions.

As a review, ask students to write balanced equations for the reactions of this activity:

$$2\,Mg + O_2 \rightarrow 2\,MgO \quad \text{(white solid)}$$
$$3\,Mg + N_2 \rightarrow Mg_3N_2 \quad \text{(magnesium nitride)}$$
$$2\,Mg + CO_2 \rightarrow 2\,MgO + C \quad \text{(some may see a black residue)}$$
$$CO_2 + Ca(OH)_2 \rightarrow CaCO_3 + H_2O$$

OPTIONAL QUESTIONS

Would carbon dioxide fire extinguishers be useful in putting out a magnesium fire?
Would combustion reactions that produce carbon dioxide contribute to acid rain?
Why should limewater test solution be freshly prepared and/or kept tightly stoppered?

Answers to Questions:

1. If the first bubbles were saved, air would contaminate the sample of the desired gas. Test results would be inconclusive.

2. a. Oxygen is the most reactive.
   b. Nitrogen is the least reactive.
3. a. Oxygen supports rusting and burning.
   b. Air contains 20.71% oxygen molecules.
   c. A higher oxygen concentration in the atmosphere would lead to more fires, and possibly to a lower general respiration rate in animals.
4. There are three components to any fire: heat, fuel, and oxidizer. Since carbon dioxide from the fire extinguisher is denser than air, the fire is smothered by preventing the less dense oxygen (the oxidizer) from reaching the fuel.
5. Blow through a straw into limewater, if the solution turns cloudy, carbon dioxide is present. Secondly, carbon dioxide may be detected by monitoring the drop in the pH of tap water while bubbling exhaled gas through the water.

DEMONSTRATION IDEA

To demonstrate that air is composed of at least two components, approximately 1/5th by volume of a gas that supports oxidation (oxygen) and 4/5th of a gas (nitrogen) that does not: wedge a piece of moistened (with vinegar) steel wool in the bottom of a graduated cylinder. Place the cylinder upside down in a beaker of water so that the internal and external water levels are even (to equalize pressure). Mark the initial level and let the apparatus sit overnight. As the steel wool rusts, oxygen is used up—assuming you used excess steel wool, the reaction will continue until approximately 1/5th of the original volume of trapped air has been consumed. A wood splint test on the remaining gas shows that it does not support combustion.

Note: Studies on the physical and chemical properties of gases gave birth to the modern science of chemistry (Cavendish, Black, Priestley, Lavoisier, etc.) and make an interesting research project for students. Two useful reference books on the history of early chemistry are: Asimov, Isaac. *Biographical Encyclopedia of Science & Technology: The Lives and Achievements of 1195 Great Scientists from Ancient Times to the Present.* New York: Avon Books, 1976. Ihde, Aaron J. *The Development of Modern Chemistry.* New York: Harper & Row, 1964.

B.2 A CLOSER LOOK AT THE ATMOSPHERE (pages 351-354)

After focusing on the individual gases and their quantities in the atmosphere, this section leads to information about the general behavior of gases, Avogadro's law, and molar volume. The point is made that the physical properties of all gases are nearly the same, provided the temperature is not too low or the pressure too high.

YOUR TURN

Avogadro's Law (page 353)

Answers to Quesitons:

a. In observation III on page 352, if all three balloons have equal numbers of molecules, they would expand the same amount under the same conditions (a 10°C temperature increase at constant pressure).
b. In observation IV, all three balloons begin with the same number of molecules. When half of the molecules are removed from each balloon, all balloons shrink to the same size, in accord with Avogadro's law.

DEMONSTRATION IDEAS

Molar Volume Display

Set up a display featuring one mole of various substances. Consider displaying 12 g carbon, 63.5 g copper, 58.5 g sodium chloride, 18 g water, and a box with the dimensions of 28.5 cm × 28.5 cm × 28.5 cm (or 22.4 L). Clearly mark the box "1 mol of any gas at STP = 22.4 L. " Contrast the simplicity of gaseous molar volume with that of liquids and solids, for which no such generalization can be made. 5 gallons is approximately 25 liters, so a large pail is roughly equal to the volume of one mole as well.

Determining the Molar Volume

To determine an approximate molar volume experimentally, fill a 1-L graduated cylinder with ice water. Invert it into a half-filled pneumatic trough. When you are ready for the demonstration, quickly rough-weigh a small piece (approximately 1 g) of dry ice, $CO_2(s)$. (Note that 1 g of dry ice will produce approximately 500 mL of gas at STP. Place the dry ice under the inverted cylinder. Measure the volume of collected gas and calculate the molar volume:

$$\frac{\text{Measured volume in L}}{\text{Mass of } CO_2(s) \text{ in grams}} \times \frac{44 \text{ g } CO_2}{1 \text{ mol } CO_2} = \text{Molar volume (L/mol)}$$

YOUR TURN

Molar Volumes and Reactions of Gases (page 354)

This activity helps reinforce an important practical consequence of Avogadro's law, namely: Since all gases (at the same temperature and pressure) have the same molar volume, the coefficients in a chemical reaction involving gases can indicate the relative volumes of reacting gases (as well as the relative number of moles).

Note: The problems can be solved by using either the ratio approach or the factor label method. The latter method is preferable for more complex questions such as #4.

Answers to Questions:

1. $(3.0 \text{ mol})(22.4 \text{ L/mol}) = 67 \text{ L } CO_2(g)$

2. $2 \text{ mol NO}(g) \quad + \; 1 \text{ mol } O_2(g) \; \rightarrow \; 2 \text{ NO}_2(g)$
    2 volumes of NO + 1 volume $O_2 \; \rightarrow \;$ 2 volumes NO_2
    $(4 \text{ L NO}) \times (1 \text{ volume } O_2/2 \text{ volumes NO}) = 2 \text{ L } O_2$
    or $\quad \dfrac{1 \text{ L } O_2}{2 \text{ L NO}} = \dfrac{x \text{ L } O_2}{4 \text{ L NO}} \quad x = 2 \text{ L } O_2$

    Note: This reaction is important in the formation of photochemical smog and will be further discussed in Part D.

3. a. $2 \text{ CO}(g) + O_2(g) \rightarrow 2 \text{ CO}_2(g)$. The equation specifies that for every 2 mol (or 2 volumes) CO, 1 mol (or 1 volume) of O_2 is needed.

    b. $(50 \text{ mol CO}) \times (1 \text{ mol } O_2/2 \text{ mol CO}) = 25 \text{ mol } O_2$

    or $\quad \dfrac{1 \text{ mol } O_2}{2 \text{ mol CO}} = \dfrac{x \text{ mol } O_2}{50 \text{ mol CO}} \quad x = 25 \text{ mol } O_2$

    c. $(1120 \text{ L CO}) \times (1 \text{ L } O_2/2 \text{ L CO}) = 560 \text{ L } O_2$

    or $\quad \dfrac{1 \text{ L } O_2}{2 \text{ L CO}} = \dfrac{x \text{ L } O_2}{1120 \text{ L CO}} \quad x = 560 \text{ L } O_2$

B.3 YOU DECIDE: ATMOSPHERIC ALTITUDE (pages 354-356)

Students explore the changes in the atmosphere on an imaginary flight into the stratosphere. The spacecraft is designed to collect data on the composition of the atmosphere at various altitudes. Students will plot the data, revealing a direct relationship between altitude and pressure; the relationship between altitude and temperature is inconsistent.

In your discussion of this activity, consider inviting two students to try to separate a pair of plungers as discussed in Question 3 on page 356.

Answers to Questions:

1. Temperature decreases linearly to 12 km, increases linearly to 50 km, decreases again linearly to 80 km. Pressure decreases in a more consistent and regular pattern, showing a "curved" (exponential) shape.
    a. Pressure.
    b. Pressure is due to the collisions of air molecules against a surface. With increasing altitude, there is a steady decrease in the number of molecules in a given volume of air.

2. a. Fall.
   b. Rise.
   c. In part A, an increase in altitude means fewer molecules in a given volume of air, which will reduce the pressure. In part B, there are more molecules at the lower altitude, and therefore more pressure.
3. a. The plungers are hard to separate because there is little pressure inside, and 14.7 lb/in² (assuming one atmosphere) pushing both together on the outside.
   b. Easier.
   c. On top of a mountain, the pressure pushing them together would be less.
4. a. (1) Mass would decrease.
      (2) Number of molecules decrease with increasing altitude.
   b. A straight line.
   c. Mass depends only on the number of molecules, and the composition of the atmosphere is fairly constant with altitude.
5. a. The temperature change graph shows three layers of temperature.
   b. Layer boundaries appear to be at the 12 km, 50 km, and 80 km altitude levels. The fourth, the thermosphere, begins around 90 km.

B.4 AIR PRESSURE (pages 356-359)

In this section, we have attempted to clarify the scientific meaning of "pressure," using SI units. After an exploration of the relationship between area and pressure, a more operational definition (using a mercury barometer) is given. This is an obvious place to show off your mercury barometer, if your school has one.

CHEMQUANDARY

Pressure Puzzles (pages 357-358)

1. The force (weight) is the same in both cases, but the area over which the constant force is exerted is different. More pressure is exerted on the floor by the stiletto-heel shoes, since they have much less surface area in contact with the floor.
2. When walking on the thin ice, the pressure is confined to the area of the soles of the shoes. Lying down on the ice, or using plywood, spreads the weight over a larger area, which may prevent falling through the ice.
3. The tires should be wide. Thin tires would tend to "cut through" the sand, slowing the vehicle's movement. The wider tires would exert less pressure because of greater surface area, and would tend to ride on the sand.

B.5 BOYLE'S LAW: PUTTING ON THE SQUEEZE (pages 359-362)

This section is an introduction to the idea that the behavior of different gases is more regular than that of liquids or solids. The mathematical form of Boyle's law is used to carry out calculations in the *YOUR TURN* exercise. Work through the examples in the text to help students grasp the nature of this inverse relationship.

Consider allowing your students to see Boyle's law in action before they are asked to work quantitative problems. Here are some ideas:

DEMONSTRATION IDEAS

1. If your school owns a vacuum pump (either electrical or mechanical hand pump), you can place a pressure gauge and a balloon or marshmallow under a bell jar and note qualitatively the relationship between pressure and volume. If a large plastic syringe is placed inside, quantitative measurements can be made and the data plotted.
2. Sargent-Welch sells a simple Boyle's law apparatus (catalog number 1077). The device is a plastic syringe/wood block system used in earlier curricula. They also sell a Boyle's law demonstrator for the overhead projector (catalog number 1077D). This is a transparent syringe/gauge system from which students can record quantitative data.
3. A much cheaper and simpler apparatus can be made by connecting a partly filled eudiometer tube to a leveling bulb with rubber tubing. When the leveling bulb is raised or lowered, the volume of the trapped gas in the tube changes, decreasing as the bulb is raised (increased pressure) and increasing when the bulb is lowered (decreased pressure).

A Volume Discount? (page 359)

The volume of a gas is not a good measure of the quantity of gas unless temperature and pressure are specified. Even though Company B's cylinder costs almost twice as much as the same size cylinder from Company A, the pressure in Company B's cylinder could be substantially higher, leading to more hydrogen molecules per liter. A better way is to buy in terms of cost per unit mass (gram or kilogram) of gas.

P-V Relationships (pages 361-362)

This activity provides qualitative and quantitative practice in applying Boyle's law to predict the pressure and volume of gas at constant temperature.

Answers to Questions:

1. a. Gases in the lungs, stomach, sinuses, and dissolved in blood will expand under conditions of reduced pressure, causing an uncomfortable feeling.
   b. Carbon dioxide inside a sealed bottle or can of soda is under pressure. When the container is opened the gas expands and rushes out due to lower external pressure. Some dissolved carbon dioxide also comes out of solution as pressure drops.
   c. If the tennis ball cans are not pressurized, the pressurized gas in the balls themselves will leak out at a faster rate during storage. Tiny pores in the structural material of the ball will leak gas, just as helium leaks from an inflated balloon.
   d. A certain volume of air is normally found in the middle ear as dictated by the pressure outside the ear. Sudden decreases in this external pressure caused by a swift elevator ascent causes the air in the middle ear to expand rapidly and rush out into the throat cavity and ear canal. (The "pop" is the releasing of the skin flap which covers the opening into the throat.)
2. $V_1 = 7100$ mL, $P_1 = 1$ atm, $V_2 = 492$ mL, $P_2 = x$
   $P_2 = (1$ atm$) \times (7100$ mL$/492$ mL$) = 14.4$ atm
3. a. $V_1 = 430$ m^3, $P_1 = 760$ mmHg, $P_2 = 596$ mmHg, $V_2 = x$
   $V_2 = (430$ m$^3) \times (760$ mmHg$/596$ mmHg$) = 548$ m^3
   b. If the windows remain closed during approach of a tornado, the sudden pressure drop and expansion of air in the room may blow out the windows.
4. a. $V_1 = x$, $P_1 = 14$ lb./in.^{2}, $V_2 = 40.0$ L, $P_2 = 44$ lb./in.^{2}
   $V_1 = (40.0$ L$) \times (44$ lb./in.$^2)/(14$ lb./in.$^2) = 126$ L
   b. Pumping by hand takes such a long time because it takes many strokes of the handle to pump such a large volume of air (126 L).

B.6 LABORATORY ACTIVITY: T-V RELATIONSHIPS (pages 362–363)

This laboratory activity introduces students to the relationship between temperature and volume of a gas sample at constant pressure. Students observe the volume of a trapped air column at different temperatures. After making a graph of their temperature/volume data, the students are asked to predict the temperature which would correspond to zero volume.

Time

One period

Materials (for a class of 24 working in pairs)

300 mL corn oil (enough to fill heating container deep enough to allow capillary tube immersion)
24 small rubber bands, orthodontist's type (2 bands/team) or cut rings from rubber tubing
12 capillary melting point tubes
12 thermometers
Paper towels
12 centimeter rulers

Advance Preparation

Obtain the rubber bands, or prepare from rubber tubing.

Pre-Lab Discussion

You may wish to distinguish Boyle's law, which involves a *constant temperature* assumption, from Charles' law, which involves a *constant pressure* assumption. (An easy way to remember this is that temperature is associated with "boyling.") Also, point out that the length of trapped air in this activity is an indirect measure of its volume, since the diameter of the tube is uniform.

Lab Tips

Heat corn oil to 130°C. As hot oil can be very hazardous, use one oil bath (under your personal control) for the entire class. Consider heating the oil in a small deep-fry appliance ("Fry Baby") if you can maintain the oil temperature at a fairly constant 130°C. If the oil is heated in a beaker on a ring stand, be sure to anchor the beaker with a ring. Place the entire apparatus in a larger pan or container to prevent hot oil from escaping if the beaker breaks.

Corn oil is specified because it has given consistently satisfactory results. If you use some other type of oil, be sure you test it in this activity first. Results have been less satisfactory, for example, with mineral oil.

The capillary tubes should be fully immersed in oil for good results.

If the procedure is followed as written, students will record temperature-length data over about at least a 90°C to 100°C range. This is highly desirable, to provide a reasonable baseline for the extrapolation to "zero volume" later. Remind students that their first data point is at the hot-oil temperature and the full length of the capillary tube.

Expected Results

The measured temperature-length data will yield reasonably linear curves. In fact, the data are probably better than they deserve to be, given the actual design of the activity. Collected experimental data over five years of teaching (involving several hundred results) have led to a median extrapolated "zero volume" temperature of –280°C, compared to the "expected" value of –273°C.

Post-Lab Activity

Focus discussion on the questions.

Answers to Questions:

1-4. The specific answers for Questions 1-4 will necessarily be based on the actual experimental data gathered.

5. The gas would not reach zero volume. It would condense to a liquid when it got sufficiently cool (when it reached its boiling point).

ALTERNATE PROCEDURE (using no oil bath)

1. Fasten a capillary tube with an oil plug in it to a thermometer, using two rubber bands. The open end of the tube should be up, and the closed end adjacent to the lowest mark on the thermometer's scale. Record the length of the air space by measuring the distance between the closed end of the tube and the *bottom* of the oil plug, using the thermometer scale as a measuring scale. (Example: closed end is at -10°C mark, bottom of oil plug is at 24°C mark. Distance is then 34°C).

2. Immerse the tube and thermometer in a test tube filled with water (the open end should be above the water level). Add the test tube/thermometer combination to a beaker 3/4 full of water. Read the temperature of the water from the thermometer, and record your results.

3. Begin heating the beaker, and every 10°C-15°C read the temperature and the length of the air space by noting the position of the bottom of the oil plug; record your data in a chart. Continue these readings until the temperature of the water is above 90°C.

4. Allow equipment to cool, and return all glassware according to teacher's instructions.

Procedure

1. Stretch the narrow end of a Beral pipet to a length of about 45 cm. Do this by grasping the bulb with your left hand while pulling the tip with pliers in a steady manner. This may take a few times— you can improve your chances for success by warming the tip of the pipet by rubbing it several times with your fingers.

2. Have a beaker of hot water (about 90°C) ready. Briefly dip the stretched stem near the bulb into the hot water, and bend it to form a J loop. Allow it to cool in that position.

3. Fill the bulb of the pipet approximately half-full with colored water (methylene blue is a suggestion). Squeeze the bulb of the pipet, then dip the open end into the solution. Avoid having droplets of liquid in the long, stretched tube.

4. Gently squeeze the bulb until the liquid is about 5 cm into the smaller tube, or until it passes the J loop. Tape the pipet to a centimeter ruler.

5. Fill a 600 mL beaker about 3/4 full of water, and place it on a ring stand with a Bunsen burner.

6. Secure the centimeter ruler with a buret clamp, with the bulb well below the surface of the water.

7. Begin heating the water gently while stirring carefully with a thermometer. Observe the column of liquid as the temperature changes. When the thermometer reaches 60°C remove the heat source.

8. As the temperature decreases begin recording the temperature and height in the tube. Take readings every 5 degrees for at least 5 readings, or more if time permits.

9. Use this data to prepare a graph of water temperature (x-axis) versus the height of the water in the pipet tube (y-axis).

B.7 YOU DECIDE: A NEW TEMPERATURE SCALE (pages 364-365)

Students are introduced to the absolute temperature scale (kelvin) and discover the mathematical convenience of converting °C to K to avoid using negative numbers. Specifically, they use their graphs from the *Laboratory Activity* B.6 to extrapolate back to the theoretical "zero volume."

Post-Activity Discussion

Help students with the correct way (by IUPAC ruling) to express the temperature using the kelvin scale. Thus 20°C is written 293 K, not 293°K. The correct way to verbalize this temperature is "293 kelvins", not "293 degrees kelvin." Point out that the pressure of a sample of gas is directly proportional to its kelvin temperature. Depending on student mathematical ability, decide whether or not to discuss the mathematical form of Charles' law.

Answers to Questions:

1. Student answers will understandably vary from the actual value of –273°C, but many should be within 25°C of this temperature.
2. Zero volume.
3. The volume is only theoretical because the gas will condense before it reaches this low temperature. Also, molecules can not disappear.
5. Student answers may vary somewhat from the actual value of 273 K for 0°C and 373 K for 100°C.
6. K = °C + 273
   °C = K – 273

CHEMQUANDARY

Behavior of Gases (page 365)

1. The carbon dioxide produced by the yeast creates pockets of gas within the dough structure that expand during the high temperature baking process.

2. Hot air is less dense (fewer molecules per given volume) than cold air.

3. The warm balloon weighs less than the volume of cold air it displaces, and so there is a net positive upward force on the balloon. Examples include smoke from a fire, steam from a tea kettle, exhaust fumes from a car, etc. Note: Archimedes' principle also explains floating and sinking in liquids.

4. If the thermostat were installed on the second floor, it would give higher readings (and therefore cause the furnace to be turned on less), than if it were on the first floor. Placing it on the first floor would result in a warmer house overall (and more costly heating bills). Note: Convection currents are most important in designing home heating systems, in cooking, for aquatic life, etc.

T-V-P Relationships (pages 365-366)

1. a. Pressure will increase by three times.
   b. Examples: pressure cooker, incineration of aerosol cans, etc.
2. a. The absolute temperature (in kelvins) will decrease by a factor of four; the absolute temperature will be 1/4th its original value.
   b. A balloon placed in a freezer.
3. Volume, like pressure, is directly proportional to the kelvin temperature. Car owners add air to their tires in winter because the decreased temperatures decrease the pressure of the enclosed volume of air. More air molecules are needed to insure proper tire pressure. Air is released when summer arrives because the pressure from the enclosed volume of air increases as the temperature increases. The increased pressure can result in a less comfortable ride and excessive wear on the tires.
4. a. Because of the large spaces between molecules in the gaseous state, they are easily affected by changes in pressure or temperature, which changes molecular motion.
   b. For practical purposes, the relatively close proximity of molecules in liquids and solids makes them non-compressible. Therefore, there are no Charles' or Boyle's laws for these states of matter. You may wish to point out that, while a given mass of water will have approximately the same volume in the liquid and solid states (slightly more in the solid state), it will increase in volume by a factor of 1000 in the gaseous state.
5. a. A given mass of warm moist air will occupy more volume than the same mass of cold air (Charles' law) and will rise above the more dense, cold air (Archimedes' principle).
   b. The two laws, Boyle's law and Charles' law, oppose each other as the balloon rises—the pressure decreases (volume should increase) and so does the temperature (which should decrease the volume). However, the data on page 355 show that the pressure changes are proportionately larger than the temperature changes. For example, from 0 km altitude to 10 km, the temperature changes 65°C, which is a 22% change, while the pressure changes from 760 mmHg to 218 mmHg, a change of 71%. Therefore, Boyle's law dominates and the volume increases.

B.8 IDEALLY, GASES BEHAVE SIMPLY (pages 366-368)

This section introduces the kinetic molecular theory as a way to understand the regularities in the behavior of gases. Analogies are introduced to help students better comprehend the unseen world of molecules in a gas.

Hammering Away at Kinetic Energy (page 366)

1. A nail would be preferred. With less mass it will attain less kinetic energy by the time it hits your foot.
2. A hammer from knee height. The final velocity of the hammer from the knee would be less than the hammer from the shoulder, and the kinetic energy would thus be reduced.

Gas Molecules in Motion (pages 367-368)

This activity helps students use the postulates of the kinetic molecular theory to explain the behavior of gaseous systems.

Answers to Questions:

1. Decreasing the volume means that there is less room for the molecules to move around in. This will cause molecules to hit the walls more often per unit time, causing an increase in pressure.
2. At higher temperatures, the kinetic energy of the molecules is greater, reflected by their higher average velocities. This increased kinetic energy will cause more energetic collisions with the walls, and higher pressure.
3. External body pressure is caused by collisions of air molecules against the skin. Since our body volume remains relatively constant, the internal body pressure must be nearly equal to the external pressure.

4. Given their high rate of motion and small size, gas molecules will eventually pass through the molecular structure of the balloon wall.

5. Helium molecules are smaller in size and exhibit a higher average velocity than molecules of air (N_2, O_2, CO_2, H_2O). Helium molecules collide more frequently with the balloon walls and fit more easily between the molecular-scale openings.

PART B: SUMMARY QUESTIONS (pages 368-369)

1. A higher atmospheric concentration of oxygen would cause more rapid corrosion (oxidation) of metals. Fires would be more common and difficult to control. Living organisms could probably gradually adapt to the metabolic effects of a higher oxygen concentration.

2. a.

   (H)(H) + (Cl)(Cl) → (H)(Cl) (H)(Cl)

   b. $H_2 + Cl_2 \rightarrow 2\ HCl$

3. No, the equal volumes of liquids (or solids) do not contain equal numbers of particles. The number of particles per volume depends on the size and packing arrangement of the units (atoms, molecules, or ions). The volume of the particles make up a significant proportion of the actual measured volume of substance. Therefore, there are no equivalent laws for liquids and solids as for gases. Different liquids and solids do not respond in the same manner as gases to changes in temperature. No constant molar volume may be given.

4. As one descends into the ocean, the pressure increases due to increasing weight of the water above. Similarly, as one descends from the outer atmosphere to the Earth's surface, the pressure increases. Note that in the first case the pressure increase is linear, and in the second case exponential.

5. As the suction cup is pressed inward, air is forced out from beneath. The atmospheric pressure acting on the outer surface is greater than the pressure beneath the cup, causing the device to remain pressed against the wall.

6. The extreme temperature, pressure, and oxygen concentration above the troposphere make life there (to understate the case) inhospitable.

7. a. constant temperature
   $V_1 = x$ $P_1 = 760$ mmHg $V_2 = 0.210$ L $P_2 = 1.30$ mmHg
   $V_1 = (0.210\ \text{L}) \times (1.30\ \text{mmHg}/760\ \text{mmHg}) = 0.00036$ L or 3.6×10^{-4} L

   b. constant volume
   (1) $T_1 = 300$ K $P_1 = 1.92$ atm $T_2 = x$ $P_2 = 5.76$ atm
       $T_2 = (300\ \text{K}) \times (5.76\ \text{atm}/1.92\ \text{atm}) = 900$ K or 627°C
   (2) $T_1 = 300$ K $P_1 = 1.92$ atm $T_2 = 1275$ K $P_2 = x$
       $P_2 = (1.92\ \text{atm}) \times (1275\ \text{K}/300\ \text{K}) = 8.16$ atm
   Yes, the tank will explode at 1275 K (1000°C).

8. Comparisons of equal volumes of warm and cold air show the warmer air to be less dense (fewer number of molecules per unit volume), with a greater average molecular velocity. In short, the less-dense air rises above the cooler, more dense air causing convection currents. Just as objects that are less dense than water rise to its surface, warm air rises and cold air falls.

EXTENDING YOUR KNOWLEDGE (page 369)

- Different environments are encountered at different depths and heights. Pressure and temperature change with depth and height. In some ways the atmosphere is more complex than the hydrosphere: (1) pressure changes exponentially, not linearly; (2) the temperature profile creates fairly permanent strata; and (3) its chemical composition varies with height and depth.

- During inhaling, the diaphragm moves down and out allowing the volume of the lungs to increase. This results in a pressure decrease within the lungs and the greater atmospheric pressure forces air into the lungs. Exhaling is the reverse process.

- Boyle's law and diving
  a. Pressure at 10.3 m = 1 atm (from air) + 1 atm equivalent (10.3 m water), or 2 atm pressure.
  b. Pressurized tanks are needed for two purposes: to withstand the increased external pressure without being crushed, and to enable sufficient oxygen to be transported in a reasonable volume.
  c. It would be decreased (crushed).
  d. If the diver does not exhale and rise slowly, the volume of the various gases trapped in the lungs and sinuses will rapidly expand, causing extreme pain. In addition, at decreased pressure the gases dissolved in the blood, particularly nitrogen, are released, causing "bends" or even rupturing of blood vessels.

e. A diver experiences increased pressures. A pilot experiences reduced pressures at high altitudes. In both cases, Boyle's law applies; the body must be protected with a pressurized suit or cabin.

- Mass of a roomful of air:
a. $(1.28 \text{ g/L})(2 \times 10^5 \text{ L}) = 2.56 \times 10^5 \text{ g}$
b. If the temperature of the air increases, the increased molecular motion would cause some of the air molecules to spread out to adjacent rooms. The density and mass of air in the room would decrease.

- Kelvin temperatures
a. Absolute zero has not been reached (but has been closely approached);
b. References include *McGraw-Hill Encyclopedia of Science and Technology*, and recent citations in the *Readers' Guide to Periodical Literature* to popular science magazines.

C: ATMOSPHERE AND CLIMATE

Part C begins to focus on the technological issues associated with the atmosphere. The nature of energy input from the sun to the Earth is presented, and the reasons for the greenhouse effect are explained. The energy exchange at the Earth's surface clarifies some of the reasons for climatic conditions, and the factors influencing carbon dioxide levels leads to a *You Decide* on CO_2 levels. Students gain some idea of the relative quantities of CO_2 produced in respiration and combustion, and propose some changes which might lower CO_2 levels. Students also learn how the chemistry of the ozone layer protects us from ultraviolet radiation, and how certain chemicals of human origin may threaten the ozone layer.

C.1 THE SUNSHINE STORY (pages 371-372)

This section provides information on radiation that showers the Earth during daylight hours. Properties of electromagnetic radiation and its interaction with molecular structure are briefly presented.

C.2 EARTH'S ENERGY BALANCE (pages 372-374)

The reasons for the relatively stable temperatures on the Earth are explained by showing the balance between incoming and outgoing radiation. The molecules in the atmosphere responsible for absorption are listed, and the potential for the greenhouse effect noted.

CHEMQUANDARY

Grab Another Blanket! (page 373)
Clouds limit the radiation of thermal energy from Earth back to space.

YOUR TURN

Earth's Energy Balance (page 373)
These questions are designed to reinforce the relationship between electromagnetic frequency and energy, interaction of light with matter, and the consequences of radiation-matter interactions on the temperature of the atmosphere.

Answers to Questions:
1. Since the energy of electromagnetic radiation (EMR) is directly proportional to its frequency, higher frequency radiation such as ultraviolet light is more energetic (per photon) than infrared photons. Ultraviolet radiation is high enough in energy to disrupt chemical bonds and harm biological organisms—far ultraviolet radiation can be ionizing.
2. First, photons of visible light allow the eye to see and the photosynthesis reaction of green plants to proceed. Second, the re-radiation of visible light into the lower energy infrared frequencies warms our atmosphere, and drives the hydrological and meteorological systems.
3. Increased concentrations of carbon dioxide may result in a warming trend and subsequent side effects such as the melting of polar ice caps and the shifting of sites suitable for agriculture.

4. If the atmosphere were thinner (contained fewer molecules), we would have less of an insulating blanket to provide a thermal balance. As a result:
   a. The average daytime temperature would be higher.
   b. The average nighttime temperature would be lower.

C.3 AT THE EARTH'S SURFACE (pages 374-375)

This section explains the physics of reflectivity and heat capacity, two properties of materials which affect the quantitites of energy absorbed. You may want to point out that materials that quickly become hot or cold generally have a lower heat capacity than those that become hot or cold slowly. (Contrast how quickly a metallic object loses temperature in a refrigerator compared to a similar mass of water.) Water's unique role in moderating temperatures and influencing wind patterns, due to its heat capacity, is pointed out.

CHEMQUANDARY

Some Reflections on Dust (page 374)

1. The snow would melt because the dust would absorb more heat—this could raise the level of the oceans.
2. If large quantities of dust of high reflectivity were to enter the atmosphere, there would be a cooling of the Earth's atmosphere as less solar radiation would reach the surface.

YOUR TURN

Thermal Properties of Materials (page 375)

Answers to Questions:

1. a. Concrete since it reflects more sunlight than the black asphalt.
   b. Sheepskin seat
2. Lower in the winter with large quantities of snow, since the snow would reflect back more solar energy.
3. A given mass of water can store more heat energy and release it over a longer period of time, than alcohol or other fluids having lower heat capacities.
4. Since the higher reflectivity of sand would tend to make it cooler than the grass, heat capacity must be responsible for the higher temperature of sand.
5. Clouds limit the radiation of thermal energy from Earth back to space.
6. The city with asphalt roads and concrete buildings would be hotter, because the black asphalt has low reflectivity/high absorptivity. In the city away from the water, the temperature-moderating effect of water is absent.

DEMONSTRATION IDEA

To show the different reflectivities of black and white: Paint one metal can black and another white, fill with equal quantities of water, place a thermometer in each and place equidistant from a heat lamp. Within an hour you will note that the water in the black can is warmer. If well insulated lids are used, you may want to repeat the experiment using empty (filled only with air) cans.

C.4 CHANGES ON THE EARTH'S SURFACE? (pages 375-379)

The greenhouse effect is not a new theory, as the text points out. However, several factors which have led recently to higher levels of CO_2 are listed, including deforestation and combustion. Temperature data on global warming prompt an exploration of the CO_2 levels in the *You Decide* activity. This section points out that the variables that control climate are not well understood, preventing confident predictions concerning the effect of carbon dioxide concentrations on climate.

C.5 YOU DECIDE: TRENDS IN CO_2 LEVELS (pages 379-380)

This activity emphasizes the use of past data to determine the future of carbon dioxide concentrations. Students plot the known carbon dioxide concentrations by years to predict the carbon dioxide concentration by extrapolation to the year 2050. Be sure to emphasize the uncertainties of such extrapolations during the post-activity discussion.

Point out that the predictions here are similar to those on page 144 in the Resource unit. If your students do not yet have a good grasp of the concept of parts per million (ppm), you may wish to have them consider the following analogies: 1 ppm is equivalent to: (a) one penny in 10,000 dollars (10^6 pennies); and (b) one millimeter in 1 kilometer (10^6 millimeters). Such analogies can give students a better sense of being able to detect in the ppm (or smaller) range when discussing molecules.

Answers to Questions:

Part 1 (Based on student-prepared graphs)

Data from "CO₂ levels"

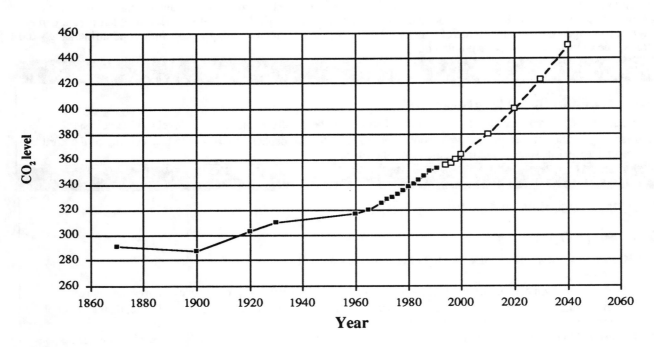

Part 2

1. Carbon dioxide levels have increased.
2. Carbon dioxide levels:
   a. 1991 = 353 ppm
   b. 2000 = 364 ppm
   c. 2050 = 450 ppm
3. No.
4. The predictions nearest the last measured data are more accurate than the one for 2050. The rate of carbon dioxide buildup in the atmosphere may change from previous rates. The further into the future we extrapolate, the greater the change in rate can be (the more uncertain our prediction). Some scientists do, however, predict that by the year 2050, the level will have doubled since 1870.
5. Some possible factors: change in the burning of fossil fuels, the amount of land we clear annually, forest fires, and a change in climatic temperature due to the beginning of a hypothesized ice age.
6. The assumption is that there will be no drastic changes in the processes that have produced the known data.

C.6 LABORATORY ACTIVITY: CO₂ LEVELS (page 380)

Students qualitatively measure the amount of carbon dioxide present in normal, combusted, and exhaled air by monitoring the change in acidity caused by CO_2 absorption in a solution. The monitor is bromthymol blue.

Time

One class period

Materials (for a class of 24 working in pairs)

12 small glass (not plastic) funnels (7-cm diameter)

12 500-mL filter flasks
12 1-hole stoppers to fit filter flasks
12 aspirator adapters for faucets
12 small (30 cm) pieces of rubber aspirator hose
24 straight glass tubes
24 glass bends
24 ring stands and clamps
36 rubber connectors
36 test tubes (18 × 150 mm)
12 large candles
12 straws
Bromthymol blue solution
12 250-mL Erlenmeyer flasks

Advance Preparation

Bromthymol blue can be purchased as either a 0.04% solution or a powder to be made up to a solution that indicates pH in the range between 6.0 (yellow) through green to 7.6 (blue). For convenience, the made-up soultion is preferred.

You may wish to pre-insert the glass into the rubber stoppers. If you allow students to do this, be sure to emphasize proper technique, using glycerin and a twisting motion.

Check the water system at your school; some systems cannot tolerate the influx of many groups all using aspirators at the same time. If so, do this activity as a demonstration.

Pre-Lab Discussion

Draw student attention to Figure 14 on page 380 and display the proper setup. Point out that, for hygienic reasons, only one student from each group should use the straw for the testing of exhaled air.

Lab Tips

Heavier walled rubber tubing is preferred when using an aspirator.

Expected Results

See Answers to Questions

Post-Lab Discussion

Emphasize that all combustion processes (which involve hydrocarbons or carbon based fuels or foods) produce carbon dioxide. This type of pollution is inevitable, though it can be lowered if we switch to energy alternatives such as nuclear, solar, wind, or geothermal power.

Point out to your students that environmental analytical chemists routinely make quantitative analyses on the order of ppm (and in many cases ppb are becoming more common and parts per trillion, ppt, possible) using automated, sensitive, instrumental techniques. Depending on the sensitivity required, different procedures may be used. You may wish to challenge your students to design an alternative carbon dioxide detection system that would allow quantitative measurements. One possibility could be based on the reaction of carbon dioxide with limewater and the subsequent increase in mass. Practical difficulties include measuring a mass difference on the order of 0.1 g, and the need to capture the water vapor that is also generated in the combustion process. (Draw the combustion gases through a drying tube filled with anhydrous calcium chloride before they enter the flask with limewater.)

Understanding combustion processes has fascinated chemists since the beginning of the discipline. The French chemist Antoine Lavoisier was the first to note the connection between combustion and human respiration. Michael Faraday delivered a whole lecture series on the history of the candle (many modern day chemistry texts have an entire laboratory activity patterned around observations of a burning candle). Much of present day chemical research is focused on ways of extracting more energy (via increased burning and thermodynamic efficiency) and less pollution from fossil fuels. Both the history and present day research make interesting library projects.

Answers to Questions:

1. Most CO_2: burning candle. Least CO_2: plain air.
2. Because combustion of paraffin (a hydrocarbon) produces CO_2, less air had to be drawn through the apparatus to effect a color change.
3. Exhaled air.
4. a. Green plants use CO_2—time should increase.

b. Persons produce CO_2—time should decrease.
c. Better ventilation reduces the concentration of exhaled CO_2—time should increase.
d. Smoking produces CO_2—time should decrease.

OPTIONAL CHEMQUANDARIES

1. Of the various gases that contribute to the greenhouse effect (CO_2, NO_2, CH_4, chlorofluorocarbons, etc.), carbon dioxide accounts for about 50% of the adverse atmospheric impact. List and briefly discuss some ways of limiting this impact. Is there a single right or best answer to the problem?
   Answer: (a) Increase energy efficiency, so that less fuel would need to be burned and less carbon dioxide produced to accomplish a given task. (b) Promote conservation measures that result in an overall lowered demand for energy. (c) Switch to solar, wind, geothermal, and nuclear energy. (d) Halt the destruction of forest and begin a reforestation plan, etc. Given our high dependence on carbon-based fuels and given the variability of local natural and technological resources, a variety of "solutions" probably are necessary.
2. How does changing the global climate via the greenhouse effect relate to issues raised in the previous *ChemCom* units?
   Answer: Water: Increased temperatures would affect both the demand and supply of water on a regional basis. **Resources:** Much of mineral production and processing (mining, roasting of ores, electrolysis, etc.), depends on the combustion of fossil fuels. **Petroleum:** The cleanest, most efficient burning of petroleum inevitably produces carbon dioxide and water—if less petroleum was burned, more would be available for building purposes well into the future. **Food:** Climate changes would trigger shifts in agricultural productivity—some presently productive areas would become deserts.
3. "There have been calls for recovery of carbon dioxide from smokestacks, especially from coal-burning electric utilities. According to one estimate, if every power plant in the world installed equipment to remove carbon dioxide from combustion gases, the annual addition of carbon dioxide to the atmosphere would be cut 30%." (*Chemical and Engineering News*, November 24, 1986, 45). Do you think this would be both a practical and effective measure? Why or why not?
   Answer: Probably not, the energy and materials costs would probably be prohibitive. Also, the carbon dioxide waste captured as some type of slurry or solid would create both environmental and economic problems of its own. Producing less carbon dioxide seems to be a wiser approach.

C.7 YOU DECIDE: REVERSING THE TREND (pages 381-382)

This activity considers what students could do to reverse a warming trend from increased carbon dioxide concentration levels. The objective is to list possible courses of actions and decide on those which would gain the strongest support, given economic, political, and social considerations.

Time

This activity can be assigned as homework. On the following day, you may want to divide students into groups to discuss their answers to the questions. Bring the students back together as a class; list the various options on the chalkboard or overhead projector. Analyze each option in terms of economic, political, and social problems.

Answers to Questions:

1. Reduce the need for fossil fuel combustion by relying more on conservation (to decrease energy demand) and alternative energy sources (solar, wind, nuclear, etc., to replace the energy presently supplied via combustion). Find better ways to prevent and control forest fires. Halt the rapid elimination of forest and land clearing and reverse the trend by planting more green plants.
2. a. Purely opinion, but a number of practical and political problems would probably limit international cooperation. Some of these include: (1) Not all scientists agree on the seriousness of the problem. (2) Less technologically advanced nations may be less capable of switching to alternative technologies and may feel that the real burden should be borne by the industrial nations that contribute more heavily to the problem. (3) A rapid reduction in the consumption of fossil fuels would cause major economic problems for both individuals and nations, especially those with economies "run" on petroleum. (4) The demand for increased food production may run counter to the "solutions."
   b. Students should realize that public support is most likely for measures that do not interfere with safety or comfort, such as lowering the number of forest fires, or providing high-quality mass transit for commuters.

C.8 OFF IN THE OZONE (pages 382-384)

The formation of ozone and its decomposition is explained.

In what respects is the problem of controlling the greenhouse effect qualitatively and quantitatively different than limiting the depletion of the ozone layer?

Answer: The greenhouse effect is created by a wide variety of manmade pollutants including carbon dioxide, nitrogen oxides, methane, and chlorofluorocarbons (CFCs). The single, major contributor, carbon dioxide, is produced whenever a carbon-based fuel is burned—short of switching to nuclear, solar, wind, or geothermal energy, people cannot avoid producing carbon dioxide. And presently, economic, political, and technical issues block any significant shift in our energy consumption patterns away from carbon-based fuels. On the other hand, ozone depletion is believed to be governed almost exclusively by CFCs that are emitted in much smaller quantities from a much smaller number of sites. Substitutes for CFCs either already exist or can be developed for most applications (witness the 1978 ban on using CFCs as propellants in aerosol cans). Also, in terms of economic development of developing countries, use of CFCs is far less critical than use of carbon-based fuels.

PART C: SUMMARY QUESTIONS (page 384)

1. a. The order of relative energy per photon is infrared (least), visible (intermediate), and ultraviolet (highest).
   b. Infrared is important in heating our atmosphere, visible enables us to see, and a minimal exposure to ultraviolet is essential to vitamin D production (see discussion in Food unit).
2. a. High energy solar radiation penetrates the atmosphere, but low energy infrared radiation is temporarily trapped.
   b. This helps maintain a comfortable average temperature.
3. The stratosphere contains the ultraviolet-absorbing ozone, protecting organisms. It also serves as a barrier that prevents the escape of molecules from the troposphere. The latter is accomplished by a temperature inversion between the troposphere and the stratosphere that prevents molecules from passing between the layers.
4. The color white reflects all frequencies of visible light, whereas colored materials absorb some of the frequencies. Absorbed visible light is converted into molecular motion (heat). Thus, wearing white will help keep one cooler.
5. a. Beach sand heats up faster during the day and cools off faster at night than water does.
   b. The heat capacity of water is much greater than that of beach sand. Also sand is stationary, while water can dissipate some of the heat by convection and evaporation.
6. Carbon dioxide and water trap lower-energy infrared radiation that is reemitted by the Earth. As a result, the Earth's average temperature is maintained at a higher level (15°C) than would be the case without this insulating blanket. This is called the greenhouse effect.
7. Burning fossil fuels, and removing green plants from the surface of the Earth, increase the amount of carbon dioxide in the atmosphere. Burning appears to be the major factor.
8. a. Increased concentrations of carbon dioxide and water.
   b. Decreased concentrations of carbon dioxide and water.
9. Ozone molecules in the stratosphere absorb high-energy ultraviolet photons that would be capable of breaking the covalent bonds in biomolecules if they were allowed to reach the Earth's surface. If the ozone layer is depleted, sunburn, cancer, and other biological problems would increase.

EXTENDING YOUR KNOWLEDGE (page 384)

• See supplemental readings and the *Readers' Guide to Periodical Literature* (for more recent articles).

D: HUMAN IMPACT ON AIR WE BREATHE

Part D explores human impact on air pollution by comparing natural and human sources of common pollutants. The different sources of pollutants are tabulated, as well as United States pollution standards. Various industrial controls are explained, some demonstrated. Smog and acid rain are shown to be products of pollution. This Part concludes with information concerning pH.

D.1 TO EXIST IS TO POLLUTE (pages 385-387)

Primary air pollutants are identified, and their natural and human-generated sources and amounts are compared. You may wish to point out some of the effects of air pollution: (1) minor nuisance and aesthetic insult (slight odor and reduced visibility); (2) property damage (discoloration of buildings, accelerated weathering, corrosion of metals); (3) damage to plants (lowered crop yields); (4) damage to animals/humans (eye and respiratory irritation, emphysema, bronchitis, lung cancer, heart disease); (5) genetic and reproductive effects; and (6) major ecosystem disruption.

D.2 YOU DECIDE: WHAT IS AIR POLLUTION? (page 387)

Here students become involved in critically analyzing data on the annual emissions of major, primary air pollutants. The questions focus on the data identifying amounts and sources of air pollutants. Students are often surprised to see the relatively small amounts of human pollution, but Question 2a makes the point that overload of nature's system of removal can occur.

Answers to Questions:

1. a. The pollutants are listed in decreasing order of annual emissions from human sources.
   b. We can control the production of these pollutants only.
   c. No.
   d. Some substances have a disproportionately negative impact on human health (carbon monoxide is highly toxic, whereas carbon dioxide is considerably less so), or on the environment (for example, discounting the uncertainties of the greenhouse effect, many scientists feel that sulfur dioxide pollution and its contribution to acid rain is a much more immediate and acute threat than is carbon dioxide). Also, the figures are global emissions; the need for specific control measures vary by location. For instance, sulfur dioxide could create a greater problem in coal-burning regions; nitrogen oxides are a concern in regions with high concentrations of cars and sunlight; hydrogen sulfide might be of greater concern near petroleum refineries.

2. a. No. Nature, including living organisms such as humans, has probably evolved mechanisms for dealing with many natural emissions. Given the recent development of industrial activity (in terms of historic and prehistoric time) and its associated pollution, nature has not had time to respond to the increased amounts of "natural " pollutants and the presence of new ones.
   b. Sulfur dioxide from human sources exceeds (by more than a factor of 10) that from natural sources.
   c. This suggests that modern society is capable of significantly affecting the amount of trace components in the atmosphere.

3. CO_2, NH_3, CH_4.

4. Combustion, especially of fossil fuels.

5. As mentioned above, nature may not have mechanisms for converting new pollutants into harmless byproducts. A case in point are the chlorofluorocarbons (Freons) which are inert in the lower atmosphere and are not decomposed until they reach the stratosphere where they may pose a threat to the ozone layer.

OPTIONAL DISCUSSION TOPIC

Indoor air pollution: As a result of the energy crisis and increased incentives to conserve energy in home heating, houses have been constructed with more insulation and less natural ventilation. Older homes typically exchanged all the interior air with exterior air about twice every hour. Today, the average United States home exchanges air about once every hour and some newer homes may be ten times "tighter." Natural radon seepage (see the Nuclear unit), gases given off by some building materials, and, most especially, cigarette smoke can significantly lower interior air quality in such homes. A study of this problem and potential solutions makes for an interesting research project. A starting point might be the article: "Indoor Air Pollution," *Science 80*, March-April, 30-33.

D.3 YOU DECIDE: MAJOR POLLUTANTS (pages 387-388)

This activity focuses on sources of pollution in the United States. Some common scapegoats, such as industry, are seen to be causing less pollution than transportation does. Students should realize that all of us, by our lifestyles based on high energy and matter consumption, contribute to the pollution problem.

Answers to Questions:

1. a. No.
   b. Transportation.

2. a. Industry contributes 35.7% of the total suspended particulates (TSP) and 42.3% of the hydrocarbons (HC).
   b. Transportation contributes 30.6% of the HC, 43.1% of the NO_x, and 66.3% of the CO.
   c. Fuel combustion (for space heating and electricity) accounts for 52.8% of the NO_x, and 80.4% of the SO_x. Emphasize that industries do not exist to pollute—but rather to make products and deliver services that society wants.

D.4 SMOG: HAZARDOUS TO YOUR HEALTH (pages 388-390)

The Air Quality Index has become an important part of weather reports. The text presents the index used, which includes the different components and the possible effects on people.

If you have had students keep track of the Air Quality Index for your area, this would be a good time to examine the data for possible correlations between weather conditions and air quality.

D.5 POLLUTION CONTROL (page 390)

The section provides a brief overview of air pollution control options. Some proposals for controlling pollution are offered, while the two questions concerning cost and benefit are raised.

CHEMQUANDARY

Airing Some Pollution Solutions (page 390)

1. Given the rapid mixing caused by diffusion, convection, and winds, pollutants are typically diluted and spread throughout rather large volumes of air. Attempting to filter such large volumes would be impractical from an energy and cost perspective.

2. Cloud seeding might clean the air over a small area, but would merely transfer the water-soluble pollutants to the ground where they could still cause damage (for example SO_x and $NO_x \rightarrow$ acid rain). Successful cloud seeding typically means that other locations may receive less than their normal share of rainfall.

   Providing all persons with masks that would prevent intake of all harmful pollutants (simple mechanical filters would not be adequate) would be expensive, repulsive, and, from a practical viewpoint, absurd. Not polluting in the first place, or controlling emissions at the source before release, are clearly more reasonable options.

D.6 INDUSTRIAL EMISSION OF PARTICULATES (pages 390-391)

Industry, which is third on the list of sources of pollution, has developed several techniques for controlling plant emissions. Four techniques are presented here. It is interesting to note that the Cottrell Corporation, which developed the most widely used technique of electrostatic precipitation, uses the revenue from its patent rights to support the Petroleum Research Corporation Fund, a very large source of research monies in the United States.

D.7 LAB DEMONSTRATION: CLEANSING AIR (page 392)

Electrostatic Precipitator Demonstration

Cottrell electrostatic precipitators use a central wire connected to a source of direct current at high voltage (30,000 V or more). As dust or aerosols pass through the strong electric field, the particles attract ions that have been formed in the field, become strongly charged themselves, and are attracted to the electrodes. The collected solid grows larger and heavier and falls to the bottom, where it is collected. Industrial electrostatic precipitators remove up to 99.5% of the total mass of particulates, but very little of the extremely fine particles (diameters less than about 10-6 m).

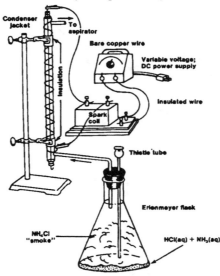

Setup for Electrostatic Precipitator Demonstration

Materials

5 mL 12.0 M hydrochloric acid [concentrated HCl]
5 mL 14.8 M ammonium hydroxide [concentrated $NH_3(aq)$]
Liebig condenser
Copper tube (1/4 in. outer diameter, 50-cm length)
2 rubber stoppers to fit condenser tube
Aspirator adapter for sink faucet
Spark coil box (tesla coil)
Power supply
4 leads
500-mL flask
2-hole stopper to fit flask
Thistle tube
Ring stand
2 rubber-ended condenser clamps (must have rubber for insulation)
Piece of rubber tubing (30-cm length) from flask to condenser
Piece of rubber tubing (50-cm length, thick walled) from condenser to aspirator
50-cm copper wire to wrap 8 times around condenser

Procedure

1. Set up the apparatus illustrated on the previous page.
2. Add 10 mL of concentrated hydrochloric acid to the flask. Stopper it.
3. Add 10 mL of concentrated ammonium hydroxide through the thistle tube and immediately turn on the aspirator. The aspirator will draw the ammonium chloride smoke into the condenser tube.
4. Allow the condenser tube to become densely filled with white smoke, ammonium chloride . Turn off the aspirator and immediately apply the electric potential. Observe the smoke disappear as the ammonium chloride collects on the walls and metal rod of the condenser. You may also notice the chemical reaction between the ammonium chloride and the copper rod which produces the greenish copper(II) chloride.

Wet Scrubbing Demonstration

Industrial wet scrubbers remove up to 99.5% of the particulates (but again few of the very fine particles) and 80-99.5% of the SO_x by passing exhaust gas through a liquid such as water.

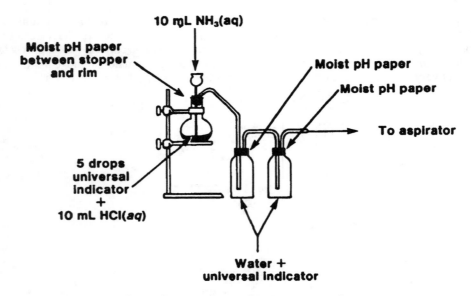

Setup for Wet Scrubbing Demonstration

Materials

Package Hydrion (pH) paper
50 mL universal indicator solution
10 mL 12.0 M hydrochloric acid [concentrated HCl]
10 mL 14.8 M ammonium hydroxide [concentrated $NH_3(aq)$]
2 500-mL flasks
500-mL filter flask (500-mL flask can be used with 2-hole stopper). If using filter flask: 1-hole stopper to fit; two
   2-hole stoppers for other flasks
U-shaped glass tube
2 straight pieces of glass tubing (to fit in first two flasks)
2 pieces rubber tubing; first to second flask (10 cm); final flask to aspirator (40 cm)
Aspirator adapter for sink faucet

Procedure

1. Add 5-10 drops of universal indicator to each collecting bottle (or flasks). Then fill each with water up to the neck, allowing just a small space for the inlet tube.
2. Place a moistened piece of pH paper between the stopper and glass rim so it hangs in the space between the stopper and the water level. (This will indicate the pH of the gases hovering above the water level.)
3. Add 10 mL of concentrated hydrochloric acid to the smoke generating flask. Immediately stopper it.
4. Pour 10 mL concentrated ammonium hydroxide through the thistle tube and immediately turn on the aspirator.
5. Direct students' attention to the color change in the water and pH paper for each flask.

Demo Tips

Note that since the electrostatic precipitator works on the principle of static electricity (which is diminished by moisture), the condenser tube and stoppers must be totally dry between demonstrations for best results.
Caution: Avoid breathing fumes from the concentrated hydrochloric acid and ammonium hydroxide bottles. Work in an area with adequate ventilation. Don't forget to wear your own protective goggles during these demonstrations. Treat the electrical circuit used in the precipitator demonstration with caution and respect. Do not touch any "live" wires. Keep students at a safe distance.

Answers to Questions:

1.　　$NH_3(g) + HCl(g) \rightarrow NH_4Cl(s)$
2.　　Universal indicator gives the pH of the solution of smoke in water as the ammonium chloride enters the flasks.
3.　　The pH paper indicates the pH of the gases hovering above the water layer just before being exhausted.
4.　　The pH of the water solution and gas above the water layer approaches neutrality as the reaction proceeds to the second flask and on to the aspirator. Scrubbing can effectively remove acidic or basic compounds from exhaust gases prior to their emission.
5.　　Wet scrubbing can remove particulate matter and lower the acidity (or basicity) of exhaust gases.
6. a. Electrostatic precipitators yield a solid waste product, whereas wet scrubbers result in a water solution that needs to be concentrated.
   b. Precipitators require expensive electricity to operate and do not remove gases such as sulfur dioxide and nitrogen dioxide that can be removed by wet scrubbers.

CHEMQUANDARY

Steps Toward Clean Air (page 392)

(1) Visible particulate matter in the air is both an eyesore and a potential health problem. Therefore its removal is desirable.
(2) Just because a power plant no longer has a visible, dirty plume does not necessarily mean that smaller, more harmful (since the body cannot rid itself of them as well) particles are also being removed. Thus, the general public could gain a false sense of security.

Previous *ChemCom* units have discussed chemical contamination of water and food. In what respects is the problem of air pollution different from these other types of pollution?

Answer: (1) At the local level individuals typically have more choices about whether or not to drink water or eat food from a specific source than they do with respect to breathing the air. (2) A much larger quantity of air (approximately 20,000 L/day or 14 kg/day) passes through the body on a daily basis than water or food. Thus, even modest contamination of the air can result in appreciable doses of a given pollutant.

D.8 PHOTOCHEMICAL SMOG (pages 392-393)

The discovery of the causes of smog is outlined, along with the chemical reactions which amplify the problem. Students are often confused by the information concerning ozone, whose disappearance was a source of concern in Part C.

Remind them that the actual concentration in the troposphere is very small, and far removed from human–generated materials.

D.9 YOU DECIDE: AUTOS AND SMOG (pages 393-394)

This in-class activity involves students in graphical analysis to uncover some of the complex chemistry involved in photochemical smog. Students can work either as individual detectives or in teams.

Answers to Questions:

1. a. Nitrogen oxides and hydrocarbons peak between about 6 A.M. and 10 A.M.
   b. These times correspond to the morning rush hours when traffic is at its peak.
2. A given pollutant may decrease because it is being physically dispersed, adsorbed onto surfaces, or chemically converted to other chemicals.
3. Nitrogen monoxide is the source of nitrogen dioxide. Most of the NO_x emissions from automobiles is in the form of nitrogen monoxide. Once released into the atmosphere, nitrogen monoxide is further oxidized to form nitrogen dioxide.
4. a. Ozone is maximized at the same time NO_x is minimized.
   b. This might suggest that NO_x is involved in ozone formation.

D.10 CONTROLLING AUTOMOBILE EMISSIONS (pages 394-396)

The Environmental Protection Agency has set emissions standards for automobiles, and the automobile industry has developed several techniques to reduce emissions. The main technique, catalytic conversion, is familiar to most students in name only. The chemistry behind the converter leads to an introduction of catalysts and their role in reducing activation energies in chemical reactions.

Point out that such control technologies are not a panacea but rather pose a number of difficult issues: (1) increased reliance on foreign countries for import of precious metals; (2) the need to recycle such metals; (3) the need for government inspection and the associated costs; and (4) the fact that air pollution levels can still increase if consumers buy more cars and/or travel more miles.

Controlling Air Pollution (page 396)

It is easier and cheaper to install, maintain, and update control technologies for stationary sites, which are relatively fewer in number than the highly mobile and numerous sites represented by the 140 million automobiles on United States roads. Additionally, governmental inspection for adherence to standards is more easily accomplished at stationary sites.

D.11 ACID RAIN (pages 398-399)

The history and causes of acid rain expand on previous material concerning acids. The oxides of sulfur and nitrogen, by-products of our demand for energy, are the main culprits, and the corrosive reaction with carbonate compounds is presented.

Upon completion of the following laboratory activity, students will have seen that both carbon dioxide and sulfur dioxide form acidic solutions when dissolved in water. You may wish to demonstrate that a third nonmetallic oxide, nitrogen dioxide, also forms acidic solutions. Given that the orange-brown gas is denser than air, a small quantity can be poured into an Erlenmeyer flask containing slightly basic water with phenolphthalein indicator. The acid formed will cause the pink color to disappear. Nitrogen dioxide can be generated by the action of nitric acid on a small quantity of copper. Generating the nitrogen dioxide in a 2-L clear plastic soft drink container, and leaving the gas inside, causes profound shrinkage and fogging of the plastic, a graphic demonstration of the destructive nature of air pollution.

OPTIONAL CHEMQUANDARIES

1. Electric utilities are responsible for about 68% of the sulfur dioxide emissions (which total about 21.4 million metric tons/yr) in the United States. (See "Acid Rain," an EPA Journal special supplement, Sept. 1986, OPA-86-009.) Would use of low sulfur coal from the western states be a viable solution? Why or why not?

   Answer: Probably not, since: (a) Such coal typically has a lower heat content than coal from the East and more would have to be burned to obtain the same quantity of energy. (b) The increased transportation (and the pollution generated in transit) costs would probably be prohibitive. (c) Mining requires water, and increasing mining in western states could exacerbate the water shortage problem there, etc.

2. Transportation (especially the private automobile) accounts for about 44% of the nitrogen oxides emissions (which total about 19.7 million metric tons/yr). Would electric cars be a practical solution to nitrogen oxide emissions? Why or why not?

   Answer: Probably not, since: (a) To date, the prototypes have been extremely limited in their size, driving range, and the working lifetime of their batteries. (b) Even if working vehicles that the public would buy were available, the electric batteries would need to be recharged by drawing on power generated by preexisting electric power plants, which generate nitrogen oxides and may generate sulfur dioxide (if coal burning) as well.

D.12 LABORATORY ACTIVITY: ACID RAIN (pages 399-400)

Students burn sulfur to produce sulfur dioxide and make a solution of SO_2 in water, testing its effects on single-celled organisms, fruit, magnesium and calcium carbonate.

Time

One class period

Materials (for a class of 24 working in pairs)

24 g sulfur powder (2 g/team)
1-2 red apples
12 cm magnesium ribbon (1 cm/team)
Paramecium (or mixed) culture
Package Hydrion (pH) paper
24 marble chips, calcium carbonate ($CaCO_3$)(2 chips/team)
Package red litmus paper
Package blue litmus paper
Paper towels
Distilled water
12 500-mL gas jars or glass bottles
12 combustion (deflagrating) spoons
12 glass plates (to fit over top of jar or bottle)
12 Bunsen burners
3-5 microscopes
12 microscope slides
12 stirring rods
12 test tubes (standard 18 × 150 mm)
12 paring knives
12 tongs
24 droppers (dropping pipets)

Advance Preparation

The red apple skins should be wiped with acetone to dissolve any protective wax, then rinsed and dried with water. *Caution: Acetone is highly flammable.*

Lab Tips

Be sure to have good room ventilation during this activity. A fume hood is a good idea.

Caution students about possible splattering when sulfur is placed under the dripping water faucet inside the fume hood.

Orange peel may be substituted for apple peel.

Marble chips sometimes do not give good results: substitute small pieces of chalk instead.

Post-Lab Discussion

Focus on the experimental results and the questions.

Answers to Questions: (page 400)

1. $S(s) + O_2(g) \rightarrow SO_2(g)$
2.   The dissolved gas increased the acidity of the water. The reaction of sulfur dioxide with the water forms sulfurous acid:

$$SO_2(g) + H_2O(l) \rightarrow H_2SO_3(aq)$$

3. a. The *Paramecium* culture dies due to a change in pH.
   b. A steady drop in the pH of lake would eventually kill the aquatic community.
4.   Since the marble in statues reacts with acid, and since the iron in the steel girders reacts with acid, the statue and bridge would deteriorate.

D.13 pH (pages 400-401)

The pH concept, originally part of the Riverwood fish kill information, is explored more thoroughly in this section. The following articles from the *Journal of Chemical Education* feature entitled "Chemical Principles Revisited" serve as excellent review for the teacher who wishes to extend the discussion presented in the text: "Acids and Bases." *J. Chem. Ed.* July 1978, 459-64. "The pH Concept" *J. Chem. Ed.* Jan. 1979, 49-53. See also the abstracts of other relevant articles found in the *J. Chem. Ed.* "Something New from the Past" feature, April 1982, 305.

YOUR TURN

pH (page 401)

1. d. A solution of household lye, 13 (most basic)
   g. Household ammonia, 11 (basic)
   b. A solution of baking soda, 9 (basic)
   f. Sugar dissolved in pure water, 7   (neutral)
   e. Milk, drinking water, 6 (acidic)
   c. A cola drink, 3 (acidic)
   a Stomach fluid, 1 (most acidic)
2.   According to the data presented in Question 1, a cola drink is 1000 (10^3) times more acidic than milk.

PART D: SUMMARY QUESTIONS (pages 401-402)

1.   Oxides of carbon (CO and CO_2), nitrogen (NO and NO_2) and sulfur (SO_2 and SO_3) particulates, hydrocarbon derivatives, and ozone.
2.   The combustion reaction for the burning of a hydrocarbon fuel even under the best circumstances, produces carbon dioxide and water. Also, in reality, high temperature combustion will always result in the production of nitrogen oxides ($N_2 + O_2 \rightarrow NO_x$), and catalytic conversions are not 100% efficient. Additionally, from an engineering perspective, combustion is rarely complete.
3.   Photochemical smog is produced by the interaction between sunlight, auto exhaust components (HC, CO, and NO_x), and naturally occurring oxidizing agents. This interaction produces a mixture of ozone, nitrogen oxides, complex organics, carbon dioxide, and water. Smog originating with other combustion processes (such as from industries, electric utilities, smelting operations) usually contains particulates and sulfur dioxide.

4. Synergism refers to the phenomenon whereby the environmental and health effects of two or more pollutants or drugs are greater than their simple additive effect. An example related to air pollution is the health impact of sulfur oxides and particulate matter. This phenomenon works against simple pollution control strategies that aim at reducing only one pollutant.

5. Because of wind patterns, acid rain is the most prevalent in the northeastern United States. The burning of high-sulfur coal in states such as Ohio is believed to be the principal source.

6. a. Nitrogen oxides and sulfur oxides are the main contributors to acid rain.
   b. Human-generated SO_x is a result of the burning of sulfur containing fuels and smelting processes, whereas NO_x is a result of high-temperature combustion processes.

7. Given the law of conservation of matter, technologies which capture air pollutants before emission into the atmosphere generate solid matter or aqueous solutions which must be recycled or "thrown away." If the material is discarded improperly, it can become a source of solid waste or water pollution.

8. Ozone is a resource in the stratosphere and a serious pollutant in the troposphere. Also, sulfur is a resource when extracted from the ground in pure form and a pollutant in fuels such as coal and oil.

9. $2 NO_2 + H_2O \rightarrow 2 H^+(aq) + NO_3^-(aq) + NO_2^-(aq)$

10. Acids: HNO_3, H_2SO_3
    Bases: $NaOH$

11. Lowest pH = (b) stomach fluid; highest pH = (c) drain cleaner.

12. The control costs of pretreating fuel to remove sulfur or of treating the exhaust gases are expensive; therefore there is resistance to changing present practices. Also, the real damage costs of acid rain are often far removed from the source, and so the benefits of control efforts may not seem worth the costs locally.

EXTENDING YOUR KNOWLEDGE (page 402)

- There is a strong positive correlation between carbon monoxide levels and traffic accidents. In heavy traffic, sustained levels of 100 or more ppm are common. A few hours of breathing such carbon monoxide levels will result in 17% or more of the body's hemoglobin molecule's being tied up with carbon monoxide. The subsequent cellular oxygen deprivation results in a lowering of body efficiency and probably contributes to some loss.

- The summary below is organized according to three headings: Engine—Advantages—Disadvantages given the present technology:
  Electric Engine—clean (at use site), quiet—limited range between charging, limited speed-weight capacity, requires expansion of electric power generation and associated problems, would involve displacing many workers
  Gas Turbine—lightweight, compact volume, long life—poor efficiency at part load and idle conditions, poor acceleration from idle, high production costs (uses cobalt and nickel) (ceramics may solve the efficiency and production problems)
  Stratified Charge—cleaner, more energy efficient, already in use
  Wankel—high power/weight ratio, lower vibration-noise, uses variety of fuels, less parts—higher pollution, lower mileage, maintenance problems
  Stirling—an external combustion engine, cleaner, less vibration and noise, little if any oil consumption, can burn variety of liquid and gaseous fuels without modification—needs a larger radiator, more appropriate to larger vehicles
  Diesel—operates at a higher temperature and is therefore more energy efficient and produces less unburned hydrocarbons, does not need spark plugs, can burn a wider variety of petrofuels—produces more particulates and NO_x, noisier, heavier, more costly, require a larger battery, more expensive to build
  Expanded Mass Transportation—cleaner, more space and energy efficient—may be less convenient for some people, initial development cost/time high

- The April 1983 issue of *CHEM MATTERS* focused on acids and bases in everyday living and included (on page 7, "Experimenter's Corner") a procedure for making a red-cabbage juice indicator. Simply, one needs to extract the pigment by boiling the cabbage in water.

E: PUTTING IT ALL TOGETHER: IS AIR A FREE RESOURCE?

Does air pollution control work? Is it worth the cost? In this final activity students complete a table of emission patterns for various pollutants from 1975-1987, in order to discover for themselves how successful efforts to control air pollution have been. The data indicate that there is progress in society's battle against air pollution. Students examine the cost and benefits of further control, after considering two views on the rights and responsibilities of citizens concerning the atmosphere.

E.1 AIR POLLUTION CONTROL: A SUCCESS? (pages 403-404)

This activity involves the students in analyzing graphical data concerning the effectiveness of pollution control measures to date.

Answers to Questions:

(Note: The following answers are based on the source data from which the graphs in Figure18 on page 404 were drawn. Do not expect such precise estimates of change in student answers, since the positions of the graph lines must be estimated. The direction of change and general magnitudes of the changes can be found on the graphs, which is the major objective of the activity.)

1. a. Particulates: $(7.0 - 10.4)/10.4 \times 100\% = -33\%$
   Sulfur oxides: $(20.4 - 25.6)/25.6 \times 100\% = -20.3\%$
   Nitrogen oxides: $(19.5 - 19.2)/19.2 \times 100\% = +1.56\%$
   Volatile organics: $(19.6 - 22.8)/22.8 \times 100\% = -14.0\%$

   Note: Additional data not found in student text are shown below.
   Carbon monoxide: $(67.6 - 80.5)/80.5 \times 100\% = -16.0\%$ (in 10^6 metric tons/yr)
   Lead: $(40.7 - 147.0)/147.0 \times 100\% = -72.3\%$ (in 10^3 metric tons/yr)

   You may wish to point out, or ask students to deduce, that the dramatic reduction in lead emissions are due to the ban on the use of leaded gasoline that was necessitated by the development of catalytic converters, as well as concerns over the health effects of lead.

   b. All pollutants decreased by at least 10% with the exception of nitrogen oxides which students may estimate as essentially remaining unchanged, based on the graphed data. Actually, it increased by about 1.6%. Thus, unless one assumes that nitrogen oxides account for a disproportionate amount of the overall air pollution problem, air quality has clearly improved.

2. If energy use stayed constant, improved pollution control technology would immediately translate to improved air quality. However, given the law of conservation of matter and the nature of the combustion process, increased energy consumption means there will be an increase in the amount of air pollutants released. Thus, air quality may either improve, decrease, or stay the same, depending on the level of increase in energy consumption and the efficiency of the controls.

3. SO_x: fuel combustion
   Particulates: industrial processes
   NO_x: fuel combustion
   VOC: transportation

4. a. Evidence that would be helpful in assessing the effects of pollution control technologies would include: frequency data on pollution alert days; comparative photographs; hospital records on air-pollution related emergencies; data on the viability of certain pollution sensitive plant species across time; etc.

   b. Both the EPA and environmental organizations agree that significant progress has been made in control technologies. Some argue, however, that we must halt the increase and must eventually decrease our dependence on fossil fuels for both ecological (acid rain and greenhouse effect) and political reasons. They argue that solar-based technologies offer a viable, safer, practical alternative.

5. Information is needed on the extent to which the best available technologies are presently being employed and present laws are being enforced, medical studies indicating that present air pollution levels are significantly and negatively impacting human health, cost-benefit data, etc.

E.2 YOU DECIDE: JUST ANOTHER RESOURCE? (pages 404-405)

Students develop questions in support of the two viewpoints, which differ in their approach to air as a resource.

E.3 PAYING THE PRICE (pages 405-407)

This section involves students in critically examining cost/benefit analyses as a means of resolving pollution-control debates. The questions posed range from factual (the first five) to open-ended opinion items (the second five). Spend most of the discussion time on the latter.

Answers to Questions:

1. a. At low pollution levels (left side of graph), control costs are highest.
   b. The control costs curve is exponential. As one approaches the theoretical goal of zero pollution, the control costs rise at an ever increasing rate. For example, let's suppose it costs an electric utility $1 million to decrease the SO_x emission by 60%—to gain the next 10% (still based on initial emission) may cost $20 million, and the next 10% may cost $40 million, and so on.

2. a. At high-pollution levels (right side of graph), damage costs are highest.
   b. Little effort is being made in this case to control emissions, therefore pollution damage is maximized.

3. a. No.
   b. No.
   c. According to the graph, zero pollution would imply infinite control costs. Also, technology cannot bypass the law of conservation of matter—if fossil fuels are burned, the absolute minimum level (assuming complete removal of sulfur, complete catalytic conversion of carbon monoxide and nitrogen oxides to carbon dioxide, nitrogen, and oxygen) of pollution would include the emission of carbon dioxide, which contributes to the greenhouse effect.

4. Improved control technologies can shift the entire control costs curve downward. They do not, however, change its basic shape or its inverse relationship to the damage costs curve.

5. a. Control costs.
   b. Control costs can be determined very precisely, since specific industries know how much it will cost to buy, install, and maintain a given technology over its useful lifetime. These costs are typically quite direct; they are actually paid by industries and ultimately by consumers. (Exceptions are possible; there might be a rise in unemployment if a plant needs to shut down, for example.)
   c. Damage costs are more difficult to assess because many costs (including both human health and environmental quality) are delayed over time (the full impact of acid rain may not be known for years) and not all individuals place the same dollar value on clean, healthy air. These costs are rarely experienced directly. Generally, epidemiological evidence can suggest the human health costs of high pollution levels, but it is more difficult to assess improvements in human health if the air quality is reasonable to start with.

6. a. Assumptions include that a dollar value can be placed on all the damage costs (what is the aesthetic value of a blue sky?); that reliable and valid measures exist for counting and comparing costs; that politically powerful groups are not operating to subvert the analysis or its associated recommendations; that the government regulation power is able to enforce recommendations, etc.
   b. None of these assumptions are valid if taken to the extreme case. This does not mean that such efforts should not be attempted, but rather that we must recognize that this will always be an imprecise and partially subjective science.

7. Control costs are direct and immediate. Damage costs are indirect and longer term.

8. a. For most local air pollution, the benefits of cleaner air are equally shared by all. A noticeable exception is acid rain where locally the costs in, say, the Ohio River Valley may be high, but the benefits are mainly experienced by locations in a northeasterly direction (due to prevailing wind patterns).
   b. Pollution control may make least "cents" to individuals or industries bearing the direct costs of pollution control technology. This is especially true for stockholders who may live far from the plant location and thus do not directly experience the adverse effects of the initial pollution, but do experience the costs of control.

9. The questions are very important—they will influence how one votes for candidates and on certain specific environmental issues. However, this does not imply that one must have actual dollar figures in mind. In fact, many environmentalists would agree with the old adage: "the most important things in life/nature are priceless" and that attempts to put a dollar figure on everything only reflects our inability to understand the real value of life/nature.

10. a. Probably not.
    b. The calculation of damage costs is limited by our lack of knowledge of long-term chronic environmental effects. Even more acute effects are often not counted in our present economic system that treats resources such as the atmosphere as "commons" and does not place a specific value on long-term environmental integrity and sustainability.

E.4 LOOKING BACK (page 407)

The final section is a synopsis of the topics covered, leaving students with a sense of the importance of their future decisions to the quality of air we all must have.

SUPPLY LIST
Expendable Items

Alka Seltzer	B.1	40 tablets
Ammonium hydroxide, conc. (demo)	D.7	15 mL
Apples	D.12	2
Balloons, large (demo)	A.2	10
Bottle, soft drink (demo)	A.2	1
Bottle, plastic, with lid (demo)	A.2	1
Bromthymol blue indicator (demo)	A.3, C.6	100 mL
Can, aluminum, soft drink (demo)	A.2	1
Candles	C.6	12
Candles, birthday (demo)	A.2	24
Capillary melting point tubes	B.6	24
Carbon dioxide gas, cylinder (opt)	A.2	1
Cardboard or waxpaper (demo)	A.2	
Cotton balls (demo)	A.2	50
Copper wire (demo)	D.7	50 cm
Dry Ice (demos)	A.2, B.2	0.5 lb.
Elodea plant (demo)	A.3	4
Helium gas, cylinder (opt; demo)	A.2	1
Hydrion paper (demos)	D.7, D.12	2 packages
Hydrochloric acid, conc. (demo)	A.2, B.1	65 mL
Hydrogen gas, cylinder (opt; demo)	A.2	1
Hydrogen peroxide, 3% solution	B.1	1.2 L
Iodine crystals (demo)	A.2	0.5 g
Limewater	B.1	60 mL
Litmus paper, red and blue	D.12	1 package
Magnesium ribbon	B.1, D.12	100 cm
Manganese dioxide	B.1	10 g
Marble chips, calcium carbonate	D.12	100 g
Methane gas	A.2	
Nitrogen (optional)	B.1	small cylinder
Nitric acid, conc. (demo)	A.2	10 mL
Oil, vegetable	B.6	300 mL
Paramecium (or mixed) culture	D.12	1
Perfume (demo)	A.2	1 bottle
Phenolphthalein (demo)	A.2	5 mL
Rubber bands, small	B.6	24
Soap (demo)	A.2	1 oz
Splints, wooden	B.1	36
Steel wool	B.1	20 g
Sulfur powder	D.12	40 g
Sulfur hexafluoride gas, cylinder (demo, opt)	A.2	1
Universal indicator	B.1	45 mL

SUPPLEMENTAL READINGS

"Acid Rain": An *EPA Journal* Special Supplement, OPA-86-009, September 1986.

"Why Isn't My Rain as Acidic as Yours?" *Journal of Chemical Education*, **62**, 158-59, 1985. Discusses how chemistry is often improperly applied to the acid rain issue.

"Balloon", Calculating Chemistry—Gases", Gas Laws and Scuba Diving", and "Getting a Lift", *CHEM MATTERS*, February, 1983.

"European Concern about Acid Rain Is Growing." *Chemical and Engineering News*, January 28, 1985, 12-18.

"Acid Rain." *CHEM MATTERS*, April 1983, 10-12. The issue focuses on acids and bases in everyday life.

Acid Rain. October 1982. Free 14 p. pamphlet of the American Chemical Society, GRASP Office, 1155 16th St., NW, Washington, DC 20036.

"Acid Rain—How Great a Menace? " *National Geographic*, November 1981, 652-81.

"Acid Rain." *Scientific American*, October 1979, 43-51.

"Chlorofluorocarbons and Stratospheric Ozone." *Journal of Chemical Education*, **64**, 387-91, 1987. Fairly technical article for teachers.

"Build a Hot Air Balloon", "Hot Air Adventure", and How the Right Professor Charles Went Up", *CHEM MATTERS*, December, 1983.

"Our Fragile Atmosphere: The Greenhouse Effect and Ozone Depletion." *EPA Journal*, December 1986, **12**.

"The Changing Atmosphere: Tending the Global Commons." *Chemical and Engineering News*, November 24, 1986, 14-64. Special report (reprint available from ACS) focusing on the struggle to understand and check the global warming and depletion of stratospheric ozone trends.

"Bubble Control", The Pumphouse Incident", and "What's That Fizz?", *CHEM MATTERS*, February, 1984.

"Are We Close to the Road's End?" A special report on carbon dioxide and other gases affecting our climate. *Discover*, January 1986, 28-50.

"Carbon Dioxide and World Climate." *Scientific American*, August 1982, 35-43.

"Special Report on the Greenhouse Effect", *Newsweek*, July 11, 1988, p. 16.

"Are We Poisoning Our Air?" *National Geographic*, **170**, 502, April, 1987.

"The Silent Summer: Ozone Loss and Global Warming: A Looming Crisis." *Newsweek*, June 23, 1986, **107**, 64.

"Tropospheric Chemistry." *Chemical & Engineering News*, October 4, 1982, 39-52. Special report (reprint available from ACS).

"Global Climatic Change." *Scientific American*, **260**, 36, 1989.

"The Planets: Between Fire and Ice." *National Geographic*, January 1985, 5-51. This is a profile of the planets emphasizing the uniqueness of Earth for supporting life.

"Your Personal Greenhouse." *CHEM MATTERS*, December, 1990.

"The Great Climate Debate." *Scientific American*, **263**, 36, 1990.

"Balloon Pressure." "Blimp Flight", "Bringing Helium Down to Earth", and "Hydrogen and Helium", *CHEM MATTERS*, October, 1985.

"Alternative Automobile Engines." *Scientific American*, July 1978, 39-49.

"Air Quality in the Home." *EPRI Journal*, March 1982, 7-14.

"Indoor Air Pollution." *Science 80*, March-April, 30-33.

American Chemical Society. *Cleaning Our Environment: A Chemical Perspective.* Washington, DC: American Chemical Society, 1978. Includes chapters on chemical analysis and monitoring, toxicology, air, water, solid waste, pesticides, and radiation.

HEALTH: YOUR RISKS AND CHOICES

PLANNING GUIDE

Section	Laboratory Activity	You Decide
A. Risk and Personal Decision–making		
A.1 Making Judgments about Risk A.2 Studying Human Disease		
B. Your Body's Internal Chemistry		
B.1 Balance and Order: Keys to Life B.2 Elements in the Human Body B.3 Cellular Chemistry B.5 How Enzymes Work B.6 How Energy is Released and Stored	B.4 Lab Demonstration: Enzymes B.7 Enzymes	
C. Acids, Bases, and Body Chemistry		
C.1 Structure Determines Function C.2 Strengths of Acids and Bases C.3 Acids, Bases, and Buffers in the Body C.5 pH in Balance	C.4 Buffers	
D. Chemistry at the Body's Surface		
D.1 Keeping Clean with Chemistry D.2 Skin: It's Got You Covered D.3 Protecting Your Skin from the Sun D.5 Getting a "D" in Photochemistry D.6 Our Crowning Glory D.8 Hair Styling and Chemical Bonding	D.4 Sunscreens D.7 Chemistry of Hair	
E. Chemical Control: Drugs and Toxins in the Human Body		
E.1 A Glimpse of Drug Function E.4 Notes on Some Other Drugs E.5 Foreign Substances in the Body E.6 Drugs in Combination E.7 Cigarette Use	E.8 Cigarette Use	E.2 Pros and Cons of Aspirin E.3 Effects of Alcohol E.9 Smoking?
F. Putting It All Together: Assessing Risk		
F.1 Risks from Alcohol and Other Drugs F.2 Personal Control of Risk F.3 Looking Back		

TEACHING SCHEDULE

	DAY 1	DAY 2	DAY 3	DAY 4	DAY 5
Class Work	Discuss pp. 410-414 YT pp. 411-412 CQ p. 412	Discuss pp. 415-417 YT pp. 417-418	YT p. 422 CQ p. 423		SQ p. 425
Laboratory		LD pp. 418-419		LA pp. 423-425	
Homework	Read pp. 410-419 SQ p.414	Read pp. 419-423	Read pp.423-425		Read pp. 426-429

	DAY 6	DAY 7	DAY 8	DAY 9	DAY 10
Class Work	Review Parts A & B Quiz Parts A & B Discuss pp. 426-429 YT p. 428		Discuss p. 432 YT pp. 432-433	Review Part C Quiz Part C Discuss pp. 434-438 CQ p.438	Discuss pp. 438-443 CQs p.443
Laboratory		LA pp. 430-431			LA pp. 441-442
Homework	Read pp. 430-431	Read pp. 432-433	SQ p. 433 Read pp. 434-438	YT pp. 435-436 Read pp. 438-442	Read pp. 443-446

	DAY 11	DAY 12	DAY 13	DAY 14	DAY 15
Class Work		Discuss pp. 447-449 CQ p. 449	Review Part D Quiz Part D Discuss pp. 450-452	YD pp. 453-454 Discuss pp. 454-459	
Laboratory	LA pp. 444-446				LA pp. 460-461
Homework	Read pp. 447-449	SQ p.449 Read pp. 450-453	YD pp. 452-453 Read pp. 453-459	Read pp. 459-462	YD pp. 461-462

	DAY 16	DAY 17	DAY 18	DAY 19	
Class Work	SQ p.462	Discuss pp. 463-464 PIAT pp. 464-465	Review unit	Exam	
Laboratory					
Homework	Read pp. 463-465				

LD= Lab Demonstration; **LA** = Laboratory Activity; **CQ** = ChemQuandary; **YT** = Your Turn; **YD** = You Decide; **PIAT** = Putting It All Together.

Our society has become extremely health conscious. Stores are full of exercise equipment and videos on working out, and health spas and exercise clubs are commonplace across the country. Increased technology in medicine has wiped out some diseases, and tremendously improved the cure rate in others.

This *ChemCom* unit presents some chemical reasons for good health and tells what individuals can do to improve their personal health. The main message of the Risk unit is that maintaining good health is largely a matter of making informed personal choices. A knowledge of body chemistry will help students understand why certain choices promote good health, while others do not. A related topic, the chemistry of personal grooming, is also explored in this unit.

OBJECTIVES

Upon completion of this unit the student will be able to:

1. Provide examples of correlation, and determine the casual relationship between the members of a given pair of events. [A.1]
2. Define epidemiology, and describe some benefits and limitations of epidemiological studies. [A.2]
3. Define homeostasis and give examples of how it is related to maintaining good health. [B.1]
4. Describe the major elements of the human body and their function in maintaining health. [B.2]
5. Explain how enzymes work and list several factors that may alter their effectiveness. [B.3-B.7]
6. Describe cellular energy production and storage, including the role of ATP. [B.6]
7. Define and give examples of acids and bases, and use net ionic equations to describe the neutralization reaction. [C.1-C.5]
8. Describe the components of a buffer and explain how it prevents acidosis and alkalosis. [C.3-C.5]
9. Apply the concept of "like dissolves like" to skin cleansing and the function of soap. [D.1, D.2]
10. Sketch the parts of human skin and describe their functions. [D.2]
11. Describe the effect of sunlight on skin and the effectiveness of PABA in sunscreens. [D.3, D.4]
12. Describe hair structure, the types of bonding in hair protein, and the effects of various hair treatment chemicals on hair. [D.6-D.8]
13. Distinguish between drugs and toxins, and describe circumstances where a substance's usual effect on homeostasis may be reversed. [E.1, E.6]
14. Use the concept of receptors to account for drug specificity and for the action of narcotic analgesics. [E.1]
15. Contrast the benefits and burdens associated with aspirin use. [E.2]
16. Outline the effect of common drugs on the human body, and the body's chemical defenses against these drugs. [E.3-E.5]
17. Discuss the role of antigen-antibody complexes in protecting the body against infectious organisms, and contrast the AIDS virus to other viruses. [E.5]
18. Use the concept of synergism to explain the hazards of combining drugs and medicines. [E.6]
19. Evaluate the products of cigarette smoking. [E.7-E.9]
20. Assess personal control of risks in terms of the maintenance of good health and well-being. [F.1, F.2]

SOURCES ON HEALTH ISSUES

1. Consumer Information Center-B, c/o R. Woods, P.O. Box 100, Pueblo, CO 81002. A distribution center for a variety of free or low cost pamphlets/booklets produced by various agencies of the federal government. Free titles related to this unit include: *Nutrition and Your Health: Dietary Guidelines* (23 p/1985/USDA/#519R); *Quackery-The Billion Dollar "Miracle" Business* (4 p/1985/FDA/#529R); *Do-It-Yourself Medical Testing* (7 p/1986/FDA/#543R); *Food and Drug Interactions* (4 p/1984/FDA/#535R); *Herbs—Magic or Toxic?* (7 p/1984/FDA/#538R); *Some Things You Should Know about Prescription Drugs* (4 p/1982/FDA/#540R); *Cancer Prevention: Good News, Better News, Best News* (20 p/1984/NIH/#547R); *Clearing the Air: A Guide to Quitting Smoking* (32 p/1984/NIH/548R); *Diet, Exercise, and Other Keys to a Healthy Heart* (8 p/1986/FDA/#552R); etc. Write for a free catalog (or pick up one in your local library).

2. Food and Drug Administration, 5600 Fisher Lane, Rockville, MD 20857. 301-443-1544. A branch of the U.S. Department of Health and Human Services that publishes a wide variety of free pamphlets and the monthly *FDA Consumer* magazine.

3. Government Printing Office, Retail Sales Outlet, 8660 Cherry Lane, Laurel, MD 20707-4980. Main GPO warehouse. Request *U.S. Government Books Catalog* and *Subject Bibliography Index.*

4. National Cancer Institute, Office of Cancer Communication, Bldg. 31, Room 10A18, Bethesda, MD 20205. 301-496-6792. Posters, pamphlets, slides, tapes, etc.

5. National Health Information Clearinghouse, P.O. Box 1133, Washington, DC 20013-1133. 800-336-4797. NHIC is a information and referral service of U.S. Public Health Service.

6. National Institute of General Medical Services, Bldg. 31, Room 4A-52, 9000 Rockville Pike, Bethesda, MD 20205. 301-496-7301. One of 12 NIH branches, NIGMS provides information and referrals to health related questions. Single free copies of: *Medicines and You* (62 p), *Inside the Cell* (96 p), *New Human Genetics* (48 p), and *Then and Now: Biomedical Science in 1887 and Today.*

7. Action on Smoking and Health, 2013 H Street, NW, Washington, DC 20006. 202-659-4310. Nonprofit, antismoking lobby and education group. Sells a variety of materials.

8. National Academy Press, 2101 Constitution Avenue, NW, Washington, DC 20418. 202-334-3313. Request free catalog of books and publications. Publications of the National Academy of Sciences range in price.

9. World Health Organization, Publications Center, 49 Sheraton Avenue, Albany, NY 12000. 518-436-9686. Request free publications directory.

10. Chemical Manufacturers Association, Publications Service, 2501 M Street, NW, Washington, DC 20037. 202-887-1223. Write for catalog.

11. Dow Chemical USA, Health and Environmental Sciences, Midland, MI 48640. Offer free booklet (21 p with bibliography), *Life in the Balance—Weighing the Questions of Risk and Benefit in Today's World* (Form No. 233-10-83).

12. E. I. du Pont de Nemours & Co., Public Relations Dept., 1007 Market St., Wilmington, DE 19898. Request publications catalog.

13. Eli Lilly & Co., Public Relations Dept., 307 East McCarty St., Indianapolis, IN 46285. Request educational programs catalog.

14. Monsanto Company, Dept A3NA-N1, 800 N. Lindbergh Blvd., St. Louis, MO 63166. Request free booklet, *The Chemical Facts of Life* (13 p)—A discussion of chemical substances, their history, use and misuse, testing procedures, benefits and risks to humans. Especially useful background for Part E.

15. Pharmaceutical Manufacturers Association, 1100 15th St., NW, Washington, DC 20005. 202-835-3400. Request publications catalog.

16. Proctor & Gamble Co., Public Relations Dept., P.O. Box 599, Cincinnati, OH 45201. Request educational publications catalog.

17. Soap and Detergent Association, 475 Park Avenue, South, New York, NY 10016. 212-725-1262. Request publications catalog.

INTRODUCTION

Students are introduced to some of the topics in this unit with information concerning health and medicine in this country. The contrast between internal and external chemistry is made, and the major point about decisions influencing our well-being sets the stage for the rest of the unit.

A: RISK AND PERSONAL DECISION-MAKING

Part A sets the stage for the rest of the unit by introducing students to questions of correlation and cause-effect.

A. 1 MAKING JUDGMENTS ABOUT RISK (page 411)

Magazines and newspapers are fond of reporting events which have improbable connections. This section begins to clarify correlated and noncorrelated events, and tests students' understanding in the *YOUR TURN* that follows.

YOUR TURN

Events: Related or Not? (pages 411-412)

Students are asked to check the correlation in several pairs of everyday events, to help clarify the difference between correlation and causation. The former is a necessary condition for the latter, but is not proof of a definite cause-effect relationship. Also, correlation is generally easier to establish than is causation.

It is helpful to check for understanding by working through a few pairs with the entire class. Excellent background reading material includes the materials mentioned at the beginning of the Teacher's Guide for this unit.

Answers to Questions:
1. Correlates with—directly causes
2. Correlates with
3. Is not related to
4. Correlates with
5. Correlates with—directly causes (sometimes)
6. Correlates with—directly causes
7. Correlates with—directly causes (with other factors)
8. Correlates with—directly causes
9. Correlates with
10. Correlates with—directly causes
11. Correlates with
12. Correlates with
13. Correlates with

CHEMQUANDARY

Cigarette Warnings (page 412)

1., 2. The change in wording reflects stronger, more compelling evidence for a cause-effect relationship between smoking and various health problems.

A.2 STUDYING HUMAN DISEASE (page 414)

Information concerning the process of medical research is given, which supports the exercise on correlation. The pros and cons of animal studies and controlled experimental studies on human subjects, that could more directly establish causal relationships, are given.

PART A: SUMMARY QUESTIONS (page 414)

1. a. Position of the moon—height of tides.
   b. Amount of exercise—body weight.

2. a.–d. Student answers may refer to decisions about smoking, exercise, dieting, etc.
3. a. Both
   b. Correlation
   c. Neither
   d. Neither
   e. Correlation

B: YOUR BODY'S INTERNAL CHEMISTRY

Part B analyzes the function of cells, the remarkable chemical factories in our bodies, starting with the elements contained in the body, and then analyzing the mechanism of enzyme action in the cells. Students learn the role of ATP in energy storage, and study the effect of temperature and pH on enzyme activity.

B.1 BALANCE AND ORDER: KEYS TO LIFE (pages 415-416)

The remarkable balance maintained by the human body is a guideline for all of life, a reminder of the need for a balanced diet given in the Food unit. Of particular interest in this section is the information about bacterial balance: some bacteria are essential for life, while others are life-threatening.

B.2 ELEMENTS IN THE HUMAN BODY (pages 416-418)

If the body is viewed as a chemical factory, a key question concerns the elements found in this factory, and the roles they play in promoting the body's proper functioning. This information also correlates with the Food unit, and student understanding is tested in the *YOUR TURN* activity.

YOUR TURN

Elements in the Body (pages 417-418)

This activity allows students to review the composition, location, and function of elements in the body. The concept of limiting reactant arises in dealing with the effects of overconsumption and underconsumption of trace minerals and macrominerals.

Answers to Questions:

1.

Element	Location in body	Health Role
Oxygen	In hemoglobin (as O_2), in molecules of carbohydrate, protein, fat, and water	Oxidizing material for fuel, component of body substance
Hydrogen	In carbohydrate, protein, fat, and water, in stomach acid and blood , buffer systems	In body substances, to aid digestion, transferred between ions and molecules in pH regulation
Carbon	In carbohydrate, protein, and fat molecules	For body structure and storage molecules
Nitrogen	In protein	For body structure and enzymes
Calcium	Bones, teeth, nerve and muscle cells	For bone, teeth structure, nerve transmission, muscle contraction
Sulfur	Some proteins	—
Phosphorus	Bones, genetic molecules, energy transfer molecules (ATP, ADP)	—
Sodium	Body fluids, cells	Needed for memory, appetite, fluid balance
Potassium	Body fluids, cells	Aids nerve function, regular heartbeat
Chlorine	Stomach acid (HCl), body fluids	—
Magnesium	Cells	Maintains a heartbeat
Iron	In hemoglobin and myoglobin molecules	Transports oxygen gas

2. No. A trace element lacking in the diet can act as a limiting reactant to stop important body reactions or functions despite the presence of adequate supplies of macrominerals. While the human body needs larger amounts of calcium and phosphorus than iron and iodine, this does not mean that iron and iodine are less important.

3. a. Eating a normal range of foods containing adequate total calories (food energy) usually ensures that sufficient amounts of these are also consumed, since these are found in a wide variety of plant and animal sources. By contrast, sources of trace minerals are limited, suggesting that a restricted diet will more likely lead to a deficiency of trace minerals.

   b. Any changes in the types and quantities of foods consumed could produce a large shift (large percent change) in the intake of certain trace minerals. Therefore, overdoses of trace minerals are more common than of macrominerals.

B. 3 CELLULAR CHEMISTRY (page 418)

The speed and precision of the body can be attributed largely to the functioning of enzymes. The examples given in this section about enzymes give some sense of the enormous extent of enzyme catalysis, which is partially apparent in the lab demonstration.

B. 4 LAB DEMONSTRATION: ENZYMES (pages 418-419)

This demonstration illustrates how enzymes function in the body to speed up certain body reactions selectively. Students find that the enzyme catalase speeds up the decomposition of hydrogen peroxide to water and oxygen gas. When the liver emulsion (the source of the catalase) is heated, students note that the enzyme (protein) is inactivated.

Question 4 on page 419 refers to a common misconception about hydrogen peroxide and cuts. H_2O_2 is such a good antiseptic because of the reactive nature of the oxygen in hydrogen peroxide, which spells death for anaerobic bacteria.

Time

One-half of a class period.

Materials

80 mL 3% hydrogen peroxide (H_2O_2) in brown bottle
2 large test tubes (25 × 200 mm)
2 pieces raw liver (size of quarter)
Ring stand and clamp for test tube
0.5 g manganese dioxide (MnO_2) (optional)
Wooden splints
Bunsen burner
Rubber or plastic glove

Advance Preparation

None

Pre-Demo Discussion

Begin by asking students why hydrogen peroxide solutions purchased in drug stores are contained in brown bottles. As the bottle labels indicate, hydrogen peroxide decomposes with age and exposure to light, heat, or dust. The "decomposition" reaction is summarized by this equation:

$$2\ H_2O_2(aq) \rightarrow 2\ H_2O(l) + O_2(g)$$

This reaction (in the absence of external factors) proceeds slowly at room temperature. A suitable catalyst (such as manganese dioxide or the enzyme catalase) causes this decomposition reaction to occur at a much faster rate. In this demonstration, formation of oxygen gas is used to estimate the extent (and rate) of hydrogen peroxide decomposition. Raw liver will be used as the source of catalase.

Procedure

1. Pour 40 mL 3% H_2O_2 into large (25 × 200 mm) test tube.
2. Pull a plastic or rubber glove onto your hand. Place a small piece of raw liver (size of a quarter) in the tube. Quickly place your gloved thumb over the tube mouth and invert the tube once. Then quickly attach the tube to a ring stand. (Students observe gas bubbles form in the tube.)

3. Test the emerging gas by thrusting a glowing wooden splint partway into the tube. (Students observe that the splint relights, consistent with the behavior of oxygen gas.)
4. Place another piece of raw liver into a second large tube. Add 40 mL of water. Boil the contents for two minutes, then drain off the water. Cool the "cooked" liver by rinsing it in a second room-temperature water sample and draining.
5. Add 40 mL of 3% H_2O_2 to the tube. Place a rubber or plastic glove on you hand. Place your gloved thumb over the tube mouth, and invert once to mix. Attach the tube to a ring stand. Perform the glowing splint test described in Step 3. (Students will observe no bubbles or evidence splint re-lighting.)

Demo Tips

Consider adding a **small** quantity (less than 0.5 g) of manganese dioxide to 20 mL of 3% H_2O_2 in a beaker to demonstrate the effect of this catalyst on the decomposition of the substrate. (This is related to the reason that dusts are listed as possible contaminants on the label of the hydrogen peroxide solution bottle—they may contain substances that can catalyze this decomposition reaction.)

Post-Demo Discussion

The molecular activity of catalase is about 5,600,000. This means that 5,600,000 molecules of hydrogen peroxide are decomposed by one enzyme molecule every minute, or about 90,000 molecules are decomposed per second.

Answers to Questions:

1. A chemical change had taken place—the decomposition of hydrogen peroxide.
2. The catalase enzyme is inactivated by heat. The thermal energy causes irreversible changes, called denaturation, in the structure of the enzyme.
3. Hydrogen peroxide solutions are unstable—they tend to react or decompose. Acetanilide or similar organic compounds are added to make the solution more stable.
4. The foaming is caused by the catalase, an enzyme found in blood.
5. a., b. Hydrogen peroxide acts as an antiseptic by killing bacteria, and as a cleansing agent due to the foaming action.

B.5 HOW ENZYMES WORK (pages 419-420)

The text presents information concerning the mechanism of enzyme action, which involves the interaction with the substrate at the "active site" of the enzyme. Often another, smaller molecule, called a coenzyme, is needed to assist the bond-breaking or bond-forming that occurs. Research has clearly demonstrated the perfect fit of substrates with enzymes, supporting the statement concerning the specificity of enzymes.

All enzyme reactions are reversible, which means that the product molecule in Figure 1 (page 420) can be broken apart into the substrate molecules.

B.6 HOW ENERGY IS RELEASED AND STORED (pages 420-422)

The role of ATP as an energy-storage molecule is explained. Many enzymes are needed for respiration, which decomposes glucose molecules according to the equation shown on page 421. There are actually 26 different steps to this equation, each step of which requires a different calalyst. Some energy bookkeeping is covered in the *YOUR TURN* exercise.

YOUR TURN

Enzymes and Energy in Action (page 422)

This activity provides practice in solving problems involving the use of body energy and its production.

Answers to Questions:

1. a. $100 \text{ mol ATP} \times \dfrac{1 \text{ mol glucose}}{38 \text{ mol ATP}} = 2.63 \text{ mol glucose}$

   b. $2.63 \text{ mol glucose} \times \dfrac{180 \text{ g glucose}}{1 \text{ mol glucose}} = 473.4 \text{ g glucose}$

2. $454 \text{ g muscle} \times \dfrac{0.001 \text{ mol ATP}}{1 \text{ g muscle}} \times \dfrac{1 \text{ mol glucose}}{38 \text{ mol ATP}} = 0.0119 \text{ mol glucose (2.1 g glucose)}$

CHEMQUANDARY

Chew on This Problem! (page 423)

1., 2. Students should be able to detect a sweeter taste after prolonged chewing. Asking for volunteers to do this is helpful.

3. Amylase, the enzyme in your saliva, has begun the process of starch digestion by converting it into glucose, which has a much sweeter taste than the starch.

B. 7 LABORATORY ACTIVITY: ENZYMES (pages 423-425)

Students observe the decomposition of protein by pepsin and starch by amylase, and note the effect of temperature and pH on enzyme action.

Time

Two class periods. Divide the class into groups of four as described in the text.

Materials (for a class of 24 working in groups of 4)

108 test tubes (16 × 125 mm)
108 stoppers for test tubes
18 10 mL graduated cylinders
6 Bunsen burners
6 250-mL beakers
6 boiling chips
6 ring stands
6 rings
6 wire gauze
6 test tube holders
240 mL pepsin solution (1.2 g of pepsin powder in 240 mL water)
Hard-boiled egg white (50 g total)
120 mL starch suspension (0.12 g starch in 120 mL water. Make fresh and keep refrigerated to retard spoilage.)
240 mL Benedict's solution
120 mL 0.5% amylase solution
1 refrigerator

Advance Preparation

Benedict's solution may be purchased from a chemical supplier or prepared by dissolving 200 g sodium carbonate monohydrate ($Na_2CO_3 \cdot H_2O$) and 173 g of sodium citrate in 800 mL distilled water. Warm the mixture slightly to speed dissolving. Dissolve 17.3 g copper(II) sulfate pentahydrate ($CuSO_4 \cdot 5H_2O$) in 100 mL distilled water, slowly pour into the first solution. Stir constantly, cool, and add enough distilled water to make one liter.

Buffer solutions are suggested for the different pH solutions. All five buffers can be made by combining different amounts of the following two solutions:

A: 12.37 g anhydrous boric acid (H_3BO_3) and 10.5 g citric acid hydrate ($C_6H_8O_7 \cdot H_2O$)/liter solution
B: 38.0 g trisodium phosphate dodecahydrate ($Na_3PO_4 \cdot 12 H_2O$)/liter solution

Make up one liter of each solution, and measure out proportional amounts of each solution. You will need 240 mL of pH 2, 4, 7, and 8 buffers and 120 mL pH 10 buffer for each class.

Buffer Solutions

pH	mL solution A	mL solution B
2	195	5
4	155	45
7	99	101
8	85	115
10	54	146

From *Journal of Chemical Education*, **38**, 559 (1961).

Commercial salivary amylase may be ordered from a chemical supplier. The original procedure for this activity asked students to contribute their own amylase and saliva, not a pleasant experience, but a possible substitute. Some people do not have amylase in their saliva.

Lab Tips

Pepsin is not tested at pH 10, because the enzyme is destroyed at that pH.

In the presence of simple sugars, Benedict's solution will exhibit a yellow, orange, or reddish precipitate of copper(I) oxide when the reagent is heated. Point out that the intensity of the color is related to the amount of glucose in the solution.

Try the activity in advance to be familiar with the expected results.

Expected Results

			pH		
	2	4	7	8	10
pepsin, warm	turbid	slightly turbid	clear	clear	—
pepsin, cold	(same results as warm, but not as pronounced)				
amylase, warm	blue	blue	orange	orange	blue
amylase, cold	(same results as warm, but fainter colors)				

Post-Lab Discussion

Pool the results from different groups. Help students recognize that these are only two of the immense number of enzymes, each of which has a specific task to perform in maintaining homeostasis in the body.

Pepsin is found in the stomach in a low pH environment, while amylase is in saliva, in the relatively neutral pH of the mouth. It is interesting that because of the pH change, starch digestion stops when swallowed material reaches the stomach, and doesn't continue until enzymes in the intestine begin breaking the starch down further. The old adage about chewing one's food thoroughly is helpful in mixing the amylase with the starch, to begin its breakdown and digestion.

Answers to Questions:

1. Pepsin pH 2, amylase pH 7
2. Enzymes work best at room temperature. If the temperature is too low, very little happens; if too high, the enzymes are denatured, as was shown in the peroxide decomposition demonstration.
3. If too much base were added, the pH would go too high and the enzyme (in this case pepsin), would be inactivated.

PART B: SUMMARY QUESTIONS (page 425)

1. Homeostasis is the maintenance of a state of balance. In the body, this state of balance is necessary for normal functioning. When such a condition of balance is upset, sickness may result. In very extreme cases, death may occur.
2. Eating a balanced diet, exercising regularly, keeping the body clean, keeping the environment as healthful as possible.
3. Hydrogen, oxygen, carbon, nitrogen, calcium.
4. a. Cells carry out chemical reactions rapidly, selectively, and efficiently.
   b. Reactions must be fast to enable the organisms to respond rapidly to safety threats (fight or flight response), and, in the case of humans, to allow mental intentions to be converted rapidly into appropriate muscular responses. Selectivity is crucial in maintaining homeostasis; the body is capable of simultaneously running a wide variety of chemical reactions to increase or decrease the concentration of particular compounds; to keep body chemistry in balance, enzymes are needed to selectively promote appropriate reactions. Efficiency enables the organism to survive without having to ingest food continuously.
5. a. Enzymes

b. These compounds act to catalyze reactions, causing them to occur faster under less extreme conditions of temperature and pressure than could be accomplished without the enzyme. Enzymes function by providing surfaces on which substrate molecules can be brought into contact with each other, by weakening bonds, and lowering the activation energy, thereby increasing speed of the reaction. Since enzymes work in a "lock-and-key" fashion, each enzyme is highly specific in terms of which reaction it catalyzes.

6. The oxidation process occurs in a sequence of reactions which allow the cells to release the energy in a controlled, step-wise fashion. For instance, the oxidation of one mole of glucose produces 38 moles of ATP. The hydrolysis of each mole of ATP releases about 8 kcal of energy.

7. $1 \text{ mol albumin} \times \dfrac{6900 \text{ g albumin}}{1 \text{ mol albumin}} \times \dfrac{17 \text{ kJ}}{1 \text{ g albumin}} \times \dfrac{1 \text{ mol ATP}}{31 \text{ kJ}} = 38,000 \text{ mol ATP}$

C: ACIDS, BASES, AND BODY CHEMISTRY

Acids and bases are defined in terms of the bond each contains: H^+ for acids, which are covalent molecules, and OH^- for bases, which are usually ionic compounds. The neutralization reaction is explained, both as a total ionic equation and a net ionic equation, in which the spectator ions (those which are unchanged in the reaction) are omitted. The final sections describe the composition and function of buffers, which are tested in a laboratory activity.

C.1 STRUCTURE DETERMINES FUNCTION (pages 426-427)

The classic Arrhenius definitions of acids and bases are given, along with equations showing the formation of the ions associated with each. Remind students of the structures of acids and bases presented in the Water unit, and the concept of "neutral", which means neither acid nor base.

C.2 STRENGTHS OF ACIDS AND BASES (page 427)

Strong acids are defined in terms of the extent of ionization, and several acids are classified as to strength.

C.3 ACIDS, BASES, AND BUFFERS IN THE BODY (page 429)

Acids and bases can denature protein, and yet the body continually handles both acids and bases without destroying enzyme activity. This section explains why, after defining buffers, by featuring the carbonic acid–hydrogen carbonate ion buffer system. This combination of weak acid and salt protects the body against drastic changes in pH. The two equations by which this happens are explained in some detail, which leads to first-hand observation of this system in the laboratory activity.

YOUR TURN

Acids and Bases (page 428)

Answers to Questions:

1. a. $HCl \xrightarrow{\text{water}} H^+(aq) + Cl^-(aq)$
   b. $\phantom{HCl \xrightarrow{\text{water}}}$ Hydrogen Chloride
2. a. $H_2SO_4 \xrightarrow{\text{water}} H^+(aq) + HSO_4^-(aq)$
   Sulfuric Acid $\phantom{\xrightarrow{\text{water}}}$ Hydrogen Hydrogen Sulfate
   b. $H_2CO_3 \xrightarrow{\text{water}} H^+(aq) + HCO_3^-(aq)$
   Carbonic Acid $\phantom{\xrightarrow{\text{water}}}$ Hydrogen Hydrogen Carbonate
3. a. $Mg(OH)_2 \xrightarrow{\text{water}} Mg^{2+}(aq) + 2OH^-(aq)$
   Magnesium hydroxide $\phantom{\xrightarrow{\text{water}}}$ Magnesium hydroxide
   b. $Al(OH)_3 \xrightarrow{\text{water}} Al^{3+}(aq) + 3OH^-(aq)$
   Aluminum Hydroxide $\phantom{\xrightarrow{\text{water}}}$ Aluminum Hydroxide
4. a. $2H^+(aq) + 2NO_3^-(aq) + Mg^{2+}(aq) + 2OH^-(aq) \longrightarrow 2NO_3^-(aq) + Mg^{2+}(aq) + 2H_2O$
   b. $2H^+(aq) + 2OH^-(aq) \longrightarrow 2H_2O$

C.4 LABORATORY ACTIVITY: BUFFERS (pages 430-431)

Students investigate the buffering ability of one of the body's two key buffer systems, the carbonic acid–hydrogen carbonate system. After making their own buffer solution, students test its effectiveness in neutralizing added acid and base.

Buffering can be a difficult concept for students, and thus it is important that they be able to see buffers in action. The marginal notes on page 430 explain the use of molarity in defining the concentration of a solution.

Time

One class period

Materials (for a class of 24 working in pairs)

250 mL 0.5 M hydrochloric acid (11 mL conc. HCl/250 mL solution)
250 mL 0.5 M sodium hydroxide (5 g NaOH/250 mL solution)
500 mL 0.1 M sodium hydrogen carbonate (4.2 g/500 mL solution)
50 mL universal indicator solution (placed in dropper bottles)
Deionized or distilled water
12 dropping bottles
24 125-mL Erlenmeyer flasks
12 50-mL burets
12 buret stands or ring stands with buret clamps
12 50-mL graduated cylinders
12 straws (each student to have own straw)
12 50 mL beakers

Advance Preparation

See solution preparations in Materials.

Pre-Lab Discussion

Briefly review the concepts associated with this activity contained in Section C.3.

Lab Tips

All references to water in this activity are to distilled (or deionized) water.

Remind students not to share the straws used to blow air into the flask to prepare the carbonate buffer system. Used straws should be discarded.

Provide universal indicator color charts or tables so students can estimate the pH values of their solutions.

If your buret supply is limited, you can reduce the number by half by asking two groups to share two burets.

Expected Results

In general, addition of acid or base to the buffered systems causes no observable pH change until the buffering capacity of the system is exceeded. Then, the pH drops or rises, approaching the pH of the added acid or base. Actual pH values will be somewhat variable.

Post-Lab Discussion

Pool the class data so all students gain the benefit of the observations of all groups. Each group should be able to observe that the buffer system strongly resists pH change in comparison with water.

Answers to Questions:

1. a. Yes.
   b. The buffer solution required much larger amounts of acid or base to change the pH.
2. a. b. In each case, more acid or base is needed with the carbonate buffer.
3. $H_2CO_3(aq) + OH^-(aq) \rightarrow HCO_3^-(aq) + H_2O(l)$
4. a. (1) Hydrochloric acid is a strong acid; potassium chloride (KCl) is the neutral salt of a strong acid (HCl) and a strong base (KOH).
       (2) NaOH is a strong base; H_2O is neutral.
       (3) $NaNO_3$ is the neutral salt of a strong acid and strong base; HNO_3 is a strong acid.
       (4) $NaC_2H_3O_2$ is the salt of a weak acid, acetic acid, $HC_2H_3O_2$.
   b. (1) This will not form a buffer in water solution.
       (2) This will not form a buffer.
       (3) This water solution will not form a buffer.
       (4) Since this pair represents a weak acid and a salt of the weak acid, it will make the best buffer system.

MICROSCALE PROCEDURE

Materials

24 well microplates
12 Beral pipets containing distilled water
12 Beral pipets containing 0.1 M sodium hydrogen carbonate.
12 Beral pipets containing 0.1 M NaOH (0.8 g of NaOH/ 200 mL solution)
12 Beral pipets containing 0.1 M HCl (1.7 mL/200 mL solution)
12 Beral pipets containing universal indicator solution
12 Beral pipet tips (for micro-straws)

Procedure

1. Add 20 drops of distilled water to one well of a 24-well plate. Add one drop of universal indicator and note its color.
2. Add 20 drops of 0.1 M sodium hydrogen carbonate to another well. Pull out the end of a clean Beral pipet, and cut it off so that you have a miniature straw about 10 cm long. Use this straw to gently blow into the $NaHCO_3$ well for 2-3 minutes: this is the buffer solution. Add one drop of universal indicator and note its color.
3. Add 5 drops of 0.1 M NaOH to the water well. Note the color produced.
4. Add 0.1 M NaOH to the buffered well until the color matches the water well color. Count the number of drops required. Record the number in your data table.
5. Repeat Steps 1 and 2 with clean, dry wells.
6. Add 5 drops of 0.1 M HCl to the water well. Note the color produced.
7. Add 0.1 M HCl to the buffered well until the color matches the water well color. Count the number of drops required. Record the number in your data table. Answer questions in the textbook (page 431).

C.5 pH IN BALANCE (pages 432-433)

The effects of deviating from a balanced pH are explained in this section. The *YOUR TURN* asks students to apply their knowledge of acids and bases to everyday situations.

YOUR TURN

Conditions that Affect pH Balance (pages 432-433)

This activity acquaints students with a number of conditions that can overload the body's buffering capacity. (The text equations on page 429 are useful in discussing these answers.) This activity also includes a review of the pH balance requirements of the body.

Answers to Questions:

1. a. Hyperventilation causes an increase in blood oxygen levels and a subsequent decrease in carbon dioxide levels.
   b. This would lower acidity and cause an increase in pH.
2. a. Holding one's breath causes a rise in carbon dioxide levels.
   b. This would cause an increase in acidity and a decrease in pH.
3. The accumulation of lactic acid would lower the blood pH to more acidic levels.
4. a. A cardiac arrest would prevent the body from eliminating carbon dioxide; thus the blood pH would decrease as the concentration of carbon dioxide (and thus carbonic acid) increases.
   b. The injection of $HCO_3^-(aq)$ from sodium hydrogen carbonate would counteract this effect, and return blood pH to more normal levels.
5. Aspirin is acetylsalicylic acid. A large excess could cause acidosis.

CHEMQUANDARY

The Shell Game (page 433)

With increasing environmental temperatures, chickens pant more rapidly to maintain their optimal body temperature. Panting lowers the concentration of carbon dioxide in their blood, making less of it available for conversion to the calcium carbonate needed for egg shells. Chickens are fed carbonated water and crushed seashells to counteract this loss.

PART C: SUMMARY QUESTIONS (page 433)

1. a. An acid contains hydrogen ions (H^+), a base contains hydroxide ions (OH^-). Many acids are molecular, covalently bonding compounds; most bases are ionic compounds.
   b. A strong acid is almost completely ionized in water, which means that it separates almost 100% into ions. A weak acid is only slightly ionized.
   c. A strong base is almost completely ionized in water, meaning it is almost 100% separated into ions. A weak base is only slightly ionized.

2. No, hydrocarbons contain the element hydrogen and are definitely not acids. Bases also contain hydrogen as an element, and yet their behavior contrasts with that of acids.

3. A buffer system contains a weak acid and the salt formed from that acid and a base. A buffer functions by reacting the salt with added acid, neutralizing the acid, and reacting the weak acid with added base, neutralizing the base.

4. (1) Hyperventilation causes decreased carbon dioxide levels and raises blood pH.
   (2) Holding one's breath causes increased carbon dioxide levels and lowers blood pH.
   (3) Strenuous muscle activity raises the lactic acid concentration and lowers blood pH.
   (4) Cardiac arrest raises carbon dioxide levels and lowers blood pH.

D: CHEMISTRY AT THE BODY'S SURFACE

Part D covers the more cosmetic aspects of health care; it focuses on skin and hair, both responsible for helping maintain the balance in our systems. How does soap work? Why should we refrain from cultivating deep suntans? How does styling change the shape (and sometimes the health) of hair? Chemistry provides the answers.

D.1 KEEPING CLEAN WITH CHEMISTRY (pages 434-437)

The beneficial effect of oil secretion in lubricating the skin contrasts with its unfortunate tendency to cause body odor. The structure of soap, and the process by which it cleanses the skin, is an application of the "like dissolves like" concept developed in the Water unit (pages 59-60). The discussion includes an important property of water molecules, hydrogen bonding, which strongly affects water's abilities to dissolve compounds. The text explains how the split personality of soap enables it to allow water and nonpolar substances (grease and oils) to mix.

YOUR TURN

Polarity and Solubility (pages 435-436)

Students are asked to connect the notion of molecular polarity to the practical property of water solubility.

Answers to Questions:

1. a. NaOH—ionic—water soluble
   b. $CH_3CH_2-O-CH_2CH_3$—nonpolar—very slightly water soluble
   c. CH_2OHCH_2OH—polar—water soluble
   d. KNO_3—ionic—water soluble

2. a. Cholesterol is practically water insoluble (0.2 mg/100 mL water). The one -OH group is only a small portion of a very large, nonpolar molecule.
   b. Given the generally nonpolar character of the molecule, one would expect it to be more soluble in nonpolar solvents.

3. a. Glucose should dissolve in water. Each glucose molecule contains several -OH groups that can form hydrogen bonds with water.
   b. The fatty oil should dissolve in gasoline. The molecule has no polar -OH groups and would probably dissolve in nonpolar solvents like gasoline.
   c. The number of polar - OH groups.

4. It is soluble in both water and oil, given that one end is polar ($-OSO_3^-$ Na^+) and the other end nonpolar (long hydrocarbon chain).

Benzoyl Peroxide and Acne (page 438)

1. Benzoyl peroxide has the ability to dissolve in the oils and then react with the bacteria, as hydrogen peroxide does.
2. The antibiotics kill the bacteria, preventing the infections that lead to pimples.

D.2 SKIN: IT'S GOT YOU COVERED (pages 438-439)

This section briefly describes the skin's biological structure and function. This background will be useful in the subsequent discussion of skin protection products.

D.3 PROTECTING YOUR SKIN FROM THE SUN (pages 439-440)

This section discusses the body's defense system against too much sun, the results of too much sun, and how sunscreens can help aid the body's defense system. The chemical changes caused by ultraviolet light are outlined, and the mechanism of PABA's shielding is explained.

D.4 LABORATORY ACTIVITY: SUNSCREENS (pages 441-442)

This activity is designed to make visible the relative effectiveness of various sunscreens solutions discussed in Section D.3, while also allowing students to act as detectives in rating an unknown oil or lotion. If you are lucky, the sun will shine at an appropriate time during this unit. Encourage students to bring a variety of samples for testing.

Time

Less than one class period

Materials (for a class of 24 working in pairs)

Bottle #2 suntan product
Bottle #6 suntan product
Bottle #10 suntan product
Bottle #15 suntan product
Bottle of oil or lotion of unknown SPF
Sunprint paper (see Lab Tips)
Acrylic sheets (such as overhead transparencies)

Advance Preparation

None, but hope for a sunny day!

Pre-Lab Discussion

Reemphasize the importance of protecting the skin from too much sun. An unnatural dark tan is not desirable. Encourage students to consider their own sunbathing habits in light of the known connection between excess sun and skin cancer.

Lab Tips

The sunprint paper is the same material used in the *Laboratory Activity* A.1 of the Nuclear unit (page 275). See those teaching notes for sources.

Care should be taken that all acrylic strips receive equal amounts of lotion, spread evenly over the slide. Spread the lotion over the slide surface with a finger.

A whole sheet of acrylic paper can be used, and the samples applied in a thin row on the acrylic. To reduce the amount of sunprint paper used, cut small strips of sunprint paper, and place them under the sunscreen samples.

Expected Results

The degree of darkening of the sunprint paper decreases as the lotion sunscreen ratings increase. Posting the sheets on the board in front of the room can simplify comparisons among groups.

Post-Lab Discussion

Focus the discussion on the results, on the questions, and on applications of student findings to their lives.

Answers to Questions:

1. Results will vary.
2. Possible trade-offs: the convenience of driving a car versus the risk of an accident, and the pleasure gained from cigarette smoking versus the increased probability of lung cancer.
3. Dark-skinned persons have a higher concentration of melanin in their skin, which screens out more of the more harmful ultraviolet rays.

D.5 GETTING A "D" IN PHOTOCHEMISTRY (pages 442-443)

Following the previous information about the potentially harmful effect of sunshine, it is important that students understand that—in moderation—sunshine is not only healthful, but essential in promoting the formation of vitamin D, necessary for bone strength.

CHEMQUANDARY

Vitamin D Deficiency (page 443, top)

Rickets is caused by diminished supplies of calcium—without a continual supply of calcium, bones become weak and brittle. A shortage of Vitamin D, which transports calcium ions, can lead to this condition even when the consumption of calcium via foods is adequate.

CHEMQUANDARY

Vitamin D Solubility (page 443, bottom)

Vitamin D is soluble in fat. Since the vitamin is practically insoluble in water, it is somewhat more difficult for the body to get rid of it, if excess quantities are ingested.

D.6 OUR CROWNING GLORY (pages 443-444)

This section describes both the macro- and micro-structure of hair as a prelude to discussions on the role of chemistry in grooming.

Ask your students if they agree with the opening premise of this section concerning the amount of time, money and concern people have with hair, an interesting discussion usually results.

D.7 LABORATORY ACTIVITY: CHEMISTRY OF HAIR (pages 444-446)

Students investigate the general characteristics of their own hair and test the effects of various hair cleansing and treating solutions on the hair properties. Students work in groups of four, and each person tests one solution: a pH 4 solution, a pH 8 solution, a permanent wave solution, or a permanent wave + neutralizer solution. In Part 2 all test their own hair.

Time

One class period

Materials (for a class of 24 working in pairs)

24 hair bundles (3-cm diameter, 15 cm long) from barber shop or hair salon (4 bundles/team)
24 wood splints (4 splints/team)
90 mL pH 4 hydrochloric acid (15 mL/team) (approximately 10 drops of 0.1 M HCl/500 mL solution)
90 mL pH 8 sodium hydroxide solution (15 mL/team) (10^{-5} g NaOH/250 mL solution)
90 mL permanent wave solution
90 mL permanent wave neutralizer
1 package Hydrion pH paper
Distilled water
24 rubber bands, orthodontist's type (4 bands/team) or cut pieces from a rubber hose
12 25-mL graduated cylinders

24 small test tubes (18 × 150 mm) (smaller than wooden splints)
6 test tube racks
6 scissors
Light source or hair dryer (optional)

Advance Preparation

Get some hair from a barber shop or beauty salon (see Lab Tips).

Pre-Lab Discussion

As a motivational device you may wish to bring in (or have students bring in) some printed advertising claims and ingredient lists for various hair care products.

Lab Tips

Arrange ahead of time with a local hair stylist to save cut hair that is at least 15 cm (6 in.) long for this experiment.
   Make sure that all tests by a given team are performed on identical hair samples.
   If the hair is left out over night, students may find textural differences.

Expected Results

In general, hair tensile strength decreases with increased pH; the samples at pH 8 and the permanent wave solution (without neutralizer) will be observed to break easiest.

Post-Lab Discussion

After discussing the questions, encourage students to apply what they have learned about hair to hair product advertising language (such as pH-balanced shampoos).

Answers to Questions:

1. Coarse hair contains more strands of protein than does thin hair.
2. The cuticle is like roof shingles. In healthy hair, rubbing one's fingers down the hair catches intact cuticle and hair bunches up. If the cuticle is damaged (broken off), the hair will be smooth and will not bunch up.
3. Answers will vary.
4. Permanent wave solution with neutralizer.
5. Permanent wave solution with neutralizer.
6. The pH 8 and permanent wave (without neutralizer) solutions make the hair most brittle.
7. High heat, extreme pH levels, and certain bleaching agents (such as chlorine) will cause the most damage to hair.
8. The sodium hydroxide solution breaks disulfide bonds and thus affects the tertiary structure of the protein. When the hair is curled after application, these bridges reform, corresponding to the shape of the curl.
9. Basic, because hair is naturally acidic and because basic solutions are better at breaking disulfide bonds.
10. Sodium hydroxide solutions are more caustic and damaging to the hair.
11. After swimming, one should at least rinse the hair or shower; it's also helpful to use an acidic rinse or conditioner to restore natural pH and oils lost by heat or chemical treatment.

D.8 HAIR STYLING AND CHEMICAL BONDING (pages 447-449)

This section describes in detail the types of bonding responsible for protein structure and curly hair. In addition to covalent bonding, hydrogen bonding, ionic bonding, nonpolar interactions and disulfide bridges contribute strongly as the protein chains interact with each other: this is called secondary and tertiary structure. Students learn how permanent wave solutions rearrange disulfide bonds, in contrast to water, which merely temporarily rearranges hydrogen or ionic bonds. Be sure to connect the discussion back to the previous (*Laboratory Activity* D. 7) observations.

CHEMQUANDARY

A Permanent Permanent? (page 449)

No, since such treatments affect only the portion of hair outside the follicle. As the hair grows, new, untreated hair will grow out from the follicle.

PART D: SUMMARY QUESTIONS (page 449)

1. a. During skin cleansing, perspiration, skin-moistening oils, dead cells, and external dirt and oil are removed.
   b. To prevent them from being oxidized to odoriferous compounds by skin bacteria.
   c. Excessive cleansing or the use of strongly basic agents can dry the skin and cause premature aging.
2. a. Ionic—water soluble
   b. Slightly polar (due to presence of –OH)—water soluble
   c. Nonpolar—need detergent
   d. Polar (presence of –OH)—water soluble
3. Soaps or detergents act to connect polar water molecules and nonpolar grease or oil molecules. They consist of long molecules with one end that is polar (and therefore soluble in water) and the other nonpolar (and therefore soluble in grease)—see Figures 4 and 5 on pages 433-437.
4. The skin acts as both a barrier and an exchange conduit between the organism and its environment. Vital body fluids are kept in and bacteria are kept out. Additionally, it is the chief organ for the sense of touch, can communicate emotions, and plays a major role in heat regulation (maintaining a constant body temperature) by sweating.
5. Tanning is the body's defense mechanism to minimize the influx of damaging UV radiation. The darkened skin consists of melanin formed by the UV-triggered oxidation of the amino acid tyrosine (see page 439).
6. Suntanning is usually associated with UV radiation in that it is energetic enough to cause disruption of some molecular bonds within skin cells. If DNA molecules are disrupted and the mutations reproduced, skin cancer may result (see Section D.3 for more detail).
7. Sunscreens contain molecules, such as PABA, that have benzene rings capable of absorbing UV photons and converting them into harmless heat before they reach the skin. The concentration of active ingredient is varied according to the desired degree of protection (see Table 3 on page 440).
8. Rickets is a disease characterized by soft and easily deformed bones. It occurs primarily in children who do not receive adequate quantities of sunshine. The lack of sunshine prevents the body from synthesizing adequate amounts of vitamin D, which is essential in Ca^{2+} absorption and incorporation into bones. Sufficient levels of sunshine and/or vitamin D supplements (in milk, tablets, etc.) can prevent the disease.
9. Hair is an elastic polymer (long-chain molecule). Rubber is also a polymer.
10. Wet hair stretches more. When hair is wet, it can be stretched to one and a half times its dry length because water (pH 7) weakens or breaks some hydrogen bonds and ionic bonds and causes swelling of the keratin.
11. Permanent wave solutions cause disulfide bonds between protein chains to break, rearrange, and then reform in a new arrangement. Water causes a temporary rearrangement of hydrogen and ionic bonding in protein chains.

EXTENDING YOUR KNOWLEDGE (page 449)

- Dark skin (containing extra melanin) is better able to prevent high energy, UV radiation from reaching the sensitive underlying skin cells. Therefore, dark-skinned persons are less likely to suffer from sunburn and skin cancer than are light-skinned persons.
- Check the Supplemental Readings on page 234 of the Teacher's Guide for references.

E: CHEMICAL CONTROL: DRUGS AND TOXINS IN THE HUMAN BODY

In this section, students learn how drugs can be used to help the body regain its chemical balance when something goes wrong, and how drugs (used improperly) and other toxins can upset the body's chemical balance. Once again, students learn that they can make certain choices to help keep their bodies in balance.

E.1 A GLIMPSE OF DRUG FUNCTION (pages 450-452)

Drugs and toxins are explained and compared, and their effects on the body explained in terms of receptors.

You should emphasize these major points: (1) The importance of dose in describing chemical toxicity. You may wish to refer back to discussions in the Water unit on pollutants, or to the Food unit discussion regarding overdoses of certain vitamins or minerals necessary in small quantities, but toxic in large quantities. (2) The fact that, although consuming (or injecting) too much of any substance may cause toxicity problems, it will not usually lead to cancer—the number of known carcinogens is a very low percentage of the total number of known substances. (3) The activity of drugs often depends on their ability to act at receptor sites. Narcotic analgesics are discussed as an example of how drugs can pass through the body's systems and mimic body chemistry to produce certain desired effects.

E.2 YOU DECIDE: PROS AND CONS OF ASPIRIN (pages 452-453)

The history of the development of aspirin is given, leading to another risk-benefit analysis. Point out that chemists, in their role as molecular architects, are continually trying to improve the effectiveness, and decrease the negative side effects, of various drugs by altering specific parts of molecules. This section considers aspirin as a specific case in which chemists started with a compound provided by nature and improved upon it.

You may wish to point out that, since its introduction in 1899 by the German Bayer Company, aspirin has grown to be the best-selling drug in the world. Over 20 billion aspirin tablets (under a variety of trade names) are consumed annually in the United States. Also point out that all drugs have side effects on a small proportion of people—in the case of aspirin some patients have minor stomach irritation, bleeding, or other symptoms. With any drug, side effects can become potentially dangerous if warnings about dose, consumption with food, use of alcohol and other drugs, or individual allergies are not followed.

This activity points out that even a common drug such as aspirin may have undesirable side effects. The choice of self-medication is best made when all the facts are known. The burden/benefit question posed by this section is: How effective is the remedy, compared to the risk that the drug will produce undesired side effects? Or, in terms of this activity: Is the pain relief from intensive use of aspirin worth the danger of its possible side effects?

Answers to Questions:

1. Student answer
2. Options that might be available to counteract the long-term side effect of hearing loss are: (1) taking another drug with the aspirin to counteract the effect; (2) wearing a hearing aid; and (3) taking a new pain-reliever that has a less negative side effect.
3. a. Effectiveness of the medication, dosage necessary, side effects, results of long-term exposure, effect on other drugs, effect on food absorption in the body, possible effects on fetuses.
   b. Manufacturers' information, personal physician, pharmacist, U.S. Public Health Service, Food and Drug Administration, publications by groups such as Consumers Union, home health encyclopedias, medical journals (*Journal of the American Medical Association*).
4. Since medications have the potential to upset the chemical balance within the body, and since many have side effects that cause or contribute to other problems, it is prudent to take medication only when absolutely necessary.

E.3 YOU DECIDE: EFFECTS OF ALCOHOL (pages 453-454)

This activity is designed to bring home the fact that alcohol is a drug with specific physiological effects. It is hoped that, without lecturing on the evils of overconsuming alcohol, we can encourage students to develop responsible attitudes regarding its use.

The ACS publication *CHEM MATTERS*, February 1985, carried an article on alcohol metabolism and dietetic and behavioral effects.

Begin by focusing the discussion on the questions. You may wish to follow this with a brief discussion of laws regarding alcohol use in your state, the nutritional effects of alcohol, or the peer pressure associated with alcohol use. As an alternative, consider inviting in an outside speaker.

Answers to Questions:

1. a. 0.05%
   b. 0.02%
2. Reaction time slowed by 15-25%, visual sensitivity reduced up to 32%, headlight recovery 7-32 seconds longer, lowered alertness, reduced coordination.
3. No, the person's blood level would be 0.095% one hour later, just under the legal limit in most states.
4. This person approaches the 0.35% blood level, thus approaching the level where death can occur.
5. The alcohol level does not exceed 0.05%, and drops to 0.02% in two hours. Given that the probability of causing an accident doubles when the driver's blood alcohol was just 0.06%, the longer the wait the better.

E.4 NOTES ON SOME OTHER DRUGS (pages 454-455)

New drugs appear and are abused continually. The intense pleasure and euphoria of cocaine, crack, ice, and designer drugs, coupled with the "promise" of physical prowess that steroids offer, continue to attract many.

E.5 FOREIGN SUBSTANCES IN THE BODY (pages 455-457)

This section discusses how the body defends itself against accidental abuse or microorganism attack through detoxification and antibody production. However, these processes are limited in scope, and continued deliberate abuse can overwhelm them. Additionally, the body is not able to protect itself from some disease-causing agents, such as the AIDS virus. It is important that students know how to reduce their possibility of exposure to this dreaded disease.

E.6 DRUGS IN COMBINATION (pages 457-459)

This section points out the high risks of experimenting with combinations of drugs. Students should understand the importance of (1) informing their doctors of all the drugs they are taking, so doctors can prescribe medications safely and effectively, and (2) carefully following the advice and warnings on the drug labels concerning dose, food, use with other drugs, etc.

E.7 CIGARETTE USE (page 459)

This section surveys some of the known negative effects of cigarette smoking on overall human health. Avoid "overkill" on the risks of smoking; let these facts, the laboratory activity, and the final activity (in Part F) provide the basis for students to make their own decisions regarding the risks of smoking.

E.8 LABORATORY ACTIVITY: CIGARETTE USE (pages 460-461)

The activity sets up a smoking machine to collect the products from cigarette smoke.
   Students then observe its effects on another living organism, *Euglena*.

Time

One period. May be done as a demonstration if time and equipment is in short supply.

Materials (for a class of 24 working in pairs)

Unfiltered, high tar cigarettes (1 cigarette/team)
Filtered low tar cigarettes (1 cigarette/team)
Box of cotton balls (drug store item)
Euglena culture
12 aspirators
24 plastic tubes (3.5 cm diameter, 20 cm length) (2 tubes/team) or drying tubes (see Advance Preparation)
12 10-cm funnels
48 pieces glass tubing (7-cm length) (4 glass bends and 1 tube per team: see Figure 22)
48 1-hole rubber stoppers (4 stoppers/team)
24 aspirator hoses
12 pairs disposable plastic gloves
12 forceps
12 50-mL beakers
24 microscope slides
Microscopes
24 ring stands and burette clamps
6-L container (optional)
Rubber tubing

Advance Preparation

The special plastic tubes (two per device) housing the smoke-trapping cotton balls (see Figure 22 on page 460) can be replaced by standard drying tubes, if desired. (Smaller balls of cotton may be needed to fit into the drying tubes, if this substitution is made.)
   If *Euglena* cultures are not available, or beyond your budget, pond water should be able to supply your students with sufficient samples to test.

Pre-Lab Discussion

In Part 2 you may wish to have half of the teams use the cotton from the nonfiltered, high tar cigarette and the other half use the cotton from the filtered low tar cigarette for comparison purposes.

Lab Tips

The large amounts of cigarette smoke generated in this activity warrant setting up one smoking device under in your fume hood, and having students collect their samples for testing from that. It also saves water.

Part 1, Step 3: Caution students to extinguish the cigarette before it reaches the filter or the tubing.

Part 2, Step 3: The *Euglena* culture should be under teacher control during the activity to prevent its contamination. Plan to dispense the culture samples to students yourself, insuring that the same dropper is always used with the culture material. (This avoids the risk that students may mix droppers, using a dropper previously in contact with the residue solution with the *Euglena* culture, and causing the entire culture to prematurely expire.)

Part 2, Step 5: Remind students to clean their dropper before the Part 2 repetition to avoid contaminating their lower trap residue with that from the upper trap.

Expected Results

- Cotton from the lower trap (unlit end) is darker than the cotton from the upper trap (lit end). One might thus conclude that the person smoking inhales more particulates than do those inhaling nearby smoke. However, some substances observed emerging from the lit end of a cigarette have led to concern among some researchers about potential dangers of "passive smoking."
- The test on the filtered cigarette yields similar though less dramatic results: the filter removes some of the particulates.
- The exposed *Euglena* exhibit irregular swimming movements and deformed body structures, eventually leading to their death.

Post-Lab Discussion

After discussing the results and the questions, you may wish to discuss the advantages and disadvantages of using *Euglena* as a test organism for the effects of smoking. The problems of extrapolating from other organisms to humans leads one to question the validity of such research, also a problem when studying food additives (see page 263 of the Food unit). You may wish to point out that nicotine is used as an agricultural insecticide.

Answers to Questions:

1. Yes, the nonfiltered cigarette exhibits the darker stain.
2. There is little difference between the cotton stains from the upper traps (lit ends) of either filtered or nonfiltered cigarettes. Students will notice that the stains in the lower trap cotton (unlit end) are darker than those from the upper (lit end) trap.
3. The treated *Euglena* exhibited irregular swimming patterns and deformed body structures, leading to death.

E.9 YOU DECIDE: SMOKING? (pages 461-462)

This activity provides an opportunity to pull together Sections E. 7 and E. 8. It is intended to give students the chance to discuss what they have learned about the health effects of smoking, and to decide for themselves if the benefits are worth the risks.

As the teacher, try to allow the students to direct the discussion. If both you and your class think that you've exhausted the topic, you may wish to discuss related issues such as: (1) the recent move for a total ban on tobacco advertising versus the right of freedom of speech; (2) the role of the government in supporting tobacco farming versus supporting cancer research; (3) the use of land for growing tobacco versus food products; and (4) the economics and ethics of exporting tobacco to developing countries.

Point out that smoking is only one (albeit a clear-cut) example of a personal habit that can decrease one's general health and life expectancy. Other examples include the foods one eats, and one's exercise patterns.

Answers to Questions:

Option 1

Positive: The monetary savings from not buying cigarettes; appreciation from nonsmokers.
Negative: Not having enjoyed the stimulation caused by smoking, and the social acceptance of other smokers.

Option 2

Positive: Temporary enjoyment of stimulation caused by smoking.
Negative: Damaged lung tissue and increased risk of emphysema, lung cancer, heart attack, and stroke; money wasted on a harmful habit; inconvenience and/or increased health risk for others with whom you live or work.

PART E: SUMMARY QUESTIONS (page 462)

1. a. Narcotic analgesics (such as morphine) are believed capable of relieving pain through their ability to block key receptor sites or nerve cells.
   b. The brain's own painkillers, endorphins and enkephalins, have several parts of their molecules which are similar to the narcotics, and would thus fit the same receptor sites.
2. a. Alcohol depresses the transmission of nerve signals by reducing the concentration of calcium ions at nerve endings, and thus prevents the release of the messenger molecules that are involved in neural transmissions.
   b. Cocaine functions as an anesthetic by reducing the sensitivity of the nerve membranes to return of messengers.
3. (1) Different medications may counteract each other; therefore the net effect of the combined treatment may be zero. (2) Different medications may interact in a manner that can harm or even kill the patient.
4. Yes, many medicines specify whether they should be taken on a full or an empty stomach to avoid gastrointestinal irritation and/or aid absorption. Alcohol interacts synergistically with many other medicines, and, if instructions are not followed, life threatening reactions can result.
5. Chemical processes in the liver include: (1) separation of useful and harmful substances; (2) storage of glucose as glycogen; (3) conversion of amino acids to proteins; and (4) the detoxification of harmful materials.
6. a. Chemical toxins may be made harmless by reactions in the liver. Biochemical agents such as viruses, bacteria, fungi, and parasites may be attacked and destroyed by antibodies.
   b. Unfortunately, the AIDS virus destroys the body's normal immune defense system.
7. Smoking (a) increases stress on the circulatory system because of blood thickening, which results from the increased production of red blood cells, needed to make up for the decreased ability of hemoglobin to carry oxygen (which is due to the increased carbon monoxide levels); it (b) scars the lungs (and possibly emphysema) as the enzyme elastase tries to clear the lungs of tar; and it (c) induces cancer.

EXTENDING YOUR KNOWLEDGE (page 462)

- See Supplemental Readings at the end of this unit.

F: PUTTING IT ALL TOGETHER: ASSESSING RISKS

Not all choices we make about health are clear cut. In this era of "carcinogen of the month", these culminating activities give students the tools to evaluate different risks for themselves. Part F describes how scientists determine whether a substance is a health hazard, and explains the uncertainties involved in making this determination. It explains how statistics can be used to help persons assess relative risks. Finally, it provides students with a method for thinking about risks where uncertainties exist regarding the dangers of these risks. A major objective is for students to question whether it is prudent to smoke tobacco, abuse alcohol, or use illegal drugs, and to realize that their health and longevity is largely dependent on decisions they make now regarding diet, exercise, and the use of drugs.

F.1 RISKS FROM ALCOHOL AND OTHER DRUGS (pages 463-464)

This section contrasts risks that are completely controllable/avoidable (use of alcohol and mind-altering drugs) to those that are somewhat less controllable. Be sure to avoid lecturing or prying into your students' private lives. Use a few examples drawn from the national news (such as, celebrities who have suffered as a result of drug abuse); let the factual data presented in the text speak for itself.

F.2 PERSONAL CONTROL OF RISK (pages 464-465)

This activity is designed to reemphasize the point that certain health hazards are within our personal control; if individuals understand the risks involved, they can make sensible choices and increase their chances of having a long, healthy life.

First have students position entries from the list of health hazards on the grid as an individual exercise. You may need to work through a couple of examples for the entire class. Then divide the class into groups of four to compare, and finally produce, a consensus grid. Students should write a brief reason for placing each hazard in the position decided upon.

Answers to Questions:

1.,2.   Possible placements:

Quadrant III (known risk, controlled)
Tuberculosis and smallpox—avoid by vaccination
Heart disease—somewhat avoidable by diet/personal habit but also genetic (also in **Quadrant IV**)
Emphysema—somewhat avoidable by not smoking
Motor vehicle accident—somewhat avoidable by wearing seat belts and safe driving (also in **Quadrant IV**)
Electrocution—avoidable by safe practices

Items falling between Quadrants III and IV
Stroke—both genetic and environmental factors involved
Homicide—odds influenced somewhat by where one chooses to live and act

Quadrant IV (known, not controlled)
Appendicitis—does not appear to be influenced by personal habits
Tornado—can only be avoided if advance notice
Lightning—an act of nature; odds can be somewhat lowered by actions
Smallpox vaccination—if an individual has a negative response (extremely unlikely)
Poison

In general, most of the 14 causes of death listed in the table given can be considered known risks (depending on one's age, sex, family history, occupation, and personal habits, one can reasonably predict the probability of dying from the given cause). Cancer in some respects falls in all four quadrants. Some habits such as smoking are known, controllable risks (**Quadrant III**). Others, such as living near a toxic waste site or eating foods containing certain additives involve a somewhat unknown and only partially controllable risk (**Quadrants I** and **II**). Some factors such as breathing polluted air or drinking polluted water represent known, mainly uncontrollable risks (**Quadrant IV**).

3.   Student answers.

4.   Answers may vary but students should recognize that enough scientific data has been collected to place both risks in **Quadrant III**.

5.   **Quadrant I:** long-term unknown, and individually uncontrollable risks of nuclear power, pesticides, recombinant DNA research, greenhouse effect.
      Quadrant II: long-term unknown, and somewhat avoidable risks of oral contraceptives, food irradiation, diagnostic X rays, water fluoridation, food additives and preservatives, etc.

6.   Focus on those categories more controlled by the individual with known risks: **Quadrant III**.

7. a.   Unknown risks become known risks only by understanding the causes of death, which may or may not be controllable by the individual. Scientific research tries to discover those conditions under which it is likely that death or disease will follow.

   b.   Those risks less controlled by the individual, such as air pollution, are best attacked through the political system in the form of legislated regulations. Individuals can do very little on their own. This is where the simple but important acts of informed voting and opinion letters to political leaders become important, effective, and responsible behaviors in a society such as ours.

8.   Answers will vary.

F.3 LOOKING BACK (page 465)

The widespread role that chemistry plays in our lives has been the focus of this unit. In keeping with the theme of *ChemCom*, students should see chemistry in their everyday lives, in this case their own bodies. Once again, the theme of making choices after weighing risks and benefits has linked chemical topics.

SUPPLY LIST
Expendable Items

Acrylic sheets	D.4	12
Amylase enzyme powder	B.7	2 g
Benedict's qualitative solution	B.7	300 mL
Boric acid	B.7	15 g
Cigarettes, unfiltered and filtered	E.8	2 packs
Citric acid	B.7	15 g
Cotton balls, large package	E.8	1 package
Eggs, hard-boiled	B.7	2
Euglena culture	E.8	230 mL
Gloves, disposable	E.8	12
Hair strands	D.7	500
Hydrochloric acid, conc.	C.4, D.7	100 mL
Hydrogen peroxide, 3% solution	B.4	80 mL
Hydrion paper AB, full scale	B.7	1 package
Liver, raw	B.4	2 pieces
Manganese dioxide	B.4	0.5 g
Pepsin enzyme powder	B.7	2 g
Permanent wave solution	D.7	180 mL
Permanent wave neutralizer	D.7	90 mL
Rubber bands, orthodontist's type	D.7	225
Sodium hydrogen carbonate	D.4	15 g
Sodium hydroxide	C.4, D.7	10 g
Splints, wooden	D.7	24
Starch, water soluble	B.7	1 g
Straws	C.4	12
Sun-sensitive paper	D.4	2 pieces
Suntan oil (SPF #2, #6, #10 and #15)	D.4	1 bottle of each
Trisodium phosphate	B.7	50 g
Universal indicator solution	C.4, D.7	50 mL

SUPPLEMENTAL READINGS

"Pills that Compete with Aspirin." *Consumer Reports*, August 1982, 395-99.

"Aspirin: Is the Warning Necessary?" *Discover*, August 1982, 18-23.

"Is Bayer Better?" *Consumer Reports*, July 1982, 347-49.

"Medicinal Aspects of Aspirin and Related Drugs." *Journal of Chemical Education*, **56**, 331-333 (1979).

"Antacids." *CHEM MATTERS*, April 1983, 6. This issue focuses on acids and bases in everyday life.

"Lead Poisoning", *CHEM MATTERS*, December, 1983.

Folk Medicine: The Art and the Science. American Chemical Society, 1986, 215 p. This is a paperbound book. Explores the medical practices of nonwestern cultures to establish a scientific basis for the successes of folk remedies. Explains why western medical researchers are increasingly turning their attention to folk medicine for new drugs.

"Drugs to Be Sought from Chinese Herbs." *Chemical and Engineering News*, July 4, 1983, 16.

"Herbs for All Seasons." *National Geographic*, March 1983, 386-409.

"Interesting Reviving in the Use of Medicinal Plants." *Chemical and Engineering News*, November 16, 1981, 88.

"Nature's Gift to Medicine." *National Geographic*, September 1974, 420-40.

"How Soap Works", "Soap", and "Soapuzzle", *CHEM MATTERS*, February, 1985.

"Household Soaps and Detergents." *Journal of Chemical Education*, **55,** 596-97 (1978). Describes the components and molecular structure of soaps and detergents and the way they work.

Nutrition and Aerobic Exercise. American Chemical Society, 1986, 160 p. This is a clothbound book. Examines the growing popularity of aerobic exercise as the focal point of daily health maintenance programs. Addresses the principal questions concerning the interaction of nutrition and aerobic exercise on muscle energy requirements, changes in nutritional needs, and long-term effects on body weight and composition. Features a discussion of the optimal plan to reduce body fat.

"The Interface Series: Energy and Exercise." *Journal of Chemical Education*, **55**, (1978). I—"How Much Work Can a Person Do?" July, 456-58; II—"Caloric Costs of Mass Transport," August, 526-28; III—"Heart Work," September, 586-87; IV—"Energy Storage Problems," October, 659-60; V—"Living Reagent," November, 726-27; VI—"Reactions for All Occasions," December, 796-97.

"Exercise." *Discover*, August 1982, 84-88.

"Pumping Oxygen." *CHEM MATTERS*, February 1984, 6-9. Exercise and oxygen metabolism.

"Alcohol." *CHEM MATTERS*, February 1985, 8-11.

"Metabolism of Alcohol." *Scientific American*, March 1976, 25-33.

"Dog Gone", "Smoking", "Test for Catalase", and "Toothpaste", *CHEM MATTERS*, February, 1986.

"Effects of Ethanol on Nutrition." *Journal of Chemical Education*, **56**, 532-533 (1979). Discusses alcohol-related malnutrition including the replacement of nutritious foods, the inhibition of vitamins, the increase in excretion of valuable minerals, and toxicity.

"Cocaine." *Scientific American*, March 1982, 128-41.

"Some Biochemistry of Sedatives." *Journal of Chemical Education*, **56**, 402-4 (1979).

"Testing Employees for Illegal Drugs." *Chemical and Engineering News*, June 2, 1986, 7-14.

"Denatured Alcohol", *CHEM MATTERS*, December, 1990.

The Journal of the American Medical Association (JAMA), **253**, #20, May 24/31, 1985. Focuses on the health effects of smoking and includes a reprint of the classic 1950 paper that first established a link between cancer and tobacco smoking; also includes current smoking trends among various age and sex groups, the effects of passive smoking, etc.

"Unwitting Guinea Pigs", *CHEM MATTERS*, October, 1985.

"The Cigarette Century." *Science 80*, Sept.-Oct. 1980, 36-43.

"Penicillin", *CHEM MATTERS*, April, 1987.

"An Anti-Cancer Diet." *Discover*, June 1984, 23+.

"Skin Deep", *CHEM MATTERS*, December, 1987.

"The Treatment of Diseases and the War Against Cancer." *Scientific American*, November 1985, 51-59.

"Abnormal Insulin", "Artificial Sweeteners", and "Dioxin [Part 1]", *CHEM MATTERS*, February, 1988.

"Special Report: Cancer." *Discover*, March 1986.

"Dioxin [Part 2]" and "Fossil Molecules", *CHEM MATTERS*, April 1988.

"Regulation of Cancer-Causing Substances." *Chemical and Engineering News*, September 6, 1982, 25-32.

"Horses and Heroin", *CHEM MATTERS*, October, 1988.

"Risk—Playing the Odds: A Worrier's Guide to the 20th Century." *Science 85*, October 1985, 29-47. Contains a wealth of useful charts, data tables, and statistics on relative risks in everyday life.

"Saving Arnold", *CHEM MATTERS*, December, 1988.

"The Science of Decision Making: How We Make Up Our Minds, and Why We're So Often Wrong." *Discover*, June 1985, 22-31.

"Distance Running", *CHEM MATTERS*, February, 1989.

"The Sun Worshippers." *CHEM MATTERS*, April 1984, 4-7.

"The Dark Side of the Sun." *Newsweek*, June 9, 1986, 60-64. Discusses the relationship between tanning and rising rates of skin cancer.

"Poison Ivy", *CHEM MATTERS*, October, 1990.

"The Immunological Functions of Skin." *Scientific American*, June 1985, 46-53.

"The New and Improved Chemistry of Cosmetics." *Science 82*, April 1982, 54-61.

"Chemistry of Cleaning." *Journal of Chemical Education*, September 1979, **56**, 610-11.

"Chemistry of Cosmetics." *Journal of Chemical Education*, December 1978, **55,** 12, 802-3. Discusses the chemistry of the skin, the structure of hair, and the chemical basis of cosmetics.

"Household Soaps and Detergents." *Journal of Chemical Education*, 1978, 55, 596-97.

"pH and Hair Shampoo." *CHEM MATTERS*, April 1983, 8-9. This issue focuses on acids and bases in everyday life.

"The pH of Hair Shampoos." *Journal of Chemical Education*, September 1977, 553-54.

"The Bald Truth about Growing Hair." *Discover*, June 1986, 72-81. Information on the science of hair and new drugs to restore hair.

"The Next Step: 25 Discoveries that Could Change Our Lives." *Science 85*, November 1985. Includes a discussion of biomedical research including replacement genes, egg development, cancer-causing genes, brain drugs, biochemical control of behavior, substitute body parts, and living drugs.

"The New Biology." *National Geographic*, September 1976 (special issue): "I. The Awesome Worlds within a Cell" (355-95); "II. The Cancer Puzzle" (396-399); "III. Seven Giants Who Led the Way" (401-7).

"Delivery of Drugs: New Systems Prolong Their Effectiveness." *Chemical and Engineering News.* April 1, 1985, 30-48. Reprint of this special report available.

"Biomaterials in Artificial Organs." *Chemical and Engineering News.* April 14, 1986, 31-48. Reprint of this special report available.

"Biochemical Studies of Aging: Fitting More Pieces into a Far-From-Finished Mosaic." *Chemical and Engineering News,* August 11, 1986, 26-39. Reprints of this special report available.

"Our Immune System: The Wars Within." *National Geographic*, June 1986, 702-36.

"Beyond Supermouse: Changing Life's Blueprint." *National Geographic,* December 1984, 818-47. Details the promise of biotechnology in agriculture, animal husbandry, medicine, and energy.

"Genetic Engineering: A Special C&EN Report." *Chemical and Engineering News,* August 13, 1984, 1-65. Discusses the science, economics, and ethics of the research to date.

"Biotechnology Moves into the Marketplace." *Chemical and Engineering News,* April 16, 1984, 11-19.

"*In-Vitro* Methods May Offer Alternatives to Animal Testing." *Chemical and Engineering News*, November 12, 1984, 25-28.

THE CHEMICAL INDUSTRY: PROMISE AND CHALLENGE

PLANNING GUIDE

Section	Laboratory Activity	You Decide
A. A New Industry for Riverwood?		
Introduction A.1 Basic Needs Met by Chemistry A.2 Industry as a Social Partner A.3 Perspectives		A.4 Products of Industry A.5 Asset or Liability?
B. An Overview of the Chemical Industry		
B.1 From Raw Materials to Products B.2 From Test Tubes to Tank Cars B.3 Close-up: The EKS Company		
C. The Chemistry of Some Nitrogen-based Products		
C.2 Fertilizer's Chemical Roles C.4 Fixing Nitrogen C.5 Nitrogen Fixation at Riverwood C.6 Nitrogen's Other Face	C.1 Fertilizer C.3 Phosphates	C.7 Food or Arms
D. Chemical Energy ↔ Electrical Energy		
D.2 Electrochemistry D.4 Industrial Electrochemistry	D.1 Voltaic Cells D.3 Electroplating	D.5 Planning for an Industry
E. Putting It All Together: Chemical Industry Past, Present , and Future		
E.1 Future Developments E.2 Looking Back and Looking Ahead		

TEACHING SCHEDULE

	DAY 1	DAY 2	DAY 3	DAY 4	DAY 5
Class Work	Discuss pp. 468-474 CA p. 474 YD p. 475	YD pp. 476-477	CA p. 474	Discuss pp. 478-484 YT pp. 480-481	SQ p. 486
Laboratory					
Homework	Read pp. 468-477 YT pp. 475-476	SQ p. 477	Read pp.478-484	Read pp. 484-486	Read pp. 487-490

	DAY 6	DAY 7	DAY 8	DAY 9	DAY 10
Class Work	Review Parts A & B Quiz Parts A & B	Discuss pp. 491-493		Discuss pp. 495-500 YT pp. 496-497	Discuss pp. 500-503 YT pp. 503-504
Laboratory	LA pp. 487-490	LA pp. 487-490	LA pp. 493-495		
Homework	Read pp. 491-493	YT p.493 Read pp. 493-495	Read pp. 495-500	Read pp. 500-505	YD p. 504

	DAY 11	DAY 12	DAY 13	DAY 14	DAY 15
Class Work	SQ pp. 504-505	Review Part C Quiz Part C	Discuss pp. 509-511 YT pp. 511-512		Discuss pp. 515-517 YD pp. 517-518
Laboratory		LA pp. 506-509		LA pp. 512-515	
Homework	Read pp. 506-509	Read pp. 509-512	Read pp. 512-515	Read pp. 515-518	

	DAY 16	DAY 17	DAY 18	DAY 19	DAY 20
Class Work	SQ p.518	PIAT pp. 519-521	PIAT pp. 519-521	Review unit	Exam
Laboratory					
Homework	Read pp. 519-521				

CA = Class Activity; **LA** = Laboratory Activity; **CQ** = ChemQuandary; **YT** = Your Turn; **YD** = You Decide; **PIAT** = Putting It All Together.

This unit of *ChemCom* addresses a variety of issues that are woven through the entire program. To reflect on the interactions between communities and the chemical industry, it is appropriate that we return to the mythical community of Riverwood, introduced in the Water unit. Using Riverwood as a backdrop, students learn about the organization and products of the EKS Nitrogen Products Company. (EKS is pronounced like the letter X.) This company serves as a model of the wide range of chemical industries and products found in the real world.

With Company EKS as an example, students are challenged to assess both the positive and the negative effects of the chemical industry on their lives, and to see their own responsibilities in an industry-society partnership.

After a walk through Company EKS and the chemistry conducted by the company, students are asked to forecast various products or processes they would like to see developed by the chemical industry to meet societal needs.

Our hope is that this *ChemCom* unit will be the beginning of a life-long interest in the interactions between chemistry and the rest of the world.

OBJECTIVES

Upon completion of this unit the student will be able to:

1. List the functions of the chemical industry and the general categories of industrial products, including present contributions and future expectations. [A.1]
2. Contrast responsibilities of the public and of industry in preserving the quality of life in a community. [A.2, A.3]
3. Outline the types of products produced by the chemical industry, and explain the importance of intermediates in production. [A.4, B.1]
4. Evaluate the potentially positive and negative impacts of a chemical industry on a community. [A.5]
5. Compare natural and synthetic products, providing examples of each. [B.1]
6. Describe the role of chemical engineers in industry, and the factors that must be considered in changing from laboratory-scale reaction levels to industrial levels. [B.2]
7. Outline the major divisions and departments of a typical chemical industry, and explain their interrelationships. [B.3]
8. Analyze a fertilizer sample for its major components, and describe their importance (particularly nitrogen compounds) in agriculture. [C.1, C.2]
9. Use colorimetry to quantify phosphate content in fertilizer samples. [C.3]
10. Apply oxidation-reduction concepts to nitrogen fixation in the Haber process. [C.4]
11. Use electronegativity values to determine oxidation states. [C.4]
12. Describe factors that must be controlled in the equilibrium synthesis of ammonia. [C.5]
13. Trace the history and development of explosives, including the contributions of Alfred Nobel. [C.6]
14. Develop and evaluate voltaic cells, using the activity series of common metals. [D.1, D.2]
15. Use the concept of half-reactions to describe commercial electrochemical cells, including their charging and discharging reactions. [D.2, D.4]
16. Demonstrate the technique of electroplating. [D.3]
17. Describe the industrial applications of electrolysis for brine decomposition and for aluminum production. [D.4]
18. Identify key considerations involved in the development of a new chemical process or product. [E.1]

SOURCES ON THE CHEMICAL INDUSTRY

1. Aluminum Association, 818 Connecticut Ave., NW, Washington, DC 20006. 202-862-5100.
2. Chemical Manufacturers Association, 2501 M Street, NW, Washington, DC 20037 202-887-1243. Write for a list of free publications. *The Chemical Balance: Benefiting People and Minimizing Risks* (32 p. booklet) is most useful.
3. *Journal of Chemical Education*—"Real World of Industrial Chemistry Series"—edited by W. Conrad Fernelius and Harold W. Wittcoff—began in February 1979, volume 56.
4. *CHEMTECH*—An ACS monthly journal of industrial chemistry; contains many highly readable and informative articles.
5. *Chemical and Engineering News.* A weekly ACS publication that often covers stories on chemistry-society interface issues—each year publishes up-to-date employment and salary information—("199X Employment Outlook"—late October, and "Salary Survey"—late June, early July), facts and figures on chemical R&D, and the top 50 chemical products, and top 100 producers.
6. American Chemical Society Education Division: Careers publications: Write 1155 16th Street, NW, Washington, DC 20036 for an up-to-date packet of free brochures.

INTRODUCTION

ChemCom returns to Riverwood, where the townspeople are considering whether to allow a chemical company to build a plant in their community. The newspaper article on pages 468-469 leads into the topics to be introduced and debated in this unit.

Your students should understand that the Riverwood decision on the ammonia plant is not intended to represent the full range of concerns that other kinds of chemical facilities or industries might evoke. Common to any industry-community decision, however, are the information-gathering and decision-making skills involved in developing informed opinions.

A: A NEW INDUSTRY FOR RIVERWOOD ?

Students learn a bit of the chemical industry's history and receive some exposure to the wide variety of areas of human activity served by this sector of our economy. The section's listing of the assets and liabilities of the plant in Riverwood (Section A.5 on pages 476-477) is not intended to be exhaustive, nor does it necessarily represent the range of assets and liabilities that might apply to such a plant in your own community.

Students should be encouraged to be constructively critical, and to use their imaginations to think about how similar kinds of issues might affect their own community, whether it be New York or Pt. Barrow.

A.1 BASIC NEEDS MET BY CHEMISTRY (pages 470-473)

This section requires no student activity other than thoughtful reading. It summarizes some past successes and future hopes of the chemical industry, providing background needed for completing the activities in Sections A.3 and A.4.

OPTIONAL DISCUSSION TOPIC/ACTIVITY

Chemistry in the Community

We hope that throughout the *ChemCom* course, you have been exposing your students via film, articles, speakers, and class field trips to the wide variety of careers that require a substantial background in chemistry. To subtly stimulate an interest in chemistry-related work, we have also incorporated two career features in each *ChemCom* unit without comment. In any case, this unit provides an ideal place to focus on scientific careers, especially those in the chemical industry.

A.2 INDUSTRY AS A SOCIAL PARTNER (pages 473-474)

This section describes the social partnership between industry and citizens. On one side, the chemical industry is obligated to produce useful products or services, while damaging the environment and public health as little as possible. On the other side, citizens should understand the requirements of the industry, and the extent of society's need for its products and services. Though such a partnership on a national scale remains more of an ideal than a reality, it underscores the responsibility citizens have in a democracy to be informed, in order to promote proper policies and regulations.

A.3 CLASS ACTIVITY: PERSPECTIVES (page 474)

This activity provides a good opportunity for students to work simultaneously on increasing their understanding of chemistry, and on developing their oral and written presentation skills. You may wish to help build interest in your chemistry program by featuring these student demonstrations and poster papers at an open-house session for younger students, or perhaps even by presenting them at a monthly PTA meeting.

In any case, you should give students ample time to prepare. Be sure to stress the importance of tying the demonstrations to important roles played by the chemical industry. *Chemical and Engineering News* and the *Journal of Chemical Education* often provide useful facts and information on the chemical industry. You may decide to ask students to submit written proposals outlining the demonstration/laboratory activity they wish to do, so that you can check on safety and supplies.

The text gives a few possibilities of chemical processes to be demonstrated. Other demonstrations may also be suitable. Such demonstration books as: *Chemical Demonstrations: A Handbook for Teachers of Chemistry*, Vols. I, II, III, and IV by Bassam Z. Shakhashiri (University of Wisconsin Press), *Chemical Demonstrations: A Sourcebook for Teachers*, Volume 1, by Lee Summerlin and James L. Ealy, Jr. , Volume 2 by Summerlin, Christie Borgford, and Ealy, and *Chemical Activities* by Borgford and Summerlin, all available from the American Chemical Society, are excellent references.

OPTIONAL BACKGROUND INFORMATION

Indicators of the Impact of the Chemical Industry:

1. Food production: According to the National Academy of Sciences, U.S. food production would be cut by at least 30% without fertilizers and pesticides. Without fumigants, preservatives and chemically based packaging materials, another 25% would spoil on the way to the market.
2. Clothing and textiles: About 70% of the fibers sent to knitting mills are composed of human-generated chemicals. Without synthetic fibers, clothing needs could be met only by diverting vast amounts of cropland from growing foods to the production of natural fibers like cotton and wool.
3. Consumer goods: More than 40% of all goods and services that America uses today rely in some way on chemicals. The chemical industry produces more than 250 million tons of chemicals annually. Chemical industry sales totaled approximately $280 billion in 1990—making chemicals the nation's fifth largest industry. (Note: Food, transportation, petroleum and coal, and machinery rank ahead of the chemical industry.)

Protecting the Environment:

By 1985, the chemical industry had invested $14 billion in pollution control to improve air and water quality and to manage wastes. Since 1961, the number of persons employed by the chemical industry on environmental problems has more than tripled and now exceeds 10,000.

Hazardous Wastes: Of the 4.5 billion tons of wastes produced in the United States per year, about 58 million tons (or about 1%) are estimated to be hazardous wastes. These wastes are produced by 22 different industries of which the chemical industry is one. The basic philosophy concerning all chemical wastes is to: (1) minimize the generation of waste through process design; (2) reclaim and recycle as much as possible; (3) treat wastes to reduce volume and hazards before any release to the environment; and (4) dispose of hard-to-treat wastes with proper safeguards.

Worker Safety: The chemical industry employs 1.1 million Americans and is ranked number one in safety among all American industries by the National Safety Council (it has the lowest incidence of recordable occupational injuries and acute and chronic illnesses). The chemical plant worker is four times safer than the average American industrial employee. Research is being carried on to measure and control the potential long-term effects of low-level exposures to certain chemicals.

Transportation of Chemicals: Despite media coverage to the contrary, most chemicals are transported safely without incident. In 1980 8,900 unintentional releases were reported from containers of five gallons or more—from drums to tank cars. This represents one incident for every 10,000 shipments, or one hundredth of one percent (0.01%). Most of these caused little personal or property damage. The chemical industry is striving to improve its record via new container designs and operator safety training.

A.4 YOU DECIDE: PRODUCTS OF INDUSTRY (page 475)

This activity helps students discover the extent of the chemical industry and its impact on daily life. In the Resource unit (page 100) students worked with products to explore renewable and nonrenewable resources. The questions here are somewhat reminiscent of that exercise.

Answers to Questions:

1. a. Purified forms of natural resources: metals (from ores), treated lumber, gasoline (from crude oil), white sugar (raw cane sugar), table salt (from brines or mines), etc.
    b. Synthesized forms of matter not found in nature: most synthetic fibers, plastics, ceramics, pesticides, and many pharmaceuticals.
2. a. Natural materials are obtained either from organic (plant/animal) or inorganic (mineral) sources.
    b. Most synthetics are fossil fuel (especially petroleum) derivatives.
3. a., b. Many different answers are possible. Consider answers based on the following table:

Natural Product	Synthetic Substitute	Advantages of Synthetic	Disadvantages of Synthetic
Animal manures	Fertilizers	More available in quantity, more concentrated, can be sold in highly specified form	Tied to the price of petroleum, more energy-intensive, if over-used can lessen the efficiency of natural processes.
Botanicals	Synthetic drugs	Can be designed to precise specification, cheaper, more readily available, usually more carefully tested	
Natural dyes	Synthetic dyes and pigments	Greater variety of colors, usually more stable	Lack natural graduations and desirable fading
Natural fibers (cotton, silk, etc.)	Synthetic fibers	Often cheaper, can be designed with specific properties (such as permanent press)	Sometimes lack absorbency, drapability, or other advantages of natural fibers
Rubber	Synthetic elastomers	Cheaper, more readily available than rubber plants, can be designed to fit precise specifications	Manufactured from a nonrenewable resource.
Soap	Synthetic detergents (surfactants)	More variety, more efficient in hard water	Manufactured from a nonrenewable resource
Wood, paper, metals	Plastics	Usually less dense, dense, can be designed to precise specifications	Made from a nonrenewable resource, usually nonbiodegradable

4. Plastics, most pharmaceuticals, etc.
5. This is a matter of opinion, but it is clear that both standards of living and average life spans have improved dramatically since the advent of the modern chemical industry. Certainly, chemistry and the products of the chemical industry pervade our lives. Many important problems in today's world require more knowledge of chemistry and its applications than we have now. It is also true, as seen in this course, that some technological applications of chemistry have created new problems for society to address.

Chemical Processing in Your Life (pages 475-476)

Many students are unaware of how many different materials come from the chemical industry. They will discover it is difficult to think of products that have not been produced or altered by the chemical industry. It is quite likely that the items they suggest actually *have* involved the chemical industry in some fashion. This can be brought out by challenging students to trace questionable products back to their origins. Examples might include wooden cutting boards (how is the wood processed? are glues involved?), natural vitamins (how is the vitamin concentrated?), and natural foods (how are they processed and packaged?).

Answers to Questions:

1. a. Unless you grow your own food, spin your own fibers, or purchase items from individuals who do, most purchased items (even if 100% natural) are packaged either in paper, cardboard, or plastics.
   b. Packaging is especially important to help maintain product quality—freshness and sanitation— during shipping and storage.
2. In most cases, synthetic products cost less than their natural alternatives (examples: nylon vs. silk, synthetic vitamins vs. natural). Typically, synthetics are at least of comparable quality if not superior (example: synthetic rubber for natural). Sometimes a blend of the two provides the best combination of cost and quality (example: clothing made by blending cotton with synthetic fibers).
3. 100% natural means only that no synthetic additives are included in the product, not that no chemical processing occurred. In fact, there has almost always been some chemical processing. Examples include 100% natural cereals and vitamins, and all-cotton or all-wool clothing items. This question also provides an opportunity to reemphasize two points: (a) all matter is made up of chemical substances; and (b) assuming purity, a given synthetic compound is identical to the natural version of the same compound and often is less expensive to obtain.

A.5 YOU DECIDE: ASSET OR LIABILITY? (pages 476-477)

The main purpose of this activity is not to have students actually make a decision concerning the proposed plant; rather, a class discussion should reveal the kinds of issues that would affect such decisions. The text outlines a number of these. Encourage students to think of others. One way to do this is to have students play the roles of either advocates or opponents for the proposed plant.

   A half-period discussion should bring out the following points: (1) Chemical industries have both assets and liabilities. (2) It is difficult, if not impossible (or at least undesirable) to put a dollar value on all risks and benefits. (3) There are always uncertainties in risk/benefit calculations. (4) People are likely to disagree on the relative risks and benefits, based on their personal values and on how they are involved with the proposed action. (5) Both short- and long-term effects need to be considered.

PART A: SUMMARY QUESTIONS (page 477)

1. a. Ideally, the relationship is analogous to biological mutualism where both partners benefit.
   b., c. Society needs industry to provide its material, energy, and employment needs. Industry needs society to buy the products and services that enable it to make a profit and remain in business.
2. The chemical industry is involved in obtaining (via growing, harvesting, or mining nature's resources), processing, packaging, and shipping nearly all the products that we use.
3. a. Possible problems include: (a) identifying all the relevant costs (or risks) and benefits, particularly the longer-term effects of newer technologies; (b) quantifying (in dollars) the cost/risk factors—this is especially true in the case of environmental degradation and decrease in the quality or length of human lives, where value judgements and subjectivity are necessarily involved; and (c) considering the politics that arise, when the benefits and costs are shared unequally between different groups.
   b. A value judgment question. Many would argue that anything affecting human life has the highest weight, with environmental considerations of second greatest value. Others would say that preserving the biosphere itself must have priority over shorter-term human considerations.

B: AN OVERVIEW OF THE CHEMICAL INDUSTRY

Part B gives students perspectives on the breadth of the chemical industry, on how chemical companies are organized, and on the types of materials used in chemical-industrial processes. Students will learn that the chemical industry touches nearly every product of our modern industrial society.

B.1 FROM RAW MATERIALS TO PRODUCTS (pages 479-483)

This section provides an overview of some processes for extracting ores and changing them into useful materials. Chemical intermediates are defined, with special emphasis on sulfuric acid, the perennial number one chemical produced by the chemical industry in the United States. Students should see the connection between producing chemical substances and meeting societal needs.

YOUR TURN

Metals from Ores (pages 480-481)

There are a couple of reasons for this exercise: (1) to review some mathematical reasoning and manipulation, and (2) to point out that during most mining operations, a tremendous quantity of unwanted matter must be removed from the wanted substance and disposed of in an environmentally sound manner. Road material is one such possibility for disposal, though since most tailings are in powder form, this idea may not work in practice.

Answers to Questions:

1. 1 kg iron $\times$ (100 kg ore/25 kg iron) = 4 kg ore.
   or 100 kg ore has 25 kg iron. 1 kg is 1/25th of that, so 100/25 kg ore = 4 kg ore.
2. 100 kg ore – 25 kg iron = 75 kg rock.
3. 10^9 kg $\times$ (1 m³/3.0 $\times$ 10³ kg) $\times$ (1/40 m) $\times$ (1/0.25 m) = 0.033 $\times$ 10⁶ m = 3.3 $\times$ 10⁴ m = 33 km
   or 10^9 kg/(3.0 $\times$ 10³ kg/m³) = 0.33 $\times$ 10⁶ m³ possible. Dividing this by the length (40 m) and the depth (0.25 m) = 33 km

CHEMQUANDARY

What Price Survival? (page 483)

This is somewhat a matter of opinion, but most chemists would probably respond that it's possible to have both material comfort and convenience and a safe environment. The law of conservation of matter tells us there will always be by-products, and thermodynamics tells us that overall disorder will tend to increase. However, by judicious use of matter and energy resources, we can maintain a high state of organization in our environment. Pollution is more a matter of misplaced resources than overconsumption. On the other hand, given limited energy resources, it may be necessary to curtail some practices that are presently considered "convenient." Possible examples may include restricting private (single-passenger) transportation, the use of throwaways, excessive packaging, and the use of some nonbiodegradable plastics.

B.2 FROM TEST TUBES TO TANK CARS (pages 483-484)

Following up on the earlier information on chemical intermediates, this section focuses on how such intermediates are produced. It discusses scientific and practical considerations that distinguish industrial-scale reactions from those conducted on a laboratory/test-tube scale. Point out the similarities and differences between the two scales.

In many ways the problems of industries are similar to those of large metropolitan areas. The analogy is as follows: in a single person there are small needs for raw materials (food and water) and small amounts of waste products, which require a simple disposal system. In a large city there are heavy demands for raw materials and vast quantities of biological waste products. Systems must be constructed to provide the food and water necessary, and safely and sanitarily dispose of the waste. Similarly, a laboratory scale reaction needs small quantities of chemicals, and produces a small amount of waste products (you might inform students of how you dispose of chemicals left over from laboratory activities). Industries need much larger amounts of chemicals and produce larger amounts of by-products, which require safe and sanitary disposal. Since students have seen the many benefits the chemical industry has produced, perhaps they will agree on the need for some manageable risks when it comes to by-products.

Other examples include a family with one cow—the manure makes good fertilizer for the garden—compared to a large dairy—there aren't enough gardens to handle all of the manure— or a farm raising 10 chickens for food and eggs as opposed to large poultry plants, with the subsequent problems of disposing of metric tons of waste products, feathers, etc.

B.3 CLOSE-UP: THE EKS COMPANY (pages 484-486)

This section outlines the organization of the EKS Company and summarizes the roles played by the major divisions and departments within a chemical company. Various groups (public relations, analytical, environmental, products) represent points of contact and interaction between the industry and society.

PART B: SUMMARY QUESTIONS (page 486)

1. Even excluding the petroleum and food industries, the remaining chemical industries earn roughly one of every ten manufacturing dollars, and employ more than one million U.S. citizens. Chemical industries are either directly or indirectly involved in nearly every product we purchase and kilojoule of energy we use.

2. a., b. (1) Sulfuric acid (H_2SO_4). Used for producing fertilizer from phosphate rock, petroleum refining, manufacturing metals, and a wide variety of substances; the typical citizen knows it as "battery acid."
   (2) Nitrogen (N_2). Used in manufacturing ammonia, as an inert atmosphere in incandescent bulbs, as a freezing agent for foods.
   (3) Oxygen (O_2). Used in oxyhydrogen or oxyacetylene welding torches, rocket propellant, submarine work by divers, oxygen tents in medicine, production of synthetic gasoline, etc.

3. a. (1) Elements. Used directly, such as iron and molybdenum for alloys, aluminum for airplane frames, copper for wiring, and oxygen in medicine.
   (2) Intermediates. Substances produced as precursors to other desired chemicals. Some examples: Ammonia is an intermediate between nitrogen and nitric acid. Chlorine is an intermediate between salt and a wide variety of plastics (including polyvinyl chloride). Sulfuric acid is used to manufacture plastics, textiles, pulp and paper, and explosives.
   (3) Petroleum-based products. These include plastics, polymers, and solvents.

4. a. Engineering—including how the size of the reaction vessel influences the control of the heat produced or consumed by a given reaction; whether a continuous flow or batch method will be used; which temperature, pressure, concentration, and catalyst conditions will be optimal.

   b. Profitability—the bottom line is whether or not the reaction can produce a product that can be sold at a price that exceeds total production costs. Small-scale reactions at the research and development level rarely have this constraint. With larger scale reactions, small adjustments in reaction conditions can mean the difference between a reaction becoming commercially feasible or not.

   c. Waste—the issue of what to do with unwanted by-products of a given reaction becomes a major issue when thousands or even billions of kilograms of product are produced. The "solutions" of dumping wastes down the drain or burying them out of sight are outrageously irresponsible. The ideal would be to find a chemical producer or consumer with a use for the waste.

5. a. Analytical department—serves as the in-house quality control unit to check on the purity of the chemicals the company buys and sells. It also may play the role of trouble shooter in helping to locate problems in the production phase.

   b. Environmental department—monitors the company's wastes and their effect on the local environment. It also oversees the installation, operation, and maintenance of pollution control technologies.

   c. Public relations department—acts as the communication link between the company and the press, general public, and government.

   d. Corporate management—oversees the work of all the other divisions and handles personnel, company policy, and finances.

   e. Research department—invents new products and processes and tries to find ways to improve existing ones.

6. Unwanted materials in one plant can become intermediates in other manufacturing efforts. For example, oxygen is a "waste" product in the production of nitrogen from air, and yet it has a wide variety of uses, rocket fuel, for welding, in medicine, etc. Sulfur dioxide, a by-product of the production of metals from ores, can be used to make sulfuric acid, number one in the "hit parade" of chemicals in the United States.

C: THE CHEMISTRY OF SOME NITROGEN-BASED PRODUCTS

Part C examines the chemistry and technological difficulties of fixing nitrogen, shows how the concept of limiting reactants applies to the use of fertilizer by plants, and describes nitrogen's role in the explosives industry.

Once again, students consider the question of who is responsible for how scientific inventions are used—this time in the context of earlier use of explosives in warfare. Alfred Nobel, the inventor of dynamite, felt anguish when his invention became a tool for war and was motivated to establish the Nobel prizes. Students should learn that as scientists or citizens, they share the responsibility for determining how new inventions are used.

C.1 LABORATORY ACTIVITY: FERTILIZER (page 487-490)

Students are asked to act as analytical chemists in performing qualitative tests to determine major ingredients in a fertilizer sample. Later, in Section C.3, they will conduct quantitative analyses.

Time

One class period.

Materials (for a class of 24 working in pairs)

Test Solutions: (see note below)

100 mL 6.0 M hydrochloric acid (50 mL conc. HCl/100 mL solution)
100 mL 3.0 M sodium hydroxide (fw = 40.0 NaOH) (12 g/100 mL solution)
100 mL concentrated sulfuric acid in dropper bottles
100 mL 0.1 M iron(II) sulfate ($FeSO_4$) (fw = 278.03 $FeSO_4 \cdot 7H_2O$) (2.78 g/100 mL solution)
100 mL 0.1 M barium chloride ($BaCl_2$) (fw = 244.3 $BaCl_2 \cdot 2H_2O$) (2.44 g/200 mL solution)

Known Solutions: 100 mL of 0.1 M solutions (see note below)

PO_4^{3-}: sodium phosphate (Na_3PO_4) (fw = 380.1 $Na_3PO_4 \cdot 12H_2O$) (3.8 g/100 mL solution)
NO_3^-: sodium nitrate ($NaNO_3$) (fw = 84.99 $NaNO_3$) (8.5 g/100 mL solution)
SO_4^{2-}: sodium sulfate (Na_2SO_4) (fw = 322.19 $Na_2SO_4 \cdot 10H_2O$) (3.22 g/100 mL solution)
K^+: potassium nitrate (KNO_3) (fw = 101.11 KNO_3) (1.01 g/100 mL solution)
NH_4^+: ammonium nitrate (NH_4NO_3) (fw = 80.04 NH_4NO_3) (0.80 g/100 mL solution)
Fe^{3+}: iron (III) nitrate, $Fe(NO_3)_3$ (fw = 404.00 $Fe(NO_3)_3 \cdot 9H_2O$) (4.04 g/100 mL solution)

Unknown "Fertilizer" Solutions:

Solutions that contain a minimum of one anion and one cation (be sure the sample does not contain both phosphate and sulfate) should be used. Concentrations of about 0.1 M are adequate to identify the ions. Examples include: (1) a solution that is 0.1 M NH_4NO_3 and 0.1 M Na_3PO_4 (testing positive for NH_4^+ and NO_3^- and PO_4^{3-}); (2) a solution of 0.1 M $Fe_2(SO_4)_3$ (testing positive for Fe^{3+} and SO_4^{2-}); etc.

Other:

100 mL 12 M (concentrated) HCl for cleaning nichrome wires
24 red litmus paper strips (2 strips/team)
12 10-cm nichrome or platinum (Pt) wire loops attached to cork or balsawood handle (300 cm of wire)
12 multiple-well spot plates
12 Bunsen burners
12 18 × 150 mm test tubes (for cleaning nichrome wires)
12 micropipets
12 squares of cobalt-blue or didinium

Pre-Lab Discussion

Remind students of the earlier work (Food unit, page 239) on the connection between limiting reactants, fertilizers, and plant growth.

Point out that sometimes more than one ion can give the same result on a given test and that, as a result, follow-up confirming tests are needed—NaOH/NaOH + litmus, $BaCl_2$/$BaCl_2$ + HCl, and flame test/flame test + cobalt glass (in the Post-Lab discussion you can point this out with specific examples).

Stress the importance of keeping a careful record of which solution is in each depression on the spot plates.

Lab Tips

The unknown should contain one anion and one cation (or, if you choose to include three ions, do not include both SO_4^{2-} and PO_4^{3-} due to some experimental complications). If more than one unknown is used for a given class, be sure to distinguish each with a code and inform the students to report the code with their results.

Be sure spot plates are rinsed with distilled water between repeated uses of the same depression.

Be sure that all reagents are made up with distilled water to avoid contamination with ions present in tap water.

Caution students about the toxicity of barium chloride.

The Bunsen burner flame should be adjusted to a hot, blue flame. Remind students that a hot needle has its own color and, unless the flame test is dramatic and obvious, to record the test as inconclusive.

Remind students that red litmus paper turns blue (or slightly violet) in a base.

Actual samples of commercial fertilizers are not recommended for the unknowns since some contain insoluble matter, time-released chemicals, or ions which complicate and confound the simple identifying chemical tests suggested in this procedure. You may however wish to **demonstrate** the presence of several ions in a actual sample of 20-20-20 commercial fertilizer.

Expected Results

Data Table—Ion Tests

Ion	PO_4^{3-}	NO_3^-	SO_4^{2-}	K^+	NH_4^+	Fe^{3+}
Color of solution	NC	NC	NC	NC	NC	Brown
NaOH	NR	NR	NR	NR	NR	Precipitate
$BaCl_2$	Precipitate	NR	Precipitate	NR	NR	NR
$BaCl_2$ + HCl	Precipitate dissolved	NR	Precipitate stayed	NR	NR	NR
NaOH + litmus	X	X	X	NR	Blue	NR
Fe^{2+}/H_2SO_4	X	Brown ring	X	X	X	X
Flame test	X	X	X	Violet	NC	Sparks
Flame test thru Co glass	X	X	X	Pink	NC	Whitish

Note: Cells marked X are tests students skip NC = No color, NR = No reaction

Post-Lab Discussion

Center your Post-Lab discussion on the logic of the data chart (the need for multiple tests on some ions), the questions provided, and any difficulties encountered in the activity. Stress the importance of the work of analytical chemists. In the particular case of fertilizers, farmers need to select a fertilizer with the proper ingredients for the needs of their soil and their plants, if they wish to avoid wasting money or even harming the environment.

Answers to Questions:

1. Potassium nitrate (KNO_3) and ammonium phosphate ($(NH_4)_3PO_4$ (or ammonium nitrate and potassium phosphate).

2. A white precipitate in the barium chloride test indicates either phosphate or sulfate. If the precipitate dissolves in 6 M hydrochloric acid, it confirms the presence of the phosphate ion.

3. The ammonium ion.

4. No, because a candle flame is not hot enough and, being highly colored itself, the flame could mask other colors. The pale blue flame of the Bunsen burner is hot enough to excite valence electrons, resulting in the emission of color, and does not pose any serious masking problem.

5. No, it would be necessary but not sufficient information. In addition to these qualitative data, quantitative (how much) analyses would be necessary to avoid needless expense or even environmental damage (by overfertilizing).

C.2 FERTILIZER'S CHEMICAL ROLES (pages 491-493)

This section considers the roles of the primary plant nutrients—nitrogen, phosphorus, and potassium. In addition to focusing on the concept of limiting reactants, remind students that fertilizers are a good example of the chemical industry's role in producing products that society needs and will buy. Chemical fertilizers, along with a climate and terrain that are conducive to farming, enable the United States to be a major food exporter. The future availability of fertilizers, (or genetic alternatives) in developing nations may constitute the difference between survival and starvation for millions of people (remind students of the discussion on world hunger in the Food unit).

Avoid getting bogged down in details of the nitrogen cycle (Figure 3 on page 492); the main point is that a natural recycling system supplies nitrogen in a usable form to plants. But, since naturally available quantities of fixed nitrogen limit crop yields, chemical fertilizers help supplement the natural system.

An interesting sidelight of the fertilizer story are some problems that can arise, such as the controversy surrounding the use of massive quantities of fertilizer on croplands. Farms are biochemical systems, not simple chemical systems. A hectare of soil contains 25 tons of living creatures which help maintain the fertility and structure of soil. Synthetic fertilizers have so successfully improved crop yields, that most U.S. farmers have abandoned old methods of spreading manure and dead plant matter on fields, and rotating crops with legumes. Yet these methods may be necessary to replace organic matter into soil; in addition, very highly concentrated doses of synthetic fertilizer may be harmful to soil creatures. The key is to learn how to use our best chemical technologies to work with nature.

YOUR TURN

Plant Nutrients (page 493)

These questions highlight the importance of using our best chemical technologies to work with nature.

Answers to Questions:

1. Legumes contain nitrogen-fixing bacteria in their roots. If such crops are rotated between harvests of plants which extract nitrogen compounds from the soil, the soil's store of nitrogen can be replenished.

2. Nonharvested parts of plants contain nitrogen, phosphorus, potassium and other nutrients which were originally extracted from the soil. By plowing the plants back into the soil these nutrients become available for the next season's crops; less fertilizer is needed than would otherwise be the case.

3. If scientists could breed plants that were capable of fixing their own nitrogen, energy costs would be reduced in that (a) little, if any, energy intensive fertilizer would be needed, and (b) less fuel would have to be burned to propel tractors.

4. a. Nitrate ions are essential precursors of amino acids and proteins. Thus, if there is an insufficient supply, plant growth will be stunted and general health reduced.

   b. Phosphorus from phosphate ions is incorporated into ADP, ATP, RNA, and DNA. A deficiency of phosphorus would result in a retardation of cellular growth and reproduction.

   c. Potassium ions activate some enzymes, help regulate the body's water balance, and are essential for the conversion of carbohydrates from one form to another. A deficiency would show up as a generalized reduction in cellular health.

C.3 LABORATORY ACTIVITY: PHOSPHATES (pages 493-495)

This activity serves as a follow-up to the qualitative *Laboratory Activity* C.1 and gives students another example (recall the *Iron in Foods Laboratory Activity* in the Food unit) of how colorimetric techniques are used by analytical chemists.

Time

One class period

Materials (for a class of 24 working in pairs)

20 g ascorbic acid
20 g commercial fertilizer (20-20-20) or as an alternative: a 1:1 molar mixture of NH_4NO_3 and Na_3PO_4
1.0 L 10.0 ppm standard phosphate buffer solution (65 mL/team)
1 g potassium phosphate, dibasic (K_2HPO_4) for preparing standard phosphate buffer solution.
   (fw = 174.18 as anhydrous K_2HPO_4)
120 mL ammonium molybdate-sulfuric acid reagent (10 mL/team)
7.5 g ammonium molybdate, $(NH_4)_2MoO_4$, crystals for preparing reagent
150 mL concentrated sulfuric acid for preparing reagent
10 L distilled water
60 test tubes (18 × 150 mm) (5 tubes/team)
12 test tube holders
12 test tube racks
12 ring stands
12 rings
12 wire screens

36 400-mL beakers
12 10-mL graduated cylinders
12 50-mL or 100 mL graduated cylinders
12 stirring rods
12 Bunsen burners
Balance

Advance Preparation

The standard phosphate solution is prepared by dissolving 1.0 g K_2HPO_4 in a small volume of water and then diluting to 1000 mL in a volumetric flask. Mix, then dilute 20 mL of this solution to 1000 mL. This produces one liter of standard (10.0 ppm) phosphate solution.

The ammonium molybdate-sulfuric acid reagent is prepared as follows. Dissolve 7.5 g of ammonium molybdate in 75 mL of distilled water. Cool the solution in an ice bath. Retain this solution for a later step.

Slowly with mixing, add 125 mL of ice-cold 18 M H_2SO_4 to 125 mL of ice-cold distilled water. *Caution: Diluting concentrated sulfuric acid generates considerable heat, and may produce splattering if not added slowly with continual stirring. Perform this dilution with considerable care.* Now slowly add the cold sulfuric acid solution (with stirring) to the ice cold ammonium molybdate solution. Keep the ammonium molybdate solution in an ice bath during this step. This makes up the 325 mL reagent solution.

Pre-Lab Discussion

Point out the connection of this activity with the earlier qualitative work with fertilizer.

Lab Tips

The actual value of phosphate as phosphorus(V) oxide (P_2O_5), using 20-20-20 fertilizer, is 20%.

The intensity of the color is affected by the length of time the solution is allowed to boil; some consistent method of heating must be agreed upon to ensure comparable results by all teams. The point at which the first "bump" of boiling occurs is an acceptable reference point.

With a spectrophotometer, data can be collected and plotted to obtain an absorption curve for the phosphate standards. This curve can then be used to compare sample values with the standards. A wavelength of 610 nm gives optimum absorption for this experiment.

Expected Results

The results will range between 17-25% phosphate [phosphorus(V) oxide].

Post-Lab Discussion

Center the discussion around the questions which emphasize the importance of colorimetric techniques, their connection to everyday observations of the color density of solutions, and the practical significance of quantitative data to a farmer.

Answers to Questions:

1.   Coffee, tea, orange juice, and powdered drinks.
2.   More easily calibrated and standardized. Able to provide more repeatable quantitative data.
3.   Since colorimetric techniques depend on comparison of an unknown to a known series of concentrations, if the standards are not accurately made, the data will be erroneous.
4.   Gravimetric techniques would involve finding the mass of precipitate. From the molar mass of the precipitate, one could calculate the mass (and original concentration) of an ion of particular interest. This technique assumes nearly complete precipitation of the ion of interest. One can also measure the turbidity produced when precipitates form.
5. a. Farmers usually analyze their soils to assess their needs for fertilizers. Matching fertilizer quantities to the needs of their soils requires knowing the amounts of each component.
   b. Over-fertilization results in a financial loss and may cause environmental damage (runoff into waterways and aquifers, killing useful soil organisms, etc.). Under-fertilization will not produce desired yields.

C.4 FIXING NITROGEN (pages 495-498)

This section develops the concept of oxidation states and redox reactions as a prelude to consideration of ammonia synthesis (Section C.5) and explosives (Section C.6). The original definitions of oxidation and reduction, first developed in the Resource unit (pages 136-137) are extended to include covalent compounds. The key to classifying these reactions is electronegativity, a measure of the ability of an element to attract shared electrons in a compound.

Note that *ChemCom* takes a fresh approach to this topic, tying the notion of positive and negative oxidation states directly to the concept of electronegativity. Students decide whether an element should be assigned a positive or negative oxidation state in a given compound by comparing electronegativities. This should enable them to follow the later development of this topic.

YOUR TURN

Electronegativity and Oxidation State (pages 496-497)

This activity connects the concept of electronegativity with the idea of oxidation numbers. The standard electronegativity chart on page 497 simplifies these problems. Encourage your students to look for patterns in the values and the arrangement of elements in the periodic table. For example, all of the metals have relatively low electronegativities.

Answers to Questions:

1.

Compound	Positive oxidation no.	Negative oxidation no.
a. SO_3	S	O
b. N_2H_4	N	H
c. H_2O	H	O
d. HCl	H	Cl
e. NaCl	Na	Cl
f. CO	C	O
g. IF_3	I	F
h. MnO_2	Mn	O

2. a.

Compound	Positive oxidation no.	Negative oxidation no.
(1) PbF_2	Pb	F
(2) NaI	Na	I
(3) K_2O	K	O
(4) NiO	Ni	O
(5) $FeCl_3$	Fe	Cl
(6) PbS	Pb	S

   b. In most binary compounds, metals have positive oxidation states and nonmetals negative oxidation states.

3.   Nickel is oxidized (since it has a lower electronegativity than sulfur and will therefore lose some control of its bonding electrons to sulfur); sulfur is reduced.

C.5 NITROGEN FIXATION AT RIVERWOOD (pages 498-500)

This section treats the synthesis of ammonia as an example of a reversible reaction harnessed for commercial production of a useful product. Most chemistry students find the information about reversible reactions new, but readily understandable. Point out that most chemical reactions are, in fact, equilibria. Le Chatelier's principle, the classic explanation of reversible reactions, is not covered here. As an option, you may wish to discuss this concept briefly.

OPTIONAL BACKGROUND INFORMATION

Nitrogen is the largest component of air, but very unreactive. Because the triple bond between the two atoms is extremely hard to break, considerable effort and energy are required for forming nitrogen compounds. Such a chemical change involving molecular nitrogen is called "fixing" nitrogen, or nitrogen fixation.

The German chemist Fritz Haber first developed a procedure for making ammonia from nitrogen and hydrogen. It required extremely high temperatures and pressures, but he succeeded by using an iron-containing catalyst. (Le Chatelier also attempted to develop a procedure for making ammonia, but was stopped by a violent explosion.)

Carl Bosch was responsible for the industrial production technique, which became extremely important for Germany's war effort during World War I. At that time, most of the nitrogen compounds needed for explosives came from potassium nitrate deposits in Chile. At the outbreak of the war this supply to Germany was cut off by the British Navy and its blockade, so the chemical industry was coverted to producing nitrogen compounds such as ammonia. Both Haber and Bosch were awarded Nobel Prizes.

The industrial production of ammonia is an excellent example of one of the goals of industry: to produce the maximum amount of product with the minimum of expense. This includes choices of starting materials, planning reaction conditions, and improving yield of product. In order to achieve these goals, the chemist determines conditions which will produce the highest yield: such conditions are called the optimum conditions.

In ammonia's case there are five optimum conditions: (1) A high concentration of starting materials, in this case nitrogen and hydrogen, must be maintained. This is partially accomplished by having high pressure. (2) The ammonia should be removed from the equilibrium reaction as it is formed. This is accomplished by cooling the products somewhat, which causes the ammonia to condense into a liquid and leave the gas phase, where reactions are occurring. The nitrogen and hydrogen do not condense under these conditions. (3) The temperature must be high enough to maintain a reasonable rate of reaction between hydrogen and nitrogen, but low enough to inhibit the decomposition of the product ammonia. (4) A catalyst is used to speed up the formation of the product ammonia, which lowers the energy needs for the reaction. (5) High pressure encourages the formation of product by increasing the collisions between nitrogen and hydrogen.

C.6 NITROGEN'S OTHER FACE (pages 500-503)

This section discusses the uses of nitrogen compounds as explosives, emphasizing the work of Alfred Nobel and the Nobel prizes as well as the chemistry of explosives. From a chemical standpoint, emphasize the following points in your class discussion: (1) the difference between molar volume of gases (22.4 L/mol at STP) and that of solids and liquids, and (2) the fact that many nitrogen-containing substances (see Figure 5 on page 503) are explosives. Note that in these explosive reactions nitrogen reverts back to its more stable zero oxidation state (in N_2).

From a social viewpoint, the following questions may be worth exploring in this and the following section: (1) Do explosives have "constructive" uses? (2) In general, how does technology relate to the Hindu saying: "Man is given the key to heaven, the same key opens the gates of hell"? (3) What role have the Nobel prizes (especially the peace prizes) had in promoting productive uses of technology?

YOUR TURN

Chemistry of Explosives (pages 503-504)

This activity is designed to reinforce these ideas: (1) Explosives are substances that either react with oxygen or self-oxidize to release large volumes of gas. (2) Compounds that contain carbon, hydrogen, oxygen, and nitrogen are often excellent explosives, since the products of the reaction—carbon monoxide, carbon dioxide, steam, and nitrogen oxides—are all highly stable, gaseous compounds. (3) Explosions are highly exothermic, rapid reactions (the products are at a lower energy state than the reactants). The rapid production of heat causes a corresponding rapid expansion of the product gases—hence the sound and force of the explosion.

Answers to Questions:

1. As written, the reaction produces 44 mol of gaseous products for each 4 moles of TNT; therefore, one mole of TNT would produce 11 moles of gases.
2. a. 11,000 "units" of gas.
   b. Since 21 moles of oxygen are used in the explosion of 4 moles of TNT, the net change in gas volume is $44 - 21 = 23$ moles of gas increase. The net volume increase would be 23,000.
3. a. $8 \times 23,000 = 184,000$-fold increase
   b. In addition to the energy produced in the reaction, this tremendous increase in volume exerts enormous pressure in the vicinity of the reaction, pushing apart any objects close to the explosion.
4. a., c. Might be explosive, since both are exothermic reactions that produce stable gaseous products from a solid or liquid substance.
   b. An endothermic reaction and does not represent a suitable explosive reaction, even though a gas is produced.

C.7 YOU DECIDE: FOOD OR ARMS (page 504)

Nitrogen provides an excellent example of a case where a particular chemical technology can be used for either constructive or destructive purposes. In this case, artificial nitrogen fixation can be used to produce fertilizers or explosives. The dilemma over whether the technology should be banned is even more complicated than the questions indicate, since explosives often find constructive non-military uses (see the photographs on pages 500 and 501, for example).

The message is that technology is not good or evil; its morality depends on how and for what purposes it is used. Banning a destructive technology that also has real benefits does not necessarily make mankind more moral; it may even result in very high societal "costs."

Answers to Questions:

1. This is partly a matter of opinion. Here are some issues that may arise:
    a. Political: If only certain countries ban ammonia, will the security of these countries be threatened by countries that continue its production?
    b. Economic: Will countries that continue production have a stronger economy due to higher crop yields and sales of fertilizers or explosives?
    c. Humanitarian: Will fewer people be fed at the lower crop yields?
    d. Practical: Will the ban (even if followed by all major powers) prevent war?

2. Over time, millions of persons would undoubtedly have starved to death, since world agricultural production would have been outstripped by population growth in nearly all countries. The economic strength of nations (such as the United States) depending on agricultural exports to maintain a favorable trade balance would be considerably lessened. And, most likely, war would still be with us.

3. Nuclear power (like ammonia technology) can be used either for constructive purposes (electric power generation; agricultural, medical and industrial applications) or destructive purposes (nuclear weapons). If the technology had been banned (even assuming a universal ban), we would not be able to benefit from its constructive uses. Whether the world would be safer or not (from the threat of nuclear war or the possible risks associated with electric power generation) is an issue to be argued.

PART C: SUMMARY QUESTIONS (pages 504-505)

1. Fertilizers typically contain a mixture of nitrogen, phosphorus, and potassium compounds, since these three elements are most commonly the limiting reactants for the growth of plants in agricultural systems. By adding these compounds to the soil, additional growth can be achieved.

2. Nitrogen in the atmosphere exists as $N_2(g)$. The stability of the triple bond in this molecule makes this compound especially unreactive, and therefore unavailable to plants. Lightning and/or some bacterial species are capable of fixing nitrogen in a form (such as nitrates or nitrites) that can be used by plants.

3. a. Oxidation state refers to hypothetical charges assigned to atoms in chemical species by assuming that all bonding electrons "belong" to the more electronegative atom in each bond. The idea is not that charges are actually formed, but that the more electronegative atom has more control over the shared electrons in each covalent bond.
    b. In redox reactions, one species gains partial control of one or more electrons (is reduced) and the other loses partial control over one or more electrons (is oxidized).
    c. In the equation given, each nitrogen atom is oxidized, since it gives up control of electrons, and each oxygen atom is reduced, as it gains control of electrons from nitrogen.

4. a. Nitrogen comes from the atmosphere; hydrogen comes from natural gas.
    b. Forward reaction is exothermic.
    c. The forward reaction is favored at low temperatures that hinder the endothermic reaction in an equilibrium.
    d. Both endo- and exothermic reactions are too slow at low temperatures, and little product is formed.
    e. A catalyst speeds up the formation of both products and reactants equally. The benefit of a catalyst is that lower temperatures are required for the reaction, and that the net time for product formation is reduced.

EXTENDING YOUR KNOWLEDGE (page 505)

*   Chemical substances found include phosphorus(V) oxide (P_2O_5), potash (K_2O), and ammonium nitrate (NH_4NO_3). Fertilizer containers typically display a series of three numbers, such as 6-12-12, which refer, respectively, to the percent (by mass) of nitrogen, available phosphorus and potassium. Grasses and other leafy materials typically have higher amounts of nitrogen. Trees and shrubs, including roses, have higher amounts of phosphorus and potassium. Sometimes 20-0-0 fertilizer is sold. This contains ammonium nitrate, which is recommended for lawns.

*   This is a discussion question, inviting student opinion and research. Direct students to examine the people involved in the Manhattan Project, which developed the first nuclear weapons; Linus Pauling, who won a Nobel prize for Peace; the current debate over genetic engineering; and Alfred Nobel, whose distress over the use of dynamite for war has already been mentioned in the text.

D: CHEMICAL ENERGY ↔ ELECTRICAL ENERGY

Part D introduces students to electrochemical cells, and helps them understand how batteries work and what the process of electroplating is. After completing Part D, students should sense the importance of electrochemistry to the chemical industry and to their lives.

D.1 LABORATORY ACTIVITY: VOLTAIC CELLS (pages 506-509)

In this laboratory activity, students construct and test various electrochemical cells. The activity applies oxidation-reduction reactions to the concrete, practical production of electrical energy.

Time

One class period

Materials (for a class of 24 working in pairs)

3.0 L 0.1 M copper(II) nitrate [$Cu(NO_3)_2$] (75 mL per pair)
 (fw = 241.60 $Cu(NO_3)_2 \cdot 3H_2O$) (72.48 g/3.0 L)
3.0 L 0.1 M zinc nitrate [$Zn(NO_3)_2$] (75 mL per pair)
 (fw = 297.49 $Zn(NO_3)_2 \cdot 6H_2O$) (89.25 g/3.0 L)
3.0 L 0.1 M magnesium nitrate [$Mg(NO_3)_2$] (75 mL per pair)
 (fw = 256.43 $Mg(NO_3)_2 \cdot 6H_2O$) (76.93 g/3.0 L)
3.0 L 0.1 M iron(III) nitrate [$Fe(NO_3)_3$] (75 mL per pair)
 (fw = 404.00 $Fe(NO_3)_3 \cdot 9H_2O$) (121.2 g/3.0 L)
12 zinc metal strips (8 cm × 15 cm × 0.25 mm)
12 copper foil strips (8 cm × 10 cm × 0.1 mm)
Distilled water
24 alligator clips with fine wire leads (2/team)
12 15-cm magnesium ribbon strips
12 iron metal strips (8 cm × 10 cm)
12 porous 75-mL cups (9 cm × 4 cm) (see note in Lab Tips)
12 400-mL beakers
12 scissors
12 centimeter scales

Optional:

12 low-current motors (1.5-3.0 V DC, 8300 rpm)
Thin cardboard or thick paper for making propeller paper towels

Advance Preparation

Prepare the solutions listed above.

Pre-Lab Discussion

You may wish to point out that the electronegativity series (Table 2 on page 497) and the activity series (Table 5 on page 507) of metals measure different properties of an element. However, in general, the lower the electronegativity of a given metal, the higher its place in the activity series (for example, sodium has an electronegativity of 0.9 versus a value of 2.4 for gold). You may also want to use the commonly invoked water-flow analogy distinguishing electrical potential in volts (analogous to water pressure in a pipe) from current in amperes (analogous to water flow rate in the pipe).

Lab Tips

The text suggests small, unglazed clay flower pots; obviously these are the cheapest way to go. Porous cups can be ordered from Sargent-Welch, catalog number 2202. Dialysis tubing (see biology teacher) may be substituted for the porous cups. The need for contact between the two solutions can be demonstrated by substituting a beaker for the porous cup: no voltage will be produced in this case.

 Iron nails may be substituted for the iron strips.

 The zinc and copper strips cannot be reused since they are cut into smaller strips and—in the case of zinc—become rather corroded. To minimize the amount of copper and zinc used, you can have the students reduce the size of plates by partially removing the plates from the solutions, and monitoring the change in current produced.

Remind students to avoid getting the electrolytic products from the plates on the skin.

Much of the equipment needed in this activity may be borrowed from the physics laboratory.

Optional: Low-current motors can be purchased at Radio Shack stores under the name Hobby Motors—catalog number 273-223. A small propeller can be fashioned from thin cardboard. The speed rate depends primarily on the current/amperes, but if all other factors are constant, cells that generate greater electric potential (volts) also produce more current. Therefore, the Mg-Cu cell will generate the highest speeds, and the Fe-Cu the slowest. Also, cutting down the size of the plates will decrease the current and the spin rate.

After students have tried different metal combinations and observed the effect on the propeller, you may want them to check the various cells with a voltmeter. Voltage can be considered the push (pressure) that drives the cell's electric current. Think of current (measured in amperes) as the number of electrons per second being released or accepted by the cell's electrodes. If the surface area is decreased, the number of electrons/second being released or accepted will also decrease. In this activity, students reduce the electrode surface area by cutting each metal strip in half until the current drops so low that it no longer turns the propeller. (You may elect to measure this current drop more quantitatively with an ammeter.)

Expected Results

From highest to lowest voltage (and propeller speeds), the cells will rank in the order Mg/Cu, Zn/Cu, and Fe/Cu. Cutting down on the size of the plates does not alter the voltage produced—that depends only on the specific combination of metals used. Students should be able to relate their findings to Figure 7 on page 509.)

Optional: All other factors being constant, cells that generate greater electric potential also produce more current, therefore causing the propeller to spin more rapidly. Cutting down the size of the plates decreases the current and the spin rate.

Post-Lab Discussion

Stress that the electrical potential (volts) of a given cell depends on the two metals and solutions used, not on electrode size. A 1.5-V carbon-zinc dry cell comes in a variety of sizes. Students might be able to speculate (correctly) that a 9-V battery is composed of six 1.5-V cells connected in series.) It is useful to have several different size cells dissected for inspection by students.

Note that the activity series does not give the standard electrode potentials for the half-reactions. Since the difference in electrical potential between successive half-reactions in this chart is not constant, cell-voltage comparisons can only be made if one electrode is used as a common reference point (Question 1). Quantitative applications of E_o values would have no additional payoff in terms of the general objectives set for this part of the *ChemCom* course. However, depending on your students, you may wish to display a standard table of electrochemical potentials with an overhead projector, and point out that quantitative measurements have been tabulated by chemists.

In addition to discussing the questions, you may decide to have students think about the chemistry of the common dry cell. (Why do "dry cells" fail to function if they are actually dry?) If done with care, common dry cells can be dissected, so students can see the functioning parts. Also, if time permits, it's possible to connect a group of homemade cells in series to play a small portable radio in class.

Materials

12 24-well microplates
0.1 M copper(II) nitrate
0.1 M zinc nitrate
0.1 M magnesium nitrate
0.1 M iron(III) nitrate
12 copper strips, 1.0 cm × 3 cm × 0.1 mm
12 copper strips, 0.5 cm × 3 cm × 0.1 mm
12 copper strips, 0.25 cm × 3 cm × 0.1 mm
12 zinc strips, 1.0 cm × 3 cm × 0.25 mm
12 zinc strips, 0.5 cm × 3 cm × 0.25 mm
12 zinc strips, 0.25 cm × 3 cm × 0.25 mm
12 magnesium ribbon strips, 3 cm
12 iron metal strips, 0.5 cm × 3 cm (iron nails may be substituted)
12 voltmeters
24 small alligator clips with fine wire leads
36 strips of filter paper, 8 mm × 3 cm
Saturated KCl solution

Procedure:

1. Add 1 mL of 0.1 M $Cu(NO_3)_2$ to each of three wells on the microplate.
2. Add 1 mL of 0.1 M $Zn(NO_3)_2$ to a well adjacent to one of the copper wells.
3. Add 1 mL of 0.1 M $Mg(NO_3)_2$ to another well adjacent to one of the copper wells.
4. Add 1 mL of 0.1 M $Fe(NO_3)_3$ to the third well adjacent to one of the copper wells.
5. Place the 0.5 cm × 3 cm × 0.1 mm copper strips in each of the copper wells.
6. Place the 0.5 cm × 3 cm × 0.25 mm zinc strip in the zinc nitrate well, the magnesium strip in the magnesium nitrate well, and the iron strip (or nail) in the iron(III) nitrate well.
7. Connect the copper and zinc wells with a piece of filter paper moistened with the saturated potassium chloride solution; make similar connections between the copper and magnesium wells, and the copper and iron wells.
8. Attach one alligator clip from the voltmeter to the copper strip in the copper well. Lightly touch the second alligator clip to the zinc strip. If the needle is deflected in the positive direction, secure the clip to the zinc. If the needle is deflected in the negative direction, reverse the positions of the clips, and attach them to the metal strips.
9. Record the reading from the voltmeter.
10. Repeat steps 8 and 9 for the copper and magnesium strips, then for the copper and iron strips.
11. Repeat the measurements, using the smaller pieces of zinc and copper in their respective solutions.
12. Repeat the measurements, using the larger pieces of zinc and copper in their respective solutions.

Answers to Questions:

1. a. Mg-Cu, Zn-Cu, and Fe-Cu.
   b. The farther apart two metals on the activity series, the greater the electrical potential of the cell.
2. a. Lower, since chromium is closer in activity to zinc than is copper.
   b. Higher, since silver is less active (or further "away" in activity) than is copper.
   c. Lower, since tin is closer in activity to copper than is zinc.
3. The electrical potential (voltage) of each battery remained the same regardless of the surface area (it depends only on the two metals used), but the current (amperes) dropped due to the decrease in surface area.
4. The Ag/Au battery is impractical due to high cost and low power output.

D. 2 ELECTROCHEMISTRY (pages 509-511)

This section provides more details regarding redox reactions, using the automobile battery as an example, and includes information about rechargable batteries.

To support this section you may wish to have a variety of dry cells disassembled. One 9-V transistor battery contains six small 1.5-V carbon/zinc cells; a 6-V lantern battery contains four 1.5-V carbon/zinc cells; and all 1.5-V carbon/zinc "batteries" (technically an incorrect term, since a battery is usually regarded as a combination of two or more cells) contain only one cell.

Students should understand that dry cells are not dry: in fact if they become totally dry they become inoperable. The term refers to the fact that the moisture is tied up as a thick slurry, and not freely moving, as in the automobile battery. The moisture is needed to allow the movement of ions throughout the cell—students have encountered this in the previous activity.

YOUR TURN

Getting a Charge from Electrochemistry (pages 511-512)

In this activity, students apply their understanding of redox reactions to electrochemical cells. Simple mnemonics which may help include: **LEO** the lion goes **GER** (Loss of Electrons is Oxidation, Gain of Electrons is Reduction) and **RC Cola** (Reduction occurs at the Cathode).

Answers to Questions:

1. a. $Al \rightarrow Sn$
    b. $Mg \rightarrow Pb$
    c. $Fe \rightarrow Cu$
2. a. $Sn^{2+}(aq) + 2 e^- \rightarrow Sn(s)$ (reduction)
        $Cd(s) \rightarrow 2 e^- + Cd^{2+}(aq)$ (oxidation)

3.   Students should draw a sketch of the electrochemical cell. The half-reactions would be:

Anode (oxidation)	Cathode (reduction)
$Ni(s) \rightarrow Ni^{2+}(aq) + 2 e^-$	$Cu^{2+}(aq) + 2 e^- \rightarrow Cu(s)$

4. a. Anode = Zn; Cathode = Cu
    b. Anode = Al; Cathode = Zn
    c. Anode = Mg; Cathode = Mn
    d. Anode = Ni; Cathode = Au

D.3 LABORATORY ACTIVITY: ELECTROPLATING (pages 512-515)

Students discover how electrical energy can be used to drive a nonspontaneous reaction (the reverse of the type of reactions discussed in Section D.2), often producing materials with enhanced appearance and/or corrosion protection.

Time

One class period

Materials (for a class of 24 working in pairs)

3.0 L 0.5 M copper(II) sulfate (250 mL/team) (fw = 249.68 $CuSO_4 \cdot 5H_2O$) (375 g/3.0 L solution)
12 pieces copper foil (3 cm × 10 cm × 0.1 mm)
12 steel wool pads
12 metallic objects to plate
12 graphite sticks
12 6-V (lantern) batteries or 6-V DC power sources
12 400-mL beakers
24 alligator clips with fine wire leads (2/team)
12 tongs
Paper towels

Advance Preparation

Prepare the 0.5 M $CuSO_4$ solution described above.

Pre-Lab Discussion

Contrast the spontaneous, exothermic reactions discussed in Section D.2 such as: $Zn(s) + Cu^{2+}(aq) \rightarrow Zn^{2+}(aq) + Cu(s)$ to the reverse of such reactions which occurs in electroplating.

Lab Tips

Be sure that the objects to be plated are cleansed of oils and rust before attempting to plate. You may wish to experiment with a range of objects to see what works best. Do not use metal objects that plate galvanically when immersed in copper(II) solution—use metals below copper in the activity series. Paper clips and keys work well.

An inexpensive carbon electrode can be crafted from a wooden soft lead pencil. Sharpen the pencil. Cut a slot in the wood near the eraser end to expose the lead. Fasten the alligator clip there. You could also cut off the pencil top and then expose the lead.

Remind students not to allow electrodes to touch each other.

Remind students to continually check the object for copper color in Part 2.

The 6-V battery can be purchased from Sargent-Welch, catalog number S-30848 or may be purchased from local stores. To lower your battery-purchase costs, consider having students work in teams of four, or ask students to bring in lantern batteries if they have them. One alternative to using batteries is to purchase a small battery charger, and connect it to Romex house wire, which can be run around the perimeter of your room to each bench. Strip a small section of insulation from the Romex at each station, which will allow students to connect their alligator clips to the exposed wire. With a small enough current, and a circuit breaker, the system should be safe and able to withstand mistakes, such as shorting out the connections.

The microscale procedure is identical, with the substitution of a U-tube for the beaker. The size of the objects to be plated must also be decreased accordingly to fit the U-tube.

Expected Results

May vary with objects being plated.

Post-Lab Discussion

Be sure students understand the difference between the spontaneous, exothermic reactions that occur in an electrochemical cell (*Laboratory Activity* D.1) and nonspontaneous, endothermic reactions that occur in electroplating. Remind the students that the same terminology is used in both cases since redox-type reactions are involved (thus reduction or gaining of electrons occurs at the cathode in both cases). One memory trick: cathode and reduction both start with consonants; anode and oxidation both start with vowels; thus the correct electrode and process become associated in both cases.

Answers to Questions :

1. a. The anode is the positive electrode (carbon electrode attached to the positive terminal of the battery).
   b. The cathode is the negative electrode (carbon electrode attached to the negative terminal of the battery).
2. The electrons are pushed by the power supply in the external circuit from the anode (+) to the cathode (–).
3. a. $Cu(s) \rightarrow Cu^{2+}(aq) + 2\ e^-$.
   b. $Cu^{2+}(aq) + 2\ e^- \rightarrow Cu(s)$.
4. a. Attach the ring to the cathode (negative) electrode, the gold metal to the anode (positive) electrode, and immerse in a gold electrolyte solution.
   b. Attach the spoon to the cathode, silver metal to the anode, and immerse in a silver electrolyte solution.
   c. Make a sample of impure copper the anode. Place a piece of pure copper (or graphite) at the cathode. Immerse both electrodes in a copper(II) ion-containing electrolyte. Pure copper metal will accumulate on the cathode; there is a net migration of copper from the impure anode to the cathode during this electrolysis.

D.4 INDUSTRIAL ELECTROCHEMISTRY (pages 515-517)

This section considers two common electrolysis reactions performed on an industrial scale. The point to emphasize is that, despite the contrast in scale between bench and industrial electrolytic reactions, the same redox chemistry applies.

D.5 YOU DECIDE: PLANNING FOR AN INDUSTRY (pages 517-518)

This assignment will require a lot of research on the part of the team members. Science and technology dictionaries, the *Chemical & Engineering News* issue on Research and Development, and encyclopedias will be necessary—alert the school librarian. If you give the assignment early enough, it may be possible to write to some of the companies listed in *Chemical & Engineering News* to ask for information relevant to the research. You could keep their materials on hand for reference work in future years.

Time

Students will need out-of-class time to research their topic. The results of their efforts should be discussed in class (one period).

PART D: SUMMARY QUESTIONS (page 518)

1. a. No.
   b. Yes.
2. Far apart. The greater the difference in the electronegativity, the electrical pressure pushing electrons from one metal to the other.
3. a. Discharging
   b. That it would consume energy (be endothermic) and is probably non-spontaneous.
   c. Endothermic
   d. Charging
   e. The extent to which a battery is charged depends on the concentration of the sulfuric acid, which has a density of about 1.84 g/mL, compared to 1.00 g/mL for pure water. Therefore the higher the density, as measured by a hydrometer, the higher the charge on the battery.
   f. When the battery is being discharged, sulfuric acid is consumed and water is produced (see equations in text). Thus, the fluid in the battery will be at maximum density when the battery is fully charged; fluid density will decrease as the battery discharges.
   g. The lead in the lead dioxide (spongy lead) is being reduced (gaining control of additional electrons); metallic lead is being oxidized (losing control of electrons).
   h. When it is being charged from an outside source, hydrogen gas may be produced as a by-product at the lead dioxide electrode, due to the reduction of $H^+(aq)$ to $H_2(g)$.
   i. If a spark is present, an explosion can occur, represented by the familiar (and dangerous) reaction:
   $$2 H_2(g) + O_2(g) \rightarrow 2H_2O + energy.$$
4. A fairly low voltage. To produce a high voltage, an active metal (easily oxidized) should be paired with an inactive metal (easily reduced) so the electrical potential generated (volts) will be large.
5. a. $Al \rightarrow Cr$
   b. $Mn \rightarrow Cu$
   c. $Fe \rightarrow Ni$

6.

Voltaic cell	**Electroplating cell**
a. exothermic	endothermic
b. spontaneous	nonspontaneous
c. from more active metal to less	from less active metal to more

7. a. Aluminum is more active than either iron or copper, suggesting that it donates electrons more readily than either of the other metals.
   b. Aluminum is found in nature in its +3 oxidation state (Al^{3+}). Compared to Fe^{3+}, Al^{3+} is far more difficult to reduce to its metallic state. The more active the metal, the harder it is to reverse the process and add electrons.
   c. The Hall process involves dissolving aluminum oxide in cryolite and electrolyzing in a cell with a carbon lining, which serves as the cathode, and carbon electrodes as the anode. The cathode donates electrons to the aluminum ions, forming metallic aluminum as a liquid, which is periodically drawn off. The carbon anodes are oxidized and must be replaced.
8. a. $4 Al^{3+}(melt) + 3 C(s) + 6 O^{2-}(melt) \rightarrow 4 Al(l) + 3CO_2(g)$

   b. $5400 \text{ g} \times \dfrac{1 \text{ mol Al}}{27 \text{ g Al}} = 200 \text{ mol Al}$

   $200 \text{ mol Al} \times \dfrac{3 \text{ e}^-}{\text{mol Al}^{3+}} = 600 \text{ mol electrons}$

E: PUTTING IT ALL TOGETHER: CHEMICAL INDUSTRY PAST, PRESENT, AND FUTURE

In some respects, the closing section attempts to pull together not only various concepts and ideas discussed in this unit, but also those in the entire *ChemCom* course. Students are asked to develop an idea for a new chemical product or process. In so doing, they will need to consider many scientific and societal issues that have pervaded this course. As before, the goal is not to attain set answers or unanimity of opinion, but rather to encourage thoughtful decision-making.

E.1 FUTURE DEVELOPMENTS (pages 519-521)

Figure 14 (page 519) summarizes key factors in the development of new materials. The four major influences on new material development combine to drive research, which will lead to new products. The normal progression is to find a product to fill a need, but occasionally a product is found first, and the need is developed to make use of that product. For example, Teflon's accidental discovery (see page 279) spurred the development of interest in non-stick cooking utensils.

After tracing some history of products of industrial research, students have the opportunity to make specific suggestions for products and processes they would like to see developed by chemical industry. This forecasting is, by its very nature, open-ended. Allow the class to divide into teams. Once groups are formed, instruct each group to spend 10 minutes brainstorming specific products or processes they would like to see developed. (Explain that the purpose of brainstorming is to allow creativity to flow and a number of ideas to be generated—evaluative or judging statements are not appropriate at this point.) Following this 10-minute period, allow five minutes for each group to decide on one specific idea as its focus. The remainder of the class period should then be used for developing this idea.

The six questions given in the text should provide some structure to this otherwise open-ended process. For homework, each team may decide to ask individual members to work up in more detail (including, perhaps, appropriate visuals) some implications of one of the questions.

On the second day of the activity, use the first 10 minutes to allow each group to organize a four- or five-minute presentation of their idea. A variety of scenarios might be employed; for example: (a) The team may act as a R & D team whose presentation is designed to convince the board of directors of their company to allocate funds to explore this new idea. (b) The team may act as an agency that has just won the advertising contract for the new idea and needs to try out a proposed campaign on a sample TV audience. (c) The team may act as a company petitioning the FDA for approval of exploratory human tests on the effectiveness and safety of a new drug or food additive.

The remainder of the class period should be devoted to the actual presentations.

Answers to Questions:

Actual responses will be specific to the products or processes being considered, but the following ideas will hopefully be built into the presentations, or evolve from class discussion:

1. a.,b. To be profitable, a new product or process must fill a consumer need. If the need does not already exist, or if the new product or process is a substitute for an existing one, consumers can be persuaded they need it, with advertising.

2. a.-c. The development of new products or processes requires a commitment of financial and human resources. Companies hope that this investment in research and development will pay off later in a profitable product or process. In that sense, R & D is similar to stock market investing: there's always the risk that, over the short run an investment will not pay off, but one hopes that over a longer time period one will be rewarded by a wise investment. Students should realize that many commercial breakthroughs in recent years have resulted from multidisciplinary teamwork. These teams include not only scientists from various disciplines, but also market analysts, advertising specialists, etc. A product is successful only if people are willing to buy it at a price that includes a reasonable profit for the manufacturer.

3. a.-d. This question touches on a theme that has been visible in several earlier units—particularly in the Resources and Petroleum units. Both practical and political factors need to be considered. In general, products or processes that use readily available materials (sand, atmospheric gases, etc.) and that are energy efficient are more desirable than those that do not.

4. a.-d. Before a new product or process can be scaled up for large-volume production, tests must ensure that it is not unacceptably harmful to humans or the environment. The potential benefits must exceed the costs/risks. This is important not only for moral reasons, but also for preserving the long-term public image of the manufacturer. If people lose faith in the products of a company, the company could fail. Also, companies need to have good public relations at the local level to address any potential questions and concerns regarding the plant.

5. Possible concerns include: Does the manufacturing of the product require resources from another country? If so, will a fair exchange be made or will the country be exploited in the process? Is the natural environment damaged in obtaining the raw materials and products or in manufacturing the product? Does the sale or use of the product run counter to attitudes and values where the product will be used? (Consider mind-altering drugs, experimental transplant procedures, genetic alterations, etc.)

6. It depends on how far removed the new idea is from the current theoretical and technological knowledge. There's a significant time interval between the initial idea, the laboratory demonstration, and the commercial production of a new product or process. In the past, the time interval was quite long (consider ideas such as powered flight, motorized transportation, etc.), but in recent years has been substantially reduced (consider the development of atomic energy, nylon, synthetic rubber and plastics, etc.).

E.2 LOOKING BACK AND LOOKING AHEAD (page 521)

The message we wish to leave students with is that learning is a lifelong process. Regardless of what career directions they take, science will continue to influence the quality of their lives. In a democratic society, citizens need to keep informed of all the issues, including those involving science and technology, in order to move in desirable directions. Many opportunities (outside of additional science courses) exist for keeping one's scientific literacy up to date, as the text suggests.

SUPPLY LIST

Expendable Items

	Section	Quantity (per class)
Ammonium molybdate	C.3	10 g
Ammonium nitrate	C.1	5 g
Ascorbic acid, vitamin C, crystals	C.3	20 g
Barium chloride dihydrate	C.1	10 g
Copper(II) nitrate trihydrate	D.1	75 g
Copper foil strips	D.1, D.3	72
Copper(II) sulfate pentahydrate	D.3	375 g
Hydrochloric acid, conc.	C.1	200 mL
Iron(III) nitrate nonahydrate	C.1 D.1	120 g
Iron(II) sulfate heptahydrate	C.1	10 g
Iron strips	D.1	12
Litmus paper, red	C.1	24 pieces
Magnesium nitrate hexahydrate	D.1	80 g
Magnesium ribbon	D.1	250 cm
Nichrome or platinum wire	C.1	300 cm
Potassium monohydrogen phosphate	C.3	2 g
Potassium nitrate	C.1	5 g
Sodium hydroxide	C.1	30 g
Sodium nitrate	C.1	5 g
Sodium phosphate dodecahydrate	C.1	10 g
Sodium sulfate decahydrate	C.1	10 g
Steel wool pads	D.3	12
Sulfuric acid, conc.	C.1, C.3	350 mL
Zinc, strips	D.1	36
Zinc nitrate hexahydrate	D.1	90 g

Nonexpendable Items

	Section	Quantity (per class)
Alligator clips with leads	D.3	24
Batteries, 6-V lantern (and/or DC power sources)	D.3	5-12
Cups, porous (75-mL)	D.1	12
Graphite sticks	D.3	24
Motors, low-current (optional)	D.1	12
Spot plates	C.1	12

SUPPLEMENTAL READINGS

"The World of Industrial Chemistry." *Journal of Chemical Education* series beginning in February 1979 issue. (Edited by W. Conrad Fernelius and Harold W. Wittcoff.) Particularly useful articles include: "The Chemical Industry: A Historical Perspective." February 1979, 114-15. "The Chemical Industry: What Is It?" April 1979, "An Acid Can be Basic—Sulfuric Acid." August 1979, 529. "Chloro-Alkali Industry." September 1980, 640-41. "Chemical Processing: Patch or Continuous." Part I September 1982, 766-68. Part II October 1982, 860-62. "A Summary Chart of the Manufacture of Important Inorganic Chemicals." May 1983, 411-13. Manufacture, properties, uses, and economic aspects of key inorganic chemicals are discussed via a flowchart used in an industrial chemistry course.

"When Push Comes to Shove", *CHEM MATTERS*, February, 1985.

"Industrial Chemistry Bibliography." April 1985, 331-32. A bibliography of articles published in the *Journal of Chemical Education* (1968-1983) that focused on industrial chemistry. Includes separate headings such as: experiments/demonstrations, chemical of the month, etc.

"Facts and Figures for the Chemical Industry." *Chemical and Engineering News.* Annual issue which includes: the top 50 chemicals and the associated production figures for the 1975-1985 period; the top 100 chemical producers; information on foreign chemical industries; and a wealth of other information and statistics.

"Facts and Figures for Chemical R & D." *Chemical and Engineering News.* Annual issue which includes: an overview of where R & D funds come from and go to; and specific information for the federal government, industry, and universities/colleges.

"Industrial Analytical Chemistry: The Eyes, Ears, and Handmaiden to Research and Development." *Journal of Chemical Education*, **63**, 237 (1986). Discusses the role of industrial analytical chemists, prerequisite training, and research challenges.

"The B.S. Chemist in Industry." *Journal of Chemical Education*, **62**, 734-735 (1985). Describes the variety of jobs available to BS chemists.

"Pharmacology: A Place for Chemists." *Journal of Chemical Education*, **59**, 231 (1982).

"Chemistry in Action: How to Plan a Visit to the Chemical Industry." *Journal of Chemical Education*, **59**, 582 (1982). Provides helpful hints on how to plan a field trip.

"Sunken Treasure", *CHEM MATTERS*, April, 1987.

"Forensic Chemistry." *Journal of Chemical Education*, **59**, 41 (1982).

"Career Choice: Chemist or Chemical Engineer." *Journal of Chemical Education*, **58**, 494 (1981).

"The Industrial Research Chemist." *Journal of Chemical Education*, **58**, 269 (1981).

"Chemistry and Public Policy." *Chemical and Engineering News*, March 9, 1987, 20-29. Special report (reprint available from ACS). A historical reflection on the role of chemistry in promoting the general welfare from the 1700s to the present.

"Leaf Jewelry", *CHEM MATTERS*, December, 1987.

The Chemical Balance: Benefiting People, Minimizing Risks. Chemical Manufacturers Association. Free booklet discusses how the chemical industry is addressing the issues of product safety, worker safety, chemical transportation safety, and managing chemical waste.

ChemEcology. Chemical Manufacturers Association. Free monthly newsletter describing the chemical industry's efforts in the environment, energy conservation, and occupational health.

"Ernie's Amazing Journey." *CHEM MATTERS*, February, 1990.

"Delaware: Who Needs to Be Big?" *National Geographic*, April 1983, 171-97.

"Electrochemistry: State of the Art Symposium." Special issue of *Journal of Chemical Education*, **60**, April 1983.

"Opportunities in Chemistry: Executive Summary." *Chemical and Engineering News*, October 14, 1985. Overview of the frontiers of chemical research.

"Harnessing Light by a Thread: Miracles of Fiber Optics." *National Geographic*, October 1979, 516-35.

"The Chip: Electronic MiniMarvel." *National Geographic*, October 1982, 421-57. "California's Silicon Valley," 459-477.

"The Laser—A Splendid Light." *National Geographic*, March 1984, 335-63.

"The Wonders of Holography." *National Geographic*, March 1984, 364-77.

TEST ITEMS

This section of your Teacher's Guide contains **Multiple-Choice Questions** and **Short Answer Questions** for each of the units. There also are nine series of **Assorted Test Items**. Each series consists of a short narrative followed by two or more related questions.

These questions should not be used as a self-sufficient set for any unit. You must make decisions about appropriate content, levels of understanding, and suitable styles of questioning. Taken as a set, the questions assess some concepts many times but omit other *ChemCom* objectives completely. Therefore, please use these questions as a guide but exercise your professional judgement to select suitable items.

The unique nature of the *ChemCom* curriculum creates special challenges in evaluating student performance. Assessment issues and new-style questioning options such as matrix questions have been addressed in several issues of *ChemComments*, a publication of the Education Division, American Chemical Society. This publication has now been superseded by *Chemunity News*, which will continue to have coverage related to the *ChemCom* curriculum. To help with end of course evaluation, a standardized examination is now available from the ACS DivCHED Examinations Institute. It incorporates novel test item formats, such as multiple-response options, linked event-decision style questions and grid questions, which permit measurement of student achievement that is difficult to assess with single-answer, multiple-choice questions.

SUPPLYING OUR WATER NEEDS

Multiple-Choice Questions

1. A vitamin supplement tablet contains 0.06 grams of vitamin C. How else can this mass be expressed?
   (A) 6 mg   (B) 60 mg   (C) 600 mg   (D) 0.006 kg

2. A "kilo" is a slang term for a kilogram, which is equal to 1,000 grams. How many milligrams are there in 5 "kilos"?
   (A) 5000   (B) 50,000   (C) 500,000   (D) 5,000,000

3. How many 250 mL servings can be poured from a 2.0-L bottle of soft drink?
   (A) 2   (B) 4   (C) 8   (D) 12

4. Which sequence reflects the decreasing abundance of water supplies on Earth?
   (A) oceans—glaciers/icecaps—groundwater—lakes—atmosphere—rivers
   (B) oceans—glaciers/icecaps—rivers—lakes—groundwater—atmosphere
   (C) glaciers/icecaps—oceans—rivers—groundwater—lakes—atmosphere
   (D) rivers—atmosphere—groundwater—lakes—oceans—glaciers/icecaps

5. If the instructions on a bottle of medicine say "shake before using", the material inside is probably a
   (A) solution.   (B) colloid.   (C) suspension.   (D) pure liquid.

6. What causes a warm bottle of soda to "explode" when placed in a freezer overnight?
   (A) The water in the soda contracts when it freezes.
   (B) The glass bottle contracts when it freezes.
   (C) The water in the soda expands when it freezes.
   (D) The glass bottle expands when it freezes.

7. Some water softening units use ion exchange resins. What occurs in such a unit?
   (A) Calcium ions are exchanged for magnesium ions.
   (B) Magnesium ions are exchanged for chloride ions.
   (C) Calcium ions are exchanged for iron ions.
   (D) Calcium ions are exchanged for sodium ions.

8. Which statement best explains the ability of the nonpolar insecticide DDT to accumulate in the environment?
   (A) DDT is very soluble in certain nonpolar fatty tissues of animals in the food chain.
   (B) DDT is very soluble in polar water contained in rivers and streams.
   (C) DDT rapidly breaks down in the environment to produce polar toxic products.
   (D) DDT is a major cause of nonpolar acid rain.

9. Which gas dissolves in atmospheric moisture to produce acid rain?
   (A) hydrogen (H_2) (B) oxygen (O_2) (C) sulfur dioxide (SO_2) (D) nitrogen (N_2)

10. How can you remove dissolved salt from water?
    (A) Filter the salt solution through sand.
    (B) Filter the salt solution through charcoal.
    (C) Distill the salt solution.
    (D) Physically separate the salt solution from the denser pure water.

11. Which item would be the most important to insure your survival for at least 3 or 4 days?
    (A) a kilogram of peanuts   (C) a package of flares
    (B) 10 liters of pure water   (D) a portable toilet

12. The largest storage place for the world's water is in
    (A) rivers.   (B) the atmosphere.   (C) the oceans.   (D) the ground.

13. Which diagram best illustrates the shape of a water molecule?

(A) H—O—H

(C)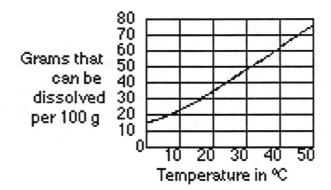

(B)

(D)

14. Farms help to pollute nearby waterways because the rain washes some of the fertilizer from the fields. The fertilizer mixes with other chemicals in the soil and often enters waterways as KNO_3. Use the solubility graph to determine the maximum number of grams of KNO_3 that could be dissolved in each 100 g of water if the temperature of the water is 20°C.

Grams that can be dissolved per 100 g

80
70
60
50
40
30
20
10
0

10 20 30 40 50
Temperature in °C

(A) 13 grams   (C) 35 grams
(B) 21 grams   (D) 46 grams

15. Distillation is not used by many water treatment plants to purify water. Why not?
(A) Distillation is too expensive.
(B) Distillation fails to remove heavy metal contamination.
(C) Distillation eventually alters the molecular structure of the water.
(D) Distillation gives off too much thermal energy.

16. Some municipalities add a soluble fluoride compound to their drinking water supply. The reason is to
(A) improve the taste of the water.
(B) eliminate the bacteria found in the water, making it safer to drink.
(C) reduce the need to clean teeth by brushing to remove bacteria.
(D) reduce tooth decay by strengthening tooth enamel.

17. Chloroform is sometimes found in water systems after purification processes have taken place. The chloroform results primarily from chemical reactions between
(A) fluoride ions and chlorine gas.
(B) organic substances and fluoride ions.
(C) chlorine and hypochlorous acid.
(D) chlorine and organic substances.

18. In 100 grams of a 25% sugar solution by mass, there are
(A) 25 grams of sugar and 75 grams of water.
(B) 25 grams of sugar and 100 grams of water.
(C) 25 grams of water and 75 grams of sugar.
(D) 100 grams of sugar and 25 grams of water.

19. These are the structural formulas for ammonia (NH_3) and hexane (C_6H_{14}). Which statement explains what will happen when ammonia and hexane are mixed?

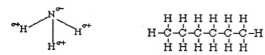

(A) They will not dissolve in each other because the molecules of one are polar while the other's molecules are nonpolar.
(B) They will not dissolve in each other because one is an ionic substance and one is a nonpolar substance.
(C) They will dissolve in each other because both are composed of molecules.
(D) They will dissolve because both are composed of ionic particles.

20. When a solid composed of ions is mixed with a liquid solvent made up of polar molecules, the solid should
(A) not dissolve because no charged particles are present.
(B) not dissolve because all the particles present carry some charge.
(C) dissolve because no charged particles are present.
(D) dissolve because all the particles present carry some charge.

21. You are marooned on a sandy island surrounded by millions of liters of ocean water. The only water available on the island is from a stagnant pond of murky-looking water. In your survival kit you have the following items.

one nylon jacket	one knife
one plastic cup	one liter-bottle of Clorox®
two plastic bags	one 5-liter glass bottle
one length of rubber tubing	one bag of salted peanuts

Which plan is likely to produce drinking water?
(A) Let a sample of pond water settle in the glass bottle for two days and then carefully pour or siphon water off the top.
(B) Obtain some sand from a high point on the island. Filter a sample of ocean water through the sand and then allow the water to settle for two days before pouring or siphoning water off the top.
(C) Pour the liter of Clorox® into the glass bottle. Fill the bottle with pond water. Allow the sample to stand for several hours before using.
(D) Filter a sample of pond water through the sand into the glass bottle. Add a capful of Clorox® and shake the bottle to mix thoroughly. Allow the sample to stand for several hours before using.

Short-Answer Questions

22. Write the formula for each compound if the name is given, and the name if the formula is given.
    a. sodium hydroxide (lye)
    b. ammonium phosphate (an important fertilizer)
    c. potassium nitrate (used in fireworks and gunpowder)
    d. NaCl (common table salt)
    e. MgO (the product formed in photoflash cubes)
    f. Fe_2O_3 (common rust)

23. Letters, words, and sentences are the basic tools for written expression. What are the chemistry equivalents to letters, words, and sentences?

24. List and briefly describe three unusual properties of water. How can you explain those properties using the concept of polarity?

25. Do you predict that oxygen is sparingly soluble or highly soluble in water? Explain your reasoning.

26. Discuss hard water in terms of its composition, difficulties of use, and treatment options.

27. Why are fish especially sensitive to thermal pollution?

28. What are the advantages and disadvantages of chlorinating drinking water rather than using untreated water?

29. What is the predicted effect of temperature and pressure on the water solubility of solids? How would these effects differ for the water solubility of gases?

30. Compare and contrast the number one use of water in the western with the number one use of water in the eastern United States. How can this difference be explained?

31. How are municipal water purification and sewage treatment similar to nature's system?

32. Is it easier to confirm the presence or to prove the absence of specific substances through chemical analysis? Explain.

33. A politician makes campaign promises that he will insure that his constituents receive 100% pure drinking water. Is this a reasonable promise? Explain.

34. Explain how the phrase "running out of water" contains both an element of truth and a mistaken notion of chemistry.

Answers to Questions:

Multiple-Choice

1. B	6. C	11. B	16. D	21. D
2. D	7. D	12. C	17. D	
3. C	8. A	13. D	18. A	
4. A	9. C	14. C	19. A	
5. C	10. C	15. A	20. D	

Short-Answer

22. a. NaOH   b. $(NH_4)_3PO_4$ c. KNO_3
    d. sodium chloride   e. magnesium oxide   f. iron(III) oxide

23. The chemistry analogs are symbols for elements, formulas for compounds, and balanced chemical equations.

24. Water has an abnormally high boiling point, a very high heat capacity, a high surface tension, and its solid state is less dense than the liquid state. The V-shape of the water molecule causes it to have a positive hydrogen "end" and a negative oxygen "end." Because water is such a polar molecule, intermolecular attractions, particularly hydrogen bonding between molecules, causes water to have these unusual properties.

25. Oxygen is only sparingly soluble in water. Remember that oxygen is nonpolar and water is polar. Therefore the "like dissolves like" rule suggests that oxygen would be either insoluble or sparingly soluble in water. Alternatively, check the actual solubility in a reference book. If oxygen is only sparingly soluble in water, its concentration will be expressed in units of mg/L or ppm. If it is highly soluble, the units g/L will be used.

26. Hard water may contain Ca^{2+}, Mg^{2+}, and/or Fe^{3+} ions dissolved in water. These ions will interfere with soap lathering by forming precipitates and forming hard deposits in situations where it is boiled. Ion-exchange systems soften hard water by substituting Na^+ ions for the hard-water ions.

27. Fish require dissolved oxygen to live. As water temperature increases, so does the rate of fish metabolism and their need for oxygen. However, the solubility of oxygen decreases with increasing water temperature, making fish particularly vulnerable to thermal pollution.

28. The chlorination of water provides some assurance that the water is bacteria free. Conversely, chlorine in water presents a health risk if the chlorine reacts with organic compounds to produce trihalomethanes, THMs. Chloroform, $CHCl_3$, is a representative trihalomethane and is a suspected carcinogen (cancer-causing agent).

29. The solubility of solids in water generally increases with increasing temperature and is unaffected by pressure. The solubility of gases in water decreases with increasing temperature and increases with increasing pressure. Therefore, these factors affecting water solubility for solids and gases are quite different.

30. In the West, irrigation accounts for some 88% of water use since rainfall is inadequate to support the region's large scale farming. In the East, with its higher population density, 51% of water use is for electric power plant cooling.

31. Water purification and sewage treatment plants start by screening large objects, allowing the water to settle, and then filtering through sand or other materials. In nature, these same steps take place through evaporation and the seepage of surface water into underground aquifers. The chlorination and flocculation steps in artificial water purification and treatment procedures do not have direct counterparts in the hydrologic cycle.

32. In general, if a substance is indicated on one or more confirming tests, one can reliably declare its presence. However, the absence of a positive test could mean several things. It may mean that the substance is not present. However, it could mean that the test is not sensitive enough to detect the low concentration of material present or that other substances are interfering with the analysis.

33. This is not a reasonable promise. Even "pure" unpolluted rainwater contains dissolved gases. Most surface or groundwater naturally contains dissolved minerals, most of which are either beneficial or at least not harmful. The cost of obtaining water without these dissolved substances would be prohibitive and a waste of taxpayers' money as there is no reason to remove many of the dissolved materials

34. The overall supply of water on our planet is constant. Water moves through its hydrologic cycle but given the law of conservation of matter and the chemical stability of water, the overall supply does not change. On the other hand, supplies of fresh water are limited and unevenly distributed. If fresh water resources are not wisely managed, the cost of producing water of sufficient quality can result in local water scarcity.

CONSERVING CHEMICAL RESOURCES

Multiple-Choice Questions

1. Which of the following processes requires the least amount of energy?
   (A) processing virgin ore
   (B) reusing a metallic article
   (C) recycling a used metal article
   (D) all require the same energy

2. A metal is usually obtained from its ore by a process called
   (A) melting.
   (B) casting.
   (C) oxidation.
   (D) reduction.

3. We are in danger of exhausting our supplies of copper because
   (A) copper atoms are being destroyed in chemical processes.
   (B) copper ores are being converted to copper metal.
   (C) copper atoms are being widely distributed in the environment.
   (D) environmentalists want stiff air pollution regulations.

4. Low-grade ores are less desirable than high grade ores because they
   (A) are more difficult to mine.
   (B) require more energy for concentration.
   (C) are scarcer than high-grade ores.
   (D) cost more per ton of ore.

5. Metals such as lithium, potassium, calcium, magnesium, and aluminum are usually obtained commercially by
   (A) mining them in their pure, elemental form.
   (B) roasting their compounds in air.
   (C) heating their compounds with carbon or carbon monoxide.
   (D) subjecting their compounds to electrolysis.

6. Which portion of our planet is the major "supply house" for the production of manufactured goods?
   (A) lithosphere
   (B) hydrosphere
   (C) atmosphere
   (D) mantle

7. Which statements about the mineral resources of our planet is true?
   (A) They are renewable and fairly evenly distributed.
   (B) They are nonrenewable and fairly evenly distributed.
   (C) They are renewable and not evenly distributed.
   (D) They are nonrenewable and not evenly distributed.

8. Which statement about reduction is not true?
   (A) Reduction is always accompanied by oxidation.
   (B) Reduction involves the gaining of electron(s).
   (C) Reduction is the process whereby metals rust.
   (D) Reduction of minerals can produce elemental metals.

9. Iron was first obtained from its ore long after copper was first obtained from its ore. What is the principal reason for this observation?
   (A) Ore containing iron is found at greater depths in the Earth's crust than is copper ore.
   (B) There is a limited supply of iron ore in the Earth's crust and an excess of copper ore.
   (C) A high temperature is necessary to obtain iron from its ore, but copper can be obtained at lower temperatures from its ore.
   (D) Until new sources were discovered, there was a scarcity of carbon which was needed to carry out the process.

10. Which process can be used to release some common metals from their compounds in ores?
    (A) oxidation
    (B) decanting
    (C) reduction
    (D) distilling

11. In the electrolysis of molten sodium chloride, sodium ions are attracted to the negative electrode and metallic sodium is formed by gain of electrons. This is
(A) oxidation.   (C) reduction.
(B) leaching.   (D) roasting.

12. In terms of the amount of energy required, aluminum is most easily obtained from
(A) clay   (B) bauxite   (C) Al_2O_3 (D) scrap aluminum

13. How many atoms of iron are formed for each formula unit of Fe_2O_3 that reacts with carbon to form metallic iron?
(A) 1   (B) 2   (C) 3   (D) 6

14. How many atoms of carbon are needed to completely react with each formula unit of Fe_2O_3 to form Fe and CO?
(A) 1   (B) 2   (C) 3   (D) 6

15. The overall reaction for the Hall-Heroult process is $2\ Al_2O_3 \rightarrow 4\ Al + 3\ O_2$. What is the greatest number of atoms of aluminum that can be obtained for each six electrons supplied to Al_2O_3?
(A) 2   (B) 4   (C) 6   (D) 8

16. For which group of starting materials will electrolytic reduction be most efficient in obtaining free metals?
(A) Ca°, Na°, Mg°, Al° (C) Li^+, K^+, Ca^{2+}, Na^+
(B) Mn°, Zn°, Cr^{2+}, Fe^{2+} (D) Pb^{2+}, Cu^{2+}, Hg^{2+}, Ag^+

17. Corrosion of metals can be prevented by sealing the surface with a nonreactive substance. Which gas is kept from reacting with the metal surface by this method?
(A) helium   (B) hydrogen   (C) oxygen   (D) nitrogen

18. What is the mass percentage of carbon in CO_2? (atomic mass of C=12, O=16)
(A) 75%   (B) 50%   (C) 33%   (D) 27%

19. What do you do at once if your hand is splashed with hot, liquid NaOH?
(A) Put on your goggles.
(B) Wash with plenty of water.
(C) Apply Vaseline over the splashed area of skin.
(D) Remove the penny from the liquid.

20. Which formula represents a major compound found in aluminum sulfide ore?
(A) AlS   (B) Al_3S_2 (C) Al_2S_3 (D) AlS_3

21. Chemists have devised a counting unit called the mole. Which of these units is most similar in concept to the mole?
(A) dozen   (B) degree   (C) gallon   (D) ton

22. A mole of sulfur atoms has a mass of approximately 32 g. How many moles of sulfur atoms are in 16 g of sulfur?
(A) 0.25 mol   (B) 0.5 mol   (C) 2.0 mol   (D) 512 mol

23. Which coefficients balance this equation?
$$\_Fe_3O_4 + \_H_2 \rightarrow \_Fe + \_H_2O$$
(A) 1, 4, 4, 3   (B) 1, 4, 3, 4   (C) 1, 1, 2, 4   (D) 2, 3, 3, 4

24. What is conserved in a balanced chemical equation?
(A) mass   (B) atoms   (C) charge   (D) all of these

25. When two metals are in contact with each other, corrosion of one of the metals may occur. A humid atmosphere accelerates any reaction. Which statement is true?
(A) Both metals react readily with oxygen in the moist air, forming two oxides.
(B) The law of conservation of matter does not allow both metals to remain unchanged in the presence of moist air.
(C) The two metals have different degrees of chemical activity and react with each other in moist air.
(D) The two metals, joined together, behave chemically as nonmetals in moist air.

26. If a strip of magnesium metal is placed into a water solution of copper(II) ions, what happens?
   (A) The magnesium strip reacts.   (C) More copper(II) ions form.
   (B) Carbon dioxide gas evolves.   (D) No reaction occurs.

27. Which gives the incorrect symbol for the corresponding element?
   (A) Na—sodium   (C) Se—selenium
   (B) H—hydrogen   (D) P—potassium

28. Which is the least important source of useful chemicals?
   (A) atmospheric gases   (B) lithosphere   (C) oil wells   (D) moon's surface

29. Which factor is least important when deciding whether to mine a specific ore compared to another ore containing the same metal?
   (A) the cost of pollution control for the processing of the ores
   (B) the concentration of metal ions in the ores
   (C) the cost of extracting the metal from the ores
   (D) the ability of the metal to combine chemically with oxygen

30. Which factor is least useful in explaining the chemical properties of substances?
   (A) the distribution of electrons in the atoms making up the substances
   (B) the relative ability of the various kinds of atoms to lose or donate electrons
   (C) the ability of chemists to modify properties through "blending"
   (D) the degree of attraction between or among atoms

31. Which statement about the future availability of chemical resources is true?
   (A) Factors such as the natural abundance, our ability to develop technologies to extract these elements, patterns of use, and the cost related to the rest of the economy will help determine future availability of chemical resources.
   (B) There will be no shortages of chemical resources in the future because atoms of elements are not destroyed during chemical changes.
   (C) All of the necessary chemical resources needed in the future will be obtained from water and air, so it will not be necessary to think about continued availability of traditional chemical resources.
   (D) Any "used up" chemical resources can always be replaced in the future by plastic substitutes, as long as consumers are willing to pay the price.

32. Suppose you are a U.S. Senator and you are considering whether to vote for or against a bill that would allow the U.S. Government to levy a tax on imported steel (iron). Which factor is least important in making your decision?
   (A) the relative chemical activity of iron compared to the activities of other metals
   (B) the known available supply of iron ore in the United States
   (C) the ability of the U.S. coal industry to supply carbon (coke) for reducing the ore
   (D) the period of time that it would take the U.S. steel industry to adapt to more efficient means of extracting iron from available iron ore resources

33. This graph represents the amount of a metal obtainable from its ore over a period of years, utilizing present technology. According to the graph, the amount of high grade ore

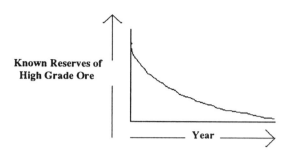

(A) decreases by the same amount each year.
(B) decreases by smaller amounts year by year.
(C) decreases by greater amounts year by year.
(D) remains the same year after year.

34.

Atomic Number	Common Oxide	Atomic Number	Common Oxide
1	H_2O	11	Na_2O
2	(He)	12	MgO
3	Li_2O	13	Al_2O_3
4	BeO	14	SiO_2
5	B_2O_3	15	P_2O_5
6	CO_2	16	SO_2
7	N_2O_5	17	Cl_2O
8	O_2	18	(Ar)
9	F_2O	19	K_2O
10	(Ne)	20	CaO

Based on the data in the table, which statement is not true?
(A) Elements with atomic number 3, 11, and 19 have similar chemical properties.
(B) Formulas of oxides can be used to show the periodic properties of elements.
(C) Elements with atomic number 2, 10, and 18 are metals so don't form oxides.
(D) Formulas of oxides of nonmetals and metals reveal periodic properties

Short-Answer Questions

35. Balance this chemical expression. (This reaction that takes place during the flocculation step of water purification.)
    __$Al_2(SO_4)_3$ + __$Ca(OH)_2$ → __$Al(OH)_3(s)$ + __$CaSO_4$

36. Compare and contrast the properties of metals and nonmetals. Consider such things as appearance, heat and electrical conductivity, ductility, and malleability.

37. What is in the raw materials used to make conventional ceramics? What are the advantages and disadvantages of ceramics for widespread future use?

38. Given two different potential mining sites, is it always true that the one containing a mineral with a higher percent metal will be more profitable to mine? Explain.

39. Briefly explain the "Spaceship Earth" analogy. In what way is the analogy useful? How is it potentially misleading?

40. Describe at least two benefits of discarding less and recycling more of our wastes.

41. In what way are phrases such as "using up" and "throwing away" inaccurate from a chemical point of view? What is the real meaning behind such phrases?

Answers to Questions:

Multiple-Choice

1. B	6. A	11. C	16. C	21. A	26. A	31. A
2. D	7. D	12. D	17. C	22. B	27. D	32. A
3. C	8. C	13. B	18. D	23. B	28. D	33. B
4. B	9. C	14. C	19. B	24. D	29. D	34. C
5. D	10. C	15. A	20. C	25. C	30. C	

Short-Answer

35. The coefficients, in order, are 1,3,2,3.

36. Most metals are solid, shiny, and grey or silver in color. Metals are usually ductile, malleable, and good conductors of both heat and electricity. Most nonmetals are colorless gases. Therefore, they are not ductile or malleable and they are poor conductors.

37. Conventional ceramics are made from abundant clay, which are silicate compounds. Silicates are composed of silicon, oxygen, and aluminum compounds and water molecules together with ions of magnesium, sodium, and potassium. Advantages of ceramics are that they are generally hard, rigid, not chemically reactive , and are resistant to wear. They have relatively high melting points and are strong at high temperatures. A disadvantage is that conventional ceramics tend to fracture if the temperature change is too sudden, but newer engineering ceramics may be able to overcome this problem by including other natural rock materials and firing the mixtures at high temperatures.

38. No, it is not always true that the mining site with the higher percent metal in the mineral will be more profitable. There are other factors to consider, such as the cost of processing, requirements of energy and water, ease of separation, and ultimate disposal of the nonuseful materials.

39. The Earth may be viewed as a spaceship since Earth and its atmosphere represent essentially a closed system. For both Earth and a spaceship, the total amount of available resources is limited since matter can neither be lost nor gained. Some consider that the analogy is a little misleading, however, since Earth continually gains radiant energy from the sun and then partially returns heat energy to space. A spaceship more closely resembles an isolated system in which neither mass nor energy is exchanged with the surroundings.

40. Discarding less waste will decrease the problem of waste storage, and recycling waste may lower the demand for nonrenewable resources. This could result in a decrease in the price of items manufactured from these resources. Other definite benefits include pollution reduction, energy savings, increased numbers of jobs, and even increased political stability in the case of some resources.

41. Atoms cannot be "used up." The law of conservation of mass states that matter cannot be created or destroyed, only converted from one form to another. When the Earth is considered as a whole, there is no "away." Discarded materials remain on the Earth. What is probably meant is that material is used up in that particular form and thrown away from the point of use.

PETROLEUM: TO BUILD ? TO BURN?

Multiple-Choice Questions

1. What is the volume of 1000 g of octane if its density is 0.7 g/mL?
   (A) 70 mL   (B) 140 mL   (C) 1.4 L   (D) 7.0 L

2. Which best describes hydrocarbons?
   (A) flammable ionic compounds that are insoluble in water and are more dense than water
   (B) flammable covalent compounds that are insoluble in water and are less dense than water
   (C) nonflammable ionic compounds that are soluble in water and less dense than water
   (D) flammable covalent compounds that are insoluble in water and more dense than water

3. Ibuprofen, like aspirin, is an anti-inflammatory drug. This is its structural formula. The functional group of this compound indicates that it is

$$CH_3-\underset{\underset{H}{|}}{\overset{\overset{CH_3}{|}}{C}}-CH_2-C_6H_4-\underset{\underset{H}{|}}{\overset{\overset{CH_3}{|}}{C}}-\underset{\underset{O}{\|}}{C}-O-H$$

   (A) an acid.   (B) an ester.   (C) a base.   (D) an alcohol.

4. Which monomer is used to make the polymer Teflon® (polytetrafluorethene)?
   (A) $CH_2=CH_2$ (C) NCH_2-CHCN
   (B) $CF_2=CF_2$ (D) $CH_2=CHCl$

5. If the human "engine" could run on methane, how many grams of methane would be needed to fill a daily need for 3000 kcal? Complete combustion of methane yields about 12 kcal/g.
   (A) 25   (B) 40   (C) 250   (D) 400

6. What was the principal fuel used in the United States prior to 1760?
   (A) coal   (B) oil   (C) wood   (D) natural gas

7. Which fossil fuel has the shortest supply of known reserves?
   (A) coal   (B) oil   (C) wood   (D) natural gas

8. Which is a true statement about the heats of combustion of the alkane series?
   (A) The molar heats of combustion are all approximately the same within the series.
   (B) There is a slight decrease in the molar heat of combustion as the number of carbon atoms increases.
   (C) The energy given off per gram of hydrocarbon increases dramatically as the number of carbon atoms increases.
   (D) Molar heats of combustion increase regularly with an increase in the number of carbon atoms in the chain.

9. Which is not a true statement about petroleum substitutes?
   (A) Production of liquid fuels from coal is a technology that has developed since the 1973 oil embargo.
   (B) Known domestic reserves of coal greatly exceed those of oil.
   (C) Shale oil processing requires large quantities of water.
   (D) Petroleum substitutes are more expensive than petroleum at the present time.

10. Which statement is true?
   (A) Bond breaking is endothermic and bond making is exothermic.
   (B) Bond breaking is exothermic and bond making is endothermic.
   (C) Bond breaking and bond making can be either endothermic or exothermic depending on the bonds.
   (D) Bond breaking is either endothermic or exothermic, but bond making does not involve a change in energy.

11. Which is not a true statement about hydrocarbons in a fractionating tower?
(A) Lighter hydrocarbons have lower boiling points and are drawn off from the top of a fractionating tower.
(B) Heavier hydrocarbons with higher boiling points tend to condense within a fractionating tower and are collected in trays at various heights in the tower.
(C) Lighter hydrocarbons have higher boiling points and are drawn off from the top of a fractionating tower.
(D) Very heavy hydrocarbons tend to remain as liquids in a fractionating tower and are drained from the bottom of the tower.

12. Which is not a true statement?
(A) Octane rating is directly proportional to the molar heat of combustion.
(B) Octane rating is directly proportional to the degree of branching.
(C) Octane rating is a measure of the antiknock characteristics of a hydrocarbon fuel.
(D) Octane rating can be increased by the addition of tetraethyl lead.

13. What is the basic composition of all hydrocarbons?
(A) water molecules and carbon atoms bonded together
(B) hydrogen and carbon atoms bonded together
(C) carbon, hydrogen, and nitrogen atoms bonded together
(D) hydrogen, carbon, and oxygen atoms bonded together

14. When bonds are broken,
(A) energy is given off.
(B) the process is termed isometric.
(C) energy is absorbed.
(D) the process is exothermic.

15. On the basis of their reactivity, which type of hydrocarbon monomers is least likely to form polymers?
(A) aromatics   (B) alkynes   (C) alkenes   (D) alkanes

16. Which is most likely to be true about the future use of petroleum in the United States?
(A) We will continue to use petroleum for fuel until all presently known oil wells in the United States become exhausted or non-productive.
(B) The United States must always depend upon foreign sources of oil.
(C) Diesel engines will disappear as it becomes more expensive to convert gasoline into the diesel oil used to fuel these engines.
(D) We will have to do with less petroleum by-products if we burn the same amount of petroleum.

17. Saturated hydrocarbons from petroleum can be converted into unsaturated hydrocarbons by a process that
(A) removes some hydrogen atoms from the hydrocarbon molecules.
(B) removes sulfur and sulfur compounds from the petroleum.
(C) bonds additional carbon atoms to the already existing saturated hydrocarbon molecules.
(D) breaks bonds and removes carbon atoms from the already existing hydrocarbon molecules.

18. Which is the correct structural formula for butane? Recall that each atom of carbon contains four valence electrons and that a dash (—) is used to represent a shared pair of electrons.

(A)
```
      H H H H
      | | | |
   H--C-C-C-C--H
      |   | |
      H   H H
      |
   H--C--H
      |
      H
```

(C)
```
      H H H
      | | |
   H--C-C-C--H
      |  ‖  |
      H     H
      |
   H--C--H
      |
      H
```

(B)
```
      H H H H
      | | | |
   H--C-C=C-C--H
      | | | |
      H H H H
```

(D)
```
      H H H H
      | | | |
   H--C-C-C-C--H
      | | | |
      H H H H
```

19. Which set of units is most appropriate for expressing the heat of combustion of a compound that burns in oxygen?
(A) g/mol   (B) J/g   (C) cal/g of O_2 (D) mol/J

20. Why did your experimentally determined heat of combustion for paraffin only approximately match the accepted value?
    (A) Different candles contain different amounts of paraffin.
    (B) Paraffin burns completely, producing only carbon dioxide and water.
    (C) The loss of heat from the metal can used as a combustion chamber was not carefully controlled.
    (D) Impurities in the water greatly affected the amount of temperature change during the investigation.

21. Which hydrocarbon has the lowest boiling point?
    (A) methane, CH_4 (C) hexane, C_6H_{14}
    (B) pentane, C_5H_{12} (D) octane, C_8H_{18}

22. As ethene is polymerized into polyethylene,
    (A) smaller molecules combine to form larger ones.
    (B) larger molecules break apart to form smaller ones.
    (C) the atoms in molecules are rearranged to produce isomers.
    (D) the different molecules in a mixture become separated into fractions.

23. The properties of hydrocarbons depend on the number and arrangement of the carbon and hydrogen atoms in their molecules. Which pair of hydrocarbons is most likely to exhibit similar properties?
    (A) methane and ethane   (C) ethane and octane
    (B) ethene and butane   (D) propane and cyclohexane

24. Below are structural formulas for four hydrocarbons. Which is an isomer of butane?

25. Which situation is least likely to result from an embargo of foreign oil?
    (A) an increase in the cost of gasoline
    (B) gas rationing
    (C) an increase in the number of petrochemicals produced
    (D) lower thermostat settings in public buildings

26. The general formula for alkane hydrocarbons is C_nH_{2n+2} and C_nH_{2n} is the general formula for another group of hydrocarbons. Which one?
    (A) alkanes   (B) alkenes   (C) aromatics   (D) napthenes

Questions 27, 28 and 29 all refer to this figure which shows the amount of energy consumed in the United States from 1850 through 1980. For example, almost 60 barrels of oil were "consumed" by each person in the United States in 1970.

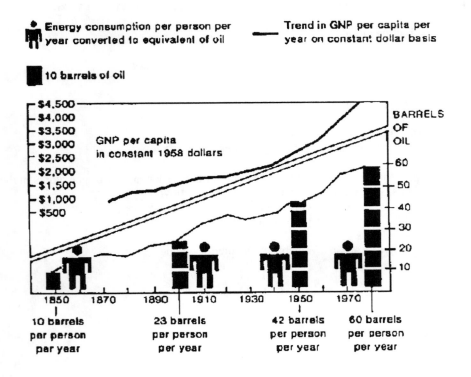

27. How many times higher was petroleum consumption in 1980 than it was in 1850?
(A) two times   (B) six times   (C) sixty times   (D) two hundred times

28. What was the GNP (gross national product) for each person (per capita) in 1950?
(A) $43   (B) $1500   (C) $2500   (D) $4500

29. How many barrels of oil were produced in the U.S. during 1980? Assume a population of 250,000,000, which is 2.5×10^8 people.
(A) 60   (B) 1500   (C) 6×10^5 (D) 1.5×10^{10}

30. Structural formulas are often more useful to chemists than are molecular formulas because structural formulas better represent
(A) the relative number of atoms in the molecules.
(B) the kind of atoms in the molecules.
(C) the three-dimensional geometry of the molecules.
(D) the distribution of the chemical bonds between the atoms in each molecule.

31. Why are there so many different hydrocarbon molecules?
(A) Carbon atoms are much larger than hydrogen atoms, allowing for many different arrangements.
(B) Carbon atoms can readily bond to each other in different ways.
(C) Carbon is found in all organic compounds.
(D) Carbon atoms form tetrahedral-shaped methane molecules.

32. Chemical compounds can be represented by formulas but mixtures cannot. Which substance, obtained from petroleum, is an example of a mixture rather than a pure compound?
(A) heptane   (B) cyclohexane   (C) gasoline   (D) octane

33. Which substance does not separate in the same fraction as the other three during the distillation of petroleum?
    (A) wax (paraffin)   (B) gasolines   (C) napthas   (D) kerosene

34. Why do the properties of petroleum vary depending on where it is mined?
    (A) Petroleum is a mixture of elements and compounds.
    (B) Petroleum has been formed under different conditions during different geological eras.
    (C) Petroleum samples contain different impurities.
    (D) All of these statements about the properties of petroleum are true.

35. Fractional distillation is based on the fact that substances found in crude oil
    (A) have different boiling points.   (C) burn in oxygen.
    (B) are used in making petrochemicals.   (D) are less dense than water.

Short-Answer Questions

36. C_2H_6O exists in two isomers, one with a boiling point of 78.3°C and the other with a boiling point of –24°C.
    a. Sketch the structural formulas for each isomer.
    b. Which isomer has the higher boiling point and explain your choice.

37. a. Write the balanced chemical equation for the combustion of propane, which is the hydrocarbon in bottled gas.
    b. If the molar heat of combustion for propane is 2200 kJ/mol, how many kJ of energy would be given off when 132 g is burned?

38. a. Write the general formula for an alkane and then for an alkene. Provide the name and formula for a specific example of each.
    b. What is the structural difference between an alkane and an alkene? How doe this difference affect chemical reactivity?
    c. Are alkanes or alkenes more likely to be used for building purposes? Why?

39. This word equation and the symbolic equation refer to the same reaction. Complete each equation and then classify the type of reaction represented.
    a   carboxylic acid + alcohol → _____ + _____

    b.   _____ + H – O – R → _____ + _____

40. a. In what sense can petroleum be considered "buried sunshine?"
    b. Explain why solar energy is classed as a renewable energy source but petroleum is not.

41. Why is petroleum so much more versatile as a building material than metallic ores?

42. Suggest the two general strategies of increasing overall energy efficiency within complex machines such as automobiles. Why is this an important objective?

Answers to Questions:

Multiple-Choice

1. C	6. C	11. C	16. D	21. A	26. B	31. B
2. B	7. B	12. A	17. A	22. A	27. B	32. C
3. A	8. D	13. B	18. D	23. A	28. C	33. A
4. B	9. A	14. C	19. B	24. B	29. D	34. D
5. C	10. A	15. D	20. C	25. C	30. D	35. A

Short-Answer

36. a. Here are the two structural formulas for C_2H_6O.

b. The first structure is polar so it will have strong forces of hydrogen bonding between molecules. It should have a much higher boiling point compared to the second structure, which is nonpolar. (The first isomer is ethanol and the second isomer is dimethyl ether.)

37. a. $C_3H_8 + 5 O_2 \rightarrow 3 CO_2 + 4 H_2O + 2200 \text{ kJ}$
    b. $132 \text{ g} \times 1 \text{ mol}/44 \text{ g} \times 2200 \text{ kJ}/1 \text{ mol} = 6600 \text{ kJ}$

38. a. The general formula for an alkane is C_nH_{2n+2} and for an alkene is C_nH_{2n}. Ethane, C_2H_6, is an example of an alkane. Ethene, C_2H_4, is an example of an alkene.
    b. In an alkane, all the bonds are single bonds. In an alkene, at least one of the carbon to carbon bonds is a double bond, providing a site for chemical reactions. Therefore, alkenes are more reactive than alkanes.
    c. Alkenes are likely to be used for building purposes because of the presence of double bonds. Alkenes are used, for example, to build polymers by means of addition reaction.

39. a. carboxylic acid + alcohol → ester + water

    b.

    This reaction, represented both by a word and a symbolic equation, is classed as a condensation reaction.

40. a. Petroleum can be considered "buried sunshine" because it was the energy of the sun that was essential to create the biomass from which petroleum was subsequently produced through the action of pressure and heat on the remains.
    b. Solar energy is a renewable resource unless, in the very long run, something extremely drastic happens to the sun such that it no longer emits energy to the Earth. Petroleum is classed as a nonrenewable resource because the climatic and environmental conditions that produced the biomass no longer exist. Even if they did, the production of petroleum through the action of pressure and heat is so slow that for all practical purpose, petroleum must be classed an a nonrenewable resource.

41. Petroleum is a complex mixture of hydrocarbons of varying length. These compounds can be easily separated by distillation and, if desired, cracked into smaller molecules or reformed into larger ones. Double and triple bonds can be added, subtracted, or substituted. Variations in bonding produces variations in properties such as hardness, flexibility, strength, and reactivity. Metallic ores typically yield one or two different forms of the metal, and do not have the great versatility of carbon atoms.

42. In a complex machine such as an automobile, some useful energy is "lost" or becomes unavailable to do useful work. This is because there are several steps involved in the conversion, and each step is less than 100% efficient. Two general strategies to improve overall energy efficiency would therefore be to decrease the number of energy conversions between fuel and its final use and/or to increase the efficiency of energy conversions.

UNDERSTANDING FOOD

Multiple-Choice Questions

1. A "high energy" food is one that is high in
   (A) vitamins.   (B) calories.   (C) proteins.   (D) minerals.

2. Food for backpacking should be high in food energy per unit weight. Therefore, it should contain a high percentage of
   (A) vitamins.   (B) fats.   (C) proteins.   (D) carbohydrates.

3. Which class of compounds provides the basic building blocks for our bodies?
   (A) minerals.   (B) fats.   (C) proteins.   (D) carbohydrates.

4. Which is not true about saturated fats?
   (A) Saturated fats are usually found as solids at room temperatures.
   (B) Saturated fats are usually derived from animal sources.
   (C) Saturated fats contain many carbon-carbon double bonds.
   (D) Saturated fats are not desirable for people on a low cholesterol diet.

5. Which vitamins are most likely to become toxic if consumed in quantities many times larger than the RDA?
   (A) A and B   (B) B and C   (C) C and D   (D) D and E

6. Which statement about the Fair Packaging and Labeling Act is not true?
   (A) Ingredients must be listed in decreasing order of abundance.
   (B) Both intentional additives and contaminants must be listed.
   (C) Foods that have defined standards of identity do not have to list their ingredients on the label.
   (D) GRAS stands for "generally recognized as safe."

7. Which is not an accurate statement about the foods we consume?
   (A) Foods supply the human body with energy.
   (B) The energy supplied by foods is usually measured in units called Calories or kilojoules.
   (C) Foods, like fossil fuels, contain mostly hydrocarbon compounds.
   (D) It takes energy to produce food, therefore food can be considered a mechanism for transforming energy.

8. Which is a correct statement about foods, their production, and their use?
   (A) A high percentage of the working population in the United States produces the country's food needs.
   (B) Comparing energy input and energy output, animal meat is the most efficient kind of food to produce in the United States.
   (C) Presently, world supplies of food energy are less than required by the world's population.
   (D) The abundance of food is determined by three factors: supply, demand, and distribution.

9. Approximately how many people die each year in the world from hunger and related diseases?
   (A) 100 million (1.0×10^8)   (C) 50 million (5.0×10^7)
   (B) 270 million (2.7×10^8)   (D) 15 million (1.5×10^7)

10. One (1) Calorie is equal to
    (A) 1000 joules.   (C) 1000 calories.
    (B) 1000 kilojoules.   (D) 1000 kilocalories.

11. Which is not considered to be intentional food additives?
    (A) sweeteners   (C) preservatives
    (B) antioxidants   (D) chemicals from packaging material

12. Iodine was used to detect vitamin C's presence in your investigation to determine the amount of the vitamin in different foods because
    (A) iodine, like oxygen, oxidizes the ascorbic acid.
    (B) iodine was the limiting factor in the reaction.
    (C) iodine reacts with vitamin C to form a blue-colored solution.
    (D) iodine also is an essential nutrient.

13. Major sources of vitamin D for humans are fish oil and butter. Which statement is most likely also true for this vitamin?
    (A) It is found in citrus fruits along with vitamin C.
    (B) Larger quantities of the vitamin (RDAs) are required for females than males.
    (C) It dissolves in fat or oil.
    (D) Lack of this vitamin is not related to malnutrition.

14. Which set of elements are included in the major biomolecules of human nutrients?
    (A) Ca, Mg, S   (C) C, H, O, N
    (B) Na, P, K   (D) Cl, Cu, Fe, Zn

15. This equation describes the reaction between magnesium metal (Mg) and hydrochloric acid (HCl).
    $$Mg + 2\,HCl \rightarrow MgCl_2 + H_2$$
    Which statement is true if you add 0.10 moles of Mg to 0.10 moles of HCl?
    (A) Mg will be the limiting reactant.
    (B) Mg will remain in excess.
    (C) Both Mg and HCl will remain in excess.
    (D) The Mg and HCl will exactly react.

16. Here is a list of ingredients needed to make a simple cake.

    | 2 cups flour | 1 cup sugar | 1/3 cup oil |
    | 2 eggs | 1.5 tbsps baking powder | 1 cup water |

    Suppose you have in storage 14 cups of flour, 26 eggs, 15 cups sugar, 15 tbsps of baking powder, 2 gallons of oil, and access to unlimited water. What is the greatest number of these simple cakes you can make?
    (A) 7   (B) 10   (C) 13   (D) 15

17. Amino acids, the building blocks for proteins, contain which set of functional groups?
    (A) —C and —H$^+$ (C) —COOH and —NH$_2$
    (B) —OH and —NH$_2$ (D) —C$_6$H$_5$ and —SO$_2$

18. How does the amount of animal protein included in diets in the United States compare to the amount in diets in the rest of the world?
    (A) People in other countries usually eat more animal protein.
    (B) People in other countries usually eat about the same amount of animal protein.
    (C) People in other countries usually eat less animal protein.
    (D) Data to answer this question are not available, so it is not possible to tell.

19. Which category of food is the body's most efficient source of energy?
    (A) fats   (C) proteins
    (B) carbohydrates   (D) vitamins

20. Which type of procedure can be used to determine the quantity of thermal energy released from food?
    (A) distillation   (C) titration
    (B) weighing   (D) calorimetry

21. A person can lose body weight if total energy intake
    (A) is greater than total energy expended or used.
    (B) is less than total energy expended or used.
    (C) is equal to total energy expended or used.
    (D) is spread out over time, no matter that total energy is expended or used.

22. The starting substance that becomes "used up" first during a chemical reaction is called a
(A) biochemical.
(B) metabolic pathway.
(C) limiting reactant.
(D) calorimeter.

23. Which set contains the primary nutrients for common agricultural crops?
(A) Ca, Mg, S
(B) N, P, K
(C) Mn, Mo, Zn
(D) B, Cl, Cu, Fe

24. Which general structural formula indicates an organic alcohol?

(A)
$$R-\overset{\overset{O}{\|}}{C}-O-H$$

(C)
$$R-O-H$$

(B)
$$R-\overset{\overset{H}{\|}}{N}-H$$

(D)
$$R-\overset{\overset{O}{\|}}{C}-O-\overset{\overset{H}{|}}{\underset{\underset{H}{|}}{C}}-R$$

25. Which statement about malnourishment is true?
(A) A malnourished person cannot build all of the required new body structures.
(B) Undernourishment always accompanies malnourishment.
(C) Inefficient burning of fuel is the major problem with malnourishment.
(D) No decomposition of H_2O molecules can take place so dehydration occurs.

26. This table indicates the percent of total calories obtainable from the protein, carbohydrate, and fat in two common foods, American cheese and peanut butter.

	percentage of total calories		
	from proteins	from carbohydrates	from fats
American cheese	24	1.7	75
Peanut Butter	17	12	71

Which is the most accurate statement about these two foods?
(A) Gram for gram, peanut butter supplies more calories from carbohydrates than does American cheese.
(B) Neither American cheese or peanut butter supplies a high percentage of calories from fats.
(C) Peanut butter supplies more calories from carbohydrates than from fat.
(D) American cheese supplies 10 times as many calories from its fat content than does peanut butter.

27. An empty, clean, dry beaker weights 34.58 g. The beaker and a milk sample weigh 51.87 g. After extracting, filtering, and drying the protein portion of the milk, the protein and filter paper weighs 1.01 g. The dry, clean filter paper weighs 0.39 g. What is the percent protein in the milk?
(A) 2.3%
(B) 3.6%
(C) 5.8%
(D) 17.3%

28. Which is not a cause of world hunger?
(A) availability of food
(B) growth in population
(C) level of income
(D) mortality rate of infants

29. What percentage of the population in the United States suffers from hunger?
(A) 2%
(B) 8%
(C) 16%
(D) 32%

30. One cup of milk contains about 300 mg of calcium. If a 16-year old female wants to meet the Recommended Daily Dietary Allowance for Calcium which is 1200 mg, how many cups of milk should she drink?
(A) 1 cup
(B) 3 cups
(C) 4 cups
(D) 8 cups

31. Given:
100 grams of wheat provide 330 Calories of energy
1 bushel of wheat has a mass of 28 kg
1 bushel of wheat costs $4

Compute the approximate cost of using only wheat to meet one person's needs for 3000 Calories (kilocalories) per day for a period of 30 days.
(A) $4
(B) $12
(C) $100
(D) $500

32. What process does this equation represent?

$$6\,CO_2 + 6\,H_2O + energy \rightarrow C_6H_{12}O_6 + 6\,O_2$$

(A) photosynthesis   (C) burning of hydrocarbons
(B) respiration   (D) reduction of metallic ores

33. How many grams of protein should a 170-pound, 16-year old boy consume per day?

Daily Recommended Dietary Allowance (RDA)
for Protein

Age Range	Grams Protein (per pound of ideal body weight)
Infants	
8 - 8.5 months	1.88
8.5 months - 1 year	0.98
Children	
1-3 years	0.81
4-6 years	0.68
7-10 years	0.55
11-14 years	0.45
15-18 years	0.39
Adults	
19 and over	0.36
Pregnant women	0.62
Nursing women	0.53

(A) 6.2 g   (B) 39 g   (C) 47 g   (D) 66 g

34. What is the total number of kilocalories of energy utilized by a student sitting in class for 5 hours, walking 2.5 mph for 30 minutes, and standing for 30 minutes?

Energy Expenditure for Typical Activities	
Activity	Kilocalories per hour
Lying down or sleeping	88
Sitting	188
Driving automobile	128
Standing	148
Walking, 2.5 mph	218
Bicycling, 5.5 mph	218
Lawn moving, power	258
Volleyball, roller skating	358
Walking, 3.75 mph	388
Swimming, 8.25 mph	388
Tennis	428
Bicycling, 13 mph	668
Skiing, 18 mph	688

(A) 550   (B) 1120   (C) 3300   (D) 2770

Short-Answer Questions

35. a. What is meant by an "essential" amino acid?
   b. What is the special significance of this term for vegetarians?

36. What groups of individuals need the highest intake of protein per body mass? Why are their needs the greatest?

37. How can you account for the immense variety of different types of proteins?

38. Why is it wise to consider both risks and benefits when discussing the banning of food additives? Use either sodium nitrite or aspartame to discuss your reasoning.

39. a. List at least three factors that influence the diet of a given nation.
    b. What common biological purpose do all national diets share?

40. Explain how the issue of world hunger can be considered either an energy crisis or a resource crisis at both the individual and societal levels.

41. a. In reference to an individual's diet, explain the wisdom in the adage "variety is the spice of life"
    b. Is simply counting calories a wise way to approach dieting?

42. The dimensions of world hunger have centered on supply, demand, and distribution concerns. Briefly discuss two factors for each of these concerns.

43. Consider the factors that determine world food supply as the "reactants in a chemical equation."
    a. What natural phenomena act as limiting reactants in the production of the total food supply?
    b. How can modern science and technology help to overcome these natural "limiting reactants?"
    c. Are there "limiting reactants" affecting food supply that cannot be overcome by science and technology alone?
    d. How can science and technology work with existing systems in society to overcome these societal "limiting reactants?"

Answers to Questions:

Multiple-Choice

1. B	6. B	11. D	16. A	21. B	26.A	31. A
2. B	7. C	12. A	17. C	22. C	27.B	32. A
3. C	8. D	13. C	18. C	23. B	28.D	33. D
4. C	9. D	14. C	19. B	24. C	29.B	34. B
5. D	10. C	15. B	20. D	25. A	30.C	

Short-Answer

35. a. Of the 20 different amino acids needed by the human body, eight cannot be internally synthesized and must occur in the diet. Their absence will cause certain key reactions to be limited even if other necessary "ingredients" are in excess. Most animal sources of protein are complete in that they contain all eight essential amino acids.
    b. If one consumes only plant-derived food, one must be sure to eat a variety of foods that "complement" each other's deficiencies so that all the amino acids are present in the diet.

36. Growing children, pregnant women and nursing women need the most protein because, in addition to needing protein for maintenance and repair of body structures, they need to provide for building new structures. This also at least partially explains why these groups are most at risk for malnutrition in times of food shortages.

37. Proteins are polymers built up from 20 different amino acid monomers. They can differ from each other in terms of their length (number of monomer units) and the type and sequencing of specific amino acids. Just as 26 letters can form thousands of words, 20 different amino acids can build an immense variety of proteins.

38. If additives serve essential health-protecting purposes, then the wisest course is to continue using the additive if the benefit is high, the risk is low, and no safer substitute is available. In the case of sodium nitrite, it serves a beneficial anti-botulism function, yet is suspected of forming carcinogenic nitrosamines. As few satisfactory alternatives are available, it continues to be used. Aspartame provides approximately 200 times the sweetening of an equivalent number of calories of sugar so can be an effective diet aid. However, a small percent of people have been found to have adverse reactions to it. Again, use continues until appropriate substitutes are found.

39. a. Key factors include the natural environment and its specific biological resources; the physical environment such as soil, rainfall, and climate; and ingrained cultural eating patterns.
    b. All foods serve the same "burning and building" functions, and are metabolized via the same biochemical processes.

40. World hunger is an energy crisis in at least two ways: (1) Large numbers of people are undernourished in the sense that they have an insufficient caloric intake to fuel bodily activities; (2) The energy needed to produce food manufacturing fertilizers and pesticides, and fuel for planting and harvesting is not available in many developing countries.

    World hunger is a societal resource crisis in that some foodstuffs are used for building materials, much like other chemical resources. Developing countries often lack chemical resources (ores, natural gas, petroleum, etc.) as much as they lack adequate supplies of protein, carbohydrates, fats, vitamins, and minerals for food.

41. a. Eating a variety of foods increases the likelihood of an adequate balance of nutrients.
    b. Simply counting calories during a weight-loss diet is not wise because the body uses foods for building as well as for burning.

42. The world's supply of food has increased by scientific and technological advances in the areas of fertilization, pest reduction, irrigation, food storage and processing, and the development of plant and animal hybrids that give higher food yields. Other factors that affect food supply include a country's socioeconomic and political forces, ecological variables such as climate and soil quality, and aid from other countries.

    The demand for food within a country is influenced by the size and age/sex distribution of the population, as well as by the dietary customs (especially the consumption of meat) and the overall standard of living. The distribution of food supplies is determined by the extent, effectiveness, and reliability of a country's transportation network, as well as by the internal economic forces (sufficient profit for the producer or seller and sufficient funds for the purchaser). In some countries, the ability to store or market foodstuffs also affects the distribution of goods.

43. a. Pest infestations, low-fertility soils, inadequate rainfall, crops with genetic inefficiencies, etc.
    b. Agrochemistry, entomology, molecular biology, and related technologies can increase crop and livestock yields by pushing back these limits.
    c. Yes, a variety of economic, social, political, and cultural factors prevent a society's poorest members from either growing their own food or having the financial resources to be able to buy it at typical market prices.
    d. The supply of natural gas and petroleum-mechanization of farming (one U.S. farmer feeds 78 people compared to one farmer in a developing nation who can just barely support his family) and fertilizer production (as much as 132 kg are applied per hectare in the United States to less than 30 kg/hectare in developing nations) are both fossil-fuel intensive. It has been estimated that if U.S. food production methods and diet were extended to the entire world, the world's known energy reserves would be exhausted in 29 years. However, it is possible to combine some aspects of modern technology with more labor-intensive farming practices in order to increase \individual and national food and economic self-sufficiency in developing countries. Work to date with such appropriate or "soft" technologies is most encouraging.

NUCLEAR CHEMISTRY IN OUR WORLD

Multiple-Choice Questions

1. Why are there relatively few naturally occuring radioisotopes on Earth?
   (A) Radioisotopes can only be produced by synthetic reactions.
   (B) Most radioisotopes originally on Earth have now decayed to stable isotopes.
   (C) All radioisotopes have half-lives on the order of only a few days or years.
   (D) There are naturally occuring radioisotopes but detection is difficult.

2. Large-scale use of transuranium metallic elements for construction purposes is unlikely because such elements
   (A) emit radiation.
   (B) are not found in nature and must therefore be synthesized via nuclear reactions.
   (C) are unstable; over time they decay to form other elements.
   (D) emit radiation, are not found in nature, and are unstable.

3. How is uranium-235 unique relative to the other isotopes of uranium?
   (A) It is the only naturally occurring radioisotope of uranium that is suitable as a fuel for fission reactors.
   (B) It is the only naturally occurring radioactive isotope of uranium.
   (C) It can be used as a fuel in nuclear power plants, but not in nuclear bombs.
   (D) It is the only radioisotope that undergoes fission.

4. Why is it impossible for a nuclear power plants to explode like a nuclear bomb?
   (A) Nuclear power plants contain far less fissionable material per unit volume than there is found in a nuclear bomb.
   (B) Nuclear power plants and nuclear bombs rely on a different type of fuel.
   (C) Nuclear power plants and nuclear bombs rely on entirely different nuclear reactions, even though the fuel is the same.
   (D) Far from being impossible, nuclear power plants often do produce explosions like nuclear bombs.

5. The half-life of plutonium-239 is 24,300 years. If a nuclear bomb released 8 kilograms of this isotope, how many years would pass before the amount was reduced to 1 kilogram?
   (A) 12,150 years   (C) 48,600 years
   (B) 24,300 years   (D) 72,900 years

6. Which lists electromagnetic radiation in order of increasing energy?
   (A) gamma rays, infrared, ultraviolet, visible, X rays
   (B) infrared, visible, ultraviolet, X rays, gamma rays
   (C) visible, infrared, ultraviolet, gamma rays, X rays
   (D) X rays, gamma rays, visible, ultraviolet, infrared

7. Which is not a true statement about nuclear radiation?
   (A) A vacuum will serve as an effective barrier to nuclear radiation.
   (B) Intensity of nuclear radiation decreases with increasing distance from the source.
   (C) Gamma rays are the most penetrating and alpha rays the least penetrating form of radiation.
   (D) Nuclear radiation can both cause cancer and be used to treat cancer.

8. Which is not a correct statement comparing nuclear power plants with coal-fired plants?
   (A) Nuclear power plants are more efficient than coal-fired plants in converting fuel to energy.
   (B) Nuclear plants produce little air pollution compared to coal-fired plants.
   (C) Nuclear plants use significantly less fuel than coal-fired plants.
   (D) Construction of nuclear plants has met with more public opposition than is the case for coal-fired plants.

9. Which process theoretically yields the greatest energy per mole of fuel?
   (A) burning natural gas (C) fission of uranium-235
   (B) burning coal   (D) fusion of hydrogen

10. Cathode rays are now known to be the same as
    (A) X rays.   (C) electrons.
    (B) visible light.   (D) alpha particles.

11. Which activity is least likely to result in any new scientific discovery?
    (A) investigating things that people do not understand
    (B) observing the effects that one substance produces on another
    (C) preparing one's mind through education
    (D) accepting all information you hear or read about as correct

12. Ernest Rutherford proved that pitchblende emitted three types of radiation in his famous lead block experiment. What was the composition of radiation that was not deflected by the magnets in this experiment?
    (A) relatively heavy, positively-charged subatomic particles
    (B) negative ions
    (C) electromagnetic radiation
    (D) electrons

13. In the classical gold-foil experiment, alpha particles were focused on a very thin film of gold. Behind the gold foil was placed a semi-circular zinc sulfide screen. Zinc sulfide fluoresces when hit by alpha particles. The pattern of the fluorescent flashes indicated that, instead of the expected spray effect, most of the alpha particles passed directly through the foil without being deflected. However, very few alpha particles bounced back or were deflected towards the alpha particle source. Which is not an interpretation of the results of this very famous experiment?
    (A) The alpha particles that bounced back must have hit heavy, positively charged particles.
    (B) Atoms of gold consist mainly of empty space.
    (C) Atoms consist of positively charged nuclei.
    (D) Atoms consist of electrons. These electrons take part in usual chemical changes.

14. Which statement about atomic structure is correct ?
    (A) The nucleus occupies about one ten-thousandth the diameter of each atom.
    (B) The molar mass of a neutron is about ten thousand grams.
    (C) Isotopes of the same element have different atomic numbers.
    (D) The mass number of an element is equal to the number of electrons outside of the nucleus of each of its atoms.

15. Which statement is true for the two symbols $^{19}_{9}F$ and $^{19}_{9}F^-$?

    (A) Both symbols represent fluoride ions.
    (B) Both symbols represent neutral atoms of fluorine.
    (C) Both symbols indicate that individual Atoms of fluorine contain 9 protons.
    (D) Each symbol indicates a different isotope of fluorine.

16. Which is the most accurate statement about the relationship between protons and neutrons in the nuclei of atoms?
    (A) Atoms of the heavy elements contain the same number of protons as neutrons.
    (B) Each atom can contain only one proton, since like charges repel each other.
    (C) The number of neutrons not protons, within atoms of a particular element determines which isotope it is.
    (D) The number of protons and neutrons contained in all atoms of any particular element never changes.

17. Which statement is correct for the symbol $^{14}_{6}C$?

    (A) It represents the isotope of carbon with 8 neutrons.
    (B) It represents the isotope of carbon with a mass number of 12.
    (C) It represents the isotope of carbon which contains 14 protons.
    (D) It represents the most common isotope of carbon.

18. The isotope with the symbol $^{238}_{92}U$ makes up over 99 percent of all uranium known to humanity on Earth. However, $^{235}_{92}U$ is the isotope of uranium most useful to nuclear scientists because it has unique properties.

    Which statement follows directly from this information??
    (A) The average atomic mass of uranium is closer to 235 than to 238.

    (B) $^{235}_{92}U$ is one of the synthetic isotopes of uranium.

    (C) Relatively large amounts of uranium ore must be mined in order to obtain small amounts of $^{235}_{92}U$.

    (D) $^{235}_{92}U$ splits into elements of lower atomic numbers when bombarded with neutrons.

19. If an element has an atomic number of 10 and an atomic mass number is 20, which statement is true?
    A particular isotope of this element must have
    (A) a relative mass of greater than 20.
    (B) an atomic number of 10.
    (C) ions that are neutral or have no charge.
    (D) atoms whose nuclei each contain 10 electrons.

20. The half-life of radium-226 is 1620 years. What fraction of an original sample of radium-226 will remain after 4860 years have passed?
    (A) one-half   (B) one-fourth   (C) one-eighth   (D) one-sixteenth

21. If a radioactive atom emits an alpha particle, the atom is changed to one with
    (A) a higher mass number and a higher atomic number.
    (B) a lower mass number and a lower atomic number.
    (C) a higher mass number and a lower atomic number.
    (D) a lower mass number and a higher atomic number.

22. Which is the most penetrating type of radiation given off by radioactive atoms?
    (A) alpha particles   (C) gamma rays
    (B) beta particles   (D) ultraviolet light

23. If a radioactive atom emits a beta particle, the atomic number of that atom increases by one. Beta emission also causes the atomic mass number of the same atom to
    (A) increase significantly.   (C) remain the same.
    (B) increase slightly.   (D) decrease significantly.

24. Which radioisotope is often used in medical treatment to destroy diseased tissue?
    (A) uranium-235   (C) cobalt-60
    (B) uranium-238   (D) strontium-90

25. Which equation represents an synthetic nuclear transformation?
    (A) $^{238}_{92}U \rightarrow {}^{4}_{2}He + {}^{234}_{90}U$

    (B) $^{1}_{1}H + {}^{9}_{4}Be \rightarrow {}^{6}_{3}Li + {}^{4}_{2}He$

    (C) $^{1}_{0}n \rightarrow {}^{1}_{1}H + {}^{0}_{-1}e + antineutrino$

    (D) $^{14}_{6}C \rightarrow {}^{14}_{7}N + {}^{0}_{-1}e$

26. Which product from nuclear fission reactions can be used to produce nuclear chain reactions?

    (A) $^{2}_{1}H$ (B) neutrons   (C) gamma rays   (D) barium ions

27. Which will not protect you from hazardous radiation exposure?
   (A) limiting the time of your exposure to radiation
   (B) staying far away from the radiation source
   (C) detecting the radiation with a rate meter
   (D) placing lead barriers between yourself and the source of radiation

28. Which nuclear particle is emitted as an atom of uranium $^{238}_{92}$U changes to an atom of thorium $^{234}_{90}$Th?

   (A) a proton   (C) an electron
   (B) a neutron   (D) an alpha particle

29. Which is not a reason that high-level radioactive wastes can be hazardous to humans?
   (A) flammability   (C) generation of heat
   (B) toxicity   (D) emission of harmful radiation

30. Which is correct about the nuclear decay process given by this equation?

$$^{210}_{82}Pb \rightarrow \quad ^{210}_{83}? \quad + \quad ^{0}_{-1}e$$

   (A) A new isotope of lead is formed.
   (B) A different element is formed.
   (C) A positive ion is formed by the loss of an electron.
   (D) A nucleus of lower atomic mass than the original nucleus is formed.

31. Which is incorrect about the nuclear decay process described by this equation?

$$^{241}_{95}Am \quad \rightarrow \quad ^{237}_{93}Np \quad + \quad ^{4}_{2}He$$

   (A) The decay produces alpha particles.
   (B) The decaying atom is of an isotope of the element americium.
   (C) Atomic number and mass are conserved.
   (D) Np is an isotope of the element Am.

32. In 1990, a large city hospital purchased 100 mg of Co-60 radioisotope. Given that the half life of Co-60 is close to five years, how much pure Co-60 would be expected to remain by the year 2000?
   (A) 250 mg   (C) 50 mg
   (B) 100 mg   (D) 25 mg

33. What percentage of all the radiation produced by nuclear power plants results in human exposure?
   (A) far more than 50%   (C) somewhere between 1 and 25%
   (B) about 25 %   (D) far less than 1%

These two statements concern a proposed nuclear power plant. Use them to answer Questions 34, 35, and 36.

Statement from a Power Company Representative
"Our company is very eager to build a new nuclear power plant. Our studies have shown that the need for electrical power in the county will increase by 28% in the next 10 years. Our present plant will be able to supply no more than 7% of that need. The people of this area will continue to use more and more electrical power over the next 30 years, and nuclear power is the best source available to them."

Statement from a Famous Nuclear Physicist
"Nuclear power is a very dangerous resource. The world has already experienced the terrible destruction that can come from nuclear bombs. The electrical power that people in this county need can be provided by means that are both safer and less expensive than nuclear power."

34. What do you conclude about the statement made by the power company's representative?
   (A) Only factual information was presented by the power company official.
   (B) The statement contains only expressions of opinion.
   (C) Both factual information and statements of opinion are presented.
   (D) The statement is only company propaganda about nuclear power.

35. What do you conclude about the famous nuclear physicist's statement?
(A) The statement contains factual information only.
(B) The opinions given are ones with which we should agree because the nuclear physicist is famous..
(C) It is possible that the scientist can have some of the same kinds of biases that a non-scientist might have.
(D) The scientist's opinions are based solely on scientific information in this case.

36. In comparing the two statements, what do you conclude?
(A) The statement from the power company representative, being a agent of business, probably contains false information.
(B) The statement from the power company representative, being a agent of business, probably contains the best economic advice for the county.
(C) The statement from the scientist, being an intellectual, should probably be the one to base action upon.
(D) Since the views expressed in both statements are different, we should seek and evaluate additional opinions on the question.

Use this table to answer Questions 37 and 38.

Isotope	Half-life	Radiation emitted
A	6.3×10^6 years	α
B	3.25 years	β
C	4.6 seconds	β

37. Which of the three isotopes is most likely to occur naturally on Earth?
(A) A   (B) B   (C) C   (D) all equally likely

38. Workers in a nuclear power plant wear film badges containing a radiation-sensitive substance to help monitor their exposure. Which isotope would emit radiation that could not be detected by a film badge?
(A) A   (B) B   (C) C   (D) none; all detected

Short-Answer

39. Fill in the rest of the information in this table. (These three isotopes were the principal radioisotopes emitted to the atmosphere during the Three Mile Island nuclear power plant accident.

Isotope	Nuclear Notation	Number of protons	Number of neutrons
krypton-88			
xenon-133			
xenon-135			

40. The three naturally-occurring decay series which begin with U-238, U-235, and Th-232 all eventually result in stable isotopes of lead (Pb-206, Pb-207, and Pb-208 respectively). Complete the following reactions that represent the last three steps in the U-235 decay series.

a) _____ $\rightarrow$ $_{-1}^{0}e + _{83}^{211}Bi$

b) $_{83}^{211}Bi \rightarrow$ _____ $+ _{84}^{211}Po$

c) $_{84}^{211}Po \rightarrow _{82}^{207}Pb +$ _____

41. Natural chlorine is a mixture of isotopes. Determine its molar mass if 75.5% of chlorine is Cl-35 and 24.5% is Cl-37.

42. Explain two basic nuclear concepts that relate to the management of nuclear waste.

43. Living adjacent to a normally functioning nuclear power plant increases a person's annual radiation dose by about 1 mrem. This is about the same as a cross-country jet flight. Do you think that this means that the risks are directly comparable and negligible? Explain your answer.

44. List the key components and explain the operation of a nuclear power plant.

45. Discuss the radon issue as an example of how attempts to resolve one problem can sometimes create new, unanticipated problems..

46. Identify at least three statements from the nuclear survey that are false and explain why they are false.

Answers to Questions:

Multiple-Choice

1. B	6. B	11. D	16. C	21. B	26. B	31. D	36. D
2. D	7. A	12. C	17. A	22. C	27. C	32. D	37. A
3. A	8. A	13. D	18. C	23. C	28. D	33. D	38. D
4. A	9. D	14. A	19. B	24. C	29. A	34. C	
5. D	10. C	15. C	20. C	25. B	30. B	35. C	

Short-Answer

39.

Isotope	Nuclear Notation	Number of protons	Number of neutrons
krypton-88	$^{88}_{36}Kr$	36	52
xenon-133	$^{133}_{54}Xe$	54	79
xenon-135	$^{135}_{54}Xe$	54	81

40. a. $^{211}_{82}Pb$ b. $^{0}_{-1}e$ c. $^{4}_{2}He$

41. $0.755 \times 35 = 26.4$
    $0.245 \times 37 = \underline{9.1}$
                  35.5 g/mol or 36 g/mol

42. One basic concept is the inverse square law. This states that the intensity of radiation decreases as the inverse of the distance squared. Therefore, the further isolated the waste is from humans, the greater the safety factor. A second concept of that of half-life. Each radioisotope will decay to half its original activity by a time determined by its half-life. The radioactivity of nuclear waste decreases over time. Therefore, the longer waste is isolated from humans the safer it becomes.

43. No, typically people can choose to take a flight or not whereas, due to economic constraints, they seldom have as much freedom about where to live. Also, a comparison of this type is often used to sidestep the tougher issues surrounding long-term nuclear waste management and the possibility of nuclear catastrophe.

44. Except for the source of heat, a nuclear power plant is much like a coal-fired plant. The most common type of nuclear reactor in the United States is called the pressurized water reactor. This type of power plant consists of uranium and plutonium fuel rods, interspersed with cadmium and boron control rods, submerged in a closed water system. This is called the core. As the neutron-absorbing control rods are withdrawn, the mass of uranium and plutonium exposed to neutrons increases. As the uranium and plutonium are bombarded by neutrons, they undergo fission and this releases tremendous heat as well as additional neutrons. The heat produced by the fissioning fuel is taken up by the water in the closed system and this water is circulated through a heat exchanger. The heat exchanger transfers the heat from the closed water system to a secondary water system. The heat converts the water in the secondary system into steam and the pressure thus generated spins a turbine which generates an electric current. It is important to note that the water in the closed system becomes highly radioactive and, for this reason, it must never come into contact with the water in the secondary system. The amount of heat produced in the core of a reactor is regulated by inserting or withdrawing the control rods. As the control rods are inserted, the cadmium and boron of which they are made absorb neutrons and this decreases the number of fission reactions that occur.

45. In our attempts to insulate our homes from the outside air and thereby conserve energy, we have sealed them so effectively that many gases, including radon, seep in from below foundations, and are trapped inside. Since the longer a person is exposed to radon, the more dangerous it is, sealing it into homes has substantially increased its ill effects.

46. Answers will vary.

CHEMISTRY, AIR, AND CLIMATE

Multiple-Choice Questions

1. Which is responsible for the largest total amount of gaseous air pollutants in the United States?
   (A) industrial processes
   (B) fuel burning for space heating and electricity
   (C) transportation
   (D) solid waste disposal

2. Which condition would result in an increase in the Earth's temperature?
   (A) removing water vapor from the atmosphere
   (B) increasing the concentration of carbon dioxide
   (C) increasing the cloud cover
   (D) covering the Earth with snow

3. Which substance bonds more strongly to hemoglobin than oxygen?
   (A) SO_2 (B) NO_2 (C) O_3 (D) CO

4. The interaction of two chemicals to give an effect greater than expected is called
   (A) catalysis.   (C) synergism.
   (B) oxidation.   (D) electrolysis.

5. Which statement best applies to the Earth's atmosphere?
   (A) All "foreign" substances entering the atmosphere are disruptive to its normal operation.
   (B) Air, in the atmosphere, is a free resource for all humans.
   (C) Natural chemistry cycles influence climate and natural recycling of the gases in the atmosphere.
   (D) In the absence of air from the atmosphere, you would live but a few hours.

6. What does the phrase "mining the atmosphere" mean?
   (A) Clean air, found in the atmosphere, is a pure substance that can be removed or mined.
   (B) Impurities from air can be removed or mined.
   (C) A decrease of carbon dioxide has occurred in the atmosphere recently due to mining.
   (D) Various elements and compounds can be separated from samples of air by fractional distillation.

7. Which statement best explains why air is gaseous?
   (A) The atmosphere exerts pressure on all objects in contact with it.
   (B) The components of air have very low boiling points.
   (C) A decrease of carbon dioxide has occurred in the atmosphere recently due to mining.
   (D) The composition of air changes from the time it enters to the time it leaves our lungs.

8. If the average adult being inhales and exhales 14 times each minute, and the average amount inhaled and exhaled each time is 500 mL, how many liters of air are inhaled and exhaled in one hour?
   (A) 840 L   (B) 420 L   (C) 60 L   (D) 0.5 L

9. Which statement about the oxygen you inhale each day is not correct ?
   (A) Only a small portion of this oxygen actually enters the bloodstream through the lungs.
   (B) At least part of the oxygen taken into the body causes $C_6H_{12}O_6$ to be converted to CO_2 and H_2O.
   (C) Oxygen usually serves as a limiting factor in the process of respiration.
   (D) Oxygen and glucose react exothermically.

10. Which statement is correct in matching properties with gases?
    (A) Both N_2 and CO_2 support the combustion of wood, but O_2 does not.
    (B) Neither N_2 nor CO_2 support the combustion of wood, but O_2 does.
    (C) Both N_2 and CO_2 form insoluble substances with limewater, but O_2 does not.
    (D) O_2 is relatively insoluble in water, but N_2 and CO_2 are quite soluble.

11. Which statement supports the idea that equal volumes of gases at the same temperature and pressure contain equal number of molecules?
    (A) Elements like argon are found in small quantities in the air.
    (B) All gases have the same density at standard conditions.
    (C) The space existing between the molecules of gases is great compared to the sizes of the molecules themselves.
    (D) Two liters of hydrogen will combine exactly with one liter of oxygen to form water.

12. Which statement is true under standard conditions of pressure and temperature for the reaction represented by this equation?

$$3 H_2(g) + N_2(g) \leftrightarrow 2 NH_3 (g)$$

    (A) N_2 is always the limiting reactant.
    (B) 6 g of H_2 ideally will react with 14 g of N_2.
    (C) 6 liters of H_2 ideally will react with 1 liter of N_2.
    (D) Two moles of NH_2 formed will occupy a volume of about 44.8 L.

13. Which statement best explains why hydrogen and helium gases, which make up about 99% of the universe's mass, are found mainly in the Earth's thermosphere rather than in the troposphere?
    (A) These are the lightest elements. Therefore, they tend to escape easily from the Earth's surface or center.
    (B) These two gases have approximately the same molar mass.
    (C) Both elements are chemically inert and therefore do not react with elements near the Earth's surface.
    (D) The two elements both contain isotopes with similar mass numbers, 2 and 3.

14. Which statement is consistent with the observation that the pressure of the Earth's atmosphere at sea level supports a column of water 10.3 m high?
    (A) The atmosphere will also support a column of mercury 10.3 m high.
    (B) A 10.3 m high column of water exerts a pressure equal to 760 mmHg so the pressure of the atmosphere may be expressed either as 10.3 m H_2O or 760 mmHg.
    (C) The great pressure of the atmosphere is produced by the relatively heavy atoms and molecules that make up the atmosphere.
    (D) The atmospheric pressure on top of a high mountain is considerably less than it is at sea level.

15. Boyle's Law is often expressed as $P_1V_1 = P_2V_2$. Which is also a correct expression?
    (A) $\dfrac{P_1}{V_1} = \dfrac{P_2}{V_2}$

    (C) $\dfrac{P_1 \cdot V_1}{P_2 \cdot V_2} = 1$

    (B) $\dfrac{P_1 \cdot V_1}{P_2} = \dfrac{P_2 \cdot V_2}{P_1}$

    (D) $\dfrac{V_1 \cdot V_2}{P_2} = P_1$

16. Which can be used to explain both Boyle's Law and Charles' Law?
    (A) atomic theory
    (C) kinetic molecular theory
    (B) law of conservation of atoms
    (D) law of conservation of charges

17. Two balloons, one filled with oxygen and the other with helium, contain the same volume of gas at the same temperature and pressure. They both are tightly sealed. By the next day, the helium-filled balloon has become much smaller than the oxygen-filled balloon. Why did this happen?
    (A) The air pressure in the room changed overnight.
    (B) The temperature of the air in the room changed overnight.
    (C) Helium molecules are smaller and lighter than oxygen molecules. Therefore, they escape more readily through the plastic or rubber of the balloon.
    (D) Helium reacts chemically with the plastic or rubber that makes up the balloon; oxygen does not react.

18. You are at the beach on a sunny day and you observe that the sand gets very hot by midday, while the water temperature remains about the same. Which is the best explanation of this observation?
    (A) The cool breeze makes the water feel colder than the sand.
    (B) Sitting on the hotter sand makes the water feel cooler.
    (C) Sand both absorbs and gives off heat more readily than water.
    (D) Hotter sand causes a sea breeze to be produced which cools the water.

19. The planet Mars has a surface temperature about forty degrees below that of Earth. Which statement helps to explain this observation?
    (A) The CO_2 and H_2O molecules in the Earth's atmosphere are good absorbers of infrared photons, trapping some of these photons and returning some of them to the Earth's surface. Mars does not have CO_2 and H_2O in its atmosphere.
    (B) CO_2 and H_2O react in the atmosphere of the Earth to produce heat energy which radiates to the Earth's surface as infrared photons but Mars does not have CO_2 and H_2O in its atmosphere.
    (C) Ultraviolet photons break single chemical bonds existing in water molecules in the atmosphere, producing H_2 and O_2 and heat which radiates to the Earth's surface but Mars does not have water molecules in its atmosphere.
    (D) Unlike the Earth's atmosphere, molecular motion in Mars' atmosphere is virtually nonexistent, thus little heat is produced.

20. Seeding clouds to produce rain would be one useful way of "supplying" rain water in areas of drought. Which statement best explains why this procedure is not considered to be too effective?
    (A) Cloud seeding is an expensive means of producing rain.
    (B) There is a limited amount of water vapor in the atmosphere. Eventually therefore, cloud seeding would not work.
    (C) Most air pollutants are not soluble in water and therefore rain would not rid the atmosphere of most pollutants.
    (D) The direction of air currents is relatively unpredictable. Therefore, the rain might not fall where most needed.

21. Electrostatic precipitators work in reducing or controlling air pollution because
    (A) many pollutants exist as charged ionic particles, or can form charged particles.
    (B) precipitators utilize water as an ionizing catalyst.
    (C) precipitators eliminate the need for waste disposal.
    (D) some useful products can be obtained from the pollutants collected by precipitators.

22. Catalytic converters in automobiles made after 1975 reduce pollution from auto exhausts. Which chemical equation best describes the main action of these converters?
    (A) $2\ NO + 2\ CO \rightarrow N_2 + 2\ CO_2$
    (B) $CH_4 + O_2 \rightarrow 2\ HCHO$
    (C) $N_2 + 2\ O_2 \rightarrow 2\ NO_2$
    (D) $Ca(OH)_2 + SO_2 \rightarrow CaSO_4 + H_2O$

23. Which matches the phrase "a hydrogen ion concentration of 10^{-7} moles/L"?
    (A) a pH of 7   (C) a basic water solution
    (B) an acidic water solution   (D) pure rain water

24. Which approach could immediately have the greatest impact on air quality?
    (A) Study the results of chemical research activity.
    (B) Reduce the miles driven by automobiles.
    (C) Vary the amount of oxygen produced through photosynthesis.
    (D) Require installation of precipitators by all industries.

25. Which is most closely related to a mechanism for reducing pollution in the atmosphere?
    (A) use of a catalyst in automobile exhaust systems
    (B) increased federal support for research in this area
    (C) purchase of fewer consumable products
    (D) increased unemployment in the United States

Short-Answer

26. Write an equation that represents how each atmospheric pollutant is formed and indicate what process is the primary source of the pollutant.
    a. CO   b. NO   c. SO_2

27. Neon is a noble gas that is used in "neon" advertising signs. What is the volume in liters of 2.0 moles of neon at STP?

28. A person states that all the concern about changing the Earth's climate is just the negative viewpoint of "pessimistic environmentalists", since any warming effect due to the greenhouse effect is likely to be offset by increased atmospheric reflectivity due to particulates. Do you agree or disagree with this opinion? Explain.

29. Why is ozone in the stratosphere a desirable gas but ozone in the troposphere an undesirable gas? Explain.

30. The burning of fossil fuels such as coal may need to be severely curtailed long before we run out of our supplies. Explain.

31. Briefly explain how geographic and weather conditions in locations such as Los Angeles can increase the severity of air pollution problems.

32. Resolution of the acid rain issue poses special problems in terms of state, regional, and even international cooperation. Explain.

33. A student in a welding class tells you he wants to purchase 100 liters of oxygen and 40 liters of acetylene. Is this sufficient information to help the student find the best buy? If not, what other information do you need and why?

34. A can of shaving cream contains the following warning: "Caution: Do not set on stove or radiator or keep where temperature will exceed 120°F". What is the scientific explanation for this warning?

Answers to Questions:

Multiple-Choice

1. C	6. D	11. D	16. C	21. A
2. B	7. B	12. D	17. C	22. A
3. D	8. B	13. A	18. C	23. A
4. C	9. C	14. B	19. A	24. B
5. C	10. B	15. C	20. D	25. A

Short-Answer

26.

	Pollutant	Equation	Primary Source
a.	CO	$C + \frac{1}{2}O_2 \rightarrow CO$	burning of fossil fuels
b.	NO	$N_2 + O_2 \rightarrow 2\,NO$	high temperature combustion
c.	SO_2	$S + O_2 \rightarrow SO_2$	burning of coal

27. 2.0 mol × 22.4 L/1 mol = 44.8

28. Although it is true that increased particulates in the upper atmosphere could increase the Earth's reflectivity and contribute to a cooling effect, the politician's logic is faulty. Many variables contribute to the Earth's energy balance; to ignore sound warnings with the hope that one pollutant might offset the effect of another is to gamble with all life on Earth. Additionally, allowing the continued unrestrained growth in fossil fuel energy consumption would contribute to other pollution and political problems.

29. In the stratosphere, ozone is an essential component that serves to shield Earth from harmful ultraviolet radiation. At the present time, holes are appearing in this thin shield due to reactions involving synthetic gases such as the freons. In the troposphere and at ground level, ozone generated by photochemical smog acts as a strong oxidizing pollutant that corrodes paint, rubber, plants, human mucous membranes, etc. The ozone issue is a good example of pollution being a resource "out of place."

30. Specific pollution problems created by the burning of coal include: (1) acid rain from SO_x and NO_x, (2) particulates, (3) greenhouse effect due to carbon dioxide, and (4) land and water pollution problems associated with mining. Some environmentalists feel that these problems will require shifting to alternative fuels (solar, wind, etc.) and increased conservation even before the price of coal becomes prohibitive.

31. Geographic regions that have many days of sunshine, frequent natural thermal inversions, a high concentration of automotive traffic, and are bounded by mountains are ideal locations for the production of photochemical smog.

32. The source of the acid rain pollution problem is often geographically removed from the site of the negative environmental effects. For example, coal-burning power plants in the Ohio River Valley, create problems for the lakes in the Adirondack Mountains in New York as well as lakes across the border in Canada. As a result, the population which generates the pollution and should pay the pollution control costs is not the same population who will experience the effects. This has posed many opportunities for negotiations between states and between the United States and Canada.

33. Volume as a measure of quantity of gas does not provide adequate information unless temperature and pressure are also specified, since volume varies with these two factors. As an alternative, gas can be purchased by mass.

34. The pressure of a fixed volume of gas will increase as the temperature increases. At some point, the pressure will cause the can to burst.

HEALTH: YOUR RISKS AND CHOICES

MULTIPLE-CHOICE QUESTIONS

1. Hair is made from
   (A) carbohydrates.   (B) cellulose.   (C) fats.   (D) proteins.

2. A drug can be definitely proven safe by:
   (A) tests on 1000 rats.   (C) tests on 1000 humans.
   (B) tests on 1000 guinea pigs.   (D) none of these choices.

3. A combination of alcohol and a barbiturate has an effect up to 200 times that of either drug taken alone. Such an effect is called
   (A) activation.   (C) synergism.
   (B) chemical dependency.   (D) synthesis.

4. Which term means the overall balance of body chemistry required for good health?
   (A) homeostasis   (C) equilibrium
   (B) ecosystems   (D) side effects

5. More than half the premature deaths in the United States can be prevented by
   (A) assuring well balanced external ecosystems of bacteria.
   (B) allowing for "population explosions" of harmful bacteria.
   (C) controlled sanitation and mass immunization.
   (D) changes in behavior related to diet, sanitation, and other personal factors.

6. Which combination of elements makes up all but about 2% of body mass?
   (A) O, C, Ca, P, S   (C) O, C, Na, H, I
   (B) O, C, H, N, Ca   (D) O, C, Si, Ca, Cl

7. Which represents a class of compounds that make up the structure of the human body?
   (A) $Na^+ + C$ (C)

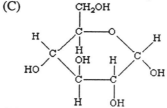

   (B) $H-N-\overset{R}{\underset{H}{C}}-\overset{}{\underset{O}{C}}-O-H$ (D) $H-O-\overset{O}{\underset{O^-}{P}}-O^-$

8. Enzymes control the rate at which chemical reactions occur in the human body. In fact, enzymes are chemical
   (A) genes.   (C) catalysts.
   (B) substrates.   (D) variations of DNA.

9. Most of the energy required in body chemistry comes from which reaction?
   (A) $C_6H_{12}O_6 + 6 O_2 \rightarrow 6 H_2O + 6 CO_2$

   (B) Energy $+ ADP + HPO_4^{2-} \rightarrow H_2O + ATP$

   (C) $H_2CO_3 \rightarrow H_2O + CO_2$

   (D) $(C_6H_{12}O_6)n$ enzyme $n C_6H_{12}O_6$

10. The change in taste of an unsalted soda cracker or a piece of white bread that is chewed for several minutes results from
    (A) the oxidation of glucose to CO_2 and H_2O
    (B) the formation of glucose from starch
    (C) the action of pepsin (enzyme) on protein to form amino acids
    (D) the interaction of amylase and pepsin

11. Which factor must be carefully controlled in the laboratory study of the enzymes that take part in digestion?
    (A) the NaCl content of saliva.
    (B) the volume of water used.
    (C) the concentration of the starch solution.
    (D) the pH.

12. Which chemical group is common to lysine, alanine, and serine?

    (A)
    $$-\overset{\underset{|}{H}}{C}-\overset{\underset{||}{H}}{\underset{O}{C}}-\overset{\underset{|}{H}}{C}-$$

    (C)
    $$-N-\overset{\underset{|}{H}}{\underset{H}{C}}-\overset{\underset{||}{H}}{\underset{O}{C}}-O-$$

    (B) —OH

    (D) HOH

13. How is it possible for hair protein to be different from skin protein or enzyme protein?
    (A) Each protein has a distinctive sequence of amino-acid units which creates a large number of possibilities for folding and interaction.
    (B) Each protein has a unique shape.
    (C) Ribosomes "direct" the order of amino acids in each protein.
    (D) Globular proteins are water soluble while fibrous proteins are tough and often ropelike in structure.

14. Which factor is the most immediate cause of normal changes in hemoglobin's ability to carry and to release oxygen within the body?
    (A) temperature   (C) sugar content of hemoglobin
    (B) pH   (D) use of an external denaturant

15. Which substance should you drink to counteract a mild and temporary alkalosis condition in your body?
    (A) carbonated soft drink
    (B) sodium bicarbonate
    (C) phosphate buffer
    (D) excessive amounts of distilled water

16. What is the main function of urine excretion?
    (A) Urine carries away the reaction products of buffer overload.
    (B) Urine produces H_2CO_3, a weak acid that neutralizes an alkalosis condition.
    (C) Urine increases the amount of CO_2 in the blood during hyperventilation.
    (D) Urine eliminates lactic acid built-up in muscles during excessive exercise.

17. Which is not likely to cause an overload of the blood's buffer systems?
    (A) hyperventilation   (C) cardiac arrest
    (B) strenuous muscular activity (D) enzymatic digestion

18. Which statement is not correct for the process of anaerobic glycolysis?
    (A) It produces energy faster than aerobic glycolysis.
    (B) It requires more oxygen than does aerobic glycolysis.
    (C) It causes lactic acid build up in the muscles.
    (D) It occurs during exercises such as weight lifting.

19. Why don't oils excreted from the body wash away with water alone?
    (A) Water molecules are strongly ionic, while oil molecules are not.
    (B) Enzymes from the body prevent the "oily dirt" from mixing with water.
    (C) Bacteria react with the secreted oils, causing the oils to break down into polar molecular substances.
    (D) Water molecules are more strongly attracted to one another than to oil molecules.

20. Control of acne through careful washing followed by moderate chemical treatment with benzoyl peroxide and antibiotics, results from
    (A) killing of harmful bacteria that live on natural oils.
    (B) the production of glycerol when soap is used to wash away the oils.
    (C) a reaction between the soap and the water.
    (D) the elimination of dead skin.

21. Why doesn't burning in living systems such as the human body occur explosively as it does in an automobile engine or a forest fire?
    (A) ADP molecules serve as a mechanism for capturing large amounts of energy.
    (B) $ATP + H_2O$ form as energy is released.
    (C) Relatively small numbers of ATP molecules are involved in storing and releasing energy.
    (D) Cells oxidize the fuel in a series of steps involving many different enzymes.

22. A suntan begins to fade if a person is not exposed to additional sunlight because
    (A) no additional melanin forms.
    (B) oxygen in the air will react with the dark-colored pigments.
    (C) the melanin dissolves in water when a person bathes.
    (D) the cells that contain melanin immediately fall from the surface of the skin.

23. Which would work best as a soap?
    (A) $CH_3-CH_2-CH_2-CH_2-CH_2-CH_2-SO_3^- Na^+$
    (B) CH_3-CH_3
    (C) glucose
    (D) gasoline

24. Why is PABA an effective screen against the ultraviolet rays from the sun and therefore good protection against sunburn?
    (A) PABA is the oil found in many suntan lotions. and therefore must be particularly effective.
    (B) The benzene ring structure of PABA absorbs ultraviolet light and dissipates its energy as heat across the molecule.
    (C) PABA is converted into melanin which protects the body from further sunburn.
    (D) The mutation of DNA molecules into vitamin D in the presence of Ca^{2+} ions is responsible for the protection by PABA.

25. Which substance plays a critical role in the liver's removal of a toxic component of certain foods?
    (A) C_6H_6 (B) $C_6H_{12}O_6$ (C) antibodies   (D) molybdenum ions

26. From a chemical viewpoint, why is it possible to style hair?
    (A) The hair cuticle consists of overlapping layers of flat cells and these are capable of sliding along each other.
    (B) Oils and natural acids coat and protect the structure of hair microfibrils, but the coating is easily lost.
    (C) The weak chemical interactions holding protein chains to one another can be shifted in location through the action of water and various chemicals.
    (D) Excess heat dries hair oils causes protein chains in hair to separate from one another.

27. How do narcotic analgesic drugs like morphine and methadone act to relieve pain?
    (A) Molecules of these drugs fit onto specific protein receptor sites on nerve cells.
    (B) The drugs reduce the sensitivity of the nervous system to all external stimuli.
    (C) The drugs cause natural painkillers to be produced within the brain.
    (D) The drugs deplete the number of calcium ions generally at nerve endings.

28. Which statement does not describe the effects of smoking cigarettes?
   (A) Carbon monoxide is one product of oxidation which reduces the ability of hemoglobin to carry oxygen.
   (B) Elastase is activated in the lungs to clear away tars. It also appears to digest protein in the lung membrane.
   (C) Cigarette smoke contains polynuclear aromatic compounds, some of which are cancer producing.
   (D) Microorganisms multiply in the cigarette smoke causing a reduction in dental decay.

29. In which case is the phrase "directly cause" used correctly?
   (A) In the United States temperature directly causes the seasons.
   (B) Catalysts directly cause changes in rates of chemical reactions.
   (C) Smoking of cigarettes directly causes lung cancer.
   (D) Changes in temperature directly cause the rate of evaporation of water to increase.

30. Research data indicates that "smokers are 10 times more likely to die from lung cancer than are nonsmokers;" and "smokers have a 70 percent greater risk of coronary heart disease than do nonsmokers." Which of these statements cannot be determined from this information.
   (A) Of the people who die from lung cancer, most are smokers.
   (B) Extensive medical research surveys have produced statistical evidence related to two major causes of death in the United States.
   (C) Evidently smoking generally is harmful in terms of shortening the lifespan of American citizens.
   (D) Coronary heart disease is a more extensive killer of humans in the United States than is lung cancer.

31. These three chemical equations represent reactions taking place in solution.

   $NaOH + HCl \rightarrow NaCl + H_2O$

   $2 NaOH + H_2SO_4 \rightarrow Na_2SO_4 + 2 H_2O$

   $LiOH + HCl \rightarrow LiCl + H_2O$

   Which statement is true?
   (A) A different acid is utilized in each reaction.
   (B) A different base is utilized in each reaction.
   (C) None of the reactants are acids or bases.
   (D) The ionic equation $H^+ + OH^- \rightarrow H_2O$ could be used to describe what happens in all three reactions.

32. What is the correct order of responses if you match the names to the formulas?

Formulas	Names
HNO_3	1. potassium hydroxide
HCl	2. hydrochloric acid
$NaOH$	3. sulfuric acid
H_2SO_2	4. calcium hydroxide
$Ca(OH)_2$	5. nitric acid
KOH	6. sodium hydroxide

   (A) 1,3,4,2,5,6   (C) 5,3,6,2,1,4
   (B) 5,2,6,3,4,1   (D) 6,5,4,3,2,1

Short-Answer

33. Define the term "homeostasis" in the context of blood pH.

34. Briefly explain the purpose of enzymes and describe how they function.

35. Briefly explain the role played by ATP in energy storage and release.

36. a) What are the four types of forces that hold protein molecules together in hair?
    b) What is the effect of water on hair?

37. Why is it true that either too much or too little exposure to the sun's rays can lead to negative health effects?

38. a) Describe how narcotic analgesics such as morphine are believed to work in the brain.
    b) How are narcotic analgesics like the brain's own painkillers?

39. a) Why is it important to follow the warnings on medication containers concerning eating and drinking?
   b) Why should you make sure your doctor is aware of all medications or drugs you are taking?

Answers to Questions:

Multiple-Choice

1. D	6. B	11. D	16. A	21. D	26. C	31. D
2. D	7. B	12. C	17. D	22. A	27. A	32. B
3. C	8. C	13. A	18. B	23. A	28. D	
4. A	9. A	14. B	19. D	24. B	29. D	
5. D	10. B	15. A	20. A	25. D	30. D	

Short-Answer

33. Homeostasis is the maintenance of balance in body systems. The pH of blood must stay constant (within a very tight range of 7.35-7.45) or death will result. The carbonate-phosphate buffer system limits or prevents acidosis or alkalosis.

34. Enzymes are very specific in the reactions they catalyze. In fact, there are more than 3000 different enzymes in every human cell, each catalyzing only one type of reaction. This specificity is a result of the way in which enzymes speed up reactions. An enzyme has the ability to bring together two or more molecules and cause them to react, or an enzyme may cause a reaction within a single molecule, such as the breaking of a bond.

35. All of the energy that the body extracts from food is converted to ATP through the great variety of biochemical pathways through which our food molecules travel. Note that not all food is used for energy production; some food is used for building tissues and other molecules as well as for other purposes. The body can hold onto energy, at least temporarily, in the form of ATP. When the body requires energy, the ATP is converted into ADP, adenosine diphosphate, or into AMP, adenosine monophosphate, releasing the energy that is tied up in the phosphate bonds.

36. a. The four types of forces that hold hair protein molecules together are covalent bonds between amino acids (peptide bonds), hydrogen bonds, nonpolar interactions, and sulfide bonds that cross link protein strands.
    b. Water, by itself, has little effect on hair protein in terms of permanent changes. A water environment can disrupt the hydrogen bonds in hair protein and allow the strands to stretch and elongate—by as much as 150%.

37. Because the sun's rays contain ultraviolet light and this radiation causes tissue damage by disrupting DNA and RNA, overexposure to the sun can cause many ill effects, including skin cancer. On the other hand, too little exposure to the sun's ultraviolet radiation can cause rickets, a result of a vitamin D deficiency. This disease manifests itself in young children as a failure of growing bones to calcify and harden. In adults, the effects of rickets, called osteomalacia, are characterized by misshapen and extremely fragile bones. It is important to note that vitamin D, cholecalciferol, is highly toxic in large doses so it is unwise to rely on supplements to replace exposure to moderate amounts of sunshine.

38. a Narcotic analgesics are believed to relieve pain by blocking key receptor sites on nerve cells in the brain.
    b These narcotic analgesics work exactly like endorphins and enkephalins, the brain's own painkillers.

39. a. Some medicines are best absorbed by the body if they come into an empty stomach, and some are more effective if they are taken with food. It is even possible that some foods can counteract or inhibit the action of a medication. For example, taking a tetracycline antibiotic with milk will ruin the bacteria-killing properties of the drug.
    b. When a patient is taking more than one drug it is important that the physician know about all of the medicines being consumed. This is because medicines may interact with each other and increase or decrease their effects. This is particularly true in the case of narcotics. Note that alcohol is a drug and should never be used while taking medications.

THE CHEMICAL INDUSTRY: PROMISE AND CHALLENGE

Multiple-Choice Questions

1. The agency that represents the public's interest in chemical safety is the
   (A) Atomic Energy Commission.
   (B) U.S. Environmental Protection Agency.
   (C) Tennessee Valley Authority.
   (D) U.S. Department of the Interior.

2. Three crucial factors when chemical reactions are scaled up to produce large quantities of high-quality products are
   (A) waste, expandibility, and costs.
   (B) engineering, risks, and costs.
   (C) engineering, profitability, and waste.
   (D) profitability, risks, and waste.

3. Which coefficients balance this equation?
   $$\_\_H_2 + \_\_N_2 \leftrightarrow \_\_\_NH_3$$
   (A) 1, 1, 2   (C) 1, 3, 2
   (B) 3, 1, 2   (D) 1, 1, 1

4. The voltage produced by an electrochemical cell depends on
   (A) the specific metals used.   (C) the type of metal and size of cell.
   (B) the size of the cells.   (D) the presence of an anode only.

5. To maximize electrical potential one should choose for the electrodes
   (A) metals that are close together on the activity series.
   (B) metals that are far apart on the activity series.
   (C) two pieces of the same metal.
   (D) one metal and one nonmetal.

6. Which would be of least direct interest to the citizens of Riverwood as they consider whether or not to allow an ammonia plant to be built in the community?
   (A) effect on air quality
   (B) possibility of chemical spills
   (C) creation of jobs for citizens of the town
   (D) increasing worldwide need for fertilizer

7. Which statement describes the major purpose of the chemical industry, assuming there is always the necessity of generating profits?
   (A) to change natural or raw materials into substances that are more useful to people
   (B) to reduce the problem of chemical wastes
   (C) to find more effective means of distilling substances like petroleum
   (D) to increase the efficiency of ammonia production for use as a fertilizer

8. Which substance is synthetic?
   (A) petroleum   (B) N_2 (C) CH_4 (D) DDT

9. Which substance is produced in the greatest amount by weight each year by the chemical industry in the United States?
   (A) H_2SO_4 (B) O_2 (C) C_2H_5COOH (D) Na_2SO_4

10. Which substance is produced in largest amount by weight each year from sea water by the chemical industry in the United States?
    (A) H_2 (B) Cl_2 (C) C_2H_5COOH (D) HNO_2

11. Which equation represents a means of fixing nitrogen so that it may be utilized as a plant nutrient?
    (A) $2 NO + O_2 \rightarrow 2 NO_2$
    (B) $N_2O_4 \rightarrow 2 NO_2$
    (C) $N_2 + 3 H_2 \rightarrow 2 NH_3$
    (D) $CO + H_2O \rightarrow CO_2 + H_2$

12. Which one of these raw materials would an ammonia company, such as EKS, be least dependent upon?
    (A) N_2 (B) H_2 (C) H_2O (D) $CaCO_3$

13. A clean nichrome wire is dipped into a solution of KNO_3. The coated wire is placed into a candle flame. The candle flame remains yellow instead of violet as expected. Which statement explains this result?
    (A) K^+ does not produce a violet flame under any circumstances, so the yellow flame is completely expected.
    (B) The candle does not burn wax completely. This causes carbon particles to incandescence, forming a yellow flame which masks the violet color.
    (C) The candle flame is too hot to produce the violet color. It will be necessary to cool the flame in some manner in order to see the expected color.
    (D) Dipping the nichrome wire into KNO_3 will always produce a yellow flame with heat. The violet flame can only be expected if Pb^{2+} ions are present.

14. Which does not add fixed nitrogen to the soil?
    (A) legumes and clover (C) nitrogen-fixing bacteria
    (B) nitrogen fertilizer   (D) liquid nitrogen from air

15. Which process does not play a role in the nitrogen cycle?
    (A) death and decay
    (B) excretion from animals
    (C) combustion or burning of CH_4
    (D) $2 NO_3^- \xrightarrow{\text{bacteria}} N_2 + 3 O_2$

16. Which step in the determination of the percentage of PO_4^{3-} in a sample of fertilizer could have led to the greatest error in your determination?
    (A) measuring 4 mL instead of 5 mL of the original solution
    (B) adding 240 mL of water instead of 245 mL of water to the fertilizer solution
    (C) pouring 21 mL instead of 20 mL of the original solution into the test tube
    (D) using 300 mL instead of 200 mL of water in the hot water bath

17. Which statement describes what happens chemically to nitrogen atoms as a result of the Haber process?
    (A) Nitrogen atoms become oxidized.
    (B) Nitrogen atoms continue to share electrons with one another.
    (C) Nitrogen atoms become reduced.
    (D) Nitrogen atoms form negatively charged ions.

18. Given that in the nitrogen molecule (N_2), each atom is in a zero oxidation state, then an atom of nitrogen in each molecule of ammonia (NH_3) is
    (A) in a positive oxidation state.
    (B) in a negative oxidation state.
    (C) in a more strongly bonded state.
    (D) still in a zero oxidation state.

19. Which pair of substances are commonly used as electrodes in a car battery?
    (A) Zn and Cu^{2+} (C) Pb and PbO_2
    (B) Pb and Mg   (D) Ag and Au

20. Which statement is correct about these half reactions?

$$Zn(s) \rightarrow Zn^{2+}(aq) + 2e^-$$
$$Cu^{2+}(aq) + 2e^- \rightarrow Cu(s)$$

(A) Zinc atoms are reduced.
(B) Zinc atoms gain electrons.
(C) Copper ions lose electrons.
(D) Copper ions are reduced.

21. Relatively large quantities of generated electricity are required to release aluminum from its minerals. Which statement best explains why?
(A) Al^{3+} ions have a very weak attraction for electrons from other chemical substances.
(B) Na_3AlF_6 is a very unstable mineral.
(C) Molten Na_3AlF_6 is a solvent for Al_2O_3.
(D) Carbon electrodes are used in the electrolytic process, which produces the aluminum from Al_2O_3.

22. Which is the most important factor in determining where to locate an aluminum producing plant?
(A) the products to be made from the aluminum
(B) the fact that aluminum readily forms an oxide coating
(C) availability of an inexpensive source of electricity
(D) the initial source of the bauxite

23. Banning the production of ammonia would not affect which of these?
(A) feeding the world
(B) producing explosives
(C) reducing a major source of heat energy (fuel)
(D) producing other nitrogen-containing intermediate compounds

24. This is the equation for the formation of NH_3 from N_2 and H_2 in the Haber process.

$$N_2 + 3\ H_2 \leftrightarrow 2\ NH_3$$

What does the double arrow symbol ($\leftrightarrow$) represent?
(A) Ammonia is a more stable substance than is N_2 or H_2.
(B) As NH_3 forms, some of it reverts back to N_2 and H_2.
(C) Cooler temperatures would produce more stable NH_3 molecules.
(D) The yield of NH_3 in this process is 100%.

25. Which equation represents an essential first step before ammonia can be produced by the Haber process?
(A) $CH_4 + H_2O \rightarrow 3\ H_2 + CO$
(B) $2\ SO_2 + O_2 \rightarrow 2\ SO_3$
(C) $4\ NH_3 + 7\ O_2 \rightarrow 4\ NO_2 + 6\ H_2O$
(D) $4\ C_3H_5(NO_3)_3 \rightarrow 12\ CO_2 + 6\ N_2 + 10\ H_2O + O_2$

26. Dynamite is not associated with which of these?
(A) the Nobel prize   (C) energy-absorbing reactions
(B) nitroglycerin   (D) non-reversible reactions

27. Which is least related to meeting societal needs through industry?
(A) availability of food for the world
(B) improved or better health
(C) development of new useful materials
(D) changes in social or cultural practices

28. The town council of Riverwood is meeting to decide if the town will issue a license to a large corporation to build a chemical plant on property in the town. Assume that the industry is one much like those considered in this unit. Which factor would be of least concern to the council members?
    (A) whether an inexpensive source of energy is available
    (B) effects of disposing of waste industrial products
    (C) effect of the industry on the town's tax revenues
    (D) income to families and individuals living in the town

29. Chemical explosives
    (A) usually contain a high percentage of the element nitrogen.
    (B) can contain built-in oxidizers.
    (C) react exothermically.
    (D) have all of these characteristics.

30. Which is a correctly balanced equation?
    (A) $Fe(s) + Fe^{3+}(aq) \rightarrow 2\ Fe^{2+}(aq)$
    (B) $NO_2(g) + H_2O(l) \rightarrow HNO_3(aq) + NO(g)$
    (C) $2\ Al(s) + 3\ Cu^{2+}(aq) \rightarrow 2\ Al^{3+}(aq) + 3\ Cu(s)$
    (D) $H_2O(l) \rightarrow H_2(g) + O_2(g)$

Short-Answer

31. If each pair of metals and solutions of their ions were put together to form a voltaic cell, in which direction would the electrons flow? (Note: You will need to consult an activity series table.)
    a) Al and Ni   b) Mn and Pb   c) Ag and Zn   d) Cu and Fe

32. Predict which element has the positive oxidation number for each of these compounds. (Note: You will need to consult a table of electronegativities.
    a) HCl c) NO_2 e) SO_2 g) CO_2
    b) H_2O d) $NaCl$ f) MnO_2 h) NH_3

33. Use a diagram to show how you would electroplate a spoon with silver metal.

34. Explain why the price of fertilizers can be strongly correlated with the price of natural gas.

35. Explain why nitrogen is a limiting reactant for plant growth, even though it makes up about 78% of the atmosphere.

36. Describe the nature of the partnership between the chemical industry and society, and list ways in which each of the partners need each other.

37. Using ammonia as an example, describe how the same chemical technology can be used for either constructive or destructive purposes.

38. List and briefly describe three factors that must be considered when a reaction is scaled up from a laboratory level to the industrial level.

Answers to Questions:

Multiple-Choice

1. B	6. D	11. C	16. A	21. A	26. C
2. C	7. A	12. D	17. C	22. C	27. D
3. B	8. D	13. B	18. B	23. C	28. A
4. A	9. A	14. D	19. C	24. B	29. D
5. B	10. B	15. C	20. D	25. A	30. C

Short-Answer

31. a. Al → Ni   b. Mn → Pb   c. Zn → Ag   d. Fe → Cu

32. a. H   c. N   e. S   g. C
    b. H   d. Na   f. Mn   h. H

33. This sketch shows how you could arrange the apparatus to electroplate a spoon with silver.

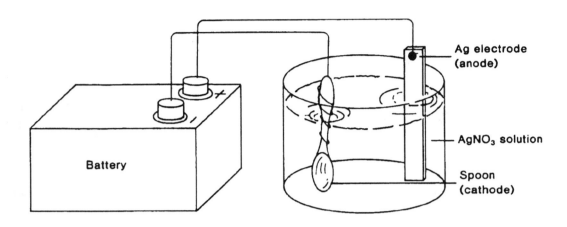

Ag electrode (anode)

AgNO$_3$ solution

Spoon (cathode)

Battery

34. The usual source of the hydrogen used in the Haber reaction is natural gas. Also, energy is needed to extract nitrogen from the atmosphere, hydrogen from methane, and to provide the startup energy to run the actual Haber reaction.

35. Nitrogen in the atmosphere exists as N$_2$(g). The stability of the triple bond in this molecule makes this compound especially unreactive and therefore unavailable to plants. Lightning and/or some bacterial species are capable of fixing" nitrogen in forms such as ammonia, nitrates, or nitrites, that can be used by plants.

36. Ideally, the relationship is analogous to biological symbiosis, where both partners benefit. Society needs industry to provide its material, energy, and employment needs. Industry needs society to buy its products and services in order to make a profit and remain in business.

37. Ammonia is the starting point for a variety of fertilizers and explosives. Judiciously applied, fertilizers raise the productivity of the soil and contribute to the resolution of the problem of world hunger. Nitrogen-based explosives can be used to take human life in wartime, but also for constructive purposes in mining and building roads. Many chemicals and technologies share this type of dual purpose. They can be used to increase or decrease the quality of human life.

38. Three factors that must be considered in scaling up a reaction to an industrial level are engineering, profitability, and waste disposal. As a reaction is scaled up, chemical engineers must deal with problems that likewise increase in scale. For example, a reaction that produces only a small amount of heat in the laboratory may produce tremendous amounts of heat at the industrial level. Unless the engineers devise a method of disposing of this heat, the reactions may go out of control, producing unanticipated products or even dangerous conditions. Chemical engineers are also concerned with profitability. They must devise the most efficient and cost-effective designs for equipment and processes and try to take advantage of chemical principles, such as Le Chatelier's principle. Waste disposal can be a colossal problem at the industrial level. The small amounts of wastes formed in a laboratory-scale reaction may be harmless, but in very large amounts they may represent substantial hazards to workers and communities.

ASSORTED TEST ITEMS

Each series consists of a short narrative followed by two or more related questions.

Series One

This is a quotation from a newspaper article.

"To lure the health-conscious consumer, soft drink makers have already removed caffeine and sugar from sodas and even added real fruit juice to the fizz. The next step is a sugar free drink that is vitamin enriched. The Company's 'Diet Squirt Plus,' which has 50 percent of the recommended daily requirement of vitamin C and 10 percent of the daily requirement of the B complex vitamins, will be introduced next year. Nutritious soft drinks are not expected to sweep the market but five or ten years from now these soft drinks could be a hot business."

Assume that you now are responsible for decisions related to the development of 'Diet Squirt Plus' and answer these questions.

1. Which substance is likely to provide the "fizz" for this drink?
   (A) O_2
   (B) N_2
   (C) CO_2
   (D) Cl_2

2. Which chemical equation describes a possible source of the "fizz" for this drink?
   (A) $2 H_2O_2 \rightarrow 2 H_2O + O_2$
   (B) $6 CO_2 + 6 H_2O \rightarrow C_6H_{12}O_6 + O_2$
   (C) $CaCO_3 \rightarrow CaO + CO_2$
   (D) $2 NaCl \rightarrow 2 Na + Cl_2$

3. "Diet Squirt Plus" must be pasteurized to prevent bacterial growth during storage. Which process is an inappropriate means of pasteurization?
   (A) adding a small quantity of antibacterial chemical to the drink
   (B) heating the bottles before adding the liquid drink
   (C) passing the sealed bottles of the drink through ultraviolet light radiation to destroy the bacteria
   (D) adding a small quantity of an alkaline substance to the drink to raise its pH

4. Which substance does this soft drink definitely not contain?
   (A) H_2O
   (B) C_2H_5OH
   (C) $C_7H_5SO_3$ (saccharin)
   (D) CO_2

5. You must decide what material to make the bottles from during a period of petroleum shortage. Which is the least likely substance to use?
   (A) glass
   (B) tin coated steel (iron)
   (C) aluminum
   (D) polystyrene

Series Two

Not long ago, thousands of gallons of petroleum spilled from a tanker into the lower end of the Mississippi River. The slowly moving current caused the oil to spread gradually across the surface of the water. Answer these questions.

1. Which cannot be an explanation of why the oil spread across the water's surface?
   (A) The speed of the water flow was slow.
   (B) The density of the oil is less than that of water.
   (C) The oil's hydrocarbon molecules are non-polar, while water molecules are polar.
   (D) The river water contained appreciable amounts of detergents.

2. Which is an incorrect statement relating to the oil spill?
   (A) Fish eventually could be killed because the oil prevents them from utilizing the oxygen atoms that make up water molecules.
   (B) Mosquito larva could be destroyed because they could not penetrate the oil surface.
   (C) Adding detergent to the water would cause the oil to emulsify with the water, reducing the oil film but perhaps causing other ecological problems.
   (D) Changes in temperature would have only a small effect on the solubility of oil in water.

3. Which describes a way to help the fish survive?
   (A) The fish would be helped by burning off the oil slick or spill instead of picking up the oil with straw.
   (B) The small amount of salt (NaCl) dissolved in the river water is reduced by the presence of the oil spill.
   (C) Removal of the oil followed by a decrease in water and air temperature would allow more oxygen to dissolve in the water.
   (D) If the oil had been left on the water, the fish could have used the oil as food.

4. Suppose that, instead of oil, a large chemical company mistakenly spilled solid caustic soda (NaOH) into the river. Which could be a result of the spill?
   (A) The river water would feel slippery to the touch where there is a high local concentration of NaOH.
   (B) The NaOH would not dissolve in the river water but would sink to the bottom of the river as a solid.
   (C) The NaOH would react with the water, forming hydrogen gas and oxygen gas with a possible danger of explosion.
   (D) An excess of H^+ ions would exist in the river water which would cause the pH of the water to decrease.

Series Three

THMs, or trihalomethanes, can be found in small amounts in "purified" water. In higher concentrations, THMs can be dangerous to human health. Answer these questions.

1. Which statement describes the formation of THM substances?
   (A) Chlorine added to water reacts with organic compounds to form THMs.
   (B) Chlorine destroys all bacteria found in the water and the result is THMs.
   (C) THMs form from the use of water softeners.
   (D) Fluorides in toothpaste produce THMs in water.

2. Which process is most closely related to the potential presence of THMs in water?
   (A) distillation   (C) oxidation   (B) filtration   (D) precipitation

Series Four

The chemical elements are the basic building blocks of all known compounds. These elements are listed in the periodic table, one of the most important sources of information to the chemist. Answer these questions.

1. Which is not true of the periodic table of the elements?
   (A) The elements on the table fall naturally into categories such as metals, non-metals, and transition elements.
   (B) The elements are listed consecutively on the table in order of increasing chemical reactivity.
   (C) The table indicates an orderly pattern of properties for the elements related to their atomic numbers.
   (D) Most, but not all, of the elements listed have been found naturally.

2. Which is a true statement about elements listed on the periodic table?
   (A) The elements He, Ne, Ar, Kr, and Xe were discovered early in the history of people because they are very reactive.
   (B) The elements Li, Na, K, Rb, and Cs are so reactive that they were easily detected by early people.
   (C) The transition metals like Fe, Cu, and Ni were discovered early in the history of people because these elements can be extracted easily from their compounds by heat.
   (D) The element H has properties typical of the halogens, and therefore it is listed on the periodic table as one of the halogens.

Series Five

You have learned that the United States uses substantial amounts of the element aluminum. This metal comes from ores found mostly in countries outside of the United States. Answer these questions.

1. Which would not help if our supply of aluminum were reduced?
    (A) conservation   (C) reuse of aluminum containers
    (B) recycling   (D) fractional distillation

2. Which set of properties makes aluminum such a useful substance?
    (A) metal oxidizes easily, low density
    (B) easily extracted from mineral, metal reduced easily
    (C) metallic color, brittle, easily liquified
    (D) magnetic, mineral has low melting point

3. Which is not a true statement about the element aluminum?
    (A) extracted from mineral with electricity
    (B) listed in group III on periodic chart
    (C) combines with Cl_2 to form AlCl and with Na to form NaAl
    (D) oxide protects metal from further oxidation

Series Six

The title "Petroleum: To Build? To Burn?" indicates the versatility of this material found in the Earth's crust. Without petroleum, our society would be very different. Answer these questions.

1. Which relates least well to petroleum?
    (A) mixture of hydrocarbons
    (B) raw material for many synthetic fibers and plastics
    (C) related chemically and geologically to natural gas
    (D) a complete substitute for coal

2. Which substance cannot be obtained or derived directly from petroleum?
    (A) CH_4 (C) $C_2H_6O_2$
    (B) $CoCl_2$ (D) C

3. C_5H_{12} is one of the many substances obtained from petroleum. Why is this substance so useful to the chemist?
    (A) Its name is pentane.
    (B) It is a hydrocarbon.
    (C) It can exist in isomeric forms.
    (D) It is an unsaturated hydrocarbon.

4. Which statement is not true for gasoline?
    (A) It is a mixture derived from petroleum.
    (B) It readily burns in the presence of oxygen and heat.
    (C) It exists in unlimited supplies.
    (D) Its cost fluctuates with the availability of petroleum.

5. Which set of substances contains the missing reactants from these three chemical equations in correct order?

    $C_2H_6 + ? \rightarrow CH_5Cl + HCl$
    $C_3H_6 + ? \rightarrow C_3H_7OH$
    $C_6H_6 + ? \rightarrow C_6H_5NO_2 + H_2O$

    (A) Cl_2, H_2O, HNO_3
    (B) HNO_2, H_2, HCl
    (C) NO_2, H_2O, HClO
    (D) H_2, Cl_2, O_2

Series Seven

Nuclear chemistry has become a very important part of our lives. You will recall that controlled nuclear reactions can be used as a source of heat energy, and the radiation of some nuclear reactions is used to control cancer and to detect other medical problems. Answer these questions.

1. Which equation represents an example of nuclear fission?

    (A) $^{235}_{92}U + ^{1}_{0}n \rightarrow ^{140}_{56}Ba + ^{93}_{36}Kr + 3^{1}_{0}n + \gamma$

    (B) $4^{1}_{1}H \rightarrow 2^{4}_{2}He + 2^{0}_{-1}e + \gamma$

    (C) $3^{4}_{2}He \rightarrow ^{12}_{6}C + \gamma$

    (D) $^{60}_{27}Co \rightarrow ^{60}_{28}Ni + ^{0}_{-1}e + \gamma$

2. Which particle is missing from both of these nuclear equations?

    $$^{239}_{94}Pu + ^{4}_{2}He \rightarrow ^{242}_{96}Cm + ?$$

    $$^{27}_{13}Al + ^{4}_{2}He \rightarrow ^{30}_{15}P + ?$$

    (A) electron   (C) alpha particle
    (B) proton   (D) neutron

3. What takes place through nuclear chemistry that was the original goal of alchemy?
    (A) fission   (C) radioactivity
    (B) transmutation   (D) neutrons

Series Eight

This graph indicates the relative estimated expenditures for major activities in the United States in 1984. These data are given in percentages of the gross national product of 3660 billion dollars. As an example, health care made up 10 percent of all of the financial activity. Use the graph and your knowledge of chemistry and math to answer these questions.

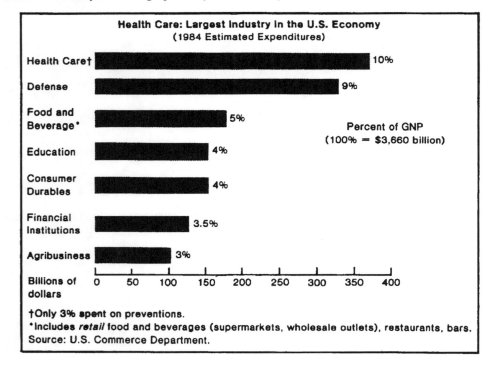

Health Care: Largest Industry in the U.S. Economy
(1984 Estimated Expenditures)

Health Care† — 10%
Defense — 9%
Food and Beverage* — 5%
Education — 4%
Consumer Durables — 4%
Financial Institutions — 3.5%
Agribusiness — 3%

Percent of GNP
(100% = $3,660 billion)

Billions of dollars: 0 50 100 150 200 250 300 350 400

†Only 3% spent on preventions.
*Includes *retail* food and beverages (supermarkets, wholesale outlets), restaurants, bars.
Source: U.S. Commerce Department.

1. How do the dollars spent on food and beverages compare to those spent on health care?
   (A) ten times more   (C) half as much
   (B) twice as much   (D) one-third as much

2. To which category of societal activity does chemistry relate?
   (A) health care   (C) agribusiness
   (B) food and beverage   (D) all of these

3. Match category of expenditure with the chemical formula most associated with that category.

Category	Chemical formula
1. Health Care	a. $C_{12}H_{22}O_{11}$
2. Defense	b. HNO_3
3. Food	c. NH_3
4. Agribusiness	d.

   (A) 1a, 2b, 3c, 4d   (C) 1d, 2b, 3a, 4c
   (B) 1d, 2c, 3b, 4a   (D) 1a, 2c, 3d, 4b

Series Nine

You have learned that the chemical industry is the nation's largest industry. This industry synthesizes, extracts, purifies, and modifies the chemical resources of the Earth's crust and atmosphere. Here are formulas for five substances directly related with this loosely structured industry. Use these choices to answer these questions.

$$A = H_2SO_4 \qquad B = C_6H_6 \qquad C = Cl_2 \qquad D = NH_3 \qquad E = CaO$$

1. Which substance is produced in greatest amount by weight each year?
   (A) A   (B) B   (C) C   (D) D   (E) E

2. Which substance is produced by electrolysis of molten NaCl?
   (A) A   (B) B   (C) C   (D) D   (E) E

3. Which substance is obtained from petroleum?
   (A) A   (B) B   (C) C   (D) D   (E) E

4. Which substance is produced by the Haber process?
   (A) A   (B) B   (C) C   (D) D   (E) E

5. Which substance is related to lime and cement?
   (A) A   (B) B   (C) C   (D) D   (E) E

6. $2 H_2 + O_2 \rightarrow 2 H_2O +$ heat energy
   This chemical equation has been considered several times in your *ChemCom* studies. Which statement does not match this equation?
   (A) Twice as many molecules of the element hydrogen as oxygen are needed to produce each two molecules of water.
   (B) $2 H_2 + O_2$ represents all together 3 molecules or 3 moles of two different elements.
   (C) Atoms of hydrogen and oxygen are rearranged to produce the product without destroying or creating atoms.
   (D) Hydrogen becomes reduced in the reaction, while oxygen becomes oxidized.

7. The United States presently houses 5% of the world's population, yet utilizes about 50% of the world's resources each year. Which statement least relates to this fact?
   (A) About 23,000 kg (25 tons) of chemical resources support the needs of each U.S. citizen each year.
   (B) Four pounds of unwanted resources are discarded each day per each U.S. citizen.
   (C) The U.S. is dependent upon obtaining certain chemical resources from other countries in order to meet the needs of its citizens.
   (D) The same atoms of elements extracted from the Earth's crust and atmosphere eventually are returned to this part of the environment.

8. You are a member of the U.S. Congress. You will soon have to vote for or against a bill which will place a new tax on oil arriving in this country in tanker loads. All of the following may influence your decision but which one would have the least effect on your vote?
   (A) the availability of U.S. oil reserves
   (B) the present economic status of the U.S. oil industry
   (C) the fraction of energy use in the United States provided for by petroleum
   (D) the level of oil production by foreign countries

9. The amount of metals we are likely to obtain from known sources of high-grade ore can be expected to decline over the foreseeable future. To reduce the difficulties associated with this fact, the most practical approach would be to
   (A) initiate a voluntary recycling program.
   (B) require mandatory recycling along with a planned program of conservation.
   (C) replace all metal parts with plastics.
   (D) begin a search for new elements to replace the currently used metals.

10. Which position is the most appropriate for a citizen to take regarding the use of the Earth's resources?
   (A) Resources are finite; thus conservation and better control of "discards" are conditions we must live with.
   (B) Scientific evidence shows that the problem has been exaggerated, and we really have nothing to be concerned about for at least another 300 years.
   (C) When the problem reaches sufficient seriousness, the scientists will find a solution to it.
   (D) This is best left to the government to deal with. There's little or nothing I can do.

Answers to Questions:

Series One
1. C
2. C
3. B
4. B
5. D

Series Two
1. D
2. A
3. C
4. A

Series Three
1. A
2. C

Series Four
1. B
2. C

Series Five
1. D
2. A
3. C

Series Six
1. D
2. B
3. C
4. C
5. A

Series Seven
1. A
2. D
3. B

Series Eight
1. C
2. D
3. C

Series Nine
1. A
2. C
3. B
4. D
5. E
6. D
7. D
8. D
9. B
10. A

BLACK LINE MASTERS

This section of your Teacher's Guide contains Black Line Masters for your use as overhead transparencies.

SUPPLYING OUR WATER NEEDS

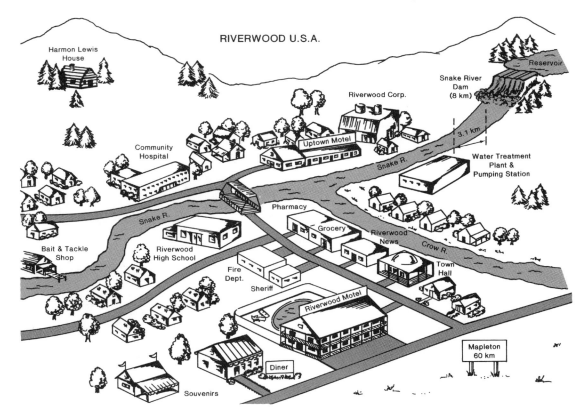

RIVERWOOD U.S.A.

Harmon Lewis House

Reservoir

Snake River Dam (8 km)

Riverwood Corp.

Uptown Motel

3.1 km

Community Hospital

Snake R.

Water Treatment Plant & Pumping Station

Pharmacy

Grocery

Riverwood News

Crow R.

Bait & Tackle Shop

Snake R.

Riverwood High School

Fire Dept.

Sheriff

Town Hall

Riverwood Motel

Mapleton 60 km

Diner

Souvenirs

W1. Riverwood Map, page 3.

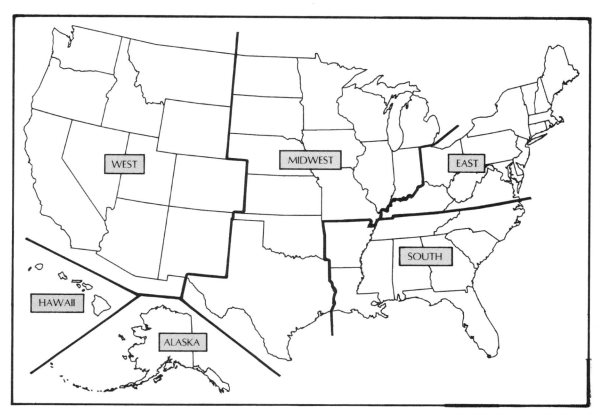

WEST

MIDWEST

EAST

SOUTH

HAWAII

ALASKA

W2. Figure 6, page 13.

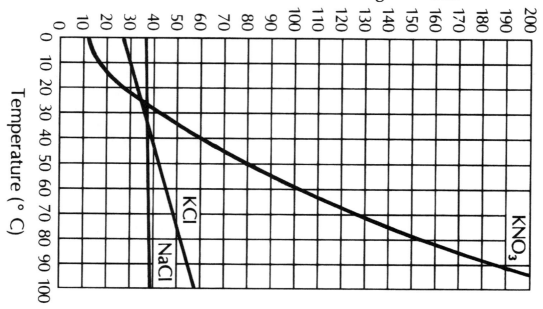

Grams of solute dissolved in 100 g of water

W3. Figure 21, page 40.

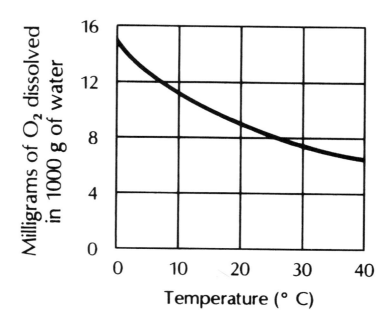

W4. Figure 22, page 41.

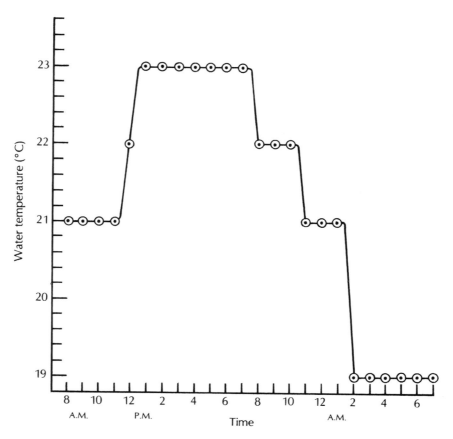

W5. Figure 25, page 50.

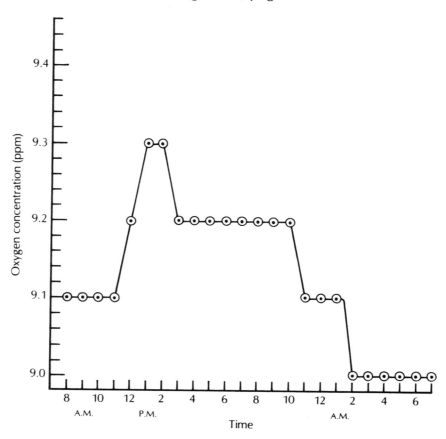

W6. Figure 26, page 50.

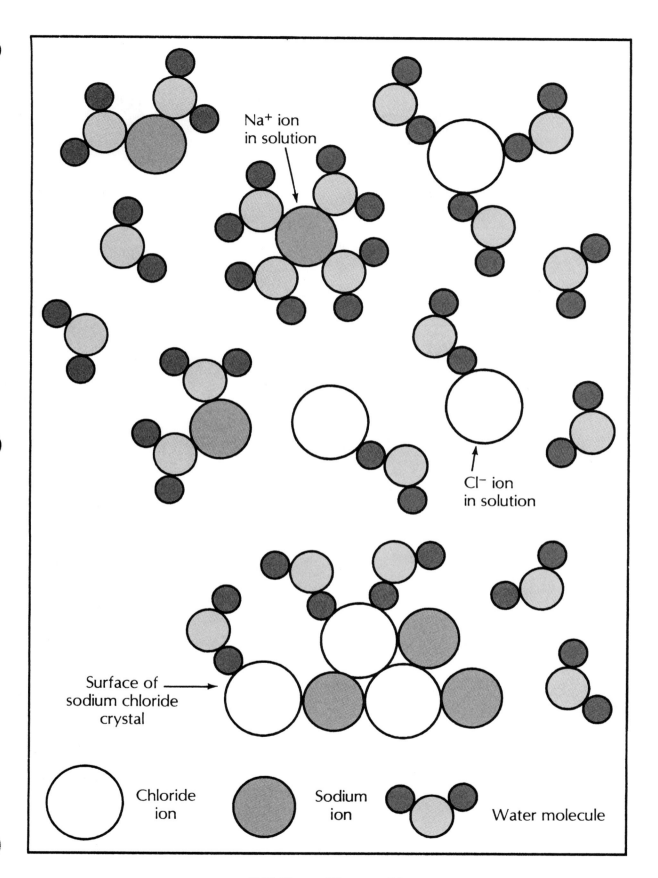

Na⁺ ion in solution

Cl⁻ ion in solution

Surface of sodium chloride crystal

Chloride ion

Sodium ion

Water molecule

W7. Figure 29, page 55

Table 12 ION CONCENTRATIONS IN THE SNAKE RIVER

Ion	Concentration Six Months Ago (ppm)	Present Concentration (ppm)	EPA Limit for Freshwater Aquatic Life (ppm)	EPA Limit for Humans (ppm)
Arsenic	0.0002	0.0002	0.44	0.05
Cadmium	0.0001	0.001	0.0015	0.01
Lead	0.01	0.02	0.074	0.05
Mercury	0.0004	0.0001	0.0041	0.05
Selenium	0.004	0.008	0.26	0.01
Chloride	52.4	51.6	No limit	250.0
Nitrate (as N)	2.1	1.9	No limit	10.0
Sulfate	34.0	35.1	No limit	250.0

W8. Table 12, page 58.

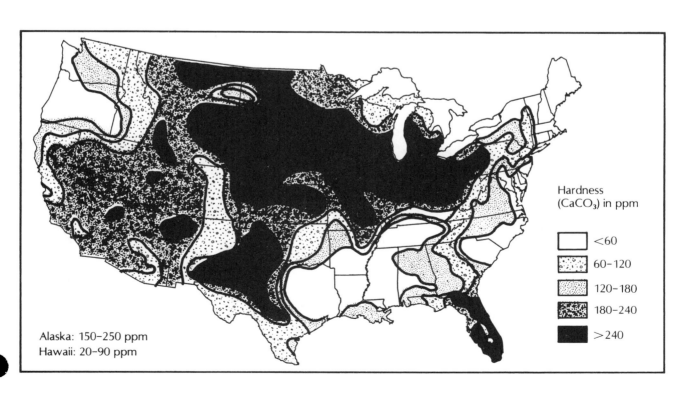

W9. Figure 31, page 67.

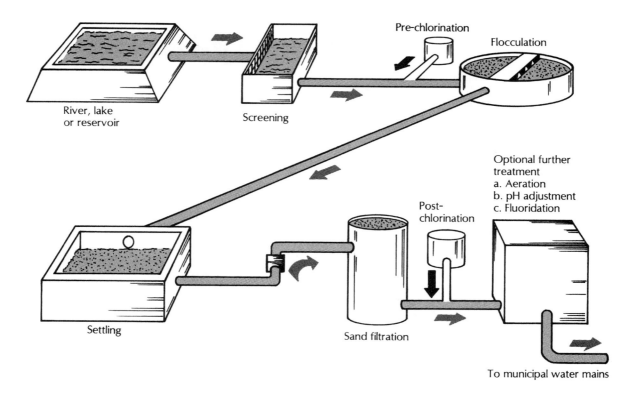

W10. Figure 34, page 70.

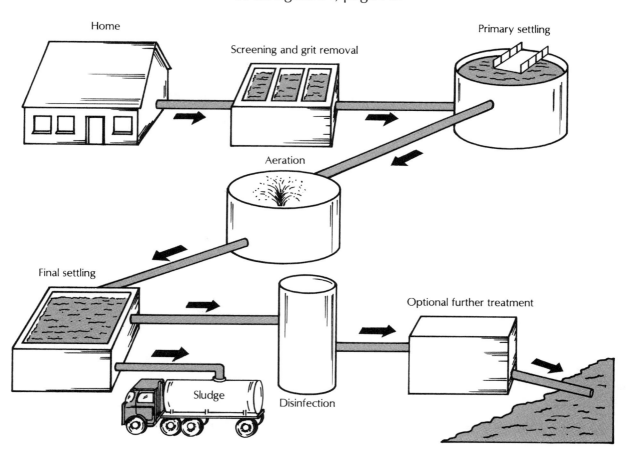

W11. Figure 35, page 72.

CONSERVING CHEMICAL RESOURCES

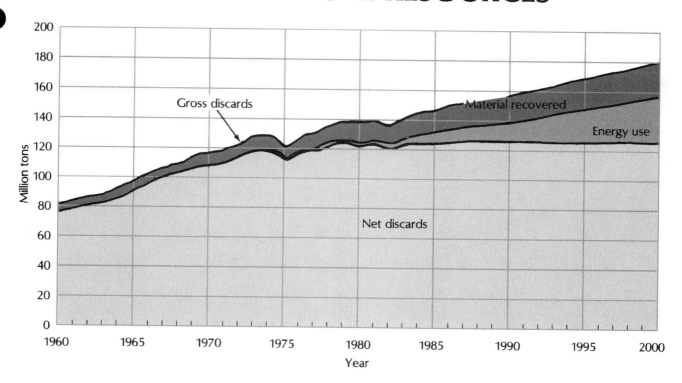

R1. Figure 5, page 98.

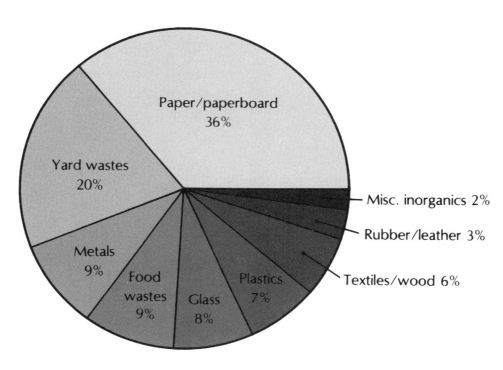

R2. Figure 6, page 98.

Table 3   **EARTH'S COMPOSITION**

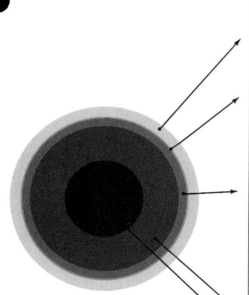

Layer of Planet	Thickness (Average)	Composition (Decreasing Order of Abundance)
Atmosphere	100 km	N_2 (78%), O_2 (21%), Ar (0.9%), He + Ne (<0.01%), variable amounts of H_2O, CO_2, etc.
Hydrosphere	5 km	H_2O, and in the oceans that cover some 71% of Earth's surface, approximately 3.5% NaCl and smaller amounts of Mg, S, Ca, and other elements as ions
Lithosphere: Crust	6400 km Top 40 km	Silicates (compounds formed of metals, Si, and O atoms). Metals include Al, Na, Fe, Ca, Mg, K, and others Coal, oil, and natural gas Carbonates such as $CaCO_3$ Oxides such as Fe_2O_3 Sulfides such as PbS
Mantle	40–2900 km	Silicates of Mg and Fe
Core	2900 km to the earth's center	Fe and Ni

R3. Table 3, page 120.

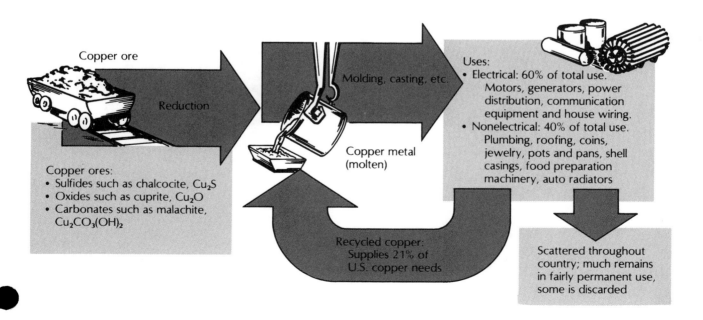

R4. Figure 11, page 132.

Table 7   **METAL ACTIVITY SERIES**

Element	Metal Ion(s) Found in Ore	Metal Obtained	Reduction Process Used to Obtain the Metal
Lithium	Li^+	$Li(s)$	Pass electric current through the molten salt (electrolysis)
Potassium	K^+	$K(s)$	
Calcium	Ca^{2+}	$Ca(s)$	
Sodium	Na^+	$Na(s)$	
Magnesium	Mg^{2+}	$Mg(s)$	
Aluminum	Al^{3+}	$Al(s)$	
Manganese	Mn^{2+}	$Mn(s)$	Heat with coke (carbon) or carbon monoxide (CO)
Zinc	Zn^{2+}	$Zn(s)$	
Chromium	Cr^{3+}, Cr^{2+}	$Cr(s)$	
Iron	Fe^{3+}, Fe^{2+}	$Fe(s)$	
Lead	Pb^{2+}	$Pb(s)$	Element occurs free or is obtained by heating in air (roasting)
Copper	Cu^{2+}, Cu^+	$Cu(s)$	
Mercury	Hg^{2+}	$Hg(l)$	
Silver	Ag^+	$Ag(s)$	
Platinum	Pt^{2+}	$Pt(s)$	
Gold	Au^{3+}, Au^+	$Au(s)$	

R5. Table 7, page 135.

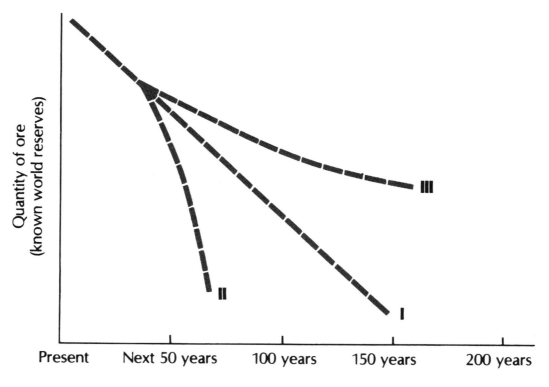

R6. Figure 14, page 144.

PETROLEUM: TO BUILD? TO BURN?

P1. Figure 1, page 154.

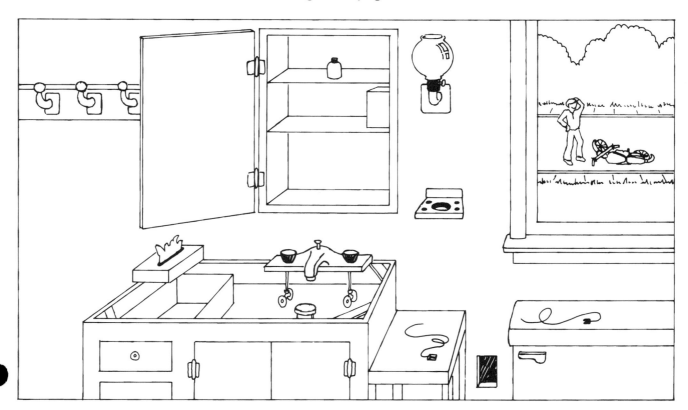

P2. Figure 1 (Teacher's Guide, page 75)

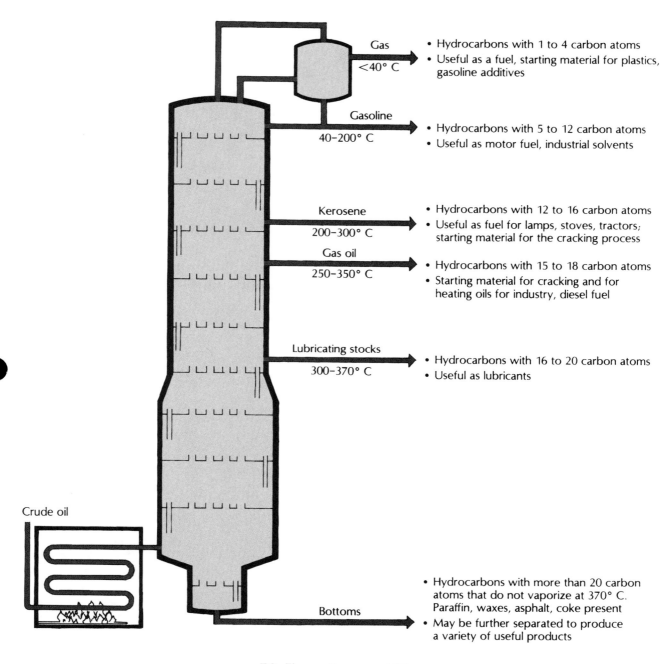

Gas
<40° C
- Hydrocarbons with 1 to 4 carbon atoms
- Useful as a fuel, starting material for plastics, gasoline additives

Gasoline
40–200° C
- Hydrocarbons with 5 to 12 carbon atoms
- Useful as motor fuel, industrial solvents

Kerosene
200–300° C
- Hydrocarbons with 12 to 16 carbon atoms
- Useful as fuel for lamps, stoves, tractors; starting material for the cracking process

Gas oil
250–350° C
- Hydrocarbons with 15 to 18 carbon atoms
- Starting material for cracking and for heating oils for industry, diesel fuel

Lubricating stocks
300–370° C
- Hydrocarbons with 16 to 20 carbon atoms
- Useful as lubricants

Crude oil

Bottoms
- Hydrocarbons with more than 20 carbon atoms that do not vaporize at 370° C. Paraffin, waxes, asphalt, coke present
- May be further separated to produce a variety of useful products

P3. Figure 5, page 159.

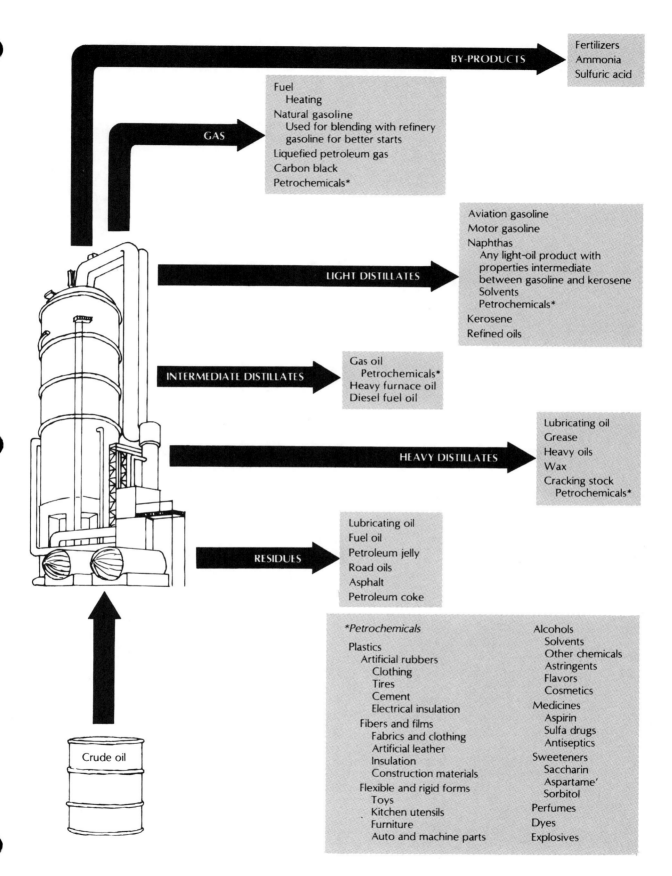

BY-PRODUCTS → Fertilizers / Ammonia / Sulfuric acid

GAS →
Fuel
  Heating
Natural gasoline
  Used for blending with refinery
  gasoline for better starts
Liquefied petroleum gas
Carbon black
Petrochemicals*

LIGHT DISTILLATES →
Aviation gasoline
Motor gasoline
Naphthas
  Any light-oil product with
  properties intermediate
  between gasoline and kerosene
  Solvents
  Petrochemicals*
Kerosene
Refined oils

INTERMEDIATE DISTILLATES →
Gas oil
  Petrochemicals*
Heavy furnace oil
Diesel fuel oil

HEAVY DISTILLATES →
Lubricating oil
Grease
Heavy oils
Wax
Cracking stock
  Petrochemicals*

RESIDUES →
Lubricating oil
Fuel oil
Petroleum jelly
Road oils
Asphalt
Petroleum coke

Crude oil

*Petrochemicals

Plastics
  Artificial rubbers
    Clothing
    Tires
    Cement
    Electrical insulation
  Fibers and films
    Fabrics and clothing
    Artificial leather
    Insulation
    Construction materials
  Flexible and rigid forms
    Toys
    Kitchen utensils
    Furniture
    Auto and machine parts

Alcohols
  Solvents
  Other chemicals
  Astringents
  Flavors
  Cosmetics
Medicines
  Aspirin
  Sulfa drugs
  Antiseptics
Sweeteners
  Saccharin
  Aspartame'
  Sorbitol
Perfumes
Dyes
Explosives

P4. Figure 6, page 165.

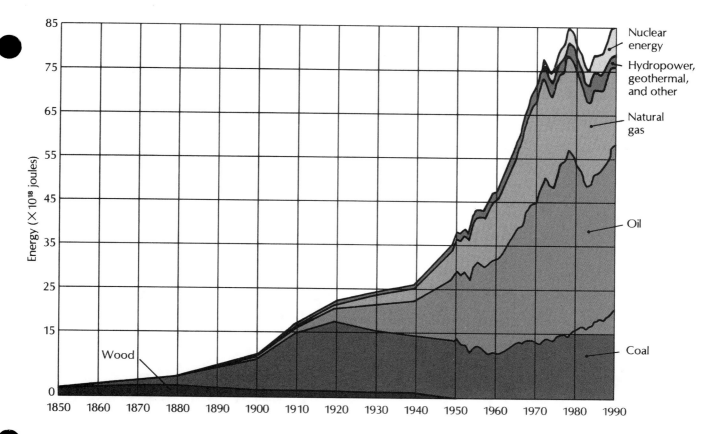

P5. Figure 9, page 179.

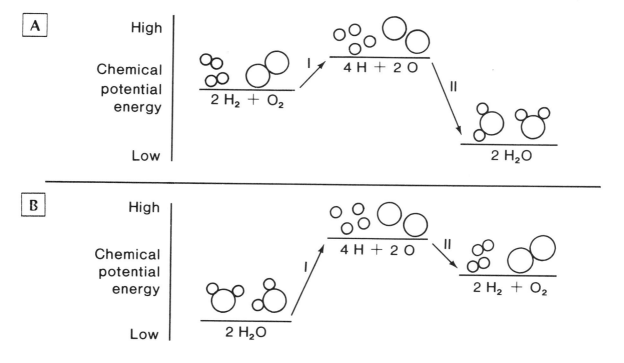

P6. Figure 10, page 182.

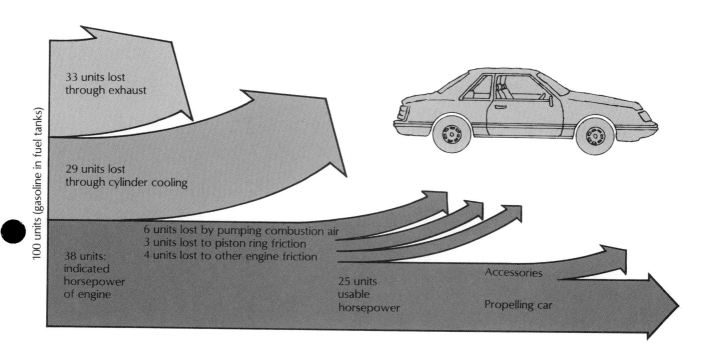

100 units (gasoline in fuel tanks)

33 units lost through exhaust

29 units lost through cylinder cooling

6 units lost by pumping combustion air
3 units lost to piston ring friction
4 units lost to other engine friction

38 units: indicated horsepower of engine

25 units usable horsepower

Accessories

Propelling car

P7. Figure 14, page 184.

UNDERSTANDING FOOD

Table 2 ENERGY EXPENDITURES

Activity	Energy Expended, Cal/hour*
Lying down or sleeping	80
Sitting	100
Driving automobile	120
Standing	140
Eating	150
Light housework	180
Walking, 2.5 mph	210
Bicycling, 5.5 mph	210
Lawn mowing	250
Golf, walking	250
Bowling	270
Walking, 3.75 mph	300
Volleyball, rollerskating	350
Tennis	420
Swimming, breaststroke	430
Swimming, crawl	520
Football, touch	530
Jogging, 11-min mile	550
Skiing, 10 mph	600
Bicycling, 13 mph	660
Football, tackle	720
Running, 8-min mile	850

*Based on a 150-pound person

F1. Table 2, page 226.

Chain form

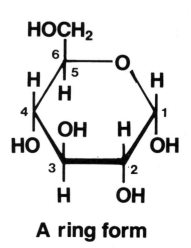

A ring form

F2. Figure 2, page 228.

F3. Figure 3, page 228.

Starch

Cellulose

F4. Figure 4, page 229.

Glycerol · Palmitic acid · Glyceryltripalmitate (a typical fat) · Water

F5. Figure 5, page 231.

(a) Palmitic acid, a saturated fatty acid

(b) Linolenic acid, a polyunsaturated fatty acid

F6. Figure 6, page 231.

Alanine (Ala)   **Cysteine (Cys)**   **Dipeptide (Ala–Cys)**   **Water**

F7. Figure 9, page 243.

Carboxyl group   **Amino group**   **Peptide bond**   **Water**

F8. Figure 10, page 243.

Table 8   **RDAs FOR PROTEIN**

Age (yr) or Condition	Median Weight (lb)	Median Height (in)	RDA (g)
Infants			
0–0.5	13	24	13
0.5–1	20	28	14
Children			
1–3	29	35	16
4–6	44	44	24
7–10	62	52	28
Males			
11–14	99	62	45
15–18	145	69	59
19–24	160	70	58
25–50	174	70	63
51+	170	68	63
Females			
11–14	101	62	46
15–18	120	64	44
19–24	128	65	46
25–50	138	64	50
51+	143	63	50
Pregnant			60
Nursing			
First 6 mo.			65
Second 6 mo.			62

Source: Food and Nutrition Board, National Academy of Sciences—National Research Council, Recommended Dietary Allowances. Revised 1989.

F9. Table 8, page 245.

Table 9   VITAMINS

Vitamin (Name)	Main Sources	Deficiency Condition
Water-soluble		
B₁ (Thiamine)	Liver, milk, pasta, bread, wheat germ, lima beans, nuts	Beriberi: nausea, severe exhaustion, paralysis
B₂ (Riboflavin)	Red meat, milk, eggs, pasta, bread, beans, dark green vegetables, peas, mushrooms	Severe skin problems
Niacin	Red meat, poultry, enriched or whole grains, beans, peas	Pellagra: weak muscles, no appetite, diarrhea, skin blotches
B₆ (Pyridoxine)	Muscle meats, liver, poultry, fish, whole grains	Depression, nausea, vomiting
B₁₂ (Cobalamin)	Red meat, liver, kidneys, fish, eggs, milk	Pernicious anemia, exhaustion
Folic acid	Kidneys, liver, leafy green vegetables, wheat germ, peas, beans	Anemia
Pantothenic acid	Plants, animals	Anemia
Biotin	Kidneys, liver, egg yolk, yeast, nuts	Dermatitis
C (Ascorbic acid)	Citrus fruits, melon, tomatoes, green pepper, strawberries	Scurvy: tender skin, weak, bleeding gums, swollen joints
Fat-soluble		
A (Retinol)	Liver, eggs, butter, cheese, dark green and deep orange vegetables	Inflamed eye membranes, night blindness, scaling of skin, faulty teeth and bones
D (Calciferol)	Fish-liver oils, fortified milk	Rickets: soft bones
E (Tocopherol)	Liver, wheat germ, whole-grain cereals, margarine, vegetable oil, leafy green vegetables	Breakage of red blood cells in premature infants, oxidation of membranes
K (Menaquinone)	Liver, cabbage, potatoes, peas, leafy green vegetables	Hemorrhage in newborns; anemia

F10. Table 9, page 252.

Table 10   RDAs FOR SELECTED VITAMINS

Sex and Age	A (µg RE)	D (µg)	C (mg)	B₁ (mg)	B₂ (mg)	Niacin (mg)	B₁₂ (µg)	K (µg)
Males								
11–14	1000	10	50	1.3	1.5	17	2.0	45
15–18	1000	10	60	1.5	1.8	20	2.0	65
19–24	1000	10	60	1.5	1.7	19	2.0	70
25–50	1000	10	60	1.5	1.7	19	2.0	80
51+	1000	10	60	1.2	1.4	15	2.0	80
Females								
11–14	800	10	50	1.1	1.3	15	2.0	45
15–18	800	10	60	1.1	1.3	15	2.0	55
19–24	800	10	60	1.1	1.3	15	2.0	60
25–50	800	5	60	1.1	1.3	15	2.0	65
51+	800	5	60	1.0	1.2	13	2.0	65

Source: Food and Nutrition Board, National Academy of Sciences—National Research Council, Recommended Dietary Allowances, Revised 1989

F11. Table 10, page 252.

Table 12   MINERALS

Mineral	Source	Deficiency Condition
Macrominerals		
Calcium (Ca)	Canned fish, milk, dairy products	Rickets in children; osteomalacia and osteoporosis in adults
Chlorine (Cl)	Meats, salt-processed foods, table salt	—
Magnesium (Mg)	Seafoods, cereal grains, nuts, dark green vegetables, cocoa	Heart failure due to spasms
Phosphorus (P)	Animal proteins	—
Potassium (K)	Orange juice, bananas, dried fruits, potatoes	Poor nerve function; irregular heartbeat; sudden death during fasting
Sodium (Na)	Meats, salt-processed foods, table salt	Headache, weakness, thirst, poor memory, appetite loss
Sulfur (S)	Proteins	—
Trace minerals		
Chromium (Cr)	Liver, animal and plant tissue	Loss of insulin efficiency with age
Cobalt (Co)	Liver, animal proteins	Anemia
Copper (Cu)	Liver, kidney, egg yolk, whole grains	—
Fluorine (F)	Seafoods, fluoridated drinking water	Dental decay
Iodine (I)	Seafoods, iodized salts	Goiter
Iron (Fe)	Liver, meats, green leafy vegetables, whole grains	Anemia; tiredness and apathy
Manganese (Mn)	Liver, kidney, wheat germ, legumes, nuts, tea	Weight loss, dermatitis
Molybdenum (Mo)	Liver, kidney, whole grains, legumes, leafy vegetables	—
Nickel (Ni)	Seafoods, grains, seeds, beans, vegetables	Cirrhosis of liver, kidney failure, stress
Selenium (Se)	Liver, organ meats, grains, vegetables	Kashan disease (a heart disease found in China)
Zinc (Zn)	Liver, shellfish, meats, wheat germ, legumes	Anemia, stunted growth

F12. Table 12, page 257.

Table 14   FOOD ADDITIVES

Additive Type	Purpose	Examples
Anticaking agents	Keep foods free-flowing	Sodium ferrocyanide
Antioxidants	Prevent fat rancidity	BHA and BHT
Bleaches	Whiten foods (flour, cheese); hasten cheese maturing	Sulfur dioxide, SO_2
Coloring agents	Increase visual appeal	Carotene (natural yellow color); synthetic dyes
Emulsifiers	Improve texture, smoothness; stabilize oil-water mixtures	Cellulose gums, dextrins
Flavoring agents	Add or enhance flavor	Salt, monosodium glutamate (MSG), spices
Humectants	Retain moisture	Glycerin
Leavening agents	Give foods light texture	Baking powder, baking soda
Nutrients	Improve nutritive value	Vitamins, minerals
Preservatives and antimycotic agents (growth inhibitors)	Prevent spoilage, microbial growth	Propionic acid, sorbic acid, benzoic acid, salt
Sweeteners	Impart sweet taste	Sugar (sucrose), dextrin, fructose, aspartame, sorbitol, mannitol

F13. Table 14, page 261.

Table 15   **MEALS AROUND THE WORLD**

	Food Energy (Cal)	Protein (g)	Iron (mg)	Vitamin B_1 (mg)
Chinese	797	36	10.5	0.66
Eskimo	872	94	19.0	0.6
Japanese	766	47	8.8	0.56
Mexican	889	27	11.0	1.14
Ugandan	828	32	6.5	0.8
U.S. citizen				
Fast-food meal	886	23	4.8	0.08
Regular meal	1212	30	6.9	0.33

Source: Adapted from S. DeVore and T. White, *The Appetites of Man: An Invitation to Better Nutrition from Nine Healthier Societies.* Anchor Books, Anchor/Doubleday, Garden City, N.Y., 1978.

F14. Table 15, page 269.

NUCLEAR CHEMISTRY IN OUR WORLD

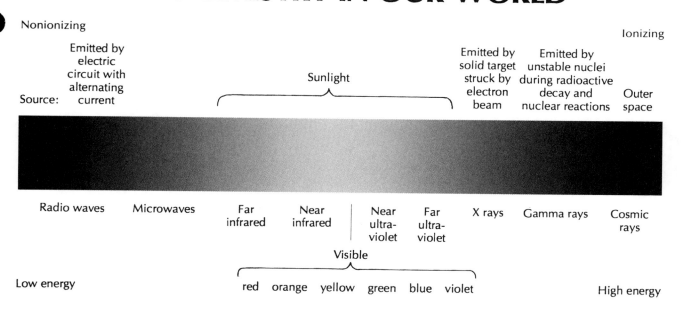

N1. Figure 1, page 276.

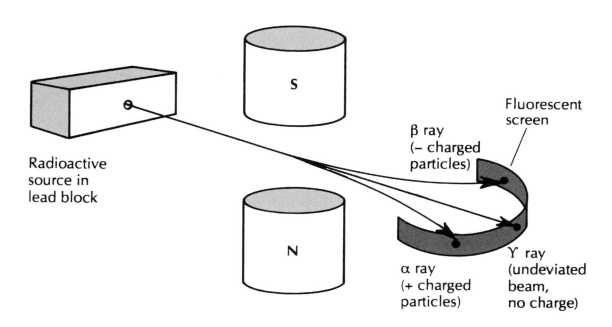

N2. Figure 4, page 280.

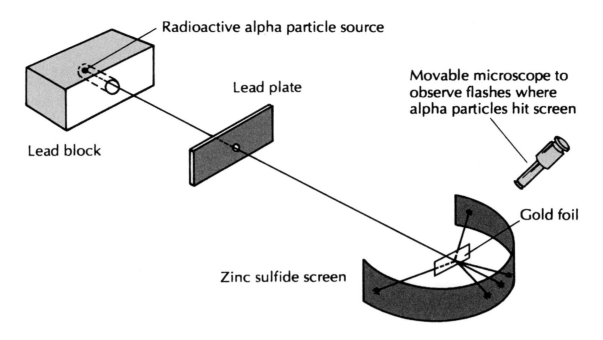

Radioactive alpha particle source

Lead plate

Movable microscope to observe flashes where alpha particles hit screen

Lead block

Gold foil

Zinc sulfide screen

N3. Figure 5, page 282.

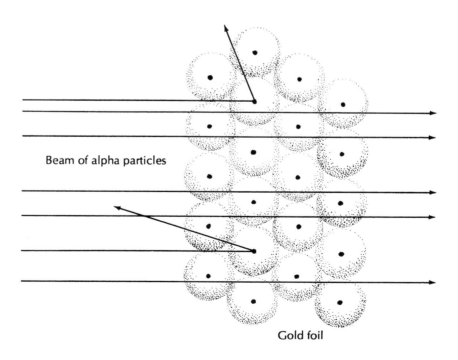

Beam of alpha particles

Gold foil

N4. Figure 6, page 282.

Table 3 SOME COMMON ISOTOPES

Symbol	Name	Total Protons (Atomic Number)	Total Neutrons	Mass Number	Total Electrons
$_{1}^{1}H$	Hydrogen-1	1	0	1	1
$_{3}^{7}Li$	Lithium-7	3	4	7	3
$_{9}^{19}F$	Fluorine-19	9	10	19	9
$_{82}^{208}Pb$	Lead-208	82	126	208	82
$_{82}^{208}Pb^{2+}$	Lead-208, (II) ion	82	126	208	80

N5. Table 3, page 284.

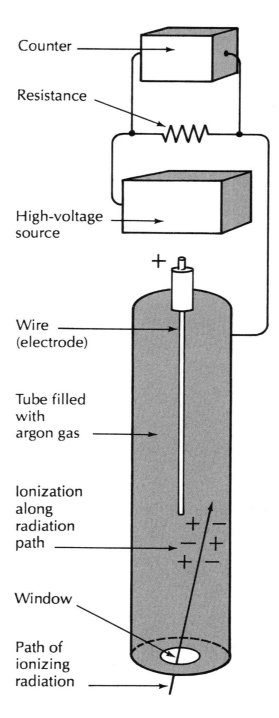

Counter

Resistance

High-voltage
source

Wire
(electrode)

Tube filled
with
argon gas

Ionization
along
radiation
path

Window

Path of
ionizing
radiation

N6. Figure 7, page 292.

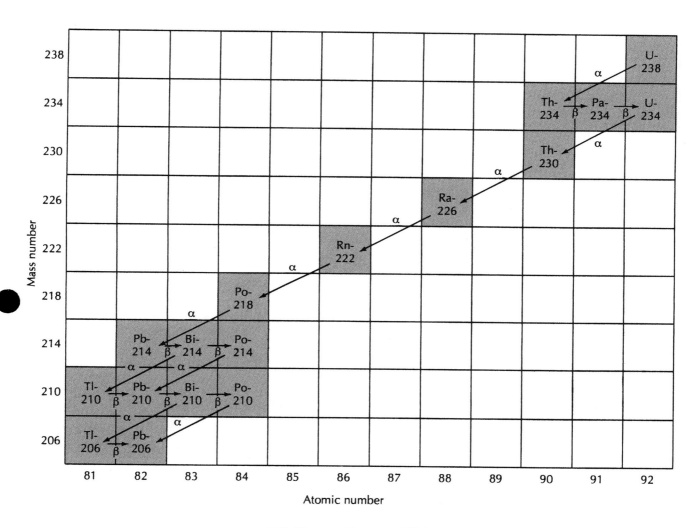

N7. Figure 14, page 297.

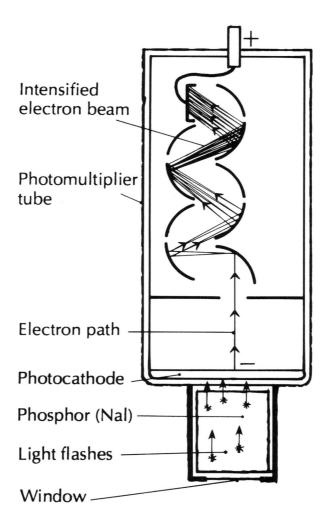

Intensified electron beam

Photomultiplier tube

Electron path

Photocathode

Phosphor (NaI)

Light flashes

Window

N8. Figure 15, page 302.

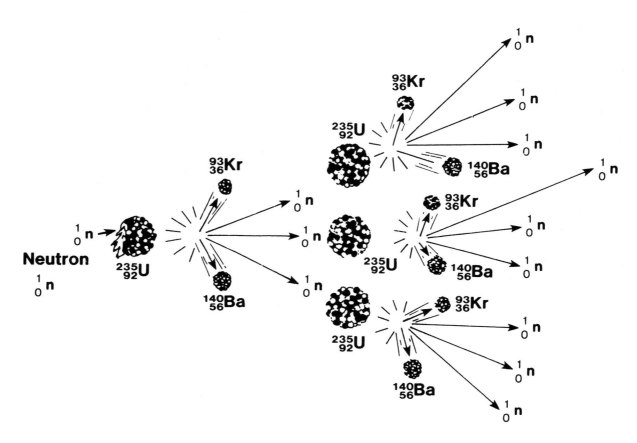

N9. Figure 17, page 310.

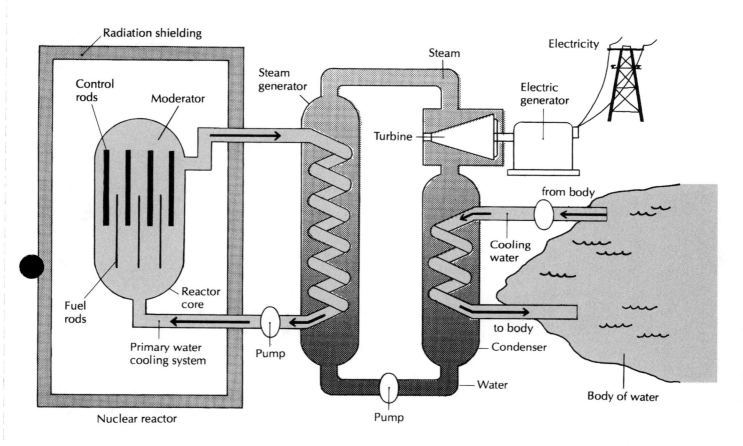

N10. Figure 20, page 312.

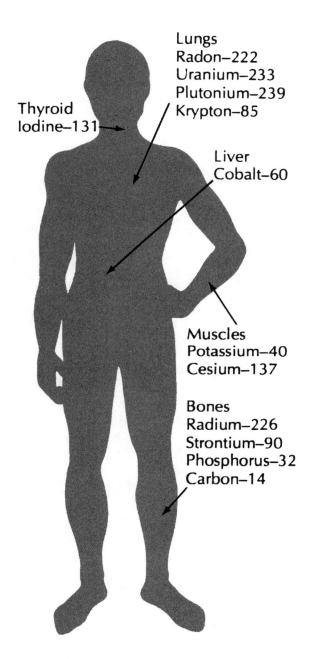

N11. Figure 25, page 325.

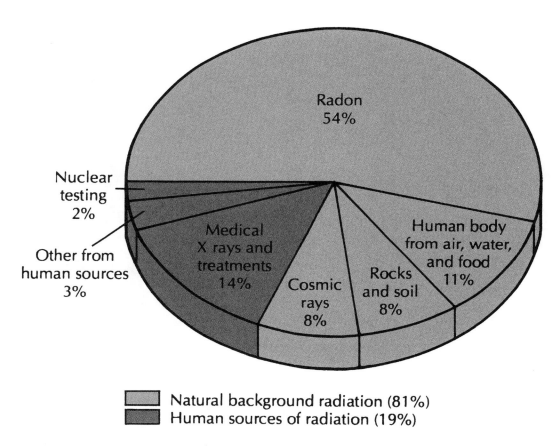

Radon
54%

Nuclear
testing
2%

Other from
human sources
3%

Medical
X rays and
treatments
14%

Cosmic
rays
8%

Rocks
and soil
8%

Human body
from air, water,
and food
11%

Natural background radiation (81%)
Human sources of radiation (19%)

N12. Figure 26, page 325.

CHEMISTRY, AIR, AND CLIMATE

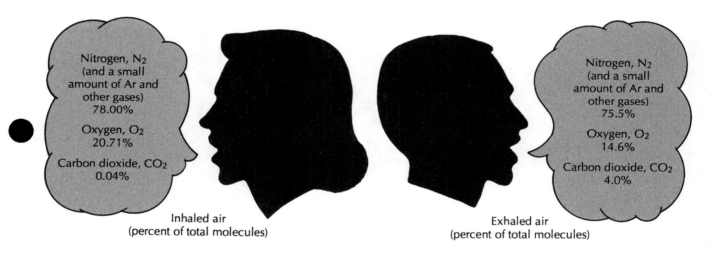

A1. Figure 1, page 344.

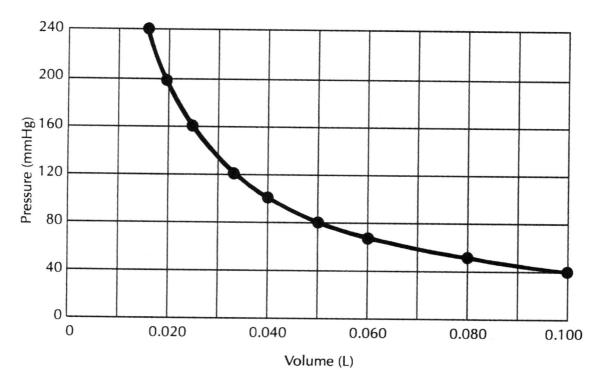

A2. Figure 6, page 360.

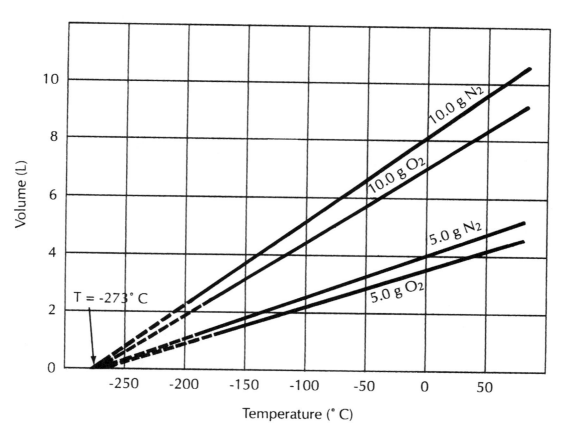

A3. Figure 9, page 364.

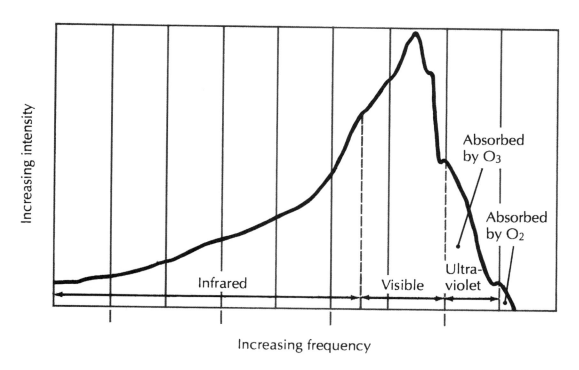

A4. Figure 10, page 371.

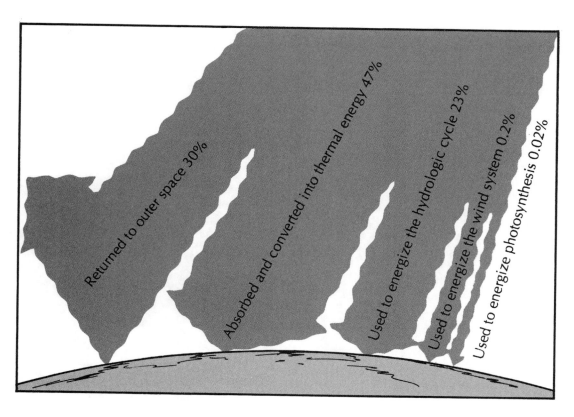

A5. Figure 11, page 372.

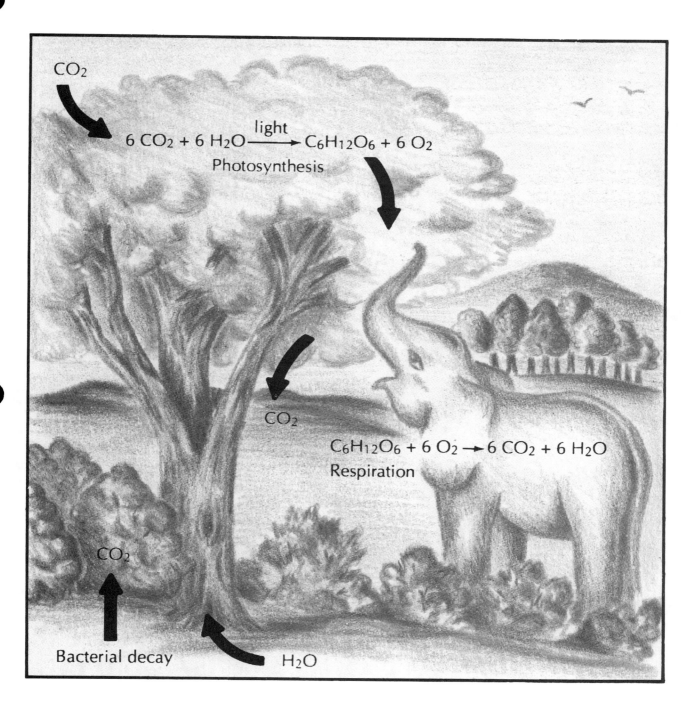

A6. Figure 12, page 376.

WORLDWIDE EMISSIONS OF AIR POLLUTANTS
Table 4
(10^{12} g/yr or 10^6 metric tons/yr)

Pollutant	Human Source	Quantity	Natural Source	Quantity
CO_2	Combustion of wood and fossil fuels	22,000	Decay; release from oceans, forest fires, and respiration	1,000,000
CO	Incomplete combustion	700	Forest fires and photochemical reactions	2,100
SO_2	Combustion of coal and oil; smelting of ores	212	Volcanoes and decay	20
CH_4	Combustion; natural gas leakage	160	Anaerobic decay and termites	1,050
NO_x	High-temperature combustion	75	Lightning; bacterial action in soil	180
NMHC	Incomplete combustion	40	Biological processes	20,000
NH_3	Sewage treatment	6	Anaerobic biological decay	260
H_2S	Petroleum refining and sewage treatment	3	Volcanoes and anaerobic decay	84

Source: Adapted from Stern et al. Fundamentals of Air Pollution, 2nd ed.: Academic Press, Inc.: Orlando, FL, 1984; pp. 30–31. Table adapted by permission of Elmer Robinson, Mauna Loa Observatory. Data based on conditions prior to 1980.

A7. Table 4, page 386.

Table 5   **U.S. POLLUTION, 1987 (in 10^6 metric tons/yr)**

Source	TSP	SO$_x$	NO$_x$	HC	CO	Total
			Pollutant			
Transportation (petroleum burning)	1.4	0.9	8.4	6.0	40.7	57.4
Fuel burning for space heating and electricity	1.8	16.4	10.3	2.3	7.2	38.0
Industrial processes	2.5	3.1	0.6	8.3	4.7	19.2
Solid waste disposal and miscellaneous	1.3	0.0	0.2	3.0	8.8	13.3
Totals	7.0	20.4	19.5	19.6	61.4	127.9

A8. Table 5, page 388.

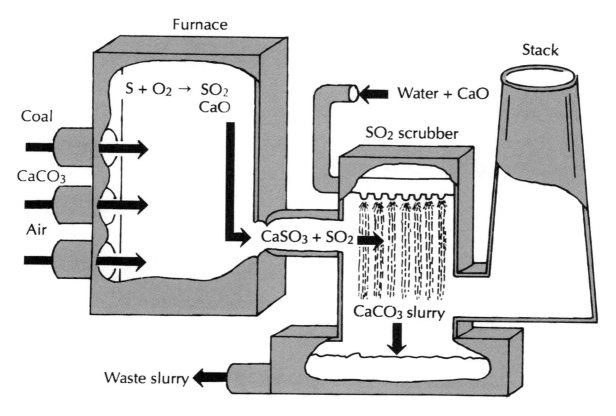

A9. Figure 15, page 391.

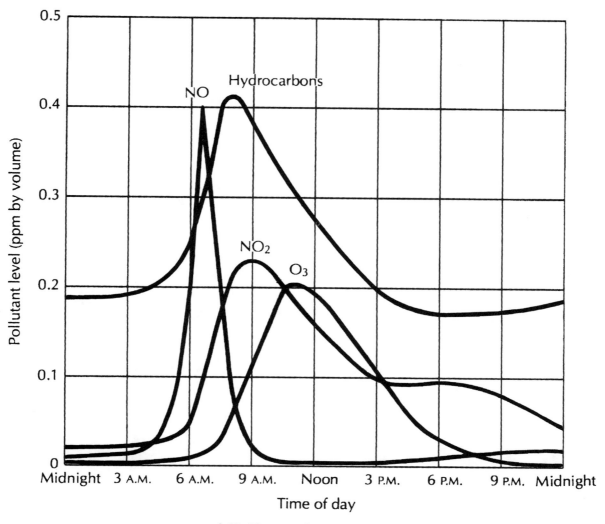

A10. Figure 16, page 394.

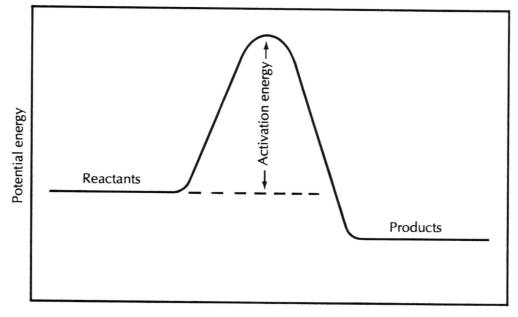

A11. Figure 17, page 395.

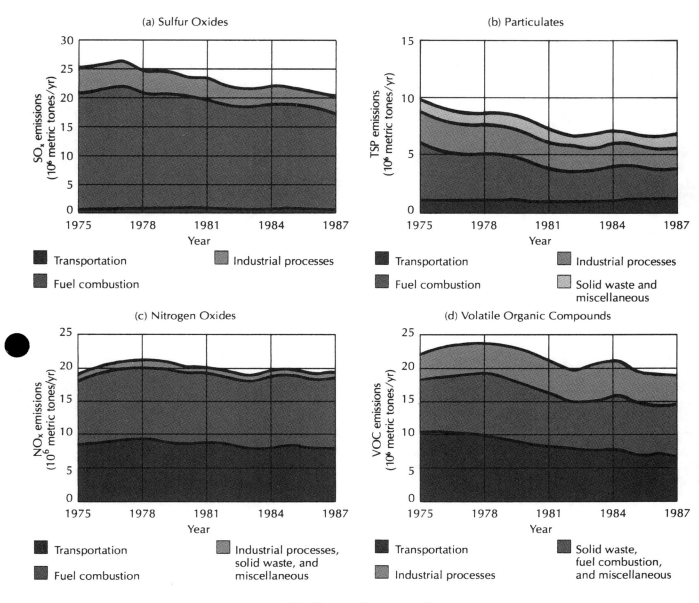

A12. Figure 18, page 404.

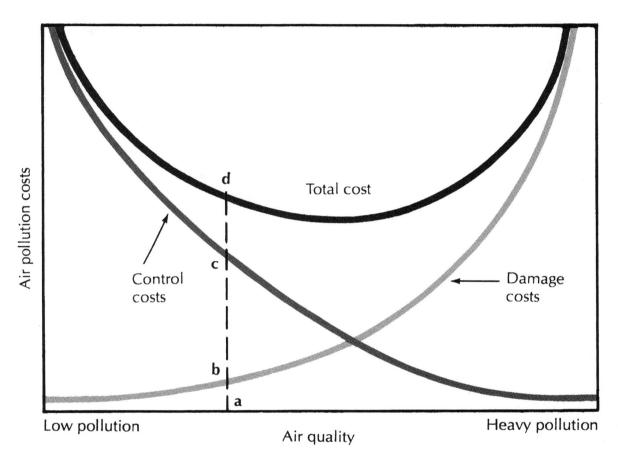

A13. Figure 19, page 406.

HEALTH: YOUR RISKS AND CHOICES

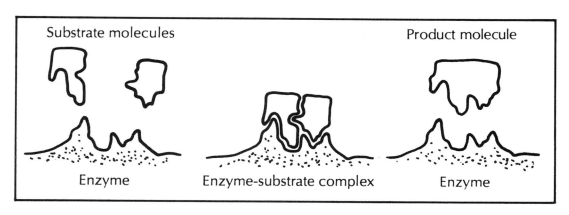

H1. Figure 1, page 420.

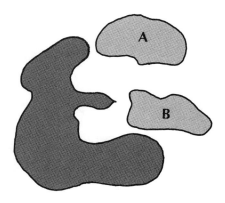

Without the coenzyme, substances A and B cannot be led to the active site of the enzyme and hence cannot react with each other.

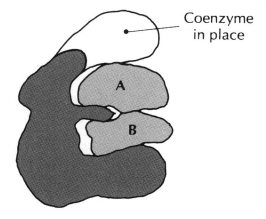

With the coenzyme, substances A and B are oriented at the active site of the enzyme. Here they can react with each other.

H2. Figure 2, page 420.

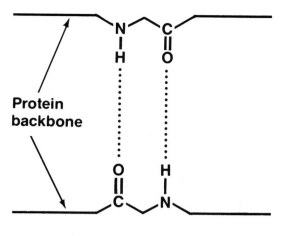

Hydrogen bonding

H3. Figure 12, page 447.

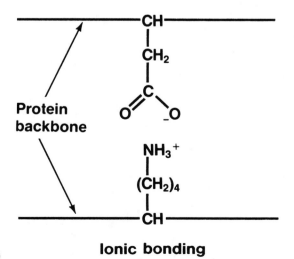

Ionic bonding

H4. Figure 14, page 447.

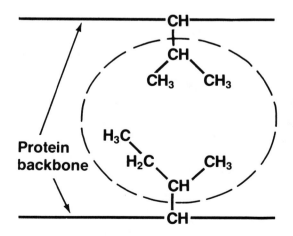

Nonpolar interactions

H5. Figure 15, page 447.

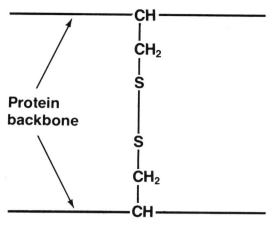

Disulfide bridge

H6. Figure 16, page 448.

(a)

(b)

(c)

(d)

H7. Figure 17, page 448.

THE CHEMICAL INDUSTRY: PROMISE AND CHALLENGE

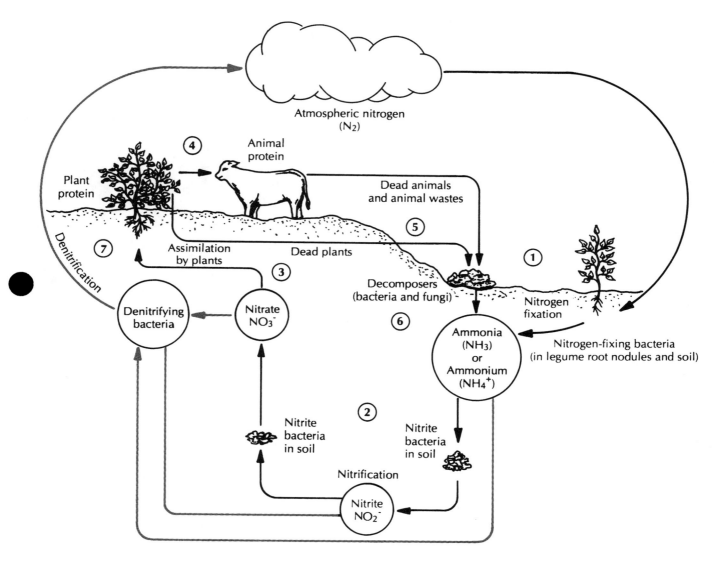

I1. Figure 3, page 492.

Table 5   **ACTIVITY SERIES OF COMMON METALS**

Metal		Products of Metal Reactivity		
Li(s)	→	$Li^+(aq)$	+	e^-
Na(s)	→	$Na^+(aq)$	+	e^-
Mg(s)	→	$Mg^{2+}(aq)$	+	$2\ e^-$
Al(s)	→	$Al^{3+}(aq)$	+	$3\ e^-$
Mn(s)	→	$Mn^{2+}(aq)$	+	$2\ e^-$
Zn(s)	→	$Zn^{2+}(aq)$	+	$2\ e^-$
Cr(s)	→	$Cr^{3+}(aq)$	+	$3\ e^-$
Fe(s)	→	$Fe^{2+}(aq)$	+	$2\ e^-$
Ni(s)	→	$Ni^{2+}(aq)$	+	$2\ e^-$
Sn(s)	→	$Sn^{2+}(aq)$	+	$2\ e^-$
Pb(s)	→	$Pb^{2+}(aq)$	+	$2\ e^-$
Cu(s)	→	$Cu^{2+}(aq)$	+	$2\ e^-$
Ag(s)	→	$Ag^+(aq)$	+	e^-
Au(s)	→	$Au^{3+}(aq)$	+	$3\ e^-$

I2. Table 5, page 507.

DISCHARGING

PbO_2 (cathode)
$PbO_2(s) + SO_4^{2-}(aq) + 4\ H^+(aq) + 2\ e^- \longrightarrow$
     $PbSO_4(s) + 2\ H_2O(l)$

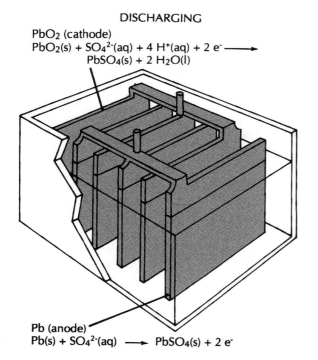

Pb (anode)
$Pb(s) + SO_4^{2-}(aq) \longrightarrow PbSO_4(s) + 2\ e^-$

CHARGING

PbO_2 (anode)
$PbSO_4(s) + 2\ H_2O(l) \longrightarrow$
     $PbO_2(s) + SO_4^{2-}(aq) + 4\ H^+(aq) + 2\ e^-$

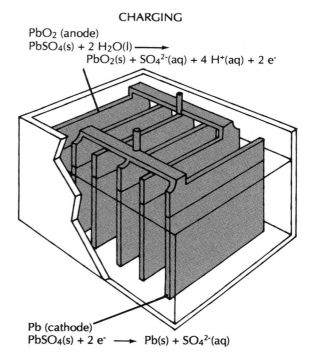

Pb (cathode)
$PbSO_4(s) + 2\ e^- \longrightarrow Pb(s) + SO_4^{2-}(aq)$

I3. Figure 9, page 511.

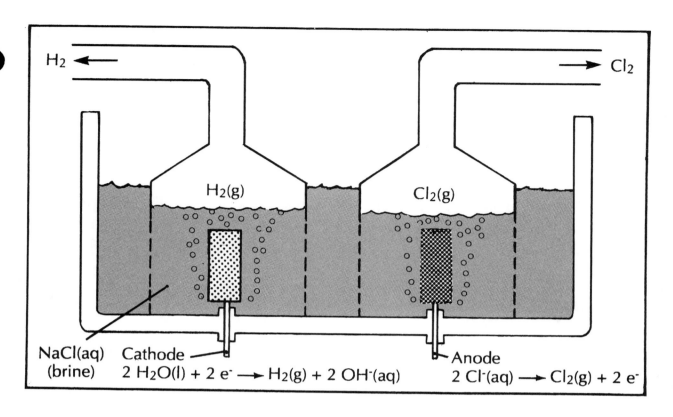

$H_2 \leftarrow$　　　　　　　　　　　　　　　　　　　$\rightarrow Cl_2$

$H_2(g)$　　　　　　　　　$Cl_2(g)$

NaCl(aq)　Cathode　　　　　　　　　　Anode
(brine)　$2 H_2O(l) + 2 e^- \rightarrow H_2(g) + 2 OH^-(aq)$　　　$2 Cl^-(aq) \rightarrow Cl_2(g) + 2 e^-$

I4. Figure 12, page 515.

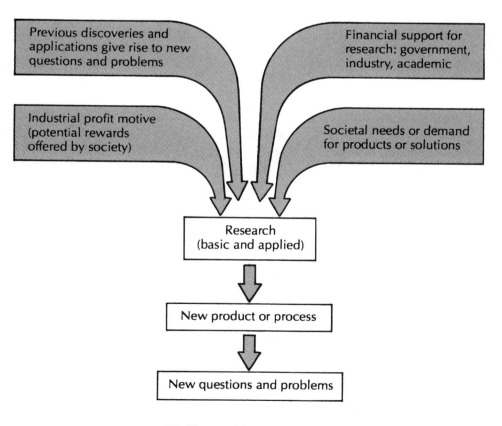

Previous discoveries and applications give rise to new questions and problems

Financial support for research: government, industry, academic

Industrial profit motive (potential rewards offered by society)

Societal needs or demand for products or solutions

Research (basic and applied)

New product or process

New questions and problems

I5. Figure 14, page 519.

Periodic Table of the Elements

1	2	3	4	5	6	7	8	9	10	11	12	13	14	15	16	17	18
H 1 1.008																	He 2 4.003
Li 3 6.941	Be 4 9.012											B 5 10.81	C 6 12.01	N 7 14.01	O 8 16.00	F 9 19.00	Ne 10 20.18
Na 11 22.99	Mg 12 24.31											Al 13 26.98	Si 14 28.09	P 15 30.97	S 16 32.07	Cl 17 35.45	Ar 18 39.95
K 19 39.10	Ca 20 40.08	Sc 21 44.96	Ti 22 47.88	V 23 50.94	Cr 24 52.00	Mn 25 54.94	Fe 26 55.85	Co 27 58.93	Ni 28 58.69	Cu 29 63.55	Zn 30 65.39	Ga 31 69.72	Ge 32 72.59	As 33 74.92	Se 34 78.96	Br 35 79.90	Kr 36 83.80
Rb 37 85.47	Sr 38 87.62	Y 39 88.91	Zr 40 91.22	Nb 41 92.21	Mo 42 95.94	Tc 43 98.91	Ru 44 101.1	Rh 45 102.9	Pd 46 106.4	Ag 47 107.9	Cd 48 112.4	In 49 114.8	Sn 50 118.7	Sb 51 121.8	Te 52 127.6	I 53 126.9	Xe 54 131.3
Cs 55 132.9	Ba 56 137.3	La★ 57 138.9	Hf 72 178.5	Ta 73 180.9	W 74 183.9	Re 75 186.2	Os 76 190.2	Ir 77 192.2	Pt 78 195.1	Au 79 197.0	Hg 80 200.6	Tl 81 204.4	Pb 82 207.2	Bi 83 209.0	Po 84 (210.0)	At 85 (210.0)	Rn 86 (222.0)
Fr 87 (223.0)	Ra 88 226.0	Ac• 89 227.0	104 (261)	105 (262)	106 (263)	107 (262)											

★ Lanthanoid series

Ce 58 140.1	Pr 59 140.9	Nd 60 144.2	Pm 61 144.9	Sm 62 150.4	Eu 63 152.0	Gd 64 157.3	Tb 65 158.9	Dy 66 162.5	Ho 67 164.9	Er 68 167.3	Tm 69 168.9	Yb 70 173.0	Lu 71 175.0

• Actinoid series

Th 90 232.0	Pa 91 231.0	U 92 238.0	Np 93 237.0	Pu 94 239.1	Am 95 243.1	Cm 96 247.1	Bk 97 247.1	Cf 98 252.1	Es 99 252.1	Fm 100 257.1	Md 101 256.1	No 102 259.1	Lr 103 260.1